A Modern Approach to Ecology of Plants

A Modern Approach to Ecology of Plants

Edited by Austin Balfour

SYRAWOOD
PUBLISHING HOUSE
New York

Published by Syrawood Publishing House,
750 Third Avenue, 9th Floor,
New York, NY 10017, USA
www.syrawoodpublishinghouse.com

A Modern Approach to Ecology of Plants
Edited by Austin Balfour

International Standard Book Number: 978-1-68286-394-7 (Hardback)

Cataloging-in-publication Data

A modern approach to ecology of plants / edited by Austin Balfour.
p. cm.
Includes bibliographical references and index.
ISBN 978-1-68286-394-7
1. Plant ecology. 2. Plant diversity. 3. Plant populations. 4. Plant biotechnology. I. Balfour, Austin.
QK901 .M63 2017
581.7--dc23

Printed in China.

TABLE OF CONTENTS

PREFACE

Plant Ecology is a vast field encompassing the study of environmental factors affecting the growth of plants. It also studies the interactions amongst plants as well as those of plants with the environment. This book contains some path-breaking studies in the field of plant ecology and provides comprehensive insights to the readers. It presents researches and studies performed by experts across the globe. This book will serve as a valuable reference to a broad spectrum of readers such as ecologists, botanists, environmentalists, agronomists, researchers, students and anyone else who is interested in exploring the field of plant ecology.

Significant researches are present in this book. Intensive efforts have been employed by authors to make this book an outstanding discourse. This book contains the enlightening chapters which have been written on the basis of significant researches done by the experts.

Finally, I would also like to thank all the members involved in this book for being a team and meeting all the deadlines for the submission of their respective works. I would also like to thank my friends and family for being supportive in my efforts.

Editor

1

Discovery and Analysis of MicroRNAs in *Leymus chinensis* under Saline-Alkali and Drought Stress Using High-Throughput Sequencing

Junfeng Zhai[1,2]◆, Yuanyuan Dong[1]◆, Yepeng Sun[1]◆, Qi Wang[3], Nan Wang[1], Fawei Wang[1], Weican Liu[1], Xiaowei Li [1], Huan Chen[1], Na Yao[1], Lili Guan[1], Kai Chen[1], Xiyan Cui[2], Meiying Yang[2], Haiyan Li[1,2]*

1 Ministry of Education Engineering Research Center of Bioreactor and Pharmaceutical Development, Jilin Agricultural University, Changchun, Jilin, China, **2** College of Life Sciences, Jilin Agricultural University, Changchun, Jilin, China, **3** High School attached to Northeast Normal University, Changchun, Jilin, China

Abstract

Leymus chinensis (Trin.) Tzvel. is a perennial rhizome grass of the Poaceae (also called Gramineae) family, which adapts well to drought, saline and alkaline conditions. However, little is known about the stress tolerance of *L. chinensis* at the molecular level. microRNAs (miRNAs) are known to play critical roles in nutrient homeostasis, developmental processes, pathogen responses, and abiotic stress in plants. In this study, we used Solexa sequencing technology to generate high-quality small RNA data from three *L. chinensis* groups: a control group, a saline-alkaline stress group (100 mM NaCl and 200 mM $NaHCO_3$), and a drought stress group (20% polyethylene glycol 2000). From these data we identified 132 known miRNAs and 16 novel miRNAs candidates. For these miRNAs we also identified target genes that encode a broad range of proteins that may be correlated with abiotic stress regulation. This is the first study to demonstrate differentially expressed miRNAs in *L. chinensis* under saline-alkali and drought stress. These findings may help explain the saline-alkaline and drought stress responses in *L. chinensis*.

Editor: Rogerio Margis, Universidade Federal do Rio Grande do Sul, Brazil

Funding: This work was supported by grants from the Program for the Special Program for Research of Transgenic Plants (Grant No. 2011ZX08010-002), the National Natural Science Foundation of China (Grant Nos. 30971804, 31101091, 31271746, 31201144), the State Key Laboratory of Crop Biology at Shandong Agricultural University, China (Grant No. 2010KF02), the Program for Young Scientific and Technological Talents & Outstanding Innovation Team (Grant No. 20111815), and the National High Technology Research and Development Program of China (863 Program) (Grant No. 2011AA100606). The funders had no role in study design, data collection and analysis, decision to publish, or preparation of the manuscript.

Competing Interests: The authors have declared that no competing interests exist.

* Email: hyli99@163.com

◆ These authors contributed equally to this work.

Introduction

Leymus chinensis (Trin.) Tzvel. is a perennial rhizome grass of the Poaceae family, which is distributed widely throughout the northern temperate areas of Eurasia [1]. *L. chinensis* is cultivated as a major grass forage product for its high protein content, productivity, palatability and nourishment. Additionally, because *L. chinensis* adapts well to drought, saline, alkaline and low temperature conditions, it is cultivated particularly on alkaline land, which means it has an important role to play in the protection of the environment [2]. However, the genome sequence of *L. chinensis* has not yet been published, and very little genetic information is publicly available. A few studies have investigated the saline-alkaline tolerance of *L. chinensis* at the molecular level, but no microRNA (miRNA) studies have been reported to date.

miRNAs are non-coding RNAs, approximately 21–26 nucleotides (nt) long, that play crucial roles in the regulation of gene expression in plants [3]. Several plant miRNAs that are involved in stress responses are regulated by abiotic stresses, including drought, high salinity, and low temperatures. For example, the expression levels of rice miR393 change under salinity and alkaline stresses and its putative target genes are related to abiotic stress [4], while miR169 plays an essential role in drought resistance in tomato [5]. Additionally, recent studies have shown that a number of other differentially expressed miRNAs are involved in stress regulation [6,7,8]. Overall, these studies suggest that miRNAs and their target mRNAs play important roles in stress tolerance. It is becoming increasingly evident that miRNAs play an important role in plant development and resistance to biotic as well as abiotic stresses.

Various methods have been used to systematically discover miRNAs in rice, wheat and maize [9,10,11,12]. To provide further insights into the role of miRNAs in *L. chinensis* tolerance to stress, we analyzed the expression patterns of miRNAs from *L. chinensis* exposed to drought and saline-alkali stresses using Solexa high-throughput sequencing technology and quantitative real time PCR (qRT-PCR). We identified and analyzed 132 known miRNAs and 16 novel miRNA candidates of *L. chinensis*. Fifteen and 19 miRNAs were differentially expressed under saline-alkali and drought stress, respectively, relative to control. These results indicated that these miRNAs were associated with drought, salinity and alkaline stress. We verified variations in the expression profiles of several miRNAs during saline-alkali and drought stresses using qRT-PCR analysis. We report the differential expression of *L. chinensis* miRNAs that may be involved in saline-alkaline and drought stress regulation. This study has important implications for gene regulation in plants under saline-alkaline and drought stress conditions.

Materials and Methods

1. Sample treatment and RNA isolation

Seeds of *L. chinensis* (Jisheng No.4 Chinese Wildrye), which has high saline-alkaline and drought resistance, were obtained from the Jilin Province Jisheng Wildrye Excellent Seed Station. The seeds were surface sterilized with 75% ethanol and reared in an artificial climate chamber in Hoagland nutrient solution. The nutrient solution was changed every 2 days. When the "3-leaves" stage was reached, the seedlings were transferred either into Hoagland nutrient solution with 100 mM NaCl and 200 mM $NaHCO_3$ to simulate salinity and alkaline stresses or into Hoagland nutrient solution containing 20% polyethylene glycol 2000 (PEG-2000) to simulate drought stress. Some plants were cultivated in Hoagland nutrient solution as a control. The seedlings were treated for 24 h, then frozen in liquid nitrogen and stored at −80°C. Total RNA was extracted with Trizol reagent (Invitrogen, CA, Carlsbad, USA) according to the manufacturer's instructions. All the experiments followed a randomized design and were conducted using three replicates of each of the samples.

2. Small RNA sequencing library construction

RNAs less than 30 nucleotides long were selected and purified by 15% denaturing polyacrylamide gel electrophoresis (PAGE). A 5' chimeric oligonucleotide adaptor primer (5′-pUCGUAUGCC-GUCUUCUGCUUGidT-3′) and a 3′ chimeric oligonucleotide adaptor primer (5′-GUUCAGAGUUCUACAGUCCGAC-GAUC-3′) were ligated to the small RNAs (sRNAs) using T4 RNA ligase and the resulting ligation products were gel purified by 10% PAGE. cDNA libraries were prepared by reverse transcription from the RNA products. The resulting libraries were sequenced on the Solexa Genome Analyzer II platform (Illumina, San Diego, CA, USA) according to the manufacturer's instructions. All the short reads were deposited in the National Center for Biotechnology Information (NCBI) and can be accessed in the Short Read Archive (SRA) under the accession number, SRP033396.

3. Bioinformatic analysis

After Solexa sequencing, low-quality reads were removed and the remaining high-quality sRNA reads were filtered to remove low quality tags and adaptor sequences. We called the filtered set of sequences the clean sequences. To identify known miRNAs, the clean sequences were searched against the non-coding RNAs (rRNA, tRNA, snRNA and snoRNA) in the Rfam 9.1 (http://rfam.janelia.org) and miRBase 19.0 (http://www.mirbase.com) databases using BLASTN with a threshold of at least 95% identity and 90% coverage (E-value ≤1e-5). To identify novel miRNAs, the sRNA reads that had matches in Rfam and miRBase were eliminated and the remaining reads were aligned to the *L. chinensis* mRNA transcript database [13] using SOAP 2.20 software (http://soap.genomics.org.cn). The SOAP2.2 output was filtered with a precursor tool to separate candidate sequences as miRNA precursors with an anchoring pattern of a block of aligned small RNAs with perfect matches. MIREAP software (http://sourceforge.net/projects/mireap/) was then used to identify candidate primary miRNAs on the genome sequence flanking the identified miRNAs. The *L. chinensis* candidate miRNAs that were detected using this approach were named according to miRBase. The RNAfold web server (http://rna.tbi.univie.ac.at/cgi-bin/RNAfold.cgi) was used to evaluate whether the predicted primary miRNAs could form suitable hairpin structures. Folding free energies (MFE) were also used to evaluate these candidate novel miRNAs. The number of reads for each identified miRNA was normalized against the total number of reads in the corresponding library to determine the expression level of the miRNA. Next, a Bayesian method was used to evaluate the statistical significance of expression differences between different samples. *P* values ≤0.01 and changes of at least two-fold were used as the threshold to confirm statistical significance. Candidate targets of the known and novel miRNAs were predicted by psRNATarget using *L. chinensis* assembled unigenes (http://bioinfo3.noble.org/psRNATarget/). The criteria used for psRNA-Target were as follows: sequences were considered to be miRNA targets when the total score was less than four mismatches (a G:U wobble pair was assigned a mismatch score of 0.5). BLASTX searches were performed using the mRNA sequences of the predicted miRNA targets to query the NCBI protein database to predict the functions of potential targets. The COG (clusters of orthologous groups) terms for the target genes were assigned by comparing the NCBI sequences that were obtained against the COG database (http://www.ncbi.nlm.nih.gov/COG).

4. miRNA validation and expression analysis by qRT-PCR

To validate expression levels of miRNAs, qRT-PCR was performed for the candidate *L. chinensis* miRNAs from the three libraries. The forward miRNA primers were designed based on the full-length miRNA sequences, and the reverse primer was the universal reverse primer for miRNAs [14]. The primers are listed in Table S1. Total RNA (1 μl) in 20 μl reaction volumes was used to synthesize reverse transcripts using a One Step PrimeScript miRNA cDNA Synthesis Kit (Takara, Japan). 5 s rRNA was used as the reference gene. The PCR conditions were as follows: an initial polymerase activation step for 3 min at 94°C, 20 cycles of 15 s at 94°C for denaturation, 20 s at 66°C for annealing, and 60 s at 72°C for elongation.

To validate expression patterns of miRNAs and their targets, qRT-PCR was performed for the predicted unigenes. Primer sequences are given in Table S2. The 25 μl reactions contained 2 μl cDNA, 12.5 μl 2× SYBR premix ExTaq (Takara, Japan), and 10 μM of the forward and reverse primers. The thermal cycling conditions were 40 cycles at 95°C for 5 s for denaturation and 60°C for 20 s for annealing and extension.

The amplification reaction was performed using the ABI7300 real-time PCR system (Applied Biosystems, Foster City, CA, USA). To calculate the relative expression of miRNAs and unigenes, the $2^{-\Delta\Delta C_t}$ value was calculated and transformed into fold-change differences.

5. 5′-RACE validation of target genes

To validate the miRNA target genes, rapid amplification of cDNA ends (RACE) assays were performed using a 5′-RACE kit (Takara, Dalian, China) [15,16]. Total RNA was treated with tobacco acid pyrophosphatase at 37°C for 1 h. A RNA Oligo adapter was then directly ligated to the treated RNAs and two gene-specific primers were used for RACE (Table S3). The PCR products were visualized on agarose gels. Bands were excised and sequenced to determine the 5′ ends of the transcripts.

Results

1. Overview of Solexa sequenced sRNA libraries from *L. chinensis* grown under stress

Solexa sequencing of the *L. chinensis* libraries generated a large number of short reads (10–40 nt long). The control library yielded 15,712,504 unfiltered raw sequence reads, and the saline-alkali and drought libraries yielded 15,076,385 and 14,725,482 raw reads, respectively. After filtering to remove low-quality reads, the control library contained 15,301,799 high-quality reads, and after filtering out the 3′-adapter, 5′-adapter and poly(A) reads, we obtained 14,825,707 clean sequences, which accounted for 96.89% of the high-quality reads. For the stress-induced libraries, we obtained 14,190,211 clean sequences from the saline-alkali library and 13,856,136 clean sequences from the drought library, which accounted for 96.90% and 96.67% of the high-quality reads in the two libraries, respectively (Table 1). Most of the sequence tags varied in length from 19 nt to 30 nt, (Figure 1A). The overall distribution of the redundant small RNAs are similar in the three samples, with one major peak at 24 nt and another minor peak at 20 nt and 21 nt. While in drought samples, the propotion of 24 nt small RNAs decreased and the 20-nt or 21-nt populations increased to a major peak.

In the control library, the 24 nt sRNAs were the most abundant, making up 39.42% of the clean sequences; in the saline-alkali and drought libraries, the 24 nt sRNAs made up 27.22% and 31.48% of the clean sequences, respectively. The 20 nt and 21 nt sRNAs were also abundant, making up 36.02%, 45.58% and 36.85% of the clean sequences in the control, saline-alkali and drought libraries respectively. It is worth noting that 24 nt, 20 nt and 21 nt are typical lengths for miRNAs (16–29 nt). Some 22 nt sRNAs were also present in the control, saline-alkali and drought libraries; however, they made up only 7.92%, 9.05% and 8.69% of the clean sequences, respectively.

There are several kinds of sRNA, including snRNAs, snoRNAs, rRNAs, tRNAs, and miRNAs (Figure S1). When all the clean sequences were searched against the GeneBank, Rfam, and miRbase databases, we found that a relatively small proportion of them mapped to known sequences. The unmatched sequences (53–67%) were characterized as unclassified sRNAs. A proportion of the clean sequences, 36.66%, 46.48% and 41.52% from the control, saline-alkali and drought libraries respectively, partially mapped to known non-coding RNA sequences (Figure 1B) and about 20% of the mapped sequences from each of the libraries were identified as miRNAs. These numbers imply that there was a relatively high abundance of miRNAs in the *L. chinensis* libraries; the unclassified sRNAs may represent either new miRNAs or other classes of non-coding RNAs.

2. Identification of conserved miRNAs expressed under saline–alkaline and drought stress

We used the clean sequences to search the miRBase 19.0 database to identify mature miRNAs, miRNA stars (a miRNA star, miRNA*, denotes an sRNA processed from the hairpin arm opposite the mature miRNA), and precursor miRNAs in the *L. chinensis* libraries. Because known miRNA sequences of *L. chinensis* are scarce, the miRNAs and pre-miRNAs of rice (*Oryza sativa* Linn.) and barley (*Hordeum vulgare*) were used as the reference sequences against which the miRNAs from *L. chinensis* were mapped. We identified 132 unique known miRNAs in the three libraries, which were distributed in 38 conserved miRNA families (Table S4). The read counts for the sequences in each of the libraries that mapped to the known miRNAs (reads number shown in figure 1B) suggested that the sequencing reads were sufficient to support an estimation of the expression levels of the identified miRNAs in the three libraries.

We found some notable quantitative differences for the known miRNAs in the stress-induced libraries. The abundance of the individual miRNAs was calculated relative to the total number of identified miRNA reads in the corresponding library. We detected 13 miRNAs for which the read counts were more than 400,000 in all three libraries. The most abundant miRNAs across the three libraries were members of the miR156 family, including lch-miR156d, lch-miR156g, lch-miR156h and lch-miR156b. Of these, the most abundant was miR156d with 2,039,447 clean sequences in the control library, followed by lch-miR156g, lch-miR156h and lch-miR156b, all with more than 200,000 read counts each. lch-miR156d is one of the most conserved miRNAs having been identified in 12 different species (*Arabidopsis thaliana, Selaginella moellendorffii, Zea mays, Sorghum bicolor, O. sativa, Vitis vinifera, Populus trichocarpa, Gossypium hirsutum, Medicago truncatula, Glycine max, Ricinus communis* and *Arabidopsis lyrata*). Similar observations were made for the members of three other miRNA families, miR168, miR166 and miR167. Each family had more than 20,000 clean sequences in each of the three libraries; among them, lch-miR168-5p and lch-miR168a, with more than 400,000 sequences each, were the most abundant. lch-miR168-5p was conserved in *Phaseolus vulgaris* and lch-miR168a was conserved in *Z. mays, S. officinarum, O. sativa, P. trichocarpa, A. lyrata, A. thaliana* and *S. bicolor*. These two high abundance miRNAs were present in similar abundances in all three libraries. The majority of the high abundance miRNAs were differentially expressed in the different libraries. For example,

Table 1. Read quality in the control and stress-induced libraries.

	Control	saline-alkaline stress group	drought stress group
Raw reads	15,712,504	15,076,385	14,725,482
high quality	15,301,799	14,644,590	14,332,735
3′adapter_null	212,414	266,964	158,349
insert_null	4,465	2,686	6,507
5′adapter_contaminants	148,528	21,684	153,276
smaller_than_18 nt	108,118	162,419	155,359
polyA	2,567	626	3,108
clean_reads	14,825,707	14,190,211	13,856,136

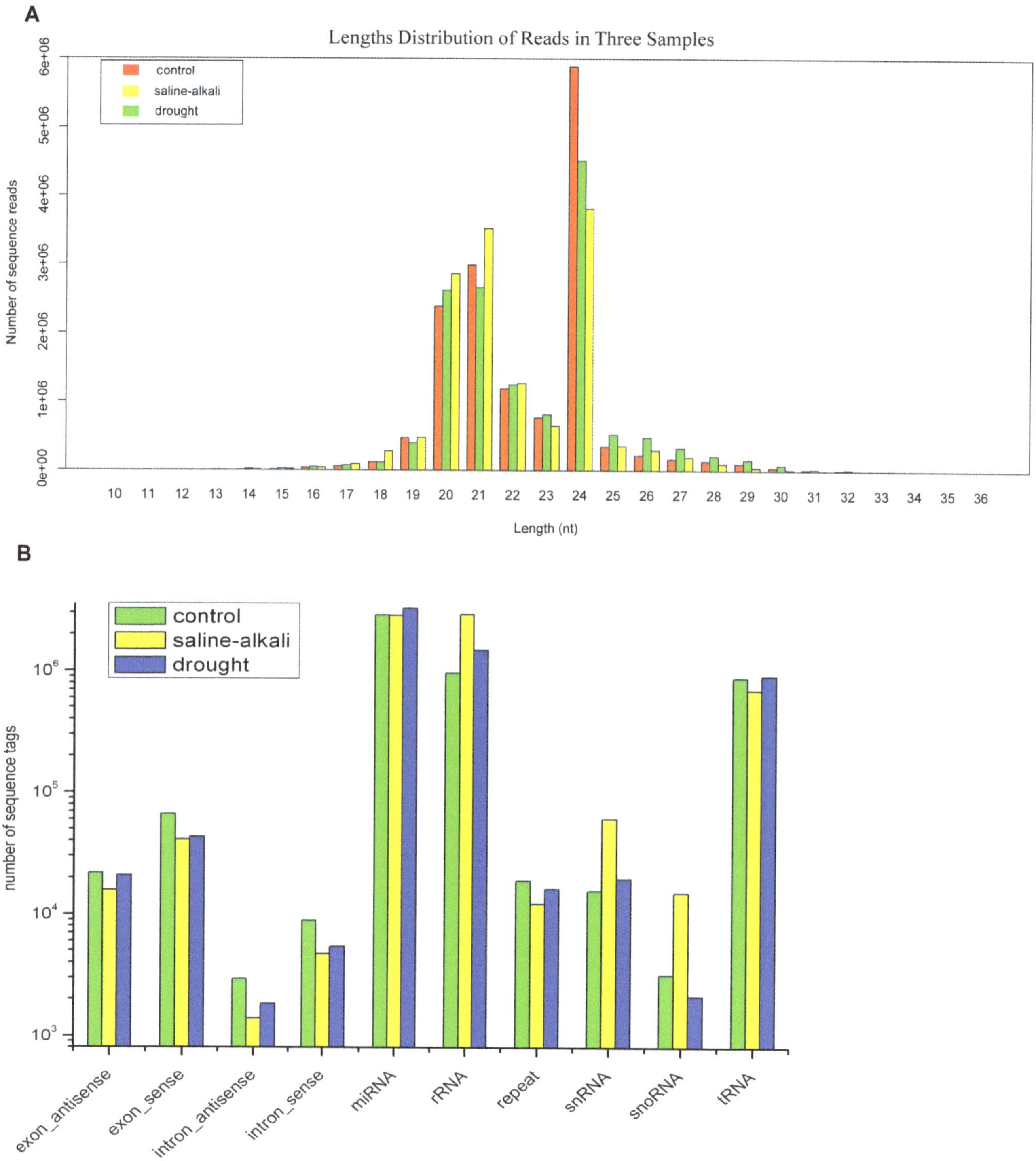

Figure 1. Length distribution and proportions of various categories of small RNAs in *L. chinensis* in the libraries from control, saline-alkaline and drought treated plants. A: Length distribution of reads in the three libraries; B: Categories of small RNAs in *the* three libraries.

miR166e with more than 56,000 clean sequences was expressed at a high level, while lch-miR166b, lch-miR166c and lch-miR169n, which also showed high expression levels, were expressed in all three libraries. Similarly, lch-miR399a, a highly expressed miRNA, was found in all three libraries. The different abundances of the different members of a miRNA family may suggest that they have varied regulatory roles under the different environmental conditions.

3. Identification of novel miRNA candidates

Because the genome of *L. chinensis* has not yet been sequenced, the identification of novel miRNAs can be a challenge. We used methods previously described to construct a non-redundant database of *L. chinensis* novel miRNA candidates by selecting the most abundant sequence as the representative for each set of similar sequences.

Sixteen novel miRNAs from the three libraries were identified as miRNA candidates (Table 2). These novel miRNAs length ranged from 20 to 23 nt and the lower minimal folding free energy (MFE) ranged from −19.1 to −110.7. These candidate sequences were among the most highly expressed sRNAs in the *L. chinensis* libraries, indicating that they are likely to play a major functional role in the plants' responses to stress. Based on their abundances and sequences, we found that these novel miRNAs displayed lower expression levels than the majority of miRNAs belonging to known families. The low abundance of the novel miRNAs might suggest specific roles for these miRNAs under the stress conditions.

4. Different expression patterns of miRNAs under saline–alkaline and drought stress conditions

miRNAs make up a large majority of the *L. chinensis* small RNA libraries when compared with other kinds of sRNA. Some identified miRNAs had more than one million sequencing reads, while others had fewer than one hundred. The most abundant miRNA was miR156. miR156d had 2,039,447 reads in the control library and 2,147,423 and 2,347,874 reads in the saline-alkaline and drought libraries, respectively. This high level of expression correlates with evolutionary conservation [17]. In contrast, some conserved miRNAs, such as miR156l, were found to be expressed at low levels compared with miR156d. It is likely that theses miRNAs have specific expression patterns in *L. chinensis* under stress conditions.

To compare differential expression patterns between control and stressed libraries, standardized treatments of miRNA abundance were performed. The common expression patterns of *L. chinensis* miRNAs under saline-alkaline and drought stress conditions are shown in Figure 2. miRNA expression varied in response to different stress-inducing conditions. Fifteen miRNAs (lch-miR160f, lch-miR171a, lch-miR171h, lch-miR319a, lch-miR319b, lch-miR390, lch-miR393, lch-miR393b, lch-miR394, lch-miR396c, lch-miR396e-3p, lch-miR396f-3p, lch-miR396g, lch-miR827a, lch-miR827b) were identified as differentially expressed with fold-changes >2 between control and saline-alkaline libraries (p value <0.1). Fourteen of them were significantly differentially expressed (p value <0.01) (lch-miR394 was not). Nineteen miRNAs were differentially expressed under drought stress with fold-changes >2 when compared with the control library (p value <0.05); seven of these (lch-miR160a, lch-miR160d, lch-miR160f, lch-miR319a, lch-miR319b, lch-miR394 and lch-miR396f-3p) were significantly differentially expressed (p value <0.01). lch-miR160a differed by a >8 fold-change between control and drought. The expression profiles of only a minority of the differentially expressed miRNAs were expressed at the same level, suggesting that differential miRNA expression might partly contribute to stress regulation among the various treatments.

5. Validation of the miRNA expression profiles by qRT-PCR

Stem-loop qRT-PCR was used to measure and validate the expression levels of seven randomly selected *L. chinensis* miRNAs (lch-miR160a, lch-miR169f, lch-miR172d, lch-miR319a, lch-miR394, lch-miR396f-3p and lch-miR397a) in the three libraries. The expression levels of these miRNAs measured by qRT-PCR were compared with those obtained by Solexa analyses of the libraries and the two sets of results were consistent (Figure 3).

6. Predicted targets of the known and novel miRNAs

Understanding the biological roles of the *L. chinensis* miRNAs requires the identification of their mRNA targets [18,19]. We used psRNATarget to predict the mRNA targets of the conserved and novel (Table S5, Table S6) miRNAs using assembled transcripts of *L. chinensis*. We used 3.5 as a cut off threshold and found 125 putative targets for 24 conserved miRNA families and 59 targets for 12 novel miRNAs.

Table 2. Novel miRNAs predicted from the three *L. chinensis* libraries.

Novel miRNA	miRNA sequence	3p/5p	strand (±)	length (nt)	folding energy (kcal/mol)
lch-MIR-01	TCTTATATTATTGGACGGAGG	3p	+	21	−72.5
lch-MIR-02	AAGTAATTTGAGACGGAGGGAGT	3p	+	23	−58.35
lch-MIR-03	AAAACATCAACAATCGGAACTTA	5p	-	23	−31.03
lch-MIR-04	GTGTGGCGGCATGGGGATGT	3p	+	20	−42.1
lch-MIR-05	TGAGAGTGCGAATACAAGGAGGT	5p	-	23	−21.7
lch-MIR-06	ATTTCTGGACGGAGGGAGTAT	5p	+	21	−19.1
lch-MIR-07	AAGAGTAGCGTTGATACACCGT	5p	-	22	−68
lch-MIR-08	AAGAATTATGGAATGGAGGGA	3p	-	21	−56.3
lch-MIR-09	AAGACAAGTAATTTGGGACGG	3p	+	21	−45.5
lch-MIR-10	CTCTTATTGAATCGCGGTAAAGT	3p	-	23	−58.3
lch-MIR-11	ATGTGTGTGTCTGTGTGTGTA	5p	-	21	−110.7
lch-MIR-12	TTCTTCGACTTGGCCATCTCCC	5p	+	21	−61.3
lch-MIR-13	AATAGGTAGGGATGACAGGATT	5p	+	22	−42.5
lch-MIR-14	GGGACGAGTTGGAAGAGGAAT	3p	+	21	−22.5
lch-MIR-15	AGGCTGTGGAGAGATGGCTGAGT	3p	-	23	−26.4
lch-MIR-16	TTTAGGAACGGAGGAAGTACA	5p	-	21	−52.6

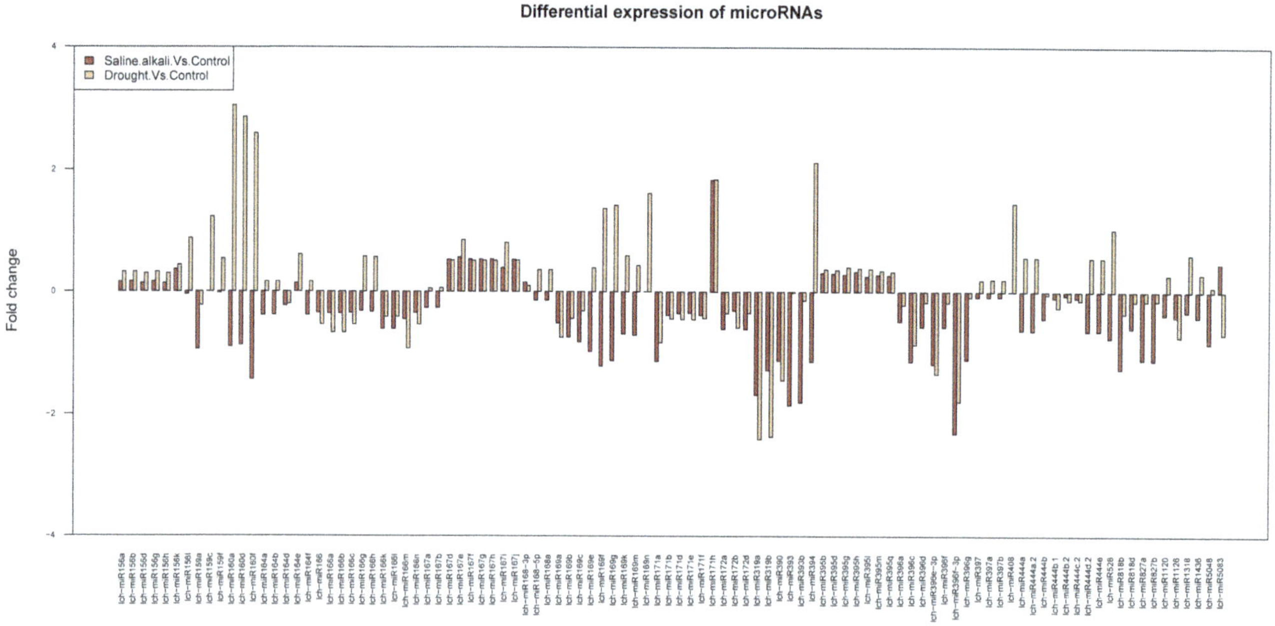

Figure 2. Differentially expressed miRNAs in the stress-induced libraries compared with the control.

The functions of all putative targets were identified by COG analysis (figure 4). Some of the predicted targets were coproporphyrinogen III oxidase, glutathione S-transferase, ubiquitin-protein ligase, MYB superfamily members, and several NAD-dependent aldehyde dehydrogenases, indicating that these predicted target genes encode a broad range of proteins correlated with metabolic and cellular processes. We also found eight putative targets for conserved miRNAs and one putative target for one of the novel miRNAs involved in abiotic stress regulation. These target genes may be involved with oxidative stress, which can damage cellular structures, in response to water deficit and pathogen attack [20,21].

We analyzed the expression of the predicted targets of known *L. chinensis* miRNAs using published transcriptome sequencing data [13]. Under saline-alkaline stress, we observed no change in the expression for four predicted targets, GW_c49847, GW_c34555,

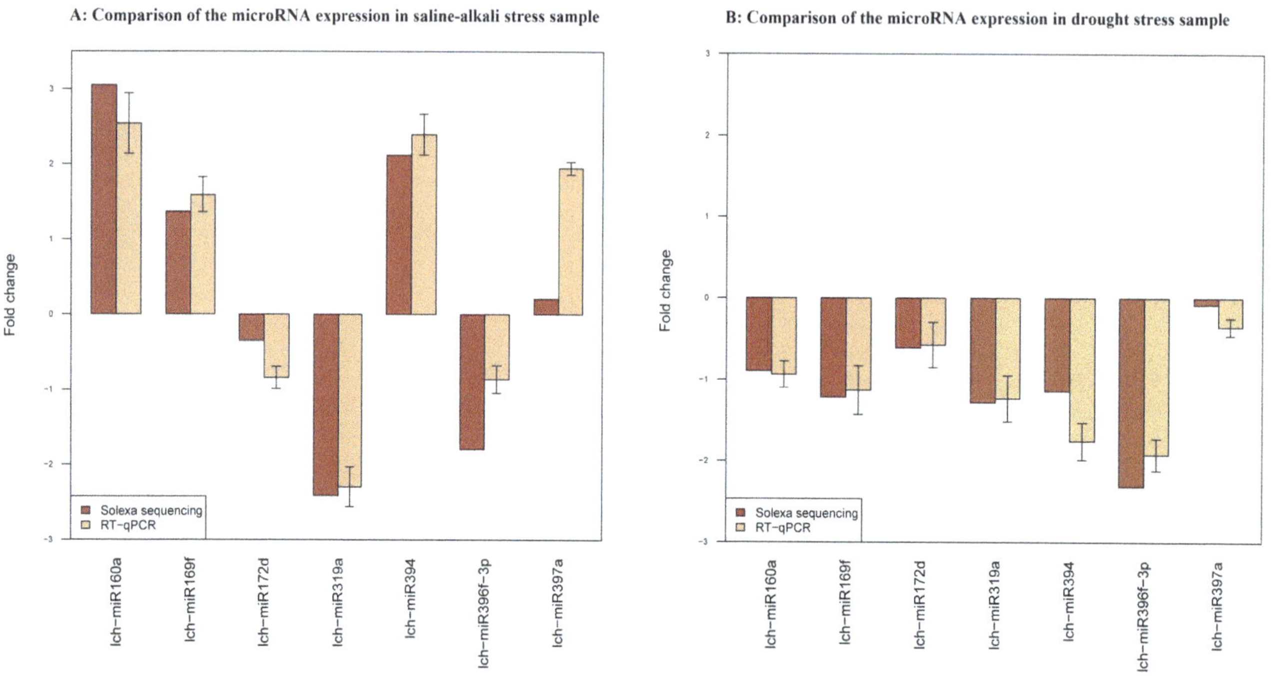

Figure 3. Comparison of the miRNA expression levels determined by Solexa sequencing and qRT-PCR. A: Comparison of miRNA expression in the saline-alkali stress sample; B: Comparison of miRNA expression in the drought stress sample.

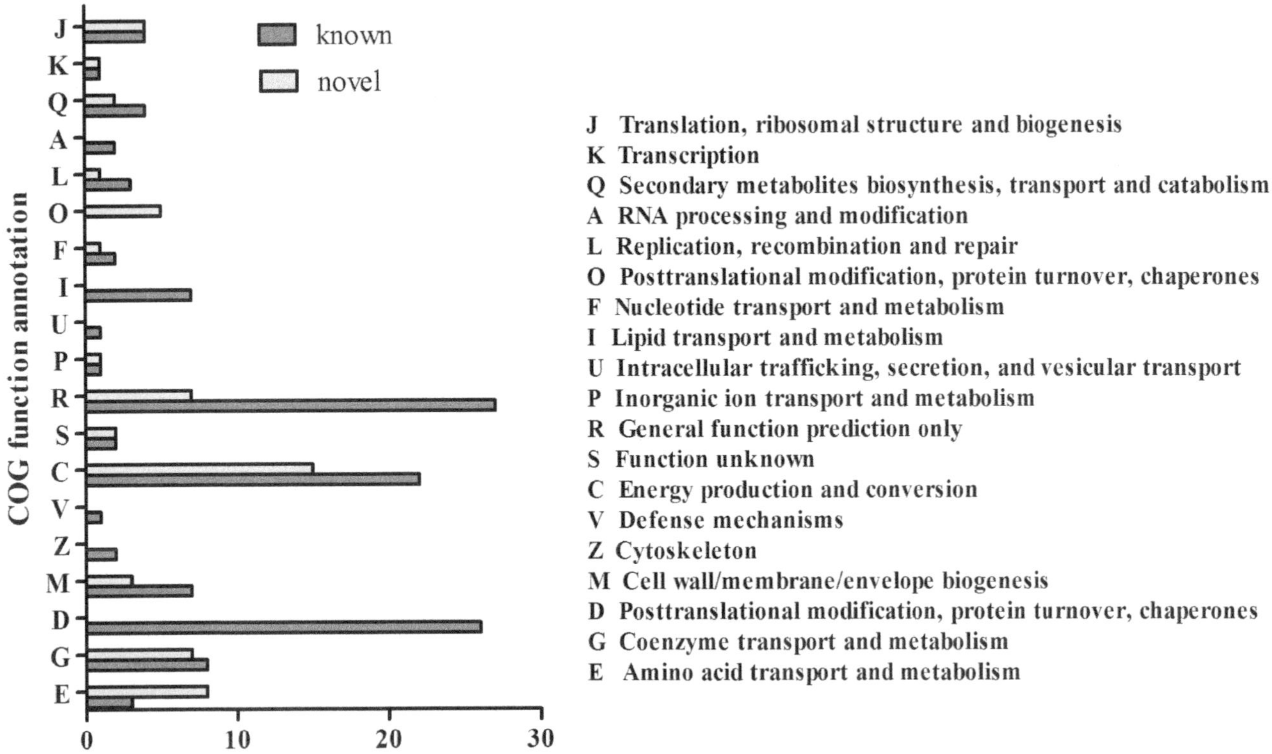

Figure 4. COG functional classification of the miRNA target genes identified in the *L. chinensis* libraries.

GW_rep_c87771 and GW_c41075, while 109 predicted targets were up- or down-regulated. Among them, 68 had similar expression patterns to the miRNAs that regulate them (table S4, table S7). We then analyzed five miRNAs (miR164a, miR172b, miR444a, miR1120 and miR1318) and their predicted targets (from the 68 described above) by qRT-PCR. After considering the miRNA expression patterns, seven target genes showed a moderate increase in expression under both saline–alkaline and drought conditions (Figure 5A, 5B). Finally, analysis of GW_c13268 by 5′-RACE revealed an mRNA fragment consistent with miRNA-guided cleavage in agreement with previous results obtained for predicted targets of known miRNAs (Figure 5C).

Discussion

Many recent studies have demonstrated that plant miRNAs are involved in environmental stress tolerance [22] and some stress-regulated miRNAs have been shown to be involved in the cell response to abiotic stresses such as salinity, cold and dehydration [23,24]. Much attention has been devoted to the identification of components involved in signal transduction pathways, such as abscisic acid signaling, because they are known to participate in responses against the adverse effects of different stresses, such as salinity and drought [25,26]. Stress response-specific miRNAs and some stress-responsive target genes were observed in previous studies [27,28]; for example, in rice, osa-miR169g was reported to be the only miRNA member of the miR169 family induced by drought [23].

Application of the newly developed high-throughput sequencing technologies has led to the identification of entire sets of miRNAs, which has delivered new insights into the role of miRNAs in plant development, and stress-related regulation. However, until now, little was known about the functions of miRNAs in abiotic stress responses in *L. chinensis*. Here, we identified 132 known miRNAs and 16 novel miRNAs after sequencing and analyzing the sRNAs of *L. chinensis*. Many miRNAs with a wide range of expression levels were found in the control, drought and saline-alkaline libraries. The most abundantly expressed miRNA family across the three libraries was miR156, which includes miR156d, miR156f, miR156h, miR156b (Table S2). miR156d, miR156f and miR156h are conserved in many species and they exhibited high expression levels in all three *L. chinensis* libraries. The most abundant was miR156d with 2,039,447 reads in the control library. Some miRNAs were differentially expressed between the two stress-induced libraries (Figure 2) and nine of them were significantly differentially expressed under the two stress conditions. Under the saline-alkaline stress conditions, four (lch-miR160a, lch-miR160d, lch-miR160f and lch-miR394) were significantly up-regulated with fold-changes >4 and two (lch-miR319a and lch-miR319b) were significantly down-regulated with fold-changes >4. lch-miR396f-3p was significantly down-regulated under the drought stress conditions. Previous reports found that members of the miR319 and miR393 families were stress inducible [29,30,31]. Therefore, we speculate that miR160, miR319a and miR396f-3p can be involved in a multi-stress response to saline-alkali conditions based on our results, and also to conditions that induce maintenance of the energy supply based on other reports [11,22]. The qRT-PCR results confirmed the up- and down-regulated trends of the differentially expressed miRNAs in the stress-induced libraries (Figure 3); in particular, lch-miR319a, lch-miR396f-3p and lch-miR397a were predicted to be associated with saline-alkaline and drought stresses.

Because the differential expression of the known miRNAs was less significant, we focused our analyses on the novel miRNAs. Deep sequencing of sRNA transcriptomes yields an incredible amount of data, which can be used to characterize not only known

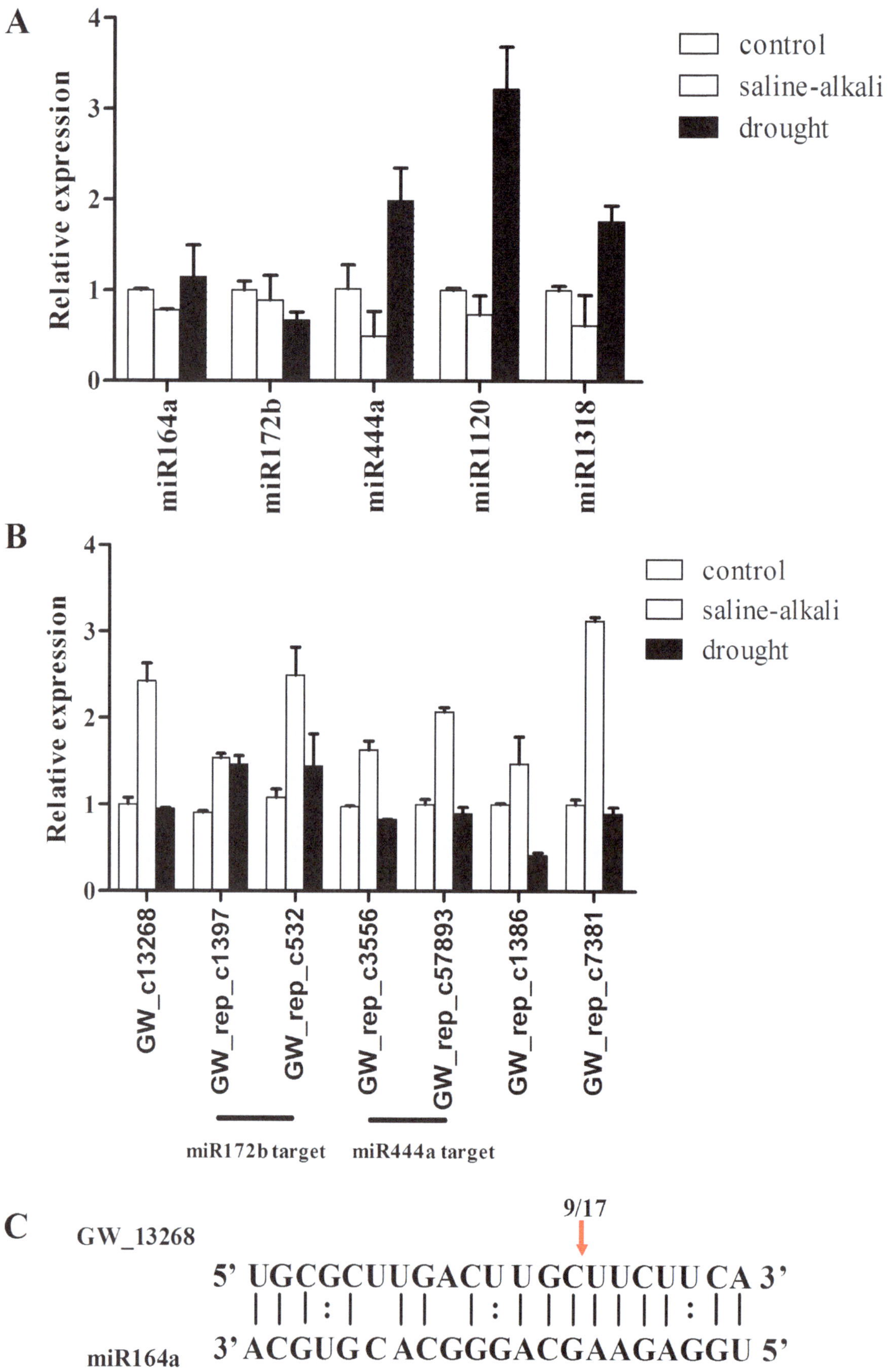
A
Relative expression
control
saline-alkali
drought
miR164a
miR172b
miR444a
miR1120
miR1318
B
Relative expression
control
saline-alkali
drought
GW_c13268
GW_rep_c1397
GW_rep_c532
GW_rep_c3556
GW_rep_c57893
GW_rep_c1386
GW_rep_c7381
miR172b target
miR444a target
C
GW_13268
9/17
5' UGCGCUUGACU U GCUUCUU CA 3'
3' ACGUG C ACGGGACGAAGAGGU 5'
miR164a

Figure 5. Analysis of potential targets in *L. chinensis*. A: Characterization of candidate miRNA expression levels under saline-alkali and drought stress by qRT-PCR; B: Comparison of predicted target expression levels under saline-alkali and drought stress by qRT-PCR; C: Cleavage site analysis of GW_13268 by 5′ RACE PCR.

miRNAs, but also novel miRNAs with high accuracy and efficiency. Because the *L. chinensis* genome has not yet been sequenced, we used complex network methods to construct a non-redundant database of miRNA candidates by selecting the most abundant sequences and aligning them with the rice genome sequence. From the three *L. chinensis* sRNA libraries, 16 miRNAs were revealed as candidate novel miRNAs. The most highly expressed sRNAs in *L. chinensis* are likely to have important functional roles in the plant's stress response.

In plants, the response to abiotic stresses is complex and involves many biochemical and molecular mechanisms that are regulated mainly through the silencing or regulation of target genes by miRNAs. miRNAs as regulators of target genes have been reported to play major roles in a plant's response to abiotic stresses, including dehydration, freezing, salinity, alkalinity [32,33]. The identification of miRNA targets is crucial for understanding the biological effects of miRNAs. The identification of the entire set of miRNAs and their targets from an organism is of fundamental importance for understanding overall gene regulation involving stress responses.

In this study, target genes for miRNAs that were differentially expressed in the three libraries were identified by a search of plant miRNA target sites, which are predominantly located in open reading frames. The predicted miRNA target genes included coproporphyrinogen III oxidase, glutathione S-transferase, ubiquitin-protein ligase, MYB superfamily members, NAD-dependent aldehyde dehydrogenases and genes involved in RNA processing. In *L. chinensis*, these target genes may participate in various aspects of plant development and stress responses. miR159 [34,35], miR160 [36,37], miR167 [38], miR319 [30,39], miR393 [31,40,41,42,43] and miR408 [44,45,46] participate in important and perhaps conserved functions, such as plant growth, development and stress responses. MYB transcription factors regulated by miR159 are involved in the biosynthesis of bioactive compounds, such as ABA [47]. In this study we identified miR159f, which may target components (ribose 5-phosphate isomerase, FAD/FMN-containing dehydrogenases) that were down-regulated under saline-alkali stress. Together with previous evidence for the targeting of MYB by miR159, these predicted targets may be up-regulated and regulated by the ABA signaling pathway. The results suggest that miRNAs may regulate ABA signaling and other stress response processes. Two miR319 members (miR319a and miR319b) were down-regulated under both saline-alkali and drought stress. miR319 was reported as a positive regulator of cold stress in rice by targeting two genes (PCF6 and TCP 21) [30]; however, these targets were not predicted from our *L. chinensis* transcript results. We predicted another two targets of miR319 (GW_rep_c24088/GW_rep_c74568, GW_rep_c58089). miR319 promotes cleavage of GW_rep_c58089 despite having a complementary site within the ORF of its target. Further qRT-PCR analysis revealed that GW_rep_c58089 is up-regulated under drought stress. Because of opposite expression patterns of miR319b and GW_rep_c58089, we suggest that GW_rep_c58089 annotated as ribulose bisphosphate carboxylase small subunit may be another target regulated in saline-alkali and drought stress in *L. chinensis*. Over-expression of miR393 in rice leads to less tolerance to salt and suggests that salt stress may cause decreased TIR1 and AFB2 mRNA degradation or translational repression [48]. miR393 is down-regulated under saline-alkali stress in our dataset. The identification of miR393, which we predict to target five unknown proteins (Table S3), will be valuable for investigating the function of miRNAs. Further evidence is needed to uncover miRNA functions in the saline-alkali stress response. Although qRT-PCR tested the expression pattern of several miRNAs and their corresponding targets, 5′-RACE PCR identified only one target of miR164a; the other six putative targets were not detected, likely reflecting low abundance and coverage of *L. chinensis* transcripts. The global identification of miRNA target genes will provide useful information to explore the functions of miRNAs in plant biogenesis

Supporting Information

Figure S1 Read abundance of various classifications of small RNAs in the *L. chinensis* libraries. A: Read abundance of various classifications of small RNAs in the control sample, B: Read abundance of various classifications of small RNAs in the saline-alkali stress sample C: Read abundance of various classifications of small RNAs in the drought stress sample.

Table S1 microRNA primers for qRT-PCR.

Table S2 microRNA and unigene primers for qRT-PCR.

Table S3 5′ RACE PCR GSP primer.

Table S4 Known miRNAs in the three small RNA libraries.

Table S5 Target genes of known miRNAs.

Table S6 Target genes of novel miRNAs.

Table S7 Expression profile of target unigenes.

Author Contributions

Conceived and designed the experiments: HL JZ. Performed the experiments: JZ YD YS QW. Analyzed the data: YD JZ YS. Contributed reagents/materials/analysis tools: NW FW WL HC NY LG KC XC MY. Wrote the paper: JZ YD YS.

References

1. Huang Z, Zhu J, Mu X, Lin J (2004) Pollen dispersion, pollen viability and pistil receptivity in Leymus chinensis. Ann Bot 93: 295–301.
2. Shu QY, Liu GS, Xu SX, Li XF, Li HJ (2005) Genetic transformation of Leymus chinensis with the PAT gene through microprojectile bombardment to improve resistance to the herbicide Basta. Plant Cell Rep 24: 36–44.
3. Mallory AC, Vaucheret H (2006) Functions of microRNAs and related small RNAs in plants. Nat Genet 38 Suppl: S31–36.
4. Gao P, Bai X, Yang L, Lv D, Pan X, et al. (2011) osa-MIR393: a salinity- and alkaline stress-related microRNA gene. Mol Biol Rep 38: 237–242.

5. Zhang X, Zou Z, Gong P, Zhang J, Ziaf K, et al. (2011) Over-expression of microRNA169 confers enhanced drought tolerance to tomato. Biotechnol Lett 33: 403–409.
6. Jeong DH, German MA, Rymarquis LA, Thatcher SR, Green PJ (2010) Abiotic stress-associated miRNAs: detection and functional analysis. Methods Mol Biol 592: 203–230.
7. Shen J, Xie K, Xiong L (2010) Global expression profiling of rice microRNAs by one-tube stem-loop reverse transcription quantitative PCR revealed important roles of microRNAs in abiotic stress responses. Mol Genet Genomics 284: 477–488.
8. Kantar M, Lucas SJ, Budak H (2011) miRNA expression patterns of Triticum dicoccoides in response to shock drought stress. Planta 233: 471–484.
9. Jin W, Li N, Zhang B, Wu F, Li W, et al. (2008) Identification and verification of microRNA in wheat (Triticum aestivum). J Plant Res 121: 351–355.
10. Zhu QH, Spriggs A, Matthew L, Fan L, Kennedy G, et al. (2008) A diverse set of microRNAs and microRNA-like small RNAs in developing rice grains. Genome Res 18: 1456–1465.
11. Ding D, Zhang L, Wang H, Liu Z, Zhang Z, et al. (2009) Differential expression of miRNAs in response to salt stress in maize roots. Ann Bot 103: 29–38.
12. McCormick KP, Willmann MR, Meyers BC (2011) Experimental design, preprocessing, normalization and differential expression analysis of small RNA sequencing experiments. Silence 2: 2.
13. Sun Y, Wang F, Wang N, Dong Y, Liu Q, et al. (2013) Transcriptome exploration in Leymus chinensis under saline-alkaline treatment using 454 pyrosequencing. PLoS One 8: e53632.
14. Chen C, Ridzon DA, Broomer AJ, Zhou Z, Lee DH, et al. (2005) Real-time quantification of microRNAs by stem-loop RT-PCR. Nucleic Acids Res 33: e179.
15. Song C, Wang C, Zhang C, Korir NK, Yu H, et al. (2010) Deep sequencing discovery of novel and conserved microRNAs in trifoliate orange (Citrus trifoliata). BMC Genomics 11: 431.
16. Zhu J, Li W, Yang W, Qi L, Han S (2013) Identification of microRNAs in Caragana intermedia by high-throughput sequencing and expression analysis of 12 microRNAs and their targets under salt stress. Plant Cell Rep 32: 1339–1349.
17. Lagos-Quintana M, Rauhut R, Lendeckel W, Tuschl T (2001) Identification of novel genes coding for small expressed RNAs. Science 294: 853–858.
18. Carlsbecker A, Lee JY, Roberts CJ, Dettmer J, Lehesranta S, et al. (2010) Cell signalling by microRNA165/6 directs gene dose-dependent root cell fate. Nature 465: 316–321.
19. Matts J, Jagadeeswaran G, Roe BA, Sunkar R (2010) Identification of microRNAs and their targets in switchgrass, a model biofuel plant species. J Plant Physiol 167: 896–904.
20. Griffiths-Jones S, Grocock RJ, van Dongen S, Bateman A, Enright AJ (2006) miRBase: microRNA sequences, targets and gene nomenclature. Nucleic Acids Res 34: D140–144.
21. Griffiths-Jones S, Saini HK, van Dongen S, Enright AJ (2008) miRBase: tools for microRNA genomics. Nucleic Acids Res 36: D154–158.
22. Jones-Rhoades MW, Bartel DP (2004) Computational identification of plant microRNAs and their targets, including a stress-induced miRNA. Mol Cell 14: 787–799.
23. Zhao B, Liang R, Ge L, Li W, Xiao H, et al. (2007) Identification of drought-induced microRNAs in rice. Biochem Biophys Res Commun 354: 585–590.
24. Sunkar R, Zhou X, Zheng Y, Zhang W, Zhu JK (2008) Identification of novel and candidate miRNAs in rice by high throughput sequencing. BMC Plant Biol 8: 25.
25. Zhu JK (2002) Salt and drought stress signal transduction in plants. Annu Rev Plant Biol 53: 247–273.
26. Seki M, Umezawa T, Urano K, Shinozaki K (2007) Regulatory metabolic networks in drought stress responses. Curr Opin Plant Biol 10: 296–302.
27. Prashanth SR, Sadhasivam V, Parida A (2008) Over expression of cytosolic copper/zinc superoxide dismutase from a mangrove plant Avicennia marina in indica rice var Pusa Basmati-1 confers abiotic stress tolerance. Transgenic Res 17: 281–291.
28. Xu DQ, Huang J, Guo SQ, Yang X, Bao YM, et al. (2008) Overexpression of a TFIIIA-type zinc finger protein gene ZFP252 enhances drought and salt tolerance in rice (Oryza sativa L.). FEBS Lett 582: 1037–1043.
29. Sunkar R, Chinnusamy V, Zhu J, Zhu JK (2007) Small RNAs as big players in plant abiotic stress responses and nutrient deprivation. Trends Plant Sci 12: 301–309.
30. Wang ST, Sun XL, Hoshino Y, Yu Y, Jia B, et al. (2014) MicroRNA319 positively regulates cold tolerance by targeting OsPCF6 and OsTCP21 in rice (Oryza sativa L.). PLoS One 9: e91357.
31. Gupta OP, Meena NL, Sharma I, Sharma P (2014) Differential regulation of microRNAs in response to osmotic, salt and cold stresses in wheat. Mol Biol Rep 41: 4623–4629.
32. Li H, Dong Y, Yin H, Wang N, Yang J, et al. (2011) Characterization of the stress associated microRNAs in Glycine max by deep sequencing. BMC Plant Biol 11: 170.
33. Song QX, Liu YF, Hu XY, Zhang WK, Ma B, et al. (2011) Identification of miRNAs and their target genes in developing soybean seeds by deep sequencing. BMC Plant Biol 11: 5.
34. Zheng B, Chen X, McCormick S (2011) The anaphase-promoting complex is a dual integrator that regulates both MicroRNA-mediated transcriptional regulation of cyclin B1 and degradation of Cyclin B1 during Arabidopsis male gametophyte development. Plant Cell 23: 1033–1046.
35. Reyes JL, Chua NH (2007) ABA induction of miR159 controls transcript levels of two MYB factors during Arabidopsis seed germination. Plant J 49: 592–606.
36. Subramanian S, Fu Y, Sunkar R, Barbazuk WB, Zhu JK, et al. (2008) Novel and nodulation-regulated microRNAs in soybean roots. BMC Genomics 9: 160.
37. Bazzini AA, Hopp HE, Beachy RN, Asurmendi S (2007) Infection and coaccumulation of tobacco mosaic virus proteins alter microRNA levels, correlating with symptom and plant development. Proc Natl Acad Sci U S A 104: 12157–12162.
38. Liu N, Wu S, Van Houten J, Wang Y, Ding B, et al. (2014) Down-regulation of AUXIN RESPONSE FACTORS 6 and 8 by microRNA 167 leads to floral development defects and female sterility in tomato. J Exp Bot 65: 2507–2520.
39. Spanudakis E, Jackson S (2014) The role of microRNAs in the control of flowering time. J Exp Bot 65: 365–380.
40. Windels D, Bielewicz D, Ebneter M, Jarmolowski A, Szweykowska-Kulinska Z, et al. (2014) miR393 is required for production of proper auxin signalling outputs. PLoS One 9: e95972.
41. Mao G, Turner M, Yu O, Subramanian S (2013) miR393 and miR164 influence indeterminate but not determinate nodule development. Plant Signal Behav 8: doi: 10 4161/psb 26753.
42. Chen ZH, Bao ML, Sun YZ, Yang YJ, Xu XH, et al. (2011) Regulation of auxin response by miR393-targeted transport inhibitor response protein 1 is involved in normal development in Arabidopsis. Plant Mol Biol 77: 619–629.
43. Si-Ammour A, Windels D, Arn-Bouldoires E, Kutter C, Ailhas J, et al. (2011) miR393 and secondary siRNAs regulate expression of the TIR1/AFB2 auxin receptor clade and auxin-related development of Arabidopsis leaves. Plant Physiol 157: 683–691.
44. Feng H, Zhang Q, Wang Q, Wang X, Liu J, et al. (2013) Target of tae-miR408, a chemocyanin-like protein gene (TaCLP1), plays positive roles in wheat response to high-salinity, heavy cupric stress and stripe rust. Plant Mol Biol 83: 433–443.
45. Mutum RD, Balyan SC, Kansal S, Agarwal P, Kumar S, et al. (2013) Evolution of variety-specific regulatory schema for expression of osa-miR408 in indica rice varieties under drought stress. FEBS J 280: 1717–1730.
46. Zhang H, Li L (2013) SQUAMOSA promoter binding protein-like7 regulated microRNA408 is required for vegetative development in Arabidopsis. Plant J 74: 98–109.
47. Li C, Lu S (2014) Genome-wide characterization and comparative analysis of R2R3-MYB transcription factors shows the complexity of MYB-associated regulatory networks in Salvia miltiorrhiza. BMC Genomics 15: 277.
48. Xia K, Wang R, Ou X, Fang Z, Tian C, et al. (2012) OsTIR1 and OsAFB2 downregulation via OsmiR393 overexpression leads to more tillers, early flowering and less tolerance to salt and drought in rice. PLoS One 7: e30039.

2

AtRD22 and AtUSPL1, Members of the Plant-Specific BURP Domain Family Involved in *Arabidopsis thaliana* Drought Tolerance

Vokkaliga Thammegowda Harshavardhan[1], Le Van Son[2,3], Christiane Seiler[1], Astrid Junker[4], Kathleen Weigelt-Fischer[4], Christian Klukas[5], Thomas Altmann[4], Nese Sreenivasulu[1,6], Helmut Bäumlein[2], Markus Kuhlmann[1]*

1 Research Group Abiotic Stress Genomics, Interdisciplinary Center for Crop Plant Research (IZN), Halle (Saale), Germany, and Leibniz Institute of Plant Genetics and Crop Plant Research (IPK) Gatersleben, Stadt Seeland, OT Gatersleben, Germany, **2** Research Group Gene Regulation, Leibniz Institute of Plant Genetics and Crop Plant Research (IPK) Gatersleben, Stadt Seeland, OT Gatersleben, Germany, **3** National Key Laboratory of Gene Technology, Institute of Biotechnology Vietnam, Academy of Science and Technology, Hanoi, Vietnam, **4** Research Group Heterosis, Leibniz Institute of Plant Genetics and Crop Plant Research (IPK) Gatersleben, Stadt Seeland, OT Gatersleben, Germany, **5** Research Group Image Analysis, Leibniz Institute of Plant Genetics and Crop Plant Research (IPK) Gatersleben, Stadt Seeland, OT Gatersleben, Germany, **6** Grain Quality and Nutrition Center, International Rice Research Institute (IRRI), Metro Manila, Philippines

Abstract

Crop plants are regularly challenged by a range of environmental stresses which typically retard their growth and ultimately compromise economic yield. The stress response involves the reprogramming of approximately 4% of the transcriptome. Here, the behavior of *AtRD22* and *AtUSPL1*, both members of the *Arabidopsis thaliana* BURP (BNM2, USP, RD22 and polygalacturonase isozyme) domain-containing gene family, has been characterized. Both genes are up-regulated as part of the abscisic acid (ABA) mediated moisture stress response. While *AtRD22* transcript was largely restricted to the leaf, that of *AtUSPL1* was more prevalent in the root. As the loss of function of either gene increased the plant's moisture stress tolerance, the implication was that their products act to suppress the drought stress response. In addition to the known involvement of AtUSPL1 in seed development, a further role in stress tolerance was demonstrated. Based on transcriptomic data and phenotype we concluded that the enhanced moisture stress tolerance of the two loss-of-function mutants is a consequence of an enhanced basal defense response.

Editor: Jin-Song Zhang, Institute of Genetics and Developmental Biology, Chinese Academy of Sciences, China

Funding: The work was funded by Vietnamese government. The junior Research Group Abiotic Stress Genomics (MK and NS) is funded by IZN (Interdisciplinary Center for Crop Plant Research, Halle (Saale), Germany. VTH and CS are funded by BMBF (GABI-GRAIN: Integrative genomics approach for exploring seed quality and yield under terminal drought; FKZ 0315041A and FKZ 0315041C) and IB-BMBF (Ind09/526). The funders had no role in study design, data collection and analysis, decision to publish, or preparation of the manuscript.

Competing Interests: The authors have declared that no competing interests exist.

* Email: kuhlmann@ipk-gatersleben.de

Introduction

Abiotic stress factors such as moisture stress, salinity, extreme temperature and variable light intensity can disturb plant metabolism and growth. It has been estimated that crop yield losses due to such stresses lie in the order of 50% [1], so increasing the resilience of crop plants will be an important contributor to yield stability. Abiotic stress affects both photosynthesis and photorespiration, as well as having an impact on the energy and redox status of the plant cell. One of the most damaging products of stress is the group of compounds referred to as reactive oxygen species (ROS). The plant response also includes the induced synthesis of certain enzymes and low molecular weight compounds associated with antioxidant activity, redox regulators, chaperones such as heat shock proteins and late embryogenesis abundant proteins, water and ion transporters, the production of compatible osmolytes to maintain cellular water content and the fine tuning of proteolysis involved in programmed cell death. Numerous attempts, with varying levels of success, have been made to genetically engineer the production of some of these components with a view to enhancing abiotic stress tolerance [2].

The phytohormone abscisic acid (ABA) is involved in the regulation of expression of many stress-responsive genes, although other stress responsive genes are known to be regulated in an ABA-independent manner [3–5]. The regulation of several stress-inducible genes is mediated by the interaction of bZIP transcription factors (ABFs) with ABA-response elements (ABRE) in target gene promoters [6].

The *Arabidopsis thaliana* gene *RESPONSIVE TO DEHYDRATION22* (*AtRD22*), originally identified as a gene which responded to dehydration [7,8], encodes a member of the BURP protein family, members of which share a highly conserved BURP domain at their C terminus [9]), sometimes also referred to as a U domain [10,11]. BURP proteins appear to be plant-specific; some examples are the *Vicia faba* unknown seed protein (USP) [10], the

Brassica napus microsporogenesis-specific protein BNM2 [12–14], the *Panicum maximum* apomixis-specific protein [15] and the wheat pollen protein RAFFTIN [16]. The soybean genome harbors 23 BURP protein encoding genes, of which 17 are responsive to stress [17–20]. The 18 *Populus trichocarpa* BURP family members fall into five recognizable sub-families [21]. There are 15 related genes in maize, one of which is specifically expressed in the root cortex parenchyma [22] while the sorghum genome harbors 11 homologs [23]. The ectopic expression of the *A. thaliana* gene *AtUSPL1* has been shown to distort seed development and to alter the morphology of seed lipid vesicles [24].

AtRD22 is up-regulated by moisture stress, salinity stress and exogenously supplied ABA [8] and its induction has been used as a marker for abiotic stress [25–29]. The heterologous expression in both *A. thaliana* and rice of the soybean gene *GmRD22* enhances salinity stress tolerance [30]. Members of the BURP family are up-regulated by stress in rice [31], soybean [30] and maize [23]. Members of the BURP family have been described in relation to stress conditions. In cotton, an RD22-like protein interacts with an α-expansin and the over expression of both proteins simultaneously promotes growth and fruit weight [32].Remarkably, the expression programmes active during stress response partially overlap with gene expression during early embryogenesis and seed desiccation [24]. Most likely ancient stress response genes have been recruited to protect seed tissue from dehydration stress in drying seeds [33,34]. Here, a combination of genetic, molecular and physiological approaches has been applied to isolate and characterize T-DNA insertion mutants of *AtRD22* and *AtUSPL1*.

Materials and Methods

Plant material and growth conditions

A. thaliana seeds (ecotype Col-0 and the three T-DNA insertion lines SALK_146066 (*rd22-1*), WiscDsLox481-484P12 (*rd22-2*) and SALK_022325 (*uspl1*)) obtained from the European Arabidopsis Stock Center (NASC) were stratified before imbibing them for three days at 4°C in the dark. The resulting germinated seedlings were grown in soil for four weeks under 60% relative humidity, a constant temperature of 22°C and under a 16 h photoperiod (light intensity of 120 $\mu mol\ m^{-2}\ s^{-1}$). Drought stress was applied by a. active dehydration under low humidity for 1–5 days and b. withholding water for 1–5 days; the relative soil water content was monitored using an HH2 moisture meter (delta-T devices, Cambridge, UK). Moisture stress was also mimicked in two week old seedlings by transferring them for three days on a MS basal medium containing one of 150 mM or 300 mM NaCl, 100 µM ABA, 4% w/v trehalose, 4% w/v sorbitol, 4% w/v glucose, 4% w/v fructose, 4% w/v sucrose, 300 mM mannitol, 15% w/v PEG 6000 or 4% w/v PEG 20000.

Detailed information is provided in Figure 1/Figure S4. Primers used for genotyping are listed in Table S1. Drought stress was applied by withdrawal of water and relative soil water content [%] was controlled with HH2 moisture meter (delta-T devices Ltd, Cambridge, UK).

Plasmid construction and plant transformation

Standard protocols [35] were used to prepare plasmids for Gateway Cloning Technology (Invitrogen GmbH, Karlsruhe, Germany). To achieve its ectopic expression, the *AtUSPL1* coding sequence was cloned into the pBENDER GATEWAY vector (a gift from B. Weisshaar, MPI, Cologne).

A 962 bp fragment harboring the *AtUSPL1* promoter was amplified, re-sequenced and inserted into the *Sph*I/*Hind*III

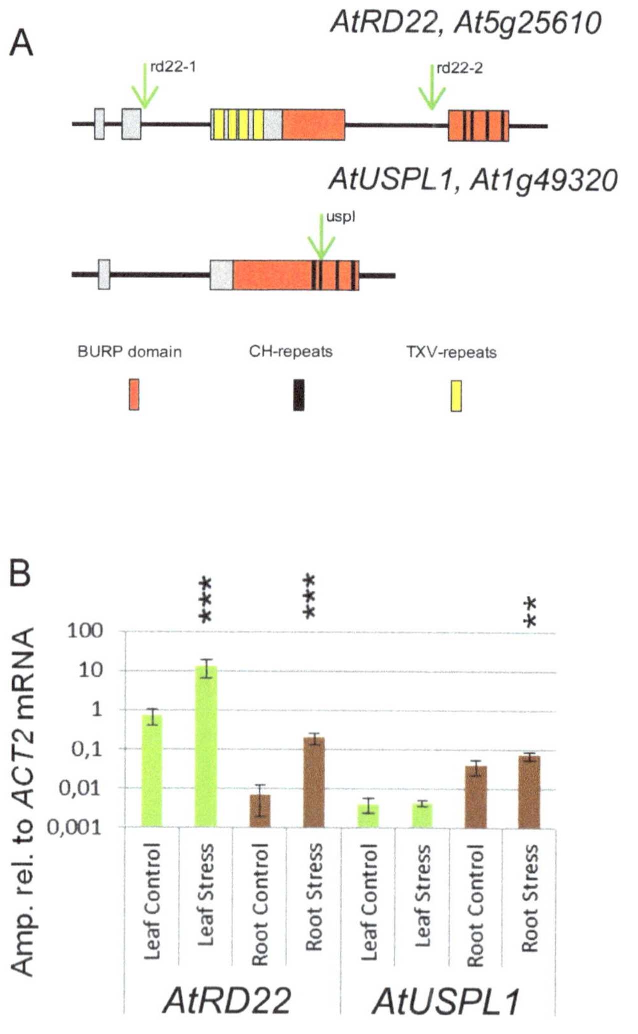

Figure 1. *AtRD22* and *AtUSPL1*, members of the BURP gene family. A. Scheme of BURP-domain containing proteins AtRD22 and AtUSPL1. BURP (named after BNM2, USP, RD22 and Polygalacturonase isozyme) proteins are identified by their C-terminal BURP-domain (red). The BURP domains contain 4 CH-repeats (black). In comparison to AtUSPL1, AtRD22 contains an additional motif, the TXV repeats (yellow), in AtUSPL1 no repetitive domain structure can be found. AGI ID is given in brackets. Position of T-DNA insertions for the used loss-of function mutants is indicated by a green arrow. B. Quantification of *AtRD22* and *AtUSPL1* mRNA in leaves and roots. The bar diagram indicates the amplification of *AtRD22* (left) and *AtUSPL1* mRNA relative to *ACT2* mRNA (amp. rel. to *ACT2* mRNA). Leaf (green) and root (brown) tissue of well watered and drought stressed (2% RWC) plants (N = 5) was analyzed. Asterisks indicate significant differences (T-test: *** $p<0.01$, ** $= p< 0.05$).

cloning site of pBIN101 (Clontech) in front of the *uidA GUS* reporter gene. Stable transformation of *A. thaliana* was performed following [36] and subsequent GUS staining to detect transgene expression following [37].

Automated plant phenotyping and image analysis

After 14 days of pre-cultivation pots (diameter 10 cm) were transferred to an automated phenotyping system composed of a conveyor belt transport system and three imaging chambers for plant imaging in the visible (390–750 nm) and NIR (1400–1510 nm with detection peak at 1452 nm) wavelength ranges as well as capturing fluorescence emission (520–750 nm) as well as a watering and weighing station (LemnaTec 3D Scanalyzer; LemnaTec, GmbH, Aachen, Germany) installed in a climate controlled plant cultivation chamber (environmental conditions: 8 h/16 h night/day, 20/22°C (night/day), relative air humidity of 60% and 240 µmol m^{-2} s^{-1} light intensity). Ten blocks (corresponding to ten biological replicates) with each 15 plants (2×5 genotypes under stress conditions, 1×5 under control conditions) were used. Half of the blocks were placed on opposite sides of the cultivation chamber and pots within one block were randomized in order to account for lane or block effects. For reasons of acclimation, plants were grown under control conditions for another week (till day 21 after sowing) before stress application was started. Automated watering was performed every 2nd day to reach defined target values of the pot weight (70% field capacity for control, no watering for soil drought stress). Relative soil water content [%] was controlled in parallel to pot weight with HH2 moisture meter (delta-T devices, Cambridge, UK). The growth rate was calculated as Relative Growth Rate (RGR) = (log(LA[tn+1]) - log(LA[tn]))/((tn+1) - tn) following [38]; Leaf area (LA), timepoint n (tn). Statistical significant growth differences were estimated by Welch t-test from values for plant area determined from top view (visual [39]). Detailed information on calculation of NIR intensities and Cameras used are provided in Table S3.

Estimation of chlorophyll content

Chlorophyll and pheophytin (a chlorophyll degradation product were extracted from the aerial tissue of 6–7 seedlings using a 1:1 acetone:DMSO mixture, and the absorbance of the extract was recorded at 663, 645 and 553 nm [40] using the UVIKON XS 60/99-90286 spectrophotometer. The experiments comprised three independent replicates. Since the chlorophyll content in the mutant plants grown under non-stressed conditions differed from that present in wild type plants, a relative value was calculated to derive the effect of the stress treatment (Figure S6C). Chlorophyll and pheophytin contents were estimated according to [41], and subjected to a one way ANOVA.

RNA isolation qRT-PCR and Northern analysis

RNA was isolated using an RNeasy kit (Qiagen, Hilden, Germany). For the purposes of Northern blotting, 10 µg RNA was loaded into each lane of a 1.2% w/v agarose, 15% v/v formaldehyde gel, electrophoresed, then transferred passively onto a Hybond N+ membrane (Amersham, GE Healthcare, Waukesha, USA) using 10× SSC as the transfer buffer. The RNA was cross-linked to the membrane by UV irradiation. An *AtUSPL1* probe was amplified from *A.thaliana* genomic DNA using primer pair USPa/b (sequences given in Table S1) and labeled with α-^{32}P dCTP via random priming (Ready Prime Labeling, Pharmacia, GE Healthcare, Waukesha, USA). Membrane/probe hybridizations were carried out in Church hybridization solution [42] at 65°C with pre-hybridization for 6 h and hybridization for 16 h. The membrane was washed for 15 min twice each in 2xSSC, 0.1% SDS, 0.5xSSC, 0.1% SDS and 0.1xSSC, 0.1% SDS at 65°C. Signals were detected and quantified with a Bio-Imaging analyser BAS2000 or X-ray film (Fuji Photo Film Co. Ltd., Tokyo, J). Quantitative real time PCRs (qRT-PCR) were run following [43], in order to assess transcript abundances in leaf and root tissue of six week old plants (five replicate RNA extractions per biological sample, and three technical replications per RNA extract). All primer sequences are given in Table S1.

RNA Isolation, target synthesis and microarray hybridization

Total RNA was isolated from the aerial tissue of two week old seedlings using the TRIzol reagent (Invitrogen) and RNAeasy columns (Qiagen) (see Figure S7 for detailed information). The integrity of the RNA was monitored by agarose gel electrophoresis, and its concentration estimated using a NanoDrop device (Peqlab, Erlangen, Germany), following the manufacturer's protocol. The integrity of the RNA was further confirmed using an Agilent 2100 Bioanalyzer in conjunction with the RNA 6000 Nano assay (Agilent Technologies, Böblingen, Germany). RNA was processed for use on an Affymetrix Arabidopsis ATH1 Genome Array (ATLAS Niolabs GmbH, Berlin, Germany).

Microarray data analysis

The ATH1 chip, which assays 22,392 unique genes, was used to contrast the transcriptomes of wild type Col-0, *rd22* and *uspl1* plants grown under both control and moisture stressed conditions. Probe set to target gene mappings were taken from the TAIR Web site: ftp://ftp.arabidopsis.org/home/tair/Microarrays/Affymetrix/affy_ATH1 _array_elements2010-12-20.txt. The microarray experiment and basic data interpretation was performed by ATLAS biolabs GmbH (Berlin, Germany) by GeneChip Operating System (GCOS) 1.4. To ensure reliability of the analyses, each GeneChip experiment was performed with two biological replicates. After logarithmic transformation of the data, the average expression for all experimental samples for this probe set was subtracted from each individual expression value, thus leading to a positive value in case of above-average expression levels and a negative value in case of below-average expression levels. GeneSpring GX software (Agilent) was used to gene-wise normalize the expression data. To identify potentially differentially expressed genes, the fold changes >2 and <2 of expression values were identified. This was done separately for the Col-0 and mutant series at different conditions. For the Col-0 and respective mutant sample, a simple moderated t-test was performed and P values were corrected using the Benjamini and Hochberg [44] false discovery rate control, applying standard limma procedures. Differentially expressed genes, between Col-0 and mutant samples were identified for both conditions using the limma nestedF procedure, applying a significance threshold of 0.5 in combination with Benjamini-Hochberg false-discovery rate control and a minimal log2-fold change value of 2. A functional categorization of the differentially transcribed genes was derived using Mapman software [45].

Results

The *BURP* gene family

The *A. thaliana* genome contains five *BURP* family genes; these include both *AtUSPL1 (At1g49320)* and *AtRD22 (At5g25610)* (Figure 1A), but also three genes encoding proteins sharing similarity with the tomato non-catalytic β-subunit of polygalacturonase. The latter have been proposed to be designated as *AtPG1* (*At1g60390*), *AtPG2* (*At1g70370*) and *AtPG3* (*At1g23760*). Based on an alignment of related sequences extracted from various species, the family can be subdivided into eight sub-families [23], according to which *AtRD22* belongs to the ATRD22-like subgroup, *AtUSPL1* to the *BNM2*-like sub-family, and *AtPG1-3* to the PG1β-like subfamily. The *A. thaliana* genome has no representative of either the V-VIII or the VfUSP like sub-groups. The N-terminal regions of *AtPG1-3* each include sequences encoding 21 FXXY–N_{9-11} repeats (of unknown function) [24]. AtRD22 contains four TXV-repeats, while AtUSPL1 has no repetitive features. Based on the domain structure and phenotype this study is restricted to the functional analysis of *AtRD22* and *AtUSPL1*.

Transcription of *AtRD22* and *AtUSPL1* during development and in response to stress

Some *Arabidopsis BURP* genes were described to be preferentially expressed in early embryogenesis [24], their involvement in stress response was less obvious. An exception of this is *AtRD22*, which was shown previously to be induced under drought treatments [8].

The transcription profiles of *AtRD22* and *AtUSPL1* were recovered from the Genevestigator database ([46] Figure S1A) and verified using qRT-PCR and Northern blotting. Archival microarray data suggested that while *AtRD22* transcript is abundant throughout plant development in the aerial part of the plant, that of *AtUSPL1* is low and is restricted to the root. In non-stressed plants, *AtRD22* transcription is highest in the leaf, and particularly so in the guard cells [46]. *AtUSPL1* transcript is most abundant in the root (Figure 1B, Figure S1B), but is also detectable in the aerial part of the plant early in development. In the root, *AtUSPL1* transcription is stimulated by exposure to either mannitol or NaCl, as is that of *AtRD22* in the leaf [47]. Here, moisture stress up-regulated *AtRD22* transcription was detected, particularly in the leaf. *AtUSPL1* transcript was detectable in the root, and its level was enhanced by the imposition of moisture stress (Figure 1B). *AtUSPL1* expression was also assayed by tracking GUS expression produced by the *pAtUSPL1::GUS* transgene. GUS activity was detected in the young leaf, in the hypocotyl and in the stem (Figure 2A). In the silique, its expression was only detectable in the mature seed funiculum (Figure 2B), while in the developing flower and stem, a low level of expression was observed (Figure 2C). GUS activity was especially strong in the root tip (Figure 2D). Compared to the strong *AtRD22* promoter activity in aerial parts of the plant [26], *ProAtUSPL1* shows strong transcriptional activity in the root tissue.

According to *in silico* data, treatment with ABA strongly induces *AtRD22* in the leaf, and *AtUSPL1* responds similarly in the root. *AtRD22* is also inducible by exposure to mannitol, glucose, nitrate, high levels of illumination, high temperature and salinity (Figure S2). The gene is down-regulated when the plant is treated with paclobutrazol, an inhibitor of gibberellin synthesis, as well as with cycloheximide, a general inhibitor of protein synthesis, and with syringolin, an inhibitor of cell proliferation.

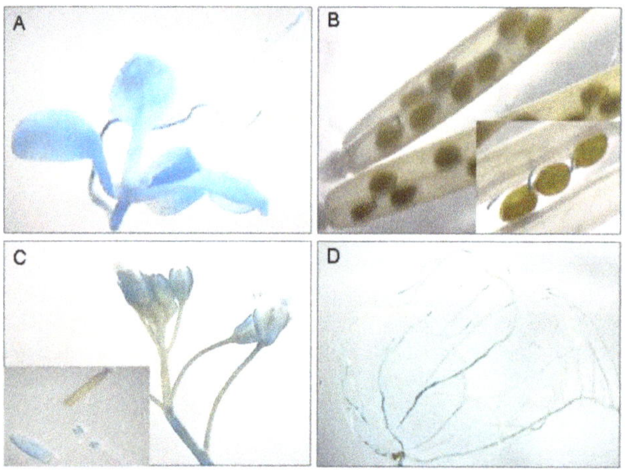

Figure 2. Histochemical *ProAtUSPL1::GUS* activity in transgenic Arabidopsis plants. The *AtUSPL1* promoter activity was determined by histochemical localisation of GUS activity derived from the transgenic *ProATUSPL1::GUS* reporter gene. Activity indicated by blue colour can be seen in A) seedling; B) in funiculus of mature seeds; C) in flowers and stems; and D) in roots.

The behavior of *AtRD22* and *AtUSPL1* T-DNA insertion mutants

In wild type plants, the expression of the two selected *BURP*-gene family members was confirmed by quantitative RT real-time PCR (Figure 1B) and microarray analysis of the aerial part and among them *AtRD22* transcript is abundant under stress treatments (Figure S3). The loss of function mutants achieved by T-DNA insertions were analysed for characterization of *AtRD22 and AtUSPL1* genes in the functional context of drought stress tolerance (Figure S4A). For *AtRD22* two independent T-DNA insertion alleles were used: *rd22-1* and *rd22-2*. For *AtUSPL1* the mutant line *uspl1* was used. To analyse the functional loss of both members of the BURP-gene family the *rd22-1/uspl1* double mutant was analysed. In the two analysed T-DNA insertion lines *rd22-1* and *uspl1* no mRNA of the respective gene was detectable by semi-quantitative PCR and Northern analysis (Figure S4B), which is corroborated by microarray analysis (Figure S4C). Therefore we assume that the used T-DNA mutants represent a loss of function mutation of the respective gene.

Loss of AtRD22 and AtUSPL1 lead to enhanced drought tolerance

The response of the loss-of-function *AtRD22 and AtUSPL1* T-DNA insertion mutants to moisture stress is summarized in Figure 3 and Figure S4/S5. Two independent *AtRD22* (*rd22-1* and *rd22-2*) and one *AtUSPL1* (*uspl1*) mutants were analysed, along with the *rd22-1/uspl1* double mutant. In both *rd22-1* and *uspl1,* transcript of the mutated gene was not detected based on either semi-quantitative PCR or Northern blotting (Figure S4B), corroborating the prediction of microarray analysis (Figure S4C). When subjected to moisture stress for 2–3 days, wild type plants became discolored as a result of an accumulation of anthocyanin and their growth ceased, whereas both the single and double mutant plants remained green and showed no evidence of any growth retardation (Figure 3A). Exposure to a longer period of moisture stress discriminated between wild type and the mutants in a similar manner (Figure S5A).

Detection of plant growth via automated phenotyping

To quantify the drought response of the mutant plants an automated phenotyping platform (LemnaTec) was used. The growth of 30 plants of control (Col-0), *rd22-1*, *rd22-2*, *uspl1* and *rd22-1/uspl1* each under defined control and drought stress condition was analysed. Drought stress was applied from 21 DAS by complete withdrawal of water. To monitor the growth of the plants, leaf area was estimated from top view images ([39,48], Figure S5B, Table S2) and used for calculation of relative growth rates [38]. An elevated growth rate of the *rd22-1* single mutant plant as well as for the double mutant plants compared to wild type could be found in the early phase of the experiment (22 DAS, Figure 3B). The first significant drought related difference in growth rate was detectable after 7 days without watering (28 DAS, Figure S5B). A significant difference in the growth rate of the *rd22/uspl1* double mutant plants compared to the wild type plants was detectable almost throughout the entire experiment initially from day 24 till day 32 after sowing under stress conditions (Figure 3B, Figure S5B).

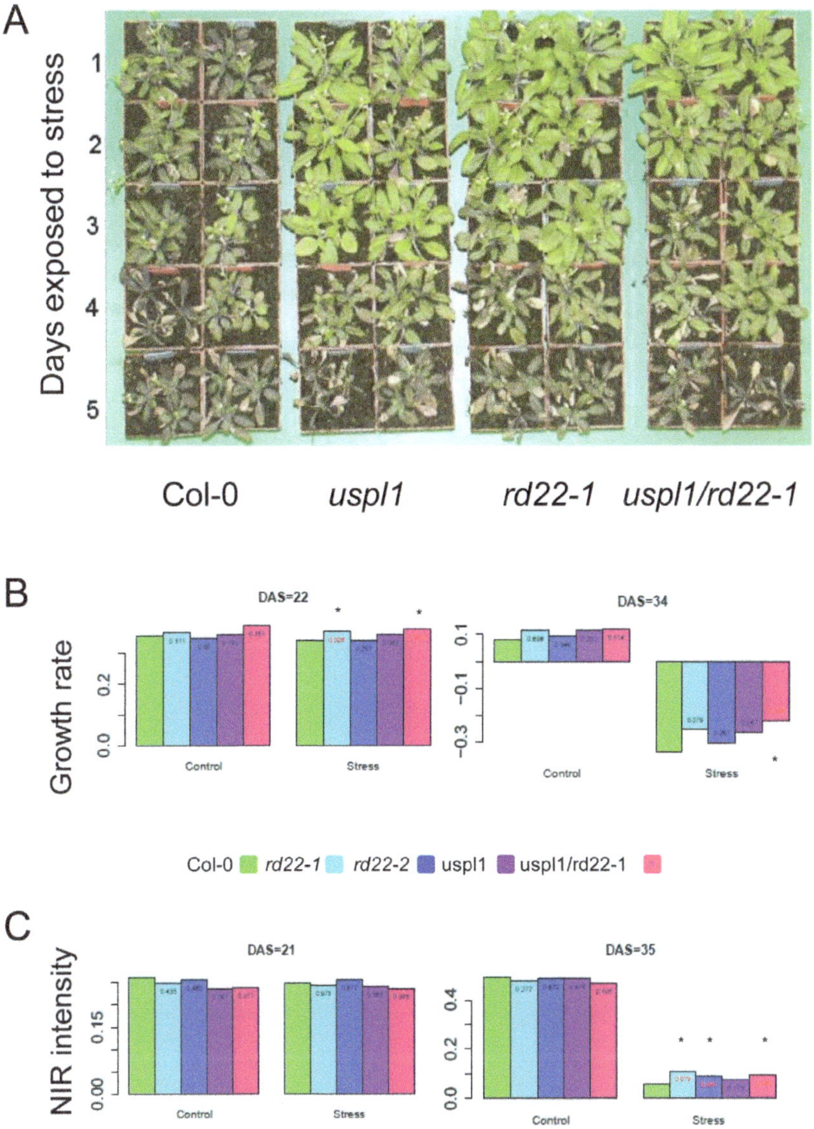

Figure 3. Influence of drought stress treatment on single and double loss of function mutants. A) Four weeks old wild type (Col-0), single and double mutant plants (*rd22-1* and *uspl1, rd22-1/uspl1*) were drought stressed for 1–5 days before they were returned to climate chamber conditions. Time of stress treatment in days is indicated left. B) Growth rates of plants under control and drought stress conditions. Bars indicate the growth rates at 22 days after sowing (DAS) for early phase of drought stress and 34 for the late phase of drought stress. For application of drought stress stop of watering started at 21 DAS. Wild type (Col-0): green bar; *rd22-1*: bright blue bar; *rd22-2*: dark blue bar; *uspl1*: purple bar; *rd22-1/uspl1*pink bar. Original data: Figure S5 B, Table S2, Asterisks indicate significant differences ($p<0.01$). C) NIR reflection as a water content-related parameter. Bars indicate the NIR intensity at 21 days after sowing (DAS) for start of experiment and at 35 DAS for the end of experiment. Wild type (Col-0): green bar; *rd22-1*: bright blue bar; *rd22-2*: dark blue bar; *uspl1*: purple bar; *rd22-1/uspl1*pink bar. $N_{control}=5$, $N_{stress}=10$ plants. Original data in Table S3.

Plant senescence as consequence of the drought stress was monitored by quantifying the ratio of yellow to green pixels. Leaves of the mutant plants *rd22-1*, *uspl1* and the double mutant showed a lower accumulation of yellow stained material, indicative for reduced senescence (33 DAS, Figure S5C). In order to analyse the relative water status of the plants the reflected near-infrared (NIR) radiation (1450 nm) from leaves was detected [49]. NIR intensity is calculated as 1 – NIR reflectance and illustrates the relation to the water content of the leaves. The drought stress related decrease of the NIR intensity was detectable at 33 DAS (Table S3). All genotypes analysed showed a similar level of detectable NIR reflectance in the beginning of the experiment and continuously under control conditions. At the end of the drought stress exposure *rd22* and *rd22/uspl1* double mutant plants showed a higher NIR intensity, indicating higher water content in the leaves (Figure 3C).

Plant growth response to salinity and osmotic stress and exposure to ABA

To further investigate the role of *AtRD22* and *AtUSPL1* during salinity stress and correlating responses on the transcriptional level, wild type and mutant plants were grown on plates with the respective treatments. Two-weeks-old seedlings were transferred for three days to MS basal medium containing one of the following stress inducing compounds: 150 mM and 300 mM NaCl, 100 µM ABA, 4% trehalose, 4% sorbitol, 4% glucose, 4% fructose, 4%

sucrose, 300 mM manitol, 15% PEG 6000 and 4% PEG 20000. Although these conditions are artificially mimicking drought, they were chosen to achieve a uniform plant response to the stimulus. When exposed to either 150 mM NaCl or 100 µM ABA, only minor signs of stress symptoms were apparent, but in the former case, the growth of both the wild type and single mutant plants was retarded (Figure S6A). Wild type and *rd22-1* plants exhibited the least extent of leaf bleaching, while the double mutant and particularly the *uspl1* single mutant, were more visibly affected. Exposure to the various sugars had only a mild effect on plant growth, but a general tendency was for the mutant plants to be more vigorous than the wild type ones. Therefore NaCl 150 mM conditions were chosen for further transcriptome and chlorophyll analysis.

Chlorophyll and pheophytin content

After a four day exposure to moisture stress, the quantity of chlorophyll a and b in the *rd22-1* mutant and the double mutant leaves was less than that in non-stressed plants (Figure S6C). Since the contents of chlorophyll a and chlorophyll b were strongly correlated with one another, subsequent measurements considered the total chlorophyll content. Pheophytin is a degradation product of chlorophyll, that accumulates during senescence, dark [50] and salt stress [51]. In plants grown in the presence of 300 mM NaCl, the content of chlorophyll (Figure 4A) fell sharply, while that of pheophytin rose (Figure 4B). However when challenged by a lesser level of stress (100 µM ABA or 150 mM NaCl), while the chlorophyll content was reduced, there was no measurable increase in pheophytin content (Figure 4C). Chlorophyll content in the *rd22-1* and *uspl1* mutants was more severely reduced in the presence of either 150 mM NaCl or 100 µM ABA. The *uspl1* mutant was the most compromised genotype with respect to chlorophyll content when the medium was supplemented with fructose, while the *rd22-1* and the double mutant plants out-performed the wild type and *uspl1* mutant plants when the stress was imparted by PEG. To clearly indicate this different chlorophyll degradation in the *rd22-1* single and double mutant plants the amount [%] of total clorophyll is indicated relative to the amount of clorophyll at control conditions (Figure 4C). Compared to the amount of chlorophyll under control conditions the *rd22-1* and *uspl1* mutant plants show the strongest reduction upon 150 mM NaCl and 100 µM ABA. In addition, the reduction of chlorophyll in the *uspl1* mutant with fructose show a major difference compared to the other genotypes. In the *rd22-1* and *rd22-1/uspl1* mutant plants the amount of chlorophyll is higher on fructose, 150 mM NaCl and PEG supplemented media compared to the wild type and *uspl1* mutant plants. Taken into account that the chlorophyll content in the mutant in the unstressed condition varies from the wild type a relative value is indicated to show the change due to the treatment (Figure S6B).

The conclusion was that AtRD22 acts to suppress chlorophyll degradation under moisture stress.

Transcriptome analysis

The transcriptional responses to exposure to 150 mM NaCl and to 4% w/v trehalose overlapped by about 10% (Figure S6A). The 150 mM NaCl treatment resulted in a changed transcript abundance for 913 of the 22,392 genes represented on the ATH1 chip. Salinity stress and both the trehalose and sorbitol treatments marginally up-regulated *AtRD22*, while the other *BURP* family genes were hardly affected (Figure S6D). The four genes showing the greatest change in transcript abundance between wild type and mutant plants grown under non-stressed conditions (Table 1) were *At4g33720* (encoding CAP, a PR protein under the control of the DREB2A tanscription factor, [52]: up-regulated by 12 fold in both *rd22-1* and *uspl1* compared to wild type); *At5g49700* (encoding a DNA-binding protein associated with moisture stress [53]: up-regulated by 56 fold in *rd22-1* and by 40 fold in *uspl1*); *At2g27550* (encoding ATC, a systemic inhibitor of floral initiation [54]: up-regulated by 23 fold in both *rd22-1* and *uspl1*); and *At2g44380* (encoding a DC1-domain containing protein, involved in the abiotic stress response, [55]: up-regulated by 22 fold in both *rd22-1* and *uspl1*). The full set of differentially transcribed genes detected in response to the various treatments is given in Table S4 (see also Figure S7). A total of 77 genes displayed differential transcription between wild type and *rd22-1* plants grown under non-stressed conditions; 31 out of these are associated with either biotic or abiotic regulatory pathways (Figure S7A) and included eight encoding a peroxidase putatively involved in H_2O_2 degradation. The absence of AtUSPL1 only re-programmed 18 genes, of which five (one encoding a peroxidase) are associated with either biotic or abiotic regulatory pathways (Figure S7B). Exposure to 150 mM NaCl resulted in the up-regulation of a number of genes associated with either moisture stress or pathogen defense (Table 2). In *rd22-1*, out of the 764 differentially transcribed genes, 231 fell into one of these two categories (Figure S7C), while in the *uspl1* mutant, only 18 genes were up-regulated, of which five were associated with either the biotic or the abiotic stress response (Figure S7D). In the latter mutant, *AtRD22* was strongly up-regulated in the presence of salinity stress. When the stress was applied by the addition of trehalose to the medium, 55 of the 171 differentially transcribed genes identified in the *rd22-1* mutant were associated with either the abiotic or the biotic stress response (Figure S7E). Over 50% (111 out of 212) of the genes responding to trehalose treatment reacted similarly to salinity stress (Figure S3A, Table S4). The set of common up- regulated genes included the transcription factor genes *AtMYB15* (*At3g23250*) and *WRKY33* (*At2g38470*), *LPT3* and *LPT4* (encoding lipid transporters), *At1g35910* (a putative trehalose-6-phosphate phosphatase), *At1g61120* (terpene synthase 04), *At1g16850* (unknown function) and *At1g78410* (VQ motif-containing protein). Although fewer genes were induced by the trehalose (212) than by the 150 mM NaCl (913) treatment, the conclusion was that the two stress agents must affect a similar class of gene. The response to both stress agents also included the down-regulation of several photosynthesis-related genes.

Discussion

RD22 and USPL1 have suppressor function during drought stress

AtRD22 and *AtUSPL1* were transcribed predominantly in, respectively, the leaf and the root. The induction of both genes by moisture stress generated a slight growth reduction. While the induction by moisture stress of *AtRD22* has been noted previously, this was not the case for *AtUSPL1*, most likely due to a concentration on the short-term stress response. Microarray-based transcriptomic analyses have shown that exposure to either NaCl or trehalose induces not only the up-regulation of salinity, trehalose and moisture stress responsive genes, but also the down-regulation of photosynthesis-related genes. The latter is implied by both the induction of leaf chlorosis and senescence and the measured fall in the chlorophyll content of the leaves of stressed plants. The present data confirmed that both salinity and trehalose activate the ABA-mediated moisture stress response. [56–58]. The sole *BURP* gene family to be up- regulated in the aerial part of the plant was *AtRD22,* consistent with observations based on the expression of the transgene *pAtRD22::GUS* [26].

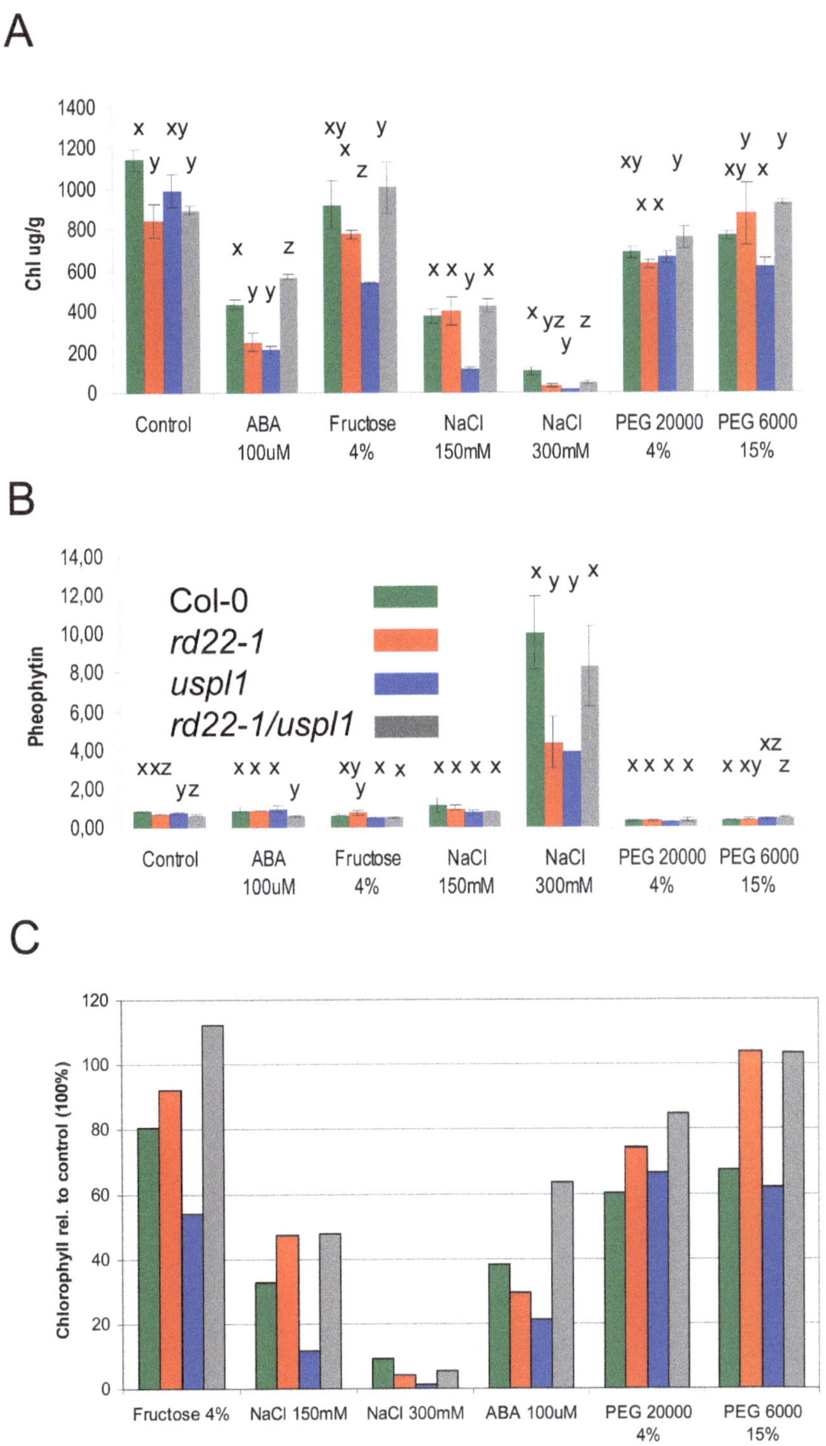

Figure 4. Chlorophyll and pheophytin content in single and double *BURP* mutants. The bars represent the total A) chlorophyll and B) pheophytin content in leaves from wild type (Col-0, green bar), single and double mutant plants (*rd22-1*, red bar, *uspl1*, blue bar and *rd22-1/uspl1*, grey bar). Chlorophyll a and b was determined separately (Figure S6A) and subsumed as total chlorophyll content. Error bar represents standard error. N = 5–6 plants in duplicate. Statistical analysis was performed by oneway ANOVA at alpa = 0.05 Tukey post hoc test: same letters indicate no difference, different letters indicate significant difference. C) The bars show the total chlorophyll content [%] relative to unstressed control plants.

The root specificity of *AtUSPL1* expression was confirmed both by the behavior of the *pAtUSPL1::GUS* transgene and by Northern hybridization experiments.

The T-DNA insertion mutants of both *AtRD22* and *AtUSPL1* produced no detectable relevant transcript, so each was taken as a genuine loss-of- function allele. In both cases, a reduced photosynthesis phenotype was exhibited, as reflected by a fall in

Table 1. Top differential expressed genes in A) *rd22-1* and B) *uspl1* at standard growth conditions in the aerial part of 2 week old seedlings.

A) *rd22*

Probe Set ID	fold change	Gene	AGI ID	put. Function	Probe Set ID	fold change	Gene	AGI ID	put. Function
257162_s_at	59		At3G24290	amm. transporter	265364_at	−45		At2G13330	Transposon
248564_at	56		At5G49700	DNA binding	246908_at	−40	RD22	At5G25610	Disease Defense
249052_at	47	PDF1.2	At5G44420	Disease Defense	251677_at	−38	ORG3	At3G56980	DNA bind
245276_at	41	ATHB-2	At4G16780		257453_at	−30		At1G65130	hydrolase
265102_at	39		At1G30870	Peroxidase	261684_at	−29		At1G47400	
259276_at	38		At3G01190	PER27 Peroxidase	265665_at	−28		At2G27420	
263096_at	38	AHB1	At2G16060		249645_at	−26	THI2.2	At5G36910	toxin receptor
264567_s_at	37		At1G05250	Peroxidase	266472_at	−26			
254044_at	35	XTR9	At4G25820	hydrolase	254170_at	−22		At4G24430	lyase
254200_at	34		At4G24110		247252_at	−22		At5G64770	

B) *uspl1*

Probe Set ID	fold change	Gene	AGI ID	put. Function	Probe Set ID	fold change	Gene	AGI ID	put. Function
247718_at	42	LTP4	At5G59310	Lipid transfer	252795_at	−38			
248564_at	40		At5G49700	DNA bind	246948_at	−23	CYP71B12	At5G25130	Cytochr. P450
261226_at	31	ATEXPA11	At1G20190	Expansin	250207_at	−22	CHS	At5G13930	
253667_at	30		At4G30170	peroxidase	256940_at	−19		At3G30720	
252374_at	24	ARR5	At3G48100	transc. Reg.	267168_at	−2		At2G37770	aldo/keto red.
254979_at	23		At4G10550	subtilase					
254832_at	23		At4G12490	protinase inhibitor					
266209_at	23	ATC	At2G27550	A.th centroradialis					
267385_at	22		At2G44380	DC domain					
247717_at	22	LTP3	At5G59320	Lipid transfer					

Complete list of differential expressed genes in Table S4.

Table 2. Differential expressed genes in A) *rd22-1* and B) *uspl1* on medium containing 150mM NaCl in the aerial part of 2 week old seedlings.

A) *rd22* on NaCl		
up regulated genes		
Probe Set ID	**Fold change**	**Gene Symbol**
247215_at	40	**PROPEP3**
266017_at	38	
252984_at	34	**ELI3-2**
252487_at	33	
256627_at	33	
246340_s_at	32	**FAMT**
264005_at	31	AGP2
257206_at	29	
267035_at	26	**AGT3**
259975_at	25	
253872_at	24	RD26
254101_at	24	**AMY1**
266267_at	23	**ATGSTU4**
256245_at	22	**HSP70**
255502_at	21	
247308_at	20	
256436_at	20	
266142_at	11	
256603_at	11	
258791_at	10	**PR4**
264415_at	6	**RAP2.6**
257517_at	4	
258277_at	4	**PAD3**
260919_at	3	
249481_at	2	
down regulated		
Probe Set ID	**Fold change**	**Gene Symbol**
246366_at	−27	
256772_at	−22	**BGAL1**
250366_at	−21	
261118_at	−20	
261684_at	−20	
267264_at	−17	**SCPL11**
258497_at	−11	**COL2**
261413_at	−11	**PLL5**
246908_at	−7	RD22
247450_at	−2	
261351_at	−2	
266363_at	−1	
B) *uspl1* on NaCl		
up regulated genes		
Probe Set ID	Fold change	Gene Symbol
261684_at	42	
258497_at	21	**COL2**
266363_at	20	
247308_at	18	

Table 2. Cont.

B) *uspl1* on NaCl		
247215_at	15	**PROPEP3**
257206_at	14	
249481_at	13	
261118_at	12	
246908_at	11	RD22
256245_at	11	**HSP70**
261351_at	11	
255502_at	11	
250366_at	11	
264415_at	10	**RAP2.6**
266267_at	10	**ATGSTU4**
256225_at	10	
256603_at	4	
258277_at	2	**PAD3**
266017_at	2	
256627_at	1	
267035_at	1	**AGT3**
down regulated		
Probe Set ID	**Fold change**	**Gene Symbol**
246366_at	−27	
266142_at	−24	
258791_at	−23	**PR4**
267264_at	−20	**SCPL11**
254101_at	−17	**AMY1**
259975_at	−15	
246340_s_at	−15	**FAMT**
256436_at	−15	
256772_at	−14	**BGAL1**
257517_at	−14	
260919_at	−13	
252487_at	−13	
252984_at	−11	**ELI3-2**
264005_at	−11	AGP2
247450_at	−11	
253872_at	−11	RD26
261413_at	−7	**PLL5**

both chlorophyll content under non-stressed growing conditions and photosynthesis-associated gene transcript abundance in stressed plants. Both under prolonged moisture stress mimicking conditions and actual moisture stress, stress tolerance was enhanced in all of the mutants. The mutants' performance (and particularly that of the double mutant), as measured by either the development of leaf area or by the rate of leaf tissue senescence, showed that they were more tolerant of moisture stress than were wild type plants. NIR intensity appeared to reliably reflect relative water content, which was higher in each of the mutants than in wild type.

Given that photosynthesis is clearly compromised by moisture stress, it was not surprising that the transcription of several photosynthesis-associated genes was altered when the plants were exposed to moisture stress. Although the mutant plants' transcriptomes were indistinguishable from that of the wild type with respect to photosynthesis-related genes, nevertheless their leaves contained less chlorophyll than the leaves of wild type plants raised under non-stressed conditions. The reduction in chlorophyll content in the mutants induced by exposure to either NaCl or PEG was less severe than that in wild type plants. Notably, under moisture stress conditions, the leaves of double mutant plants retained more chlorophyll than those of wild type ones. The retention of chlorophyll can be expected to support a higher rate of photosynthesis, so that less transpiration is required to generate a given quantity of assimilate. Since the plant's capacity to retain its water is improved, its water content under moisture stress conditions was greater (Figure 3C).

The observed up-regulation of peroxidase encoding genes in the mutants suggested a secondary effect of the loss-of-function mutations. The transcriptomic data implied that the oxidative state within the mutant plants differed from that within the wild type, with a knock-on effect on gene expression in both ABA-dependent and ABA-independent pathways [59]. Both the phenotype of the mutants as well as the transcriptional response of both genes suggested that under moisture stress conditions, their products exerted a suppressor function. Similar conclusions have been drawn regarding the effect of mutations of genes encoding SnRK2 [60] the effect of which is an almost complete abolition of the ABA response. In particular, *AtRD22* is not transcribed in these mutants when the growing medium is supplemented with ABA reflecting moisture stress condition. Under normal growing conditions, the *SnRK2* mutants exhibit reduced growth, which (along with plant survival) is improved by the addition of ABA to the growing medium. The inference is that the BURP domain containing proteins AtRD22 and AtUSPL1 act as suppressors of the ABA-mediated moisture stress response.

RD22 act as suppressor predominantly in the leaf, while USPL1 act in the root

Although both *AtRD22* and *AtUSPL1* exhibited organ-specific transcription, the loss of function of each gene induced a comparable improvement in the plant's tolerance to moisture stress, at least at the level of the vigour of the leaves. The sensitivity of the microarray platform rules out any trans-silencing effects of either T-DNA insertion on other members of the BURP gene family. Since this sort of silencing has been associated with several T-DNA insertion mutations [61], the assumption is that both *AtRD22* and *AtUSP1* are involved in a holistic moisture stress response. The microarray analysis demonstrated that the two loss-of-function mutations resulted in the induction of a partially overlapping set of genes in the aerial part of the plant, even though *AtUSPL1* was transcribed specifically in the root. Not only was the tolerance of moisture stress enhanced in both mutants, but also their chlorophyll content was reduced. As the lack of *AtRD22* and *AtUSP1* led to an enhanced tolerance to moisture stress, the proposed suppressor function of the BURP domain-containing proteins during an episode of moisture stress includes an organ-specific component. Both proteins are part of an ABA mediated moisture stress response pathway, with AtRD22 functioning mainly in the aerial part of the plant and AtUSPL1 in the root.

Increased drought resistance might be correlated to an increased defence gene response

The transcriptomic analysis of the two mutants revealed an increased transcript abundance compared to the wild type levels with respect to various genes associated with the response to biotic stress (for example, the gene *PDF1.2* was up-regulated by nearly 50 fold). This class of genes was differentially transcribed both under non-stressed and moisture stressed conditions. Several peroxidase encoding genes were also up-regulated. Peroxidases are known to represent an integral component of the plant's hypersensitive response [62]. H_2O_2 is also known as signal molecule during the stress response [63,64]. A loss of control over the production of H_2O_2 might result in the described reduction of chlorophyll content [65,66] and increased water content under drought stress conditions. Such regulation would subsequently lead to better performance of the mutant plants under drought stress.

The up-regulation of peroxidase encoding genes implies a level of linkage between the ABA-mediated moisture stress response and defense against pathogen invasion. Such a connection has been proposed in a suggested model for the function of OCP3, a homeodomain transcription factor. The loss of *OCP3* function results in an enhanced tolerance to moisture stress and at the same time an increased sensitivity to ABA. The abundance of *AtRD22* transcript in the *ocp3* mutant is not different to that in the wild type, and the plant's susceptibility to *Botrytis cinerea* infection is reduced [67]. The *ocp3* mutant was initially identified and named after the phenotype of constitutive over expression of a cationic peroxidase [68]. Here, *AtOCP3* was marginally down regulated when the plants were challenged by NaCl, but there was no transcriptional difference between wild type and either of the two mutants. The implication is that the BURP-containing proteins act to enhance the plant's moisture stress response via the up-regulation of peroxidase encoding genes. In soybean, a direct interaction between RD22 and cell wall-localized peroxidases has been described [69]. The up- regulation of H_2O_2 detoxifying enzymes enhances moisture stress tolerance [70] as well as explaining the link with the pathogen defense response.

A reduced chlorophyll content, in conjunction with an elevated level of transcription of defense response genes under non-stressed growing conditions, suggests that the mutant plants are primed to mount a stronger and/or more rapid set of measures to prevent the accumulation of ROS. A change in the accumulation of H_2O_2 would not only have an impact on the defense response [71], but also on the response to moisture stress [63,72]. Due to the overall elevated expression of H_2O_2-detoxifying enzymes such responses could be limited, leading to an enhanced drought resistance. Our study provides the first functional approach to investigate *BURP*-domain encoding gene function in addition to the previously published structural comparisons.

Supporting Information

Figure S1 Expression profile of the *Arabidopsis thaliana BURP* gene family. A) Expression profile of the *Arabidopsis thaliana BURP* gene family. Data obtained from Genevestigator database (Zimmerman *et al.*, 2004). Relative expression of *AtRD22* (red) and *AtUSPL1* (blue) is given for the different developmental stages of Arabidopsis life cycle (left to right: germinating seed, seedling, Young rosette, developed rosette, bolting, young flower, developed flower, flowers and siliques, mature siliques, senescence). B) Expression of *AtUSPL1* confirmed by Northern Blot analysis. Expression of *AtUSPL1* was determined from root, leaf, shoot, young silique and total flower tissue of *Arabidopsis thaliana* (*Col-0*) plants. For equal loading of the RNA samples probing of the membrane with specific probe against housekeeping mRNA of *AtROC1* (rotamase cyclophilin, renamed in *AtCYP1*) was performed. C) Expression of *AtRD22* obtained from Arabidopsis eFP Browser (Winter *et al.*, 2007). The expression of selected stimuli (Cold: 4°C, Osmotic: 300 mM Mannitol, Salt: 150 mM NaCl and Drought: air steam 15 min) is displayed for the aerial as well as the hypogeic part of the plant. *AtRD22* expression is induced in the aerial part of the plant after applying osmotic, salt stress and slightly increased after loss of water due to airstream treatment. D) Expression of *AtUSPL1* obtained from Arabidopsis eFP Browser (Winter *et al.*, 2007). The expression of selected stimuli (Cold: 4°C, Osmotic: 300 mM mannitol, Salt: 150 mM NaCl and Drought: air steam 15 min) is displayed for the aerial as well as the hypogeic part of the plant. *AtUSPL1* expression is induced in the hypogeic part of the plant after applying osmotic, salt stress and slightly increased after loss of water due to airstream treatment.

Figure S2 Expression of the BURP domain containing gene family in *Arabidopsis thaliana*. Expression analysis of *AtRD22* (*At5G25610*) and *AtUSPL1* (*At1G49320*) obtained from Genevestigator database (Zimmerman et al., 2004) displaying induced and reduced expression after different conditions and stresses. Displayed are only changes in expression upon stress/ treatment above threefold with a statistic significance $p<0.001$. Red indicates up-regulation; Green indicates down-regulation.

Figure S3 Comparison of gene expression in Arabidopsis wild type plants grown on 150 mM NaCl and 4% trehalose supplemented MS medium. BURP gene family mRNA in *Col-0* under selected stress conditions. Bars indicate the expression pattern obtained by microarray analysis using ATH1 chip: *AtRD22* (red): 246908_at; *AtUSPL1* (blue): 262388_at; *AtPG1* (dark green): 265131_at; *AtPG2* (green): 264277_at; *AtPG3* (bright green): 264315_at. Displayed is the rel. Abundance of mRNA.

Figure S4 Characterization of BURP mutant plants. A) Scheme of the *AtRD22* and *AtUSPL1* gene model. In blue the exon-intron structure within the coding region of the respective gene is given. The protein structure refers to Figure 1. In the encoded protein parts are given in aminoacids [aa]. And the size of the fulllenght protein is given below the gene description. Position of T-DNA insertions of used mutant plants are indicated by black lines. Mutant alleles for *rd22*: *rd22-1* (SALK_146066) and *rd22-2* (WiscDsLox481-484P12). Mutant alleles for *uspl1*: *uspl1*: (SALK_022325). The T-DNA insertion line SALK_146066 (*rd22-1*) is based on pROK2 conferring kanamycin resistance and the WiscDsLox481-484P12 (*rd22-2*) is based on pWiscDs-Lox conferring phosphinotricin (BASTA) resistance. *uspl1* T-DNA insertion lines SALK_022325 (referred to as *uspl1*, based on pROK2 conferring kanamycin resistance from Nottingham Arabidopsis Stock Centre) was analyzed. The position of the T-DNA insertion in *At1G49320 (AtUSPL1)* was determined by PCR and subsequent sequencing. The position of the T-DNA insertion are depicted and confirmed by PCR. Double mutant *rd22-1/uspl1* line was generated by crossing SALK_146066 and SALK_022325 and identified in the F_3 generation by PCR. Kanamycin and phosphinotricin resistance of the plants was tested on germination medium (1 MS salts; 10 g/l sucrose) plates with 40 mg/l kanamycin or 20 mg/l glufosinate-ammonium under long day conditions. B) Analysis of used T-DNA insertion mutants. The absence of *AtRD22* and *AtUSPL1* mRNA in homozygous *rd22* and *uspl1* mutant plants was determined by Northern Blot analysis (left) and semi quantitative RT-PCR. C) Analysis of BURP-gene family mRNA in *rd22-1* and *uspl1* mutants by microarray analysis on MS medium. Bars indicate the rel. expression signal obtained by microarray analysis using ATH1 chip from each single experiment: *AtRD22* (red): 246908_at; *AtUSPL1* (blue): 262388_at; *AtPG1* (dark green): 265131_at; *AtPG2* (green): 264277_at; *AtPG3* (bright green): 264315_at. (TIF)

Figure S5 Increased drought stress resistance of the *rd22* and *uspl1* mutant plants. A) The plants were drought stressed by withdrawal of water. Top: day 0 (80% RWC in the soil); Bottom: appearance of plants after 8 days without watering. B) Top: Projected area of wild type and mutant plants under control conditions and drought stress (dotted line) obtained by lemnatec phenotyping; Drought stress was started 21 days after sawing (DAS). Middle: Growth rates calculated based on Poorter and Lewis 1986 for individual days. Wild type (Col-0): green line; *rd22-1*: bright blue line; *rd22-2*: dark blue line; *uspl1*: purple line; *rd22-1/uspl1*pink line (+/- s.e.m.). Bottom: Statistical analysis or growth rates at 28 DAS. Asterisks indicate significant differences ($p<0.05$) between control and stress. C) Estimation of senescence after drought stress. The graph indicates the ration of yellow to green pixels in the plant area of the analysed top view images from day 33. Wild type (Col-0): green bar; *rd22-1*: bright blue bar; *rd22-2*: dark blue bar; *uspl1*: purple bar; *rd22-1/uspl1*: pink bar. $N_{control} = 5$, $N_{stress} = 10$ plants. Asterisks indicate significant differences ($p<0.05$).

Figure S6 A) Differentially expressed genes (Col-0) between 150 mM NaCl containing medium and standard growth conditions sorted by relation to pathway (Mapman). Top: Differentially expressed genes (Col-0) between 150 mM NaCl containing medium and standard growth conditions sorted by relation to pathway (Mapman). The different numbers indicate different categories/pathways and are described in the table. Bottom: Differentially expressed genes (Col-0) between standard growth conditions and 4% trehalose containing medium. The bar diagram indicates the percentage of common (red) and inverse (yellow) regulated genes upon salt (grey) and sugar (blue) treatment. Approximately half of the genes induced by 4% trehalose treatment are also reacting on 150 mM NaCl treatment. B) Influence of ABA and NaCl on single and double loss of function mutants. Growth phenotypes of wild type (Col-0), single and double mutant plants (*rd22-1* and *uspl1, rd22-1/uspl1*) on standard MS-medium and 150 mM NaCl, 300 mM NaCl and 100 µM ABA supplemented MS-Medium. Two week old seedlings were transferred for 3 days to the MS basal and supplemented medium. C) Chlorophyll and pheophytin content of wild type and *rd22-1*, *uspl1* and *rd22-1/uspl1* mutant plants on different supplemented media. Two week old seedlings were transferred to the MS basal + one of the following stress treatments for 4 days: 100 µM ABA, 4% fructose, 150 mM NaCl, 300 mM NaCl, 4% PEG 20000 and 15% PEG 6000. From Top to bottom: Chlorophyll a and b content. The error bar represents standard error. Content was estimated from 5-6 plants in duplicate. Statistical analysis was performed by oneway ANOVA at alpa = 0.05 Tukey post hoc test: same letters indicate no difference, different letters indicate significant difference. Chlorophyll a and b and pheophytin content [%] relative to unstressed control plants.

Figure S7 Differentially expressed genes in the *rd22-1* and *uspl1* plants grown on control, 150 mM NaCl and 4% trehalose containing medium. List of used categories is given in Figure 3 A. Top: Display of top regulated category (Bin 20, biotic and abiotic stress pathways) of differential regulated genes. Bottom: Overview of all differentially regulated genes mapped to categories (Bins) by MAPMAN. For RNA extraction tissue was grinded in liquid nitrogen and the homogenized powder was added to 1 ml TRIZOL and incubated at RT for 5 min. Samples were centrifuged at 10.000 rpm for 10 min and the supernatant was transferred to a new tube. 200 µl of chloroform were added and incubated at room temperature for 2–3 min. Samples were again centrifuged as described above and the aqueous supernatant was transferred to the QIAshredder column and centrifuged for 30 s at 10000 rpm. 350 µl of RLT buffer (plus β-mercaptoethanol) and 250 µl of absolute ethanol were added to the flow-through and passed through an RNAeasy spin column. All the following steps were performed as described in the manufacturer's protocol followed by in-column DNAse digestion. A) Schematic display of differentially expressed genes in *rd22-1* by

MAPMAN. 31 out of 77 differential regulated genes are mapping to biotic and abiotic stress pathways. B) Schematic display of differentially expressed genes in *uspl1* by MAPMAN. 5 out of 18 differential regulated genes are mapping to biotic and abiotic stress pathways. C) Schematic display of differentially expressed genes in *rd22-1* on 150 mM NaCl by MAPMAN. 231 out of 764 differentially regulated genes are mapping to biotic and abiotic stress pathways. D) Schematic display of differentially expressed genes in *uspl1* on 150 mM NaCl by MAPMAN. 7 out of 12 differentially regulated genes are mapping to biotic and abiotic stress pathways. E) Schematic display of differentially expressed genes in *rd22* on 150 mM NaCl by MAPMAN. 55 out of 171 differentially regulated genes are mapping to biotic and abiotic stress pathways.

Table S1 Primer information.

Table S2 Average projected plant area (mm2) from top images.

Table S3 Average near-infrared intensity as observed from top images, obtained using a Nir 300 camera from VDS Vosskühler (now Allied Vision Technologies). High values indicate relative high water content.

Table S4 List of differentially expressed genes.

Acknowledgments

We thank Elke Liemann and Jana Lorenz for excellent technical assistance, Dijun Chen for support with statistical analysis of lemnatec data, Ingo Mücke for his support using the automated phenotyping system.

Author Contributions

Conceived and designed the experiments: NS HB MK TA. Performed the experiments: VTH CS LVS KWF. Analyzed the data: AJ CK. Wrote the paper: MK AJ TA.

References

1. Bray EA, Bailey-Serres J, Weretilnyk E (2000) Responses to abiotic stresses. Biochemistry and Molecular Biology of Plants, Rockville, MD, USA: American Society of Plant Physiologists. pp.1158–1203.
2. Sreenivasulu N, Sopory SK, Kishor PBK (2007) Deciphering the regulatory mechanisms of abiotic stress tolerance in plants by genomic approaches. Gene 388: 1–13.
3. Agarwal PK, Agarwal P, Reddy MK, Sopory SK (2006) Role of DREB transcription factors in abiotic and biotic stress tolerance in plants. Plant Cell Rep 25: 1263–1274.
4. Hasegawa PM, Bressan RA, Zhu JK, Bohnert HJ (2000) Plant Cellular and Molecular Responses to High Salinity. Annu Rev Plant Physiol Plant Mol Biol 51: 463–499.
5. Yamaguchi-Shinozaki K, Shinozaki K (2006) Transcriptional regulatory networks in cellular responses and tolerance to dehydration and cold stresses. Annu Rev Plant Biol 57: 781–803.
6. Kasuga M, Liu Q, Miura S, Yamaguchi-Shinozaki K, Shinozaki K (1999) Improving plant drought, salt, and freezing tolerance by gene transfer of a single stress-inducible transcription factor. Nature Biotechnology 17: 287–291.
7. Yamaguchi-Shinozaki K, Shinozaki K (1992) A novel Arabidopsis DNA binding protein contains the conserved motif of HMG-box proteins. Nucleic Acids Res 20: 6737.
8. Yamaguchi-Shinozaki K, Shinozaki K (1993) The plant hormone abscisic acid mediates the drought-induced expression but not the seed-specific expression of rd22, a gene responsive to dehydration stress in Arabidopsis thaliana. Mol Gen Genet 238: 17–25.
9. Hattori J, Boutilier KA, van Lookeren Campagne MM, Miki BL (1998) A conserved BURP domain defines a novel group of plant proteins with unusual primary structures. Mol Gen Genet 259: 424–428.
10. Baumlein H, Boerjan W, Nagy I, Bassuner R, Van Montagu M, et al. (1991) A novel seed protein gene from Vicia faba is developmentally regulated in transgenic tobacco and Arabidopsis plants. Mol Gen Genet 225: 459–467.
11. Wohlfarth T, Braun H, Kirik V, Kolle K, Czihal A, et al. (1998) Regulation and evolution of seed globulin genes. Journal of Plant Physiology 152: 600–606.
12. Boutilier KA, Gines MJ, DeMoor JM, Huang B, Baszczynski CL, et al. (1994) Expression of the BnmNAP subfamily of napin genes coincides with the induction of Brassica microspore embryogenesis. Plant Mol Biol 26: 1711–1723.
13. Teerawanichpan P, Xia Q, Caldwell SJ, Datla R, Selvaraj G (2009) Protein storage vacuoles of Brassica napus zygotic embryos accumulate a BURP domain protein and perturbation of its production distorts the PSV. Plant Molecular Biology 71: 331–343.
14. Treacy BK, Hattori J, Prudhomme I, Barbour E, Boutilier K, et al. (1997) Bnm1, a Brassica pollen-specific gene. Plant Molecular Biology 34: 603–611.
15. Chen LZ, Guan LM, Seo M, Hoffmann F, Adachi T (2005) Developmental expression of ASG-1 during gametogenesis in apomictic guinea grass (Panicum maximum). Journal of Plant Physiology 162: 1141–1148.
16. Wang AM, Xia Q, Xie WS, Datla R, Selvaraj G (2003) The classical Ubisch bodies carry a sporophytically produced structural protein (RAFTIN) that is essential for pollen development. Proceedings of the National Academy of Sciences of the United States of America 100: 14487–14492.
17. Batchelor AK, Boutilier K, Miller SS, Hattori J, Bowman LA, et al. (2002) SCB1, a BURP-domain protein gene, from developing soybean seed coats. Planta 215: 523–532.
18. Datta N, Lafayette PR, Kroner PA, Nagao RT, Key JL (1993) Isolation and Characterization of 3 Families of Auxin down-Regulated Cdna Clones. Plant Molecular Biology 21: 859–869.
19. Granger C, Coryell V, Khanna A, Keim P, Vodkin L, et al. (2002) Identification, structure, and differential expression of members of a BURP domain containing protein family in soybean. Genome 45: 693–701.
20. Xu HL, Li YX, Yan YM, Wang K, Gao Y, et al. (2010) Genome-scale identification of Soybean BURP domain-containing genes and their expression under stress treatments. Bmc Plant Biology 10.
21. Shao YH, Wei G, Wang L, Dong Q, Zhao Y, et al. (2011) Genome-Wide Analysis of BURP Domain-Containing Genes in Populus trichocarpa. Journal of Integrative Plant Biology 53: 743–755.
22. Held BM, John I, Wang H, Moragoda L, Tirimanne TS, et al. (1997) Zrp2: a novel maize gene whose mRNA accumulates in the root cortex and mature stems. Plant Mol Biol 35: 367–375.
23. Gan D, Jiang H, Zhang J, Zhao Y, Zhu S, et al. (2011) Genome-wide analysis of BURP domain-containing genes in maize and sorghum. Mol Biol Rep 38: 4553–4563.
24. Van Son L, Tiedemann J, Rutten T, Hillmer S, Hinz G, et al. (2009) The BURP domain protein AtUSPL1 of Arabidopsis thaliana is destined to the protein storage vacuoles and overexpression of the cognate gene distorts seed development. Plant Molecular Biology 71: 319–329.
25. Abe H, Yamaguchi-Shinozaki K, Urao T, Iwasaki T, Hosokawa D, et al. (1997) Role of arabidopsis MYC and MYB homologs in drought- and abscisic acid-regulated gene expression. Plant Cell 9: 1859–1868.
26. Iwasaki T, Yamaguchishinozaki K, Shinozaki K (1995) Identification of a Cis-Regulatory Region of a Gene in Arabidopsis-Thaliana Whose Induction by Dehydration Is Mediated by Abscisic-Acid and Requires Protein-Synthesis. Molecular and General Genetics 247: 391–398.
27. Park MY, Chung MS, Koh HS, Lee DJ, Ahn SJ, et al. (2009) Isolation and functional characterization of the Arabidopsis salt-tolerance 32 (AtSAT32) gene associated with salt tolerance and ABA signaling. Physiol Plant 135: 426–435.
28. Sanchez JP, Chua NH (2001) Arabidopsis PLC1 is required for secondary responses to abscisic acid signals. Plant Cell 13: 1143–1154.
29. Song HM, Zhao RM, Fan PX, Wang XC, Chen XY, et al. (2009) Overexpression of AtHsp90.2, AtHsp90.5 and AtHsp90.7 in Arabidopsis thaliana enhances plant sensitivity to salt and drought stresses. Planta 229: 955–964.
30. Wang HM, Zhou L, Fu YP, Cheung MY, Wong FL, et al. (2012) Expression of an apoplast-localized BURP-domain protein from soybean (GmRD22) enhances tolerance towards abiotic stress. Plant Cell and Environment 35: 1932–1947.
31. Ding Z, Li S, An X, Liu X, Qin H, et al. (2009) Transgenic expression of MYB15 confers enhanced sensitivity to abscisic acid and improved drought tolerance in Arabidopsis thaliana. J Genet Genomics 36: 17–29.
32. Xu B, Gou JY, Li FG, Shangguan XX, Zhao B, et al. (2013) A Cotton BURP Domain Protein Interacts With -Expansin and Their Co-Expression Promotes Plant Growth and Fruit Production. Molecular Plant 6: 945–958.
33. Schallau A, Kakhovskaya I, Tewes A, Czihal A, Tiedemann J, et al. (2008) Phylogenetic footprints in fern spore- and seed-specific gene promoters. Plant Journal 53: 414–424.
34. Sreenivasulu N, Radchuk V, Strickert M, Miersch O, Weschke W, et al. (2006) Gene expression patterns reveal tissue-specific signalling networks controlling programmed cell death and ABA-regulated maturation in developing barley seeds (vol 47, pg 310, 2006). Plant Journal 47: 987–987.

35. Orkin S (1990) Molecular-Cloning - a Laboratory Manual, 2nd Edition - Sambrook, J, Fritsch, Ef, Maniatis, T. Nature 343: 604–605.
36. Bechthold N, Elli J, Pelletier G (1993) *In planta Agrobacterium*-mediated transfer by infiltration of adult *Arabidopsis thaliana* plants. CR Acad Sci Paris 316: 871–883.
37. Jefferson RA, Kavanagh TA, Bevan MW (1987) Gus Fusions - Beta-Glucuronidase as a Sensitive and Versatile Gene Fusion Marker in Higher-Plants. Embo Journal 6: 3901–3907.
38. Poorter H, Lewis C (1986) Testing Differences in Relative Growth-Rate - a Method Avoiding Curve Fitting and Pairing. Physiologia Plantarum 67: 223–226.
39. Klukas C, Chen D, Pape JM (2014) IAP: an open-source information system for high-throughput plant phenotyping. Plant Physiol.
40. Govind G, Harshavardhan VT, Patricia JK, Dhanalakshmi R, Senthil Kumar M, et al. (2009) Identification and functional validation of a unique set of drought induced genes preferentially expressed in response to gradual water stress in peanut. Mol Genet Genomics 281: 591–605.
41. Hiscox JD, Israelstam GF (1979) Method for the Extraction of Chlorophyll from Leaf Tissue without Maceration. Canadian Journal of Botany-Revue Canadienne De Botanique 57: 1332–1334.
42. Church GM, Gilbert W (1984) Genomic sequencing. Proceedings of the National Academy of Sciences of the United States of America 81: 1991–1995.
43. Seiler C, Harshavardhan VT, Reddy PS, Hensel G, Kumlehn J, et al. (2014) Abscisic acid flux alterations result in differential abscisic acid signaling responses and impact assimilation efficiency in barley under terminal drought stress. Plant Physiol 164: 1677–1696.
44. Benjamini Y, Hochberg Y (1995) Controlling the False Discovery Rate - a Practical and Powerful Approach to Multiple Testing. Journal of the Royal Statistical Society Series B-Methodological 57: 289–300.
45. Thimm O, Blasing O, Gibon Y, Nagel A, Meyer S, et al. (2004) MAPMAN: a user-driven tool to display genomics data sets onto diagrams of metabolic pathways and other biological processes. Plant Journal 37: 914–939.
46. Zimmermann P, Hirsch-Hoffmann M, Hennig L, Gruissem W (2004) GENEVESTIGATOR. Arabidopsis microarray database and analysis toolbox. Plant Physiology 136: 2621–2632.
47. Winter D, Vinegar B, Nahal H, Ammar R, Wilson GV, et al. (2007) An "Electronic Fluorescent Pictograph" Browser for Exploring and Analyzing Large-Scale Biological Data Sets. Plos One 2.
48. Klukas C, Pape JM, Entzian A (2012) Analysis of high-throughput plant image data with the information system IAP. J Integr Bioinform 9: 191.
49. Munns R, James RA, Sirault XR, Furbank RT, Jones HG (2010) New phenotyping methods for screening wheat and barley for beneficial responses to water deficit. J Exp Bot 61: 3499–3507.
50. Schelbert S, Aubry S, Burla B, Agne B, Kessler F, et al. (2009) Pheophytin Pheophorbide Hydrolase (Pheophytinase) Is Involved in Chlorophyll Breakdown during Leaf Senescence in Arabidopsis. Plant Cell 21: 767–785.
51. Santos CLV, Campos A, Azevedo H, Caldeira G (2001) In situ and in vitro senescence induced by KCl stress: nutritional imbalance, lipid peroxidation and antioxidant metabolism. Journal of Experimental Botany 52: 351–360.
52. Qin F, Sakuma Y, Tran LSP, Maruyama K, Kidokoro S, et al. (2008) Arabidopsis DREB2A-interacting proteins function as RING E3 ligases and negatively regulate plant drought stress-responsive gene expression. Plant Cell 20: 1693–1707.
53. Catala R, Ouyang J, Abreu IA, Hu Y, Seo H, et al. (2007) The Arabidopsis E3 SUMO ligase SIZ1 regulates plant growth and drought responses. Plant Cell 19: 2952–2966.
54. Huang NC, Jane WN, Chen J, Yu TS (2012) Arabidopsis thaliana CENTRORADIALIS homologue (ATC) acts systemically to inhibit floral initiation in Arabidopsis. Plant Journal 72: 175–184.
55. Li C, Lv J, Zhao X, Ai X, Zhu X, et al. (2010) TaCHP: a wheat zinc finger protein gene down-regulated by abscisic acid and salinity stress plays a positive role in stress tolerance. Plant Physiol 154: 211–221.
56. Barrero JM, Rodriguez PL, Quesada V, Piqueras P, Ponce MR, et al. (2006) Both abscisic acid (ABA)-dependent and ABA-independent pathways govern the induction of NCED3, AAO3 and ABA1 in response to salt stress. Plant Cell Environ 29: 2000–2008.
57. Hwang IW, Goodman HM (1995) An Arabidopsis-Thaliana Root-Specific Kinase Homolog Is Induced by Dehydration, Aba, and Nacl. Plant Journal 8: 37–43.
58. Rosado A, Schapire AL, Bressan RA, Harfouche AL, Hasegawa PM, et al. (2006) The Arabidopsis tetratricopeptide repeat-containing protein TTL1 is required for osmotic stress responses and abscisic acid sensitivity. Plant Physiology 142: 1113–1126.
59. Shinozaki K, Yamaguchi-Shinozaki K (1992) [Plant genes induced by drought stress and ABA]. Tanpakushitsu Kakusan Koso 37: 1190–1199.
60. Fujii H, Verslues PE, Zhu JK (2011) Arabidopsis decuple mutant reveals the importance of SnRK2 kinases in osmotic stress responses in vivo. Proc Natl Acad Sci U S A 108: 1717–1722.
61. Daxinger L, Hunter B, Sheik M, Jauvion V, Gasciolli V, et al. (2008) Unexpected silencing effects from T-DNA tags in Arabidopsis. Trends in Plant Science 13: 4–6.
62. Almagro L, Gomez Ros LV, Belchi-Navarro S, Bru R, Ros Barcelo A, et al. (2009) Class III peroxidases in plant defence reactions. J Exp Bot 60: 377–390.
63. Desikan R, Cheung MK, Bright J, Henson D, Hancock JT, et al. (2004) ABA, hydrogen peroxide and nitric oxide signalling in stomatal guard cells. Journal of Experimental Botany 55: 205–212.
64. Hancock JT, Desikan R, Neill SJ (2001) Role of reactive oxygen species in cell signalling pathways. Biochem Soc Trans 29: 345–350.
65. Kar RK, Choudhuri MA (1987) Possible Mechanisms of Light-Induced Chlorophyll Degradation in Senescing Leaves of Hydrilla-Verticillata. Physiologia Plantarum 70: 729–734.
66. Hynninen PH, Kaartinen V, Kolehmainen E (2010) Horseradish peroxidase-catalyzed oxidation of chlorophyll a with hydrogen peroxide Characterization of the products and mechanism of the reaction. Biochimica Et Biophysica Acta-Bioenergetics 1797: 531–542.
67. Ramirez V, Coego A, Lopez A, Agorio A, Flors V, et al. (2009) Drought tolerance in Arabidopsis is controlled by the OCP3 disease resistance regulator. Plant Journal 58: 578–591.
68. Coego A, Ramirez V, Gil MJ, Flors V, Mauch-Mani B, et al. (2005) An Arabidopsis homeodomain transcription factor, OVEREXPRESSOR OFCATIONIC PEROXIDASE 3, mediates resistance to infection by necrotrophic pathogens. Plant Cell 17: 2123–2137.
69. Wang YN, Liu C, Li KX, Sun FF, Hu HZ, et al. (2007) Arabidopsis EIN2 modulates stress response through abscisic acid response pathway. Plant Molecular Biology 64: 633–644.
70. Miao Y, Lv D, Wang P, Wang XC, Chen J, et al. (2006) An Arabidopsis glutathione peroxidase functions as both a redox transducer and a scavenger in abscisic acid and drought stress responses. Plant Cell 18: 2749–2766.
71. Mullineaux P, Ball L, Escobar C, Karpinska B, Creissen G, et al. (2000) Are diverse signalling pathways integrated in the regulation of arabidopsis antioxidant defence gene expression in response to excess excitation energy? Philos Trans R Soc Lond B Biol Sci 355: 1531–1540.
72. Neill S, Barros R, Bright J, Desikan R, Hancock J, et al. (2008) Nitric oxide, stomatal closure, and abiotic stress. J Exp Bot 59: 165–176.

3

New Insights into How Increases in Fertility Improve the Growth of Rice at the Seedling Stage in Red Soil Regions of Subtropical China

Yilin Li[1], **Weiming Shi**[1], **Xingxiang Wang**[1,2]*

1 State Key Laboratory of Soil and Sustainable Agriculture, Institute of Soil Science, Chinese Academy of Sciences, Nanjing, China, **2** Jiangxi Key Laboratory of Ecological Research of Red Soil, Yingtan, Jiangxi, China

Abstract

The differences in rhizosphere nitrification activities between high- and low- fertility soils appear to be related to differences in dissolved oxygen concentrations in the soil, implying a relationship to differences in the radial oxygen loss (ROL) of rice roots in these soils. A miniaturised Clark-type oxygen microelectrode system was used to determine rice root ROL and the rhizosphere oxygen profile, and rhizosphere nitrification activity was studied using a short-term nitrification activity assay. Rice planting significantly altered the oxygen cycling in the water-soil system due to rice root ROL. Although the oxygen content in control high-fertility soil (without rice plants) was lower than that in control low-fertility soil, high rice root ROL significantly improved the rhizosphere oxygen concentration in the high-fertility soil. High soil fertility improved the rice root growth and root porosity as well as rice root ROL, resulting in enhanced rhizosphere nitrification. High fertility also increased the content of nitrification-induced nitrate in the rhizosphere, resulting in enhanced ammonium uptake and assimilation in the rice. Although high ammonium pools in the high-fertility soil increased rhizosphere nitrification, rice root ROL might also contribute to rhizosphere nitrification improvement. This study provides new insights into the reasons that an increase in soil fertility may enhance the growth of rice. Our results suggest that an amendment of the fertiliser used in nutrient- and nitrification-poor paddy soils in the red soil regions of China may significantly promote rice growth and rice N nutrition.

Editor: Wagner L. Araujo, Universidade Federal de Vicosa, Brazil

Funding: This research was financially supported by the National Natural Science Foundation of China (No. 31471948), the Special Fund for Agro-scientific Research in the Public Interest (No. 201203050-3) and the GanPo 555 Talents Program of Jiangxi Province, China. The funders had no role in study design, data collection and analysis, decision to publish, or preparation of the manuscript.

Competing Interests: The authors have declared that no competing interests exist.

* Email: xxwang@issas.ac.cn

Introduction

Nitrification, an important component of the nitrogen cycle, involves two aerobic microbial processes. In nitrification, ammonia (NH_3) is first converted to nitrite (NO_2^-); then, NO_2^- is further converted to nitrate (NO_3^-) [1,2]. Nitrification occurs in a variety of environments, such as sediments and soils. However, in waterlogged environments, such as paddy soils and flooding waters, nitrification is limited by the low availability of oxygen, which results in low NO_3^- availability. An increased number of hydroponic experiments have demonstrated that rice absorbs and assimilates NO_3^- in preference to ammonium (NH_4^+) [3,4]. Rice growth, yield, net N acquisition and N use efficiency (NUE) have been shown to be superior using mixed NH_4^+ and NO_3^- nutrition compared with pure NH_4^+ cultivation [3,5,6]. In paddy fields, NO_3^- concentrations often fluctuate and are difficult to control compared with hydroponic culturing, primarily because of denitrification in the bulk soil and nitrification in the rhizosphere. In our previous soil culture study [7], different rice cultivars (Indica and Japonica) showed a significant difference in the nitrification activity in their rhizospheres, and rhizosphere nitrification was closely related to rice growth and N nutrition as well as rice N accumulation and NUE.

The oxidation of NH_3 is a microbial process and is thought to be the rate-limiting step for nitrification [8]. Recently, according to metagenomic [9,10] and cultivation [11] methods, another type of NH_3 oxidiser, the ammonia-oxidising archaea (AOA) group, has been found to be abundant in the natural environment [12] and has been hypothesised to be a dominant group among the ammonia-oxidising prokaryotes in the soil [13]. Because ammonia-oxidising bacteria (AOB) and AOA are both strictly aerobic chemolithoautotrophic microorganisms, nitrification can only occur in the presence of oxygen and is restricted in the oxic upper few millimetres of flooded soils [14–16].

However, rice plants form numerous adventitious roots containing aerenchyma. These roots promote radial oxygen loss (ROL) from the roots to the rhizosphere, supporting aerobic microbial nitrification [17,18]. Furthermore, ROL alleviates the phytotoxicity of reduced compounds (e.g., Fe^{2+}, Mn^{2+}, H_2S) in rice [19]. ROL is of biological significance for submerged plants and has been investigated extensively using various methods [20–23]. A Clark-type oxygen microelectrode has been used to measure the rice ROL and the detailed oxygen vertical profiles in the field, indicating that rice plants at 21 d had 20% air saturation at the root surface. Diffusion of O_2 away from the root was able to create oxic conditions up to a distance of 0.4 mm away from the root [23]. The diameter of this oxygen microelectrode was only 25 μm, which should have caused minimal damage to the soil and rice plants.

Two paddy soils (representing low and high fertilities) in red soil regions with low nitrification activity were examined by oxygen microelectrode techniques to study the relationship between rice ROL and rhizosphere nitrification and the contribution of root ROL to rice N nutrition. Red soils, which can be classified as acrisols and ferralsols according to the Food and Agriculture Organization (FAO) nomenclature [24], are widespread in tropical and subtropical China, covering 11.8% of the nation's land area. The red soil region contains abundant water resources and heat resources and is one of the principal rice production areas in China [25]. Due to intensive leaching and weathering, long-term inappropriate utilisation and management, and the inherent fragility of hill ecosystems, the red soils are characterised by low pH and fertility, as well as low nitrification activity [26,27]. When natural factors, such as parent materials, topography, and climate, are the same in the paddy soil-forming process, time and human factors, such as long-term fertilisation, are the primary influencing factors in improving soil fertility. Two paddy soils collected from paddy fields with different fertility levels were defined as low fertility (i.e., reclamation of approximately 15 years, long-term inorganic and organic fertilisation in the rice season) and high fertility (i.e., reclamation great than 50 years, long-term inorganic and organic fertilisation in the rice season and manure application in the fallow winter). Although increasing soil fertility may improve rice root growth, little is known about whether it could change the rice root aerenchyma development and ROL. Our previous study showed that oxygen concentrations in paddy soil (without rice plants) with high fertility were much lower than those measured in low-fertility paddy soil collected from red soil regions (unpublished data). In this study, we hypothesised that rice plants grown in high-fertility paddy soil would improve the root aerenchyma development and ROL, which would lead to increased oxygen in the rhizosphere soil, and, thus, enhance rhizosphere nitrification activity.

Materials and Methods

Paddy soils and rice plant cultivars

Paddy soils derived from Quaternary red clay (hydragric anthrosols according to the FAO nomenclature [24]) were collected from the Chinese Academy of Sciences Red Soil Ecological Experimental Station, located in Yingtan (28°15′20″ N, 116°55′30″ E), Jiangxi Province, which is one of the major rice production areas in China. The cultivation practice in the area is two crops per year. The Yingtan region has a typical subtropical monsoon climate with an annual precipitation of 1785 mm and a mean annual temperature of 17.8°C. Two paddy soils (0–20 cm) collected from paddy fields with different fertility levels were defined as low fertility and high fertility, and their properties are described in Table 1. The soil samples were air-dried, ground, sieved through 0.71-mm mesh, and stored for further incubation experiments. The plant residues were carefully removed by hand. One of the most popular Indica rice cultivars in Jiangxi Province, Hesheng 10, was used in the experiment.

Soil incubation and sampling

A rhizobox, as described by Li *et al.* [28], was used to collect soil samples at various distances from the rice root surface. Two pieces of nylon nets (with a 30-μm mesh size) were used to divide the rhizobox into three compartments to facilitate the isolation of the rice root from the rhizosphere soil. Three rice plants were planted in the middle compartment of each rhizobox. The mesh of the nylon net was fine enough to allow the water and nutrient elements to pass through. A 600-g sample of the paddy soil (air dried and sieved) was mixed thoroughly with urea (120 mg N kg^{-1}) and KH_2PO_4 (93 mg kg^{-1}) to fill each rhizobox (the fertiliser applications were performed according to the conventional amounts used by the local farmers to support rice growth adequately). A rhizobox without rice plants was used as the control (CK). One centimetre of surface water was maintained after rice germination, using deionised water every morning and evening throughout the incubation. The rice plants were incubated at 25°C with a 16-h photoperiod during the experiment. Three replicate rhizoboxes for each treatment were used, and soil samples were collected at 40, 50 and 60 d after rice germination; an additional six replicate rhizoboxes were used for in situ measurements of soil oxygen concentration and rice root ROL 50 d after germination.

A previous experiment in this laboratory showed that between 40 and 60 days after rice germination, the rice roots were fully developed and filled the middle compartment of the rhizobox so that the nylon nets would not influence the rice root growth. Therefore, the soil adhering to the rice roots in the inner rhizobox was defined as root surface soil, and the soil in the outer rhizobox within a 4-mm distance from the nylon nets was defined as rhizosphere soil. To prevent the water layer from interfering with the soil mineral N and nitrification activity determination, no watering occurred on the evening of the day prior to sampling to maintain a thin (1–2 mm) water layer. The sampling was conducted at 10:00 am, and each treatment (three replicates) was first frozen at −20°C for 2 h to harden the paddy soil for subsequent slicing. First, the rhizobox was carefully split along the two nylon nets into three compartments: middle, right, and left parts. Next, a piece of organic glass board (2 mm thick, 8 cm×8 cm) was inserted into the inner side of the right or left part of the rhizobox. Then, a soil sample of 2-mm thickness was pushed aside, and the soil was sliced to obtain a soil sample located 2 mm away from the root surface. Next, another piece of organic glass board (2 mm thick, 8 cm×8 cm) was inserted to obtain a layer located 4 mm away from the root surface. Using the same procedure, soil samples were obtained at distances of 6, 8, 10, 20, 30 and 40 mm from the root surface. The soil samples collected in the right and left compartments at the same distance were mixed together for the assessments. The root surface soil was separated from the rice roots using sterilised tweezers. All soil samples were collected at the same time as the plant sampling to assay for mineral N and nitrification activity.

Mineral N assays

Fresh soil samples were extracted with 2 mol L^{-1} of KCl (soil:solution ratio: 1:10), and the extracts were measured colourimetrically for NH_4^+ and NO_3^- [26] using a continuous flow autoanalyser (model Autoanalyzer 3, Bran + Luebbe, Hamburg, Germany).

Short-term nitrification activity assays

Short-term nitrification estimates are typically used for the nitrification activity assay. The principle underlying this method is based on the determination of NO_2^- after the incubation of soil

Table 1. Properties of the paddy soils used in the experiment.

Property	Low fertility	High fertility
pH (water:soil, 2.5:1)	4.75	5.02
Organic matter (g kg^{-1})	17.5	39.1
Total N (g kg^{-1})	0.85	1.74
Total P (g kg^{-1})	0.32	0.67
NH_4^+-N (mg kg^{-1})	0.72	2.27
NO_3^--N (mg kg^{-1})	0.17	0.21
Clay (%)	36.3	20.9
Average yield of double-cropping rice per year (kg hm^{-1})	7500	10000
Reclamation years	15	>50

samples with $NaClO_3$ in the absence of NH_4^+ for 24 h at 25°C [29,30]. Briefly, each soil sample (5 g) was shaken with 2.5 mL of $NaClO_3$ (75 mmol L^{-1}, which prevents oxidation of NO_2^-) at 170 rpm on a rotary shaker (model HZ-9611K, Hualida Laboratorial Equipment Co., Ltd., Taicang, China) for 30 min, and then incubated aerobically in an incubator for 24 h at 25°C. After incubation, NO_2^- was extracted from the soil samples into a total volume of 15 mL using two solutions (5 mL of deionised H_2O followed by 10 mL of 2 mol L^{-1} KCl) by shaking at 170-rpm for 30 min on a rotary shaker, as described previously. The contents were mixed thoroughly and immediately filtered. Five millilitres of the clear filtrate was pipetted into glass test tubes, followed by 3 mL of the buffer (0.19 mol L^{-1} NH_4Cl, pH 8.5) and 2 mL of the reagent (1% sulphanilamide, 0.05% N-(1-naphthyl) ethylene-diamine dihydrochloride and 10% phosphoric acid mixed solution), for NO_2^- determination. The contents were again vigorously shaken and allowed to stand for 15 min at room temperature. For a control measurement, soil samples were extracted as previously described after incubation with $NaClO_3$ at −20°C. The amount of NO_2^- produced was measured based on the absorbance at 520 nm in a spectrophotometer. Pre-experiments using the paddy soil without planted rice (waterlogged for 40 d) to observe the kinetics of ammonia oxidation in the first 24 hours were performed and revealed that nitrite was produced linearly over time ($p<0.05$).

In situ measurements of soil oxygen concentrations

An additional three replicate rhizoboxes were used for the in situ measurement of soil oxygen concentrations 50 d after germination. The incubation conditions were the same as described above except that a 1-cm water layer was maintained in the rhizoboxes for the in situ measurement.

The oxygen microelectrode was a miniaturised Clark-type oxygen electrode with a guard cathode (OXY25, ø = 20–30 µm, Unisense, Aarhus, Denmark [31]). The soil oxygen profile measurements were performed in the right and left compartments of each rhizobox and in the centre of the range at a distance of 2 mm from the nylon nets for all treatments. All measurements were performed with a 50-µm depth interval using a microelectrode with a diameter of approximately 25 µm, and the periods for "wait before measure" and "measure" were both set as 3 s.

In vitro measurements of rice root oxygen loss in an agar microcosm

Three rhizosphere boxes were split into three compartments each, and the rice plants with soil were placed into a plastic bucket filled with deionised water. The soil adhering to the rice roots was removed by repeated immersion in the deionised water. After the soil was removed, the rice roots were separated according to the individual plants and were then individually placed into beakers with International Rice Research Institute (IRRI) rice nutrient solution (pH 5.5) [32] for 2 h. The roots of each individual rice plant were then inserted into a container filled with IRRI rice nutrient solution (pH 5.5) containing 0.9% agar [23] before the agar was allowed to solidify at 38°C. The container was combined with one small culture dish (6 cm in diameter, 2.5 cm in depth) and one large culture dish (15 cm in diameter, 2.5 cm in depth). The rice root (to be measured) was inserted into the small culture dish, whereas the other roots were evenly distributed in the large culture dish. Next, paraffin oil was poured onto the surface of the solidified agar in the small culture dish to prevent air penetration. To obtain comparable results, the sample root was placed horizontally in the agar. Because the solidified agar was a clear medium, it allowed observation of the rice roots and provided fixation support for the roots. To measure the oxygen levels of the outer and inner rice roots, the microelectrode had to be sufficiently thin to longitudinally penetrate the root. Therefore, a much slimmer microelectrode (OXY10, ø = 8–10 µm, Unisense, Aarhus, Denmark [31]) was used for the measurement. Newly formed roots with lengths of approximately 45 mm were selected and measured at 25 mm from the root tip. The measurements were conducted after the rice roots had been fixed in the agar for approximately 2 h; these measurements were repeated for different rice plants with different newly formed roots with the same length as described above. All measurements were conducted with a 50-µm depth interval, and the periods for "wait before measure" and "measure" were both set to 3 s.

Because oxygen concentration is very sensitive to temperature, soil and agar temperatures were monitored and recorded during the oxygen measurements. All of the oxygen measurements, including soil oxygen concentration and root ROL, were performed in a specialised laboratory without electrical noise interference at a temperature of 20°C.

Plant dry matter and N accumulations assays

The rice tissues were separated into shoots and roots and heated at 105°C for 30 min to terminate enzyme activity. The samples were then oven-dried at 70°C for 48 h to a constant weight. The dried plant material was ground and digested for total N determination using the Kjeldahl method. A 5-mL aliquot from the 100-mL digested solution was then analysed for N using a continuous flow auto-analyser (model Autoanalyser 3, Bran + Luebbe, Hamburg, Germany).

Plant NRA and GSA assays

Fresh leaves and roots of both treatments were immediately frozen with liquid N and then stored in a refrigerator at −70°C for the later measurement of glutamine synthetase activity (GSA) and NO_3^- reductase activity (NRA).

The level of GSA was assayed using an in vitro method. Frozen plant material (0.5 g) was ground in a precooled mortar with sand and a pestle after adding liquid N and then homogenised in an extraction buffer (pH 7.2, 10 mL g^{-1} fresh weight) containing 0.5 mmol L^{-1} of EDTA and 50 mmol L^{-1} of K_2SO_4. The homogenates were centrifuged at 20,000 *g* for 20 min. Next, 1.2 mL of the clear filtrate was pipetted into a centrifuge tube, to which was then added 0.6 mL of imidazole-HCl (pH 7.0, 0.25 mol L^{-1}), 0.4 mL of sodium glutamate (pH 7.0, 0.3 mol L^{-1}), 0.4 mL of ATP-Na (pH 7.0, 15 mmol L^{-1}), 0.2 mL of $MgSO_4$ (0.5 mol L^{-1}) and 0.2 mL of hydroxylamine (1 mol L^{-1}). After the mixture was incubated at 25°C for 20 min, the reaction was terminated by adding 0.8 mL of acidic $FeCl_3$ [24% (W/V) trichloroacetic acid and 10% (W/V) $FeCl_3$ in 18% HCl]. The amount of γ-glutamyl hydroxamate was measured with a spectrophotometer at 540 nm. One unit of GS activity was defined as the amount of enzyme that catalysed the formation of 1 μmol γ-glutamyl hydroxamate per min at 25°C [33].

The level of NRA was assayed by another in vitro method. Frozen plant material (0.5 g) was ground in a precooled mortar with sand and a pestle after adding liquid N and then homogenised in an extraction buffer (10 mL g^{-1} fresh weight) containing 5 mmol L^{-1} of EDTA, 5 mmol L^{-1} of cysteine and 25 mmol L^{-1} of potassium phosphate buffer (pH 8.7). The homogenates were centrifuged at 20,000 *g* for 20 min. Next, 0.4 mL of the clear filtrate was pipetted into a centrifuge tube, and the filtrate was added to 1.0 mL of KNO_3 (0.1 mol L^{-1}) and 0.6 mL of NADH (2 mg mL^{-1}). After the mixture was incubated at 25°C for 30 min, the excess NADH was oxidised by the addition of 0.5 mL of sulphanilamide (1%). The mixture was centrifuged at 20,000 *g* for 15 min. The amount of NO_2^- produced was measured after the addition of N-(1-naphtyl) ethylene-diamine dihydrochloride (0.02%) using an absorbance of 540 nm in a spectrophotometer [34].

Plant POR assay

For this assay, 0.4–0.6 g of fresh adventitious roots was sampled according to published methods [13,35,36]. POR was determined as , where POR is the root porosity (%), P_{gr} is the mass of the pycnometer with water and ground roots (g), P_r is the mass of the pycnometer with water and roots (g), r is the mass of roots (g), and P is the mass of the pycnometer with water (g).

Data analysis

All statistical analyses were performed using SPSS version 13.0. A one-way ANOVA with a homogeneity of variance test was used, followed by a least significant difference (LSD) test to check for quantitative differences between treatments. The statistically significance level was set at $p<0.05$.

Results

Distribution of soil nitrification activity

The nitrification activities of the root surface and rhizosphere soils increased over time, whereas those in the bulk soil did not change significantly (Fig. 1A, B and C). The maximal nitrification activities were found 2 mm away from the root surface in the rhizosphere soil, and their values were 2.76 (0.33) μg kg^{-1} h^{-1} for the low-fertility treatment and 4.32 (0.92) μg kg^{-1} h^{-1} for the high-fertility treatment for the last sampling period (Fig. 1C). Minimal nitrification activities appeared in the root surface and bulk soils (20–40 mm away from the root surface). Throughout the sampling period, the nitrification activities in the paddy soil with high fertility were always significantly higher than those in the soil with low fertility, except for the root surface soil 40 d after germination (Fig. 1A). Compared with the control, rice planting generally increased the rhizosphere soil nitrification activity significantly, but no increased effect of rice planting on nitrification occurring in the root surface or bulk soil was observed in either treatment. For example, the nitrification activity 2 mm away from the root surface was 1.8, 2.8 and 5.2 times higher than that in the control in the low-fertility soils at 40, 50 and 60 d after germination, respectively, and the corresponding activity in the high-fertility soil was 1.7, 2.9 and 3.9 times higher than the control, respectively (Fig. 1A, B, C and D).

Distribution of soil nitrate concentration

The NO_3^- concentration did not change over time. Although there was a strong absorption of NO_3^- by the rice roots, the horizontal distribution of soil NO_3^- at different distances from the root surface showed no concentration gradient (Fig. 2A, B and C). This result might have been due to the strong mobility of NO_3^- in the soil. The NO_3^- concentrations in the high-fertility soil were always higher than those in the low-fertility soil. This difference which might have contributed to the higher nitrification activity of the high-fertility soil described above (Fig. 1A, B, and C). The NO_3^- concentration in the soil planted with rice did not differ significantly from that in the control (Fig. 2A, B, C and D).

Distribution of the soil ammonium concentration

The NH_4^+ concentration in the soil decreased over time (Fig. 3), suggesting that a large quantity of NH_4^+ was absorbed by the rice roots. This conclusion is partially supported by the distinct depletion of NH_4^+ in the zone near the root surface (Fig. 3A, B, and C), where the NH_4^+ concentration increased with increasing distance from the rice roots. In the period 40–60 d after germination, the NH_4^+ concentrations of the high-fertility treatment in the bulk soil were always higher than those of the low-fertility treatment, except for the measurement taken 40 d after germination (Fig. 3A). This difference increased over time (Fig. 3A, B, and C). As the NH_4^+ concentrations in the control increased slightly with time, the NH_4^+ concentrations in the control for the high-fertility treatment were significantly higher than those in the control for the low-fertility treatment (Fig. 3D). Due to the NH_4^+ uptake by rice roots, the content of NH_4^+ in the treatments with rice was significantly lower than that in the control (without rice), especially on the last sampling day. For example, the NH_4^+ concentrations 40 mm away from the rice root (bulk soil) were 19.9% and 26.9% of those in the control at 60 d after germination in the low- and high-fertility soils, respectively, and those in the root surface soil were only 9.75% and 8.86% of the control.

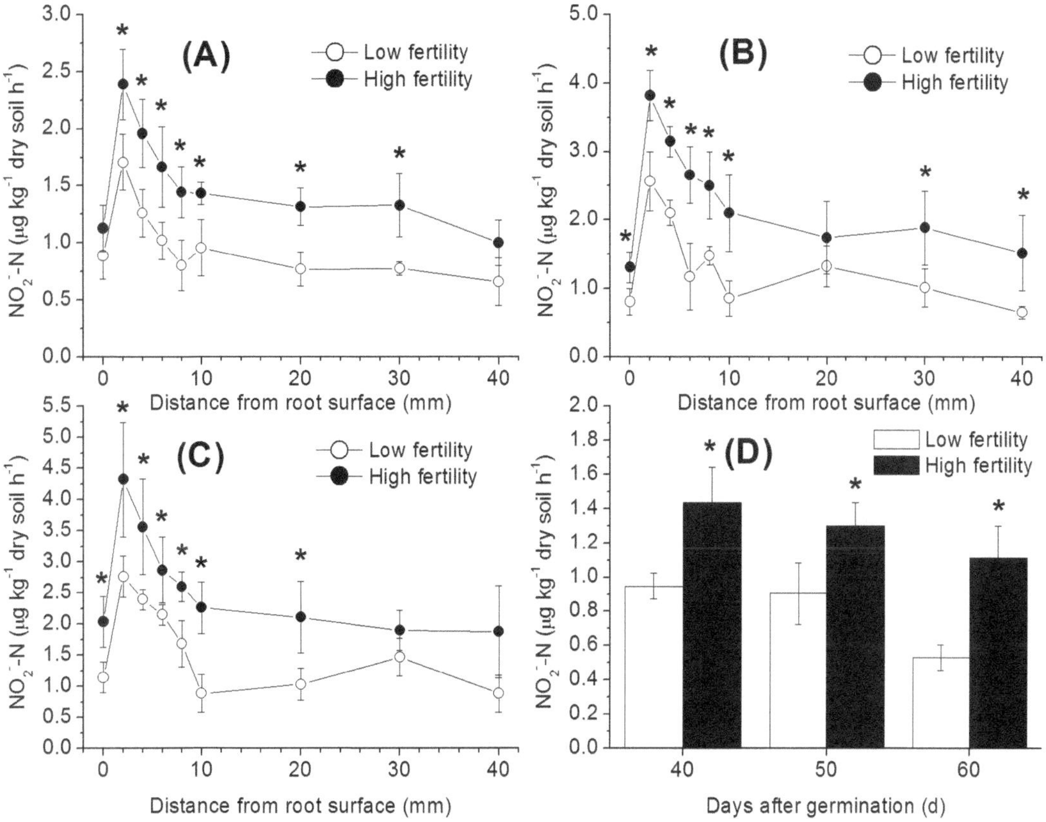

Figure 1. Nitrification activities measured at various distances from the rice root surface in paddy soils with different fertility levels (A) 40 d after germination, (B) 50 d after germination, (C) 60 d after germination, relative to the (D) control (without rice plants). The bars indicate ±SD. Dots and bars with * indicate a significant difference ($p<0.05$) between the two fertility levels.

Rice plant nitrogen assimilation, growth, and nitrogen accumulation

The NRA and GSA in the rice leaves and roots did not change over time, and the levels of NRA and GSA in the rice leaves were significantly higher than those in the roots (Table 2). In addition, high soil fertility ameliorated the NRA and GSA levels significantly; the leaf and root NRAs increased by 48.2% and 49.7% in the high-fertility soil compared to those in the low-fertility soil, whereas the leaf and root GSA levels in the high-fertility soil were, respectively, 37.6% and 103.1%, greater than those in the low-fertility soil (Table 2).

The rice biomass and N accumulation measurements differed significantly between the two fertility levels. The high-fertility soil increased the rice plant biomass by 90.4%, 65.2%, and 45.6% compared with the low-fertility soil when measured 40, 50, and 60 d after germination, respectively. The corresponding N accumulations increased by 47.9%, 48.6%, and 29.9% at the measurements taken 40, 50, and 60 d after germination, respectively (Table 2). High soil fertility markedly improved the growth and N uptake of the rice.

Rice root morphological structure

We observed large changes in the morphological structure of the rice roots and found that the major characteristics of the developmental level of the root system such as root biomass, adventitious root number and root diameter were much greater in the high-fertility soil than in the low-fertility soil (Table 3). Furthermore, the characteristics of the rice root, such as POR, which indicated the developmental degree of aerenchyma, were significantly higher in the high-fertility soil than in the low-fertility soil (Table 3). In summary, the high-fertility soil markedly improved the growth of the rice root and the root porosity.

Distribution of the soil oxygen concentration

The oxygen dissolved in the water layer remained stable. The average values were approximately 231 and 243 μmol L^{-1} in the low-fertility and high-fertility treatments ($n = 6$; Fig. 4A), respectively, and 218 and 209 μmol L^{-1} in their respective controls ($n = 6$; Fig. 4B). Rice planting significantly increased the content of dissolved oxygen in the water layer. The oxygen concentration decreased with soil depth so rapidly that it became zero at a depth of approximately 2.6–4.2 mm below the soil surface (Fig. 4A, B). Interestingly, the oxygen concentration and oxygen penetration depth in the rhizosphere soil with high fertility were significantly higher than the corresponding values in the rhizosphere soil with low fertility, whereas a reverse trend existed in the controls of these two fertility levels, although certain values at certain depths showed no significant difference due to the presence of abnormal peaks. For example, at a depth of 2 mm from the soil surface, the average dissolved oxygen concentrations in the low- and high-fertility rhizosphere soils were 33.1 and 68.9 μmol L^{-1}, respectively ($n = 6$; Fig. 4A), and their corresponding controls were 24.5 and 13.7 μmol L^{-1}, respectively ($n = 6$; Fig. 4B). Among the six repetitions, the abnormal peaks appeared more often in the rice planting treatments than in the control. This finding may be a result of the rice root ROL.

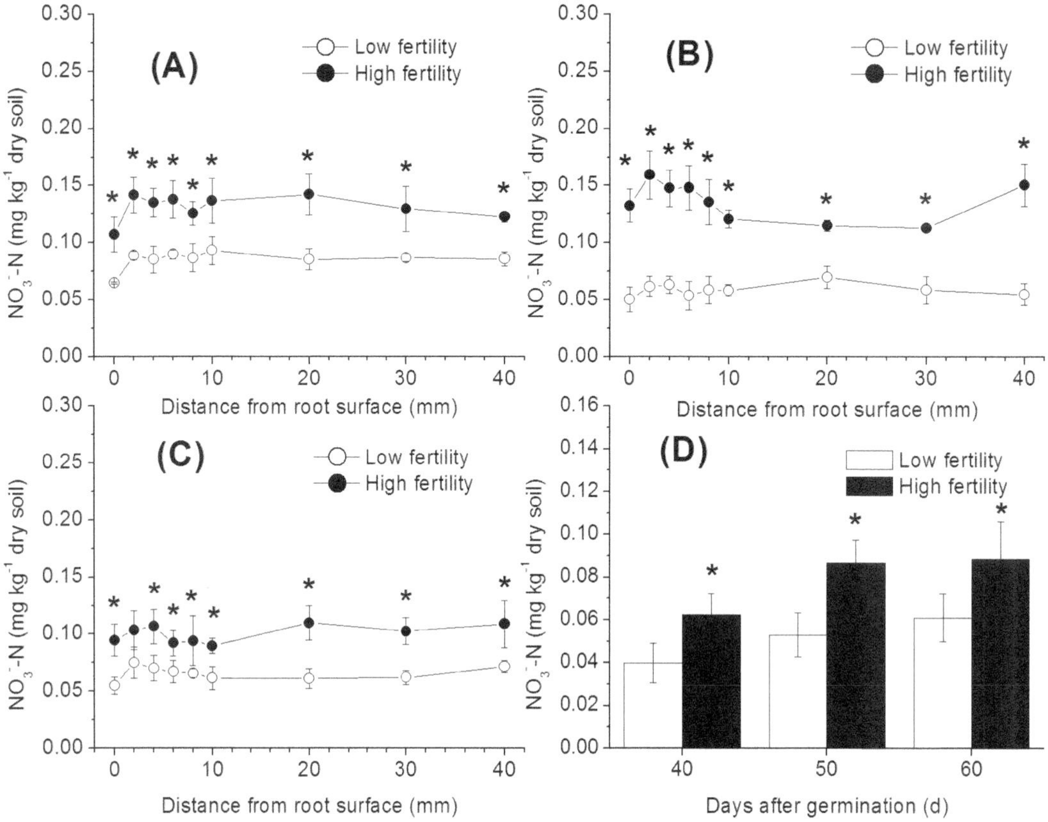

Figure 2. NO_3^--N measured at various distances from the rice root surface in paddy soils with different fertility levels (A) 40 d after germination, (B) 50 d after germination, and (C) 60 d after germination, relative to the (D) control (without rice plants). The bars indicate ±SD. Dots and bars with * indicate a significant difference ($p<0.05$) between the two fertility levels.

Root radial oxygen loss

The oxygen in the agar was detected at an average distance of 400 μm from the root surface in the low-fertility treatment (Fig. 5A), in comparison with an average distance of 600 μm from the root surface in the high-fertility treatment (Fig. 5B). This finding indicated that the root ROL level of rice in the high-fertility soil was significantly higher than that in the low-fertility soil. The oxygen concentration decreased rapidly with distance near the rice root surface. When the microelectrode was in contact with the root surface, the average oxygen concentrations were 50.2 and 97.7 μmol L^{-1} in the low- and high-fertility treatments, respectively. The oxygen concentration remained unchanged inside the rice root, and the oxygen concentrations in the roots of the low-fertility treatment were lower than those in the high-fertility treatment (Fig. 5A, B). For example, the average oxygen concentrations in the low-and high-fertility treatments were 55.4 and 103.3 μmol L^{-1}, respectively.

Discussion

The penetration depths of oxygen in the soil surface and in the rhizosphere generally depend on both rice root ROL and the oxygen consumption caused by soil organisms and chemical oxidation [37]. In fact, rice planting could significantly alter (and, most likely, improve) the cycling of oxygen in the water and soil system due to rice root ROL, and our results confirmed this hypothesis (see Fig. 4A, B). For example, the dissolved oxygen concentration in the water layer was approximately 231 and 243 μmol L^{-1} in the low- and high-fertility treatments (Fig. 4A), respectively, compared with 218 and 209 μmol L^{-1} in their corresponding controls (Fig. 4B). In planted microcosms (planting with 110-d-old rice seedlings), oxygen is still detectable to a depth of at least 40 mm [38], which appears to be the maximum depth reported to date [23]. Interestingly, our results showed that the oxygen concentration in the rhizosphere soil with high fertility was significantly higher than that in the rhizosphere soil with low fertility (Fig. 4A), whereas the soil oxygen concentration in the high-fertility soil control was significantly lower than that in the low-fertility soil control (Fig. 4B). Because the control paddy soil without rice plants had no exogenous replenishment of oxygen as occurred for the rice root ROL, except for the dissolved oxygen provided by the flooding water, the high organic matter content in the high-fertility soil (see Materials and Methods) led to more severe oxygen depletion. This result may have been due to the higher oxygen consumption required by the degradation of organic matter and reduced compounds in the high-fertility soil compared with the low-fertility soil [37]. Although the oxygen content in the control high-fertility soil was lower than that in the control low-fertility soil (Fig. 4B), the soil oxygen concentration in the high-fertility paddy soil with rice plants was higher than that in the low-fertility soil (Fig. 4A), potentially due to the high rice root ROL in the high-fertility soil (Fig. 5B).

The difference in dissolved oxygen concentrations in rhizosphere soils between the high- and low-fertility soils appeared to be related to the differences in rice root ROL in these soils. When rice plants were grown in the high-fertility soil, the rice root growth

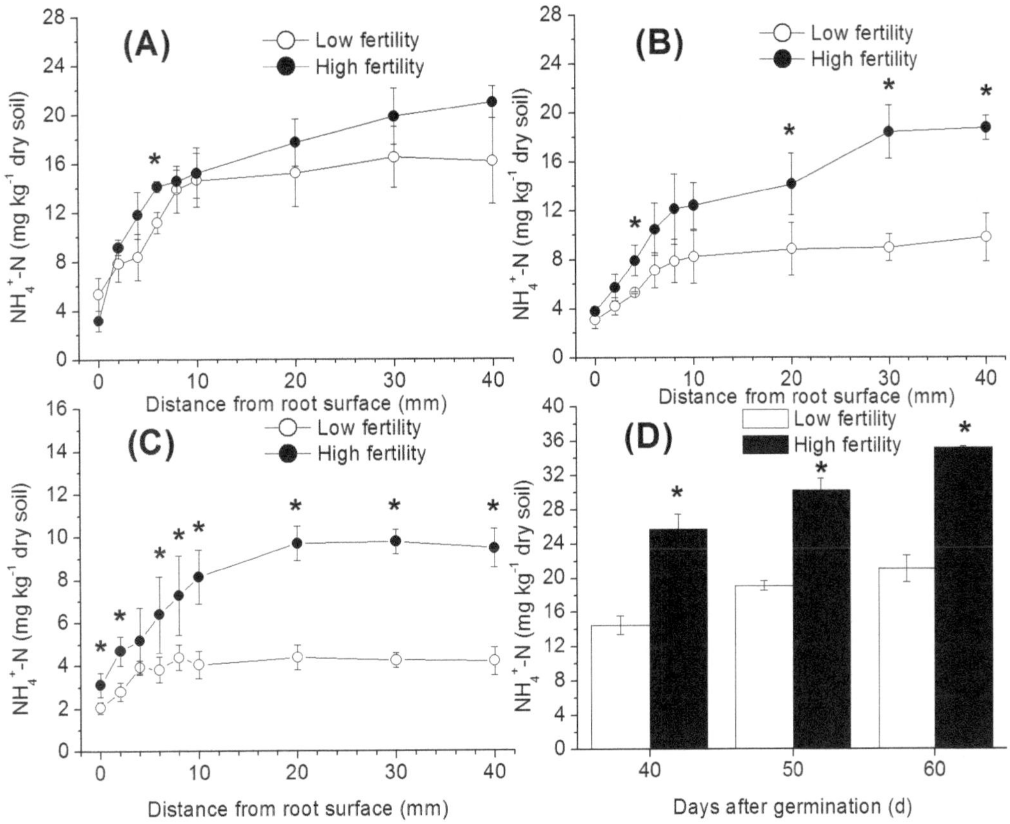

Figure 3. NH_4^+-N measured at various distances from the rice root surface in paddy soils with different fertility levels (A) 40 d after germination, (B) 50 d after germination, and (C) 60 d after germination, relative to the (D) control (without rice plants). The bars indicate ±SD. Dots and bars with * indicate a significant difference ($p<0.05$) between the two fertility levels.

was greatly improved; the root biomass, adventitious root number, and root diameter of the rice grown in the high-fertility soil were much greater than the corresponding values for the rice grown in the low-fertility soil (Table 3). A similar observation has previously been reported for wheat [39]. Not only did the rice root biomass, adventitious root number and root diameter change, but root porosity (see Table 3) was also improved by high soil fertility. This phenomenon does not appear to have been previously reported, as its positive feedback mechanism is unclear. ROL from rice roots estimated by Kludze *et al.* [19], showed a significant and positive correlation with POR. Oxygen efflux from root to soil represented 30%–40% of oxygen transportation from shoot to root [17]. Well-developed root systems and aerenchyma formation resulted in higher rice root ROL in the high-fertility soil (Fig. 5) [7] and consequently increased the oxygen content in the rhizosphere soil (Fig. 4A). Mugnai et al. [40] investigated maize roots and found that the root apex transition zone (intercalated between the apical division zone and the elongation zone [41]) plays central roles in both sensing and adapting to root hypoxia and emits the greatest amount of nitric oxide. This discovery motivated us to further investigate whether the root apex transition zone of rice plays a key role in altering the ROL and, correspondingly, the rhizosphere soil oxygen distribution.

An agar microcosm was used to further quantify the rice root ROL. The results indicated that the root ROL of rice planted in the high-fertility soil was greater than that in the low-fertility soil. The average oxygen concentrations at the root surface were 50.2 and 97.7 μmol L^{-1}, and they decreased rapidly to nearly zero at distances of approximately 400 μm and 600 μm from the root surfaces in the low- and high-fertility treatments, respectively (Fig. 5A, B). The average oxygen concentrations in the interior of the rice roots in the low- and high-fertility treatments were approximately 55.4 and 103.3 μmol L^{-1}, respectively (Fig. 5A, B). Similar results have been obtained by Revsbech *et al.* [23], who found that oxygen leakage created an oxic zone extending approximately 150 μm away from the fine lateral root (for 21-d-old rice); the inner oxygen concentration of the root (42-d-old rice) was approximately 75 μmol L^{-1}. The rice root ROL can also promote the development of aerobic niches in the rhizosphere to restrict the accumulation of phytotoxic compounds (e.g., Fe^{2+}, Mn^{2+}, H_2S) as well as denitrification [42] and maintain aerobic microbial processes, such as nitrification [14,19].

Due to long-term fertiliser changes, soil fertility increased significantly and promoted soil nitrification, as widely reported in previous studies of paddy soil [8,16] and upland soil [2,43]. A low clay content contributed to a high soil porosity, which promoted gas transport by diffusion and advection, thereby supporting soil nitrification. The high ambient NH_4^+ pool in the high-fertility soil also improved its nitrification activity (Fig. 1 and 5). The maximal nitrification activities were found in the rhizosphere soil located 2 mm away from the root surface, and decreased with distance from the root surface (Fig. 1). In our previous study, the nitrification activity gradient also revealed nitrification activity differences in different rice cultivars. It also showed that the maximum nitrification activity detected in the rhizosphere soil, i.e., the maximum nitrification activities of the Indica- and

Table 2. Rice plant N assimilation, growth, and nitrogen accumulation[a].

Fertility level	Sampling date (d)	NRA (μg g^{-1} FW h^{-1})		GSA (mmol g^{-1} FW min^{-1})		Biomass (g DW plant^{-1})	N accumulation (mg N plant^{-1})
		Leaf	Root	Leaf	Root		
Low fertility	40	7.21±0.93 ^{b}	4.06±0.33 ^{c}	1.85±0.14 ^{c}	0.08±0.02 ^{b}	0.49±0.10 ^{d}	16.8±1.77^{d}
	50	6.84±0.76 ^{b}	4.61±0.82 ^{bc}	2.09±0.05 ^{bc}	0.09±0.01 ^{b}	0.85±0.09 ^{c}	23.4±0.93 ^{c}
	60	7.38±0.30 ^{b}	4.14±1.08 ^{c}	2.35±0.15 ^{bc}	0.11±0.04 ^{b}	1.32±0.15 ^{b}	34.8±3.23 ^{b}
High fertility	40	10.6±1.38 ^{a}	5.85±0.69 ^{ab}	2.56±0.45 ^{b}	0.18±0.02 ^{a}	0.93±0.14 ^{c}	24.8±3.59 ^{c}
	50	10.7±2.41 ^{a}	6.86±0.49 ^{a}	2.58±0.25 ^{b}	0.18±0.01 ^{a}	1.41±0.23 ^{b}	34.8±3.88 ^{b}
	60	10.5±0.49 ^{a}	6.45±1.22 ^{a}	3.51±0.41 ^{a}	0.21±0.06 ^{a}	1.92±0.15 ^{a}	45.1±3.24 ^{a}

[a]Values represent means ± SDs with 3 replicates (with each replicate composed of rice plant tissue sampled from different rhizoboxes of each treatment). Different letters in the same column indicate a significant difference at the $p<0.05$ level.

Japonica- planted soils occurred 6 and 2 mm away from the rice roots, respectively [28]. The use of a rhizobox is convenient for distinguishing between the root surface and rhizosphere soil, but this design produces a dense root-zone and thus leads to a generally more oxidised environment, which may enhance nitrification. In contrast, more competition for NH_4^+ in this design may lower nitrification.

Based on our previous experience, a water layer of 1 cm that is maintained with applications of deionised water every morning and evening throughout the incubation would make it difficult to slice the soil. Additionally, mineral N, especially NO_3^-, would be dissolved in the overlying water because the oxygen concentration was much higher in the water layer than in the soil profile. If we continued to maintain a 1-cm water layer, the concentration of NO_3^- in the soil would be very low and might even remain undetected because the paddy soil that we used is derived from Quaternary red clay and has relatively low pH values (usually 4–5.5). The nitrification activity in this paddy soil is much weaker than that in neutral paddy soil [26]. The NO_3^- concentration in the paddy soil is correspondingly low. For this reason, we chose not to water on the evening of the day prior to the sampling date. The water layer was actually 1–2 mm deep when we measured the mineral N and nitrification activity. The advantage of this approach is that the soil is easy to slice, moreover, the NO_3^- concentration can be measured more effectively. However, the disadvantage of this approach is that it imperceptibly stimulates nitrification and results in overestimates of the soil NO_3^- concentration. In this experiment, the NO_3^- concentration in the soil planted with rice did not differ significantly from that in the control. Because the nitrification activity in the rice planting soil was much higher than that in the control (Fig. 1D), the majority of the NO_3^- resulting from increased nitrification might have been absorbed primarily by the rice roots. NO_3^- can be absorbed rapidly by rice roots or can diffuse rapidly to other parts of the soil. Therefore, the NO_3^- distribution in flooded paddy soil is relatively even. However, advection caused by evaporation from the leaves and following water transport to the roots increases transport processes in the soil to a greater extent than that resulting from diffusion. This pattern, was also confirmed by our previous experiment [28]. The low pH caused by acidic root exudates [44] and NH_4^+ (nitrification substrate) exhaustion in the root surface soil (Fig. 3A, B, and C) strongly inhibited root surface nitrification in both of the studied soil types (Fig. 1A, B, and C) [14]. Due to the high nitrification activity in the high-fertility soil, the NO_3^- concentrations in the high-fertility soil were higher than those in the low-fertility soil (Fig. 2). This difference, might have caused more NO_3^- uptake and assimilation by the rice plants grown in the high-fertility soil (the NRA results in Table 2 may support this hypothesis).

The NO_3^- that is converted via nitrification in the root surface and rhizosphere soils is very important for N nutrition of rice [3,4]. Extensive hydroponic experiments have demonstrated that rice growth, yield, net N acquisition, N translocation to the shoot, and NUE are greater with a mixed NH_4^+ and NO_3^- supply than with NH_4^+ alone [3,5,6]. In the present study, the NO_3^- concentration in the rice-planted soil was not significantly different from that measured in the control (Fig. 2), although the nitrification activity in the rice-planted soil was much higher than that in the control (Fig. 1), thus indicating that greater amounts of NO_3^- absorbed by the rice root might have produced the results described above. NO_3^- reductase is a key enzyme involved in NO_3^- assimilation in crops, and its activity is strongly dependent on the external NO_3^- concentration [28]. In this experiment, the rice leaf and root NRA in the high-fertility soil increased by 48.2% and 49.7%,

Table 3. Rice root growth and morphology[a].

Fertility level	Sampling date (d)	Root biomass (g DW plant^{-1})	Adventitious root number (plant^{-1})	Root diameter (mm)	POR (%)
Low fertility	40	0.11±0.02[d]	22±3[d]	0.64±0.05[d]	14.1±1.46[e]
	50	0.20±0.03[c]	42±3[c]	0.78±0.06[c]	16.2±1.13[d]
	60	0.24±0.01[c]	49±3[c]	0.90±0.02[b]	16.9±1.38[cd]
High fertility	40	0.21±0.04[c]	28±3[d]	0.75±0.04[c]	19.6±1.44[bc]
	50	0.35±0.02[b]	61±13[b]	0.90±0.03[b]	22.0±2.92[ab]
	60	0.42±0.06[a]	82±7[a]	1.06±0.09[a]	23.1±2.36[a]

[a]Values represent means (means ± SDs), and different letters in the same column indicate a significant difference at the $p<0.05$ level. Data were pooled for n = 3, except for adventitious root diameter (n = 12), which was measured at the middle of each root selected for the average length of the whole root.

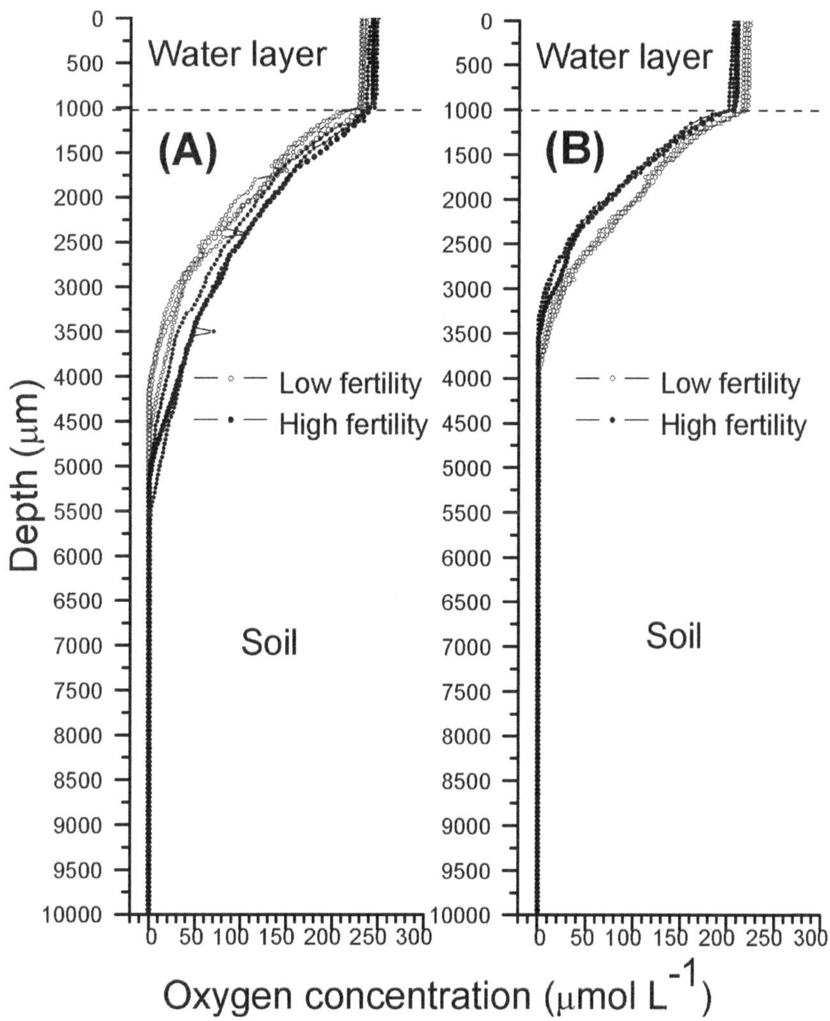

Figure 4. A representative graph showing oxygen profiles of (A) the rhizosphere soil (2 mm away from the root surface) and (B) the control (without rice plants) 50 d after germination in paddy soils with different fertility levels at 20°C. The oxygen microelectrode used was 20–30 μm in diameter (OXY25, Unisense, Aarhus, Denmark [31]). Three rhizoboxes for each treatment were used in the in situ measurements, and the oxygen concentration profile was determined in the right and left compartments of each rhizobox. All measurements were conducted in the centre of the range marked by a 2-mm distance from the nylon nets in the low- and high-fertility soils with and without the rice plant treatments. Hence, the measurements were repeated six times in each treatment. Three results (measured in the right compartment) are shown, and each curve shows one individual soil oxygen profile.

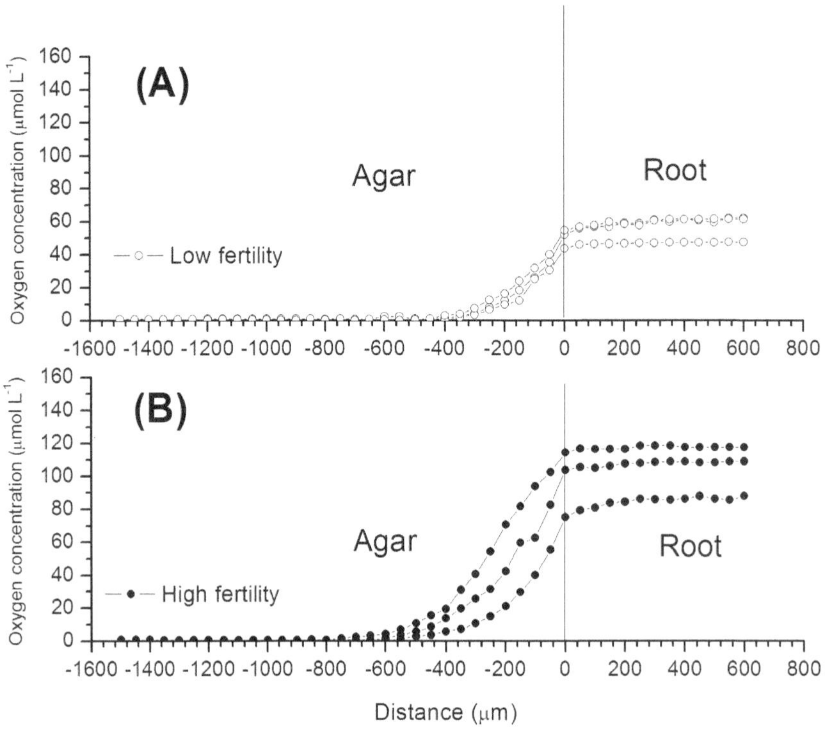

Figure 5. A representative graph showing oxygen profiles of the rhizosphere and within the rice root, 50 d after germination, growing in the (A) low- and (B) high-fertility soils at 20°C. The rice plant roots were separated from the soil and placed in agar with a nutrient solution. Newly formed roots with uniform 45-mm lengths were selected, and both were measured at a position 25 mm away from the root tips. The average diameters of the roots growing in the low- and high-fertility conditions were 0.74 and 0.93 mm at the measured points, respectively. The measurements were repeated in different rice plants and with different newly formed roots, with lengths ranging from 30 to 55 mm. An OXY10 oxygen microelectrode (Unisense, Aarhus, Denmark [31]) 8–10 μm in diameter was used to penetrate the rice root longitudinally. Three rhizoboxes were used for the individual rice root ROL measurements. Each curve shown in the figure indicates an individual rice root.

respectively, compared with those in the low-fertility soil (Table 2), which also confirmed that more NO_3^- was absorbed and assimilated by the rice planted in the high-fertility soil compared with that planted in the low-fertility soil. Greater amounts of NO_3^- absorbed and assimilated by rice plants will also accelerate NH_4^+ uptake and assimilation, perhaps through the NO_3^--specific induction of additional pathways for NH_4^+ assimilation [3], as also confirmed by the GSA results (Table 2). Hence, the rice N accumulation and rice growth of plants grown in the high-fertility soil were improved compared with those of plants grown in the low-fertility soil (Table 2).

Conclusions

The present study investigated reasons that increasing soil fertility would enhance the growth of rice. The study adopted a new perspective focusing on rice root ROL and its corresponding soil oxygen profile. The fertile soil had high porosity (because of the low clay content) and a high NH_4^+ content, which supported soil nitrification. A high rhizosphere oxygen pressure distribution caused by the rice root ROL, which was attributed to the improved rice root growth and root porosity observed in the high-fertility soil, also contributed to the promotion of rhizosphere nitrification. The mechanism accelerating the development of rice root aerenchyma in the high-fertility soil requires further exploration. Investigation of the mechanisms of N loss and the quantification of N loss were not performed in this study but will be addressed by our laboratory in future research. Long-term tillage and fertilisation (especially organic fertiliser) would significantly increase soil fertility, but a large amount of organic matter would decrease the soil oxygen content. Fortunately, rice planting altered the water layer and soil oxygen status via root ROL, and rice in the fertile soil further improved the rhizosphere soil oxidisation environment. Our results suggest that increased fertilisation could lead to increased soil aeration in red soil regions in China, which might promote rice growth and rice N nutrition.

Acknowledgments

We thank Dr. O. Pedersen at the Freshwater Biological Laboratory, Institute of Biology, University of Copenhagen, Denmark, for his patience and experienced instruction in oxygen microelectrode measurement and careful English revisions of this manuscript.

Author Contributions

Conceived and designed the experiments: YLL WMS XXW. Performed the experiments: YLL. Analyzed the data: YLL. Contributed reagents/materials/analysis tools: YLL XXW. Wrote the paper: YLL WMS XXW.

References

1. Abbasi MK, adams WA (1998) Loss of nitrogen in compacted grassland soil by simultaneous nitrification and denitrification. Plant and Soil 200: 265–277.
2. He J Z, Shen JP, Zhang LM, Zhu YG, Zheng YM, et al. (2007) Quantitative analyses of the abundance and composition of ammonia-oxidizing bacteria and

ammonia-oxidizing archaea of a Chinese upland red soil under long-term fertilization practices. Environmental Microbiology 9: 2364–2374.
3. Kronzucker HJ, Siddiqi MY, Glass ADM, Kirk GJD (1999) Nitrate-ammonium synergism in ice: a subcellular flux analysis. Plant Physiology 119: 1041–1046.
4. Kronzucker HJ, Glass ADM, Siddiqi MY, Kirk GJD (2000) Comparative kinetic analysis of ammonium and nitrate acquisition by tropical lowland rice: implications for rice cultivartion and yield potential. New Phytologist 145: 471–476.
5. Raman DR, Spanswick RM, Walker LP (1995) The kinetics of nitrate uptake from flowing nutrient solutions by rice: Influence of pretreatment and light. Bioresourse Technology 53: 125–132.
6. Duan YH, Zhang YL, Ye LT, Fan XR, Xu GH, et al. (2007) Responses of rice cultivars with different nitrogen use efficiency to partial replaced nitrate nutrition. Annals of Botany 99: 1153–1160.
7. Li YL, Fan XR, Shen QR (2008) The relationship between rhizosphere nitrification and nitrogen-use efficiency in rice plants. Plant, Cell and Environment 31: 73–85.
8. Chen XP, Zhu YG, Xia Y, Shen JP, He JZ (2008) Ammonia-oxidizing archaea: important players in paddy rhizosphere soil? Environmental Microbiology 10: 1978–1987.
9. Venter JC, Remington K, Heidelberg JF, Halpern AL, Rusch D, et al. (2004) Environmental genome shotgun sequencing of the Sargasso Sea. Science 304: 66–74.
10. Schleper C, Jurgens G, Jonuscheit M (2005) Genomic studies of uncultivated archaea. Nature Reviews Microbiology 3: 479–488.
11. Könneke M, Bernhard AE, de la Torre JR, Walker CB, Waterbury JB, et al. (2005) Isolation of an autotrophic ammonia-oxidizing marine archaeon. Nature 437: 543–546.
12. Francis CA, Roberts KJ, Beman JM, Santoro AE, Oakley BB (2005) Ubiquity and diversity of ammonia-oxidizing archaea in water columns and sediments of the ocean. Proceedings of the National Academy of Sciences of the United States of America 102: 14683–14688.
13. Leininger S, Urich T, Schloter M, Schwark L, Qi J, et al. (2006) Archaea predominate among ammonia-oxidizing prokaryotes in soils. Nature 442: 806–809.
14. Arth I, Frenzel P (2000) Nitrification and denitrification in the rhizosphere of rice: the detection of processes by a new multi-channel electrode. Biology and Fertility of Soils 31: 427–435.
15. Briones AM, Okabe S, Umemiya Y, Ramsing NB, Reichardt W, et al. (2002) Influence of different cultivars on populations of ammonia-oxidizing bacteria in the root environment of rice. Applied and Environmental Microbiology 68: 3067–3075.
16. Wang YN, Ke XB, Wu LQ, Lu YH (2009) Community composition of ammonia-oxidizing bacteria and archaea in rice field soil as affected by nitrogen fertilization. Systematic and Applied Microbiology 32: 27–36.
17. Armstrong W (1979) Aeration in higher plants. Advances in Botanical Research 7: 225–332.
18. Armstrong W, Justin SHFW, Beckett PM, Lythe S (1991) Root adaptation to soil waterlogging. Aquatic Botany 39: 57–73.
19. Kludze HK, DeLaune RD, Patrick WH (1993) Aerenchyma formation and methane and oxygen exchange in rice. Soil Science Society of America Journal 57: 386–391.
20. Armstrong W (1964) Oxygen diffusion from the roots of some British Bog plants. Nature 204: 801–802.
21. Armstrong W (1971) Radial oxygen losses from intact rice roots as affected by distance from the apex, respiration and waterlogging. Physiologia Plantarum 25: 192–197.
22. Kludze HK, DeLaune RD, Patrick WH (1994) A colorimetric method for assaying dissolved oxygen loss from container-grown rice roots. Agronomy Journal 86: 483–487.
23. Revsbech NP, Pedersen O, Reichardt W, Briones A (1999) Microsensor analysis of oxygen and pH in the rice rhizosphere under field and laboratory conditions. Biology and Fertility of soils 29: 379–385.
24. FAO (1998) FAO, ISRIC, ISSS World Reference Base for Soil Resources. World Soil Resources Reports 84. Rome, Italy.
25. Zhong WH, Cai ZC (2007) Long-term effects of inorganic fertilizers on microbial biomass and community functional diversity in a paddy soil derived from quaternary red clay. Applied Soil Ecology 36: 84–91.
26. Yuan F, Ran W, Shen QR, Wang DZ (2005) Characterization of nitrifying bacteria communities of soils from different ecological regions of China by molecular and conventional methods. Biology and Fertility of Soils 41: 22–27.
27. Zhang MK, Xu JM (2005) Restoration of surface soil fertility of an eroded red soil in southern China. Soil and Tillage Research 80: 13–21.
28. Li YL, Zhang YL, Hu J, Shen QR (2007) Contribution of Nitrification Happened in Rhizospheric Soil Growing with Different Rice Cultivars to N Nutrition. Biology and Fertility of Soils 43: 417–425.
29. Berg P, Rosswall T (1985) Ammonium oxidizer numbers, potential and actual oxidation rates in two Swedish arable soils. Biology and Fertility of Soils 1: 131–140.
30. Alef K, Nannipieri P (1995) Methods in Applied Soil Microbiology and Biochemistry. In: Berg P, Rosswall T, Editors.Assay of nitrification (short-term estimations).New York: Harcourt Brace Academic Press. pp.241–242.
31. Revsbech NP (1989) An oxygen microsensor with a guard cathode. Limnology and Oceanography 34: 474–478.
32. Fan XR, Jia LJ, Li YL, Smith SJ, Miller AJ, et al. (2007) Comparing nitrate storage and remobilization in two rice cultivars that differ in their nitrogen use efficiency. Journal of Experimental Botany 58: 1729–1740.
33. Zhang CF, Peng SB, Peng XX, Chavez A Q, Bennett J (1997) Response of glutamine synthetase isoforms to nitrogen sources in rice (*Oryza sativa* L.) roots. Plant Science 125: 163–170.
34. Ogawa T, Fukuoka H, Yano H, Ohkawa Y (1999) Relationships between nitrite reductase activity and genotype-dependent callus growth in rice cell cultures. Plant Cell Reports 18: 576–581.
35. Jensen CR, Luxmoore RJ, Van Gundy SD, Stolzy LH (1969) Root air-space measurements by a pycnometer method. Agronomy Journal 61: 474–475.
36. Kim JD, Jugsujinda A, Carbonell-Barrachina AA, DeLaune RD, Patrick WH (1999) Physiological functions and methane and oxygen exchange in Korean rice cultivars grown under controlled soil redox potential. Botanical Bulletin of Academia Sinica 40: 185–191.
37. Ratering S, Schnell S (2001) Nitrate-dependent iron (II) oxidation in paddy soil. Environmental. Microbiology 3: 100–109.
38. Frenzel P, Rothfuss F, Conrad R (1992) Oxygen profiles and methane turnover in a flooded rice microcosm. Biology and Fertility of Soils 14: 84–89.
39. Wang YF, Yu ZW, Li SX, Yu SL (2003) Effects of nitrogen rates and soil fertil ity levels on root nitrogen uptake and assimilation and grain protein content of winter wheat. Plant Nutrition and Fertilizer Science 9: 39–44.
40. Mugnai S, Azzarello E, Baluška F, Mancuso S (2012) Local root apex hypoxia induces NO-mediated hypoxic acclimation of the entire root. Plant and Cell Physiology 53: 912–920.
41. Baluška F, Mancuso S, Volkmann D, Barlow PW (2010) Root apex transition zone: a signaling-response nexus in the root. Trends in Plant Science 15: 402–408.
42. Sorrell BK (1999) Effect of external oxygen demand on radial oxygen loss by *Juncus* roots in titanium citrate solutions. Plant, Cell and Environment 22: 1587–1593.
43. Chu HY, Fujii T, Morimoto S, Lin XG, Yagi K, et al. (2007) Community structure of ammonia-oxidizing bacteria under long-term application of mineral fertilizer and organic manure in a sandy loam soil. Applied and Environmental Microbiology 73: 485–491.
44. Colmer TD, Bloom AJ (1998) A comparison of NH_4^+ and NO_3^- net fluxes along roots of rice and maize. Plant, Cell and Environment 21: 240–246.

4

New Insights into Regulation of Proteome and Polysaccharide in Cell Wall of *Elsholtzia splendens* in Response to Copper Stress

Tingting Liu, Chaofeng Shen, Yi Wang, Canke Huang, Jiyan Shi*

Institute of Environmental Science and Technology, College of Environmental and Resource Sciences, Zhejiang University Hangzhou, P.R. China

Abstract

Background and Aims: Copper (Cu) is an essential micronutrient for plants. However, excess amounts of Cu are toxic and result in a wide range of harmful effects on the physiological and biochemical processes of plants. Cell wall has a crucial role in plant defense response to toxic metals. To date, the process of cell wall response to Cu and the detoxification mechanism have not been well documented at the proteomic level.

Methods: An recently developed 6-plex Tandem Mass Tag was used for relative and absolute quantitation methods to achieve a comprehensive understanding of Cu tolerance/detoxification molecular mechanisms in the cell wall. LC–MS/MS approach was performed to analyze the Cu-responsive cell wall proteins and polysaccharides.

Key Results: The majority of the 22 up-regulated proteins were involved in the antioxidant defense pathway, cell wall polysaccharide remodeling, and cell metabolism process. Changes in polysaccharide amount, composition, and distribution could offer more binding sites for Cu ions. The 33 down-regulated proteins were involved in the signal pathway, energy, and protein synthesis.

Conclusions: Based on the abundant changes in proteins and polysaccharides, and their putative functions, a possible protein interaction network can provide new insights into Cu stress response in root cell wall. Cu can facilitate further functional research on target proteins associated with metal response in the cell wall.

Editor: Raffaella Balestrini, Institute for Sustainable Plant Protection, C.N.R., Italy

Funding: The work was supported by the National Natural Science Foundation of China (11179025, 21177109) and Program for New Century Excellent Talents in University (NCET-11-0455). The funders had no role in study design, data collection and analysis, decision to publish, or preparation of the manuscript.

Competing Interests: The authors have declared that no competing interests exist.

* Email: shijiyan@zju.edu.cn

Introduction

Copper (Cu) is an essential micronutrient for plants because it is a structural and catalytic component of several proteins or enzymes involved in various physiological metabolisms [1]. However, excess Cu in plants is harmful; it induces phytotoxicity [2] and disturbs the physiological metabolisms, including carbohydrate metabolism, protein metabolism, mineral nutrition, cell respiration, and photosynthesis [3]. Given its redox properties, Cu^{2+} can catalyze the generation of reactive oxygen species (ROS) and Fenton reactions that damage cellular components or interfere with cellular transport processes [4]. Plants cannot avoid the stress conditions from Cu-contaminated areas. Thus, several resistance mechanisms for Cu tolerance have been developed [5], including exclusion, compartmentalization, chelation, and binding to organic ligands, which enable maintenance of Cu homeostasis for plant survival even in a highly contaminated environment [6].

Plant cell wall is the primary site for signal perception and defense response, which is significant in responding to environmental stresses [7]. The cell walls of plants that grow on heavy metal-contaminated areas provide physical barriers against toxic heavy metals and actively participate in plant defense response. The cell wall serves as a repository for high contents of metals; it senses stress signals and transmits them to the cell interior, thereby affecting the cell fate decision [8]. The main chemical components of a plant cell wall include polysaccharides and proteins. Polysaccharides represent up to 90% of the plant cell wall and constitute three different kinds of polymers, namely, cellulose, hemicelluloses, and pectins. Cell wall proteins (CWPs) of plants comprise less than 10% of the cell wall dry weight but have major roles in cell wall structure, metabolism, cell enlargement, signal transduction, defense responses, and many other physiological events [9].

Certain plants can accumulate high content of metal ions in their cell wall through various compounds. Cell wall polysaccharides possess crucial roles in heavy metal binding and accumulation. Binding of metal ions depends on the number of functional groups in the cell wall polysaccharides. Previous studies confirmed that the essential capacity of the cell wall for binding metal ions depends mainly on the abundance of polysaccharides in the carboxyl groups [10]. Current information about the regulation of cell wall polysaccharides under metal stress remains limited.

Abiotic stress induces expression of stress-responsive proteins, which lead to cellular adaptation in plant growth under stress conditions. Therefore, studies have been carried out to identify various CWPs involved in stress response, particularly those of regulatory or targeting function proteins [11–17]. Most reports have focused on leaf cell wall proteomes under abiotic stress, such as salt in tobacco [11]; wounding in *Medicago* [12]; dehydration in chickpea, rice, and maize [9,14,15]; zinc toxicity in *Hordeum* [18]; nickel toxicity in barley [19]; manganese toxicity in cowpea [20,21]; boron deficiency in *Lupinus albus* [22]; and flooding in soybean [23]. Few studies have focused on the root cell wall proteomes under abiotic stress [24–26]. All of these studies broadened our understanding of the complicated regulation of apoplast proteins. An in-depth proteome study of the cell wall can provide extensive information on the intrinsic mechanism of stress response attributed to a possible relationship between protein abundance and plant stress tolerance [27].

Elsholtzia splendens is a plant that can survive in a highly Cu-polluted area. This species can accumulate more than 1000 mg/kg of Cu in the body [28], and the major proportion (70%) of Cu ions is stored in the root cell wall [29]. Nevertheless, little is known about the defense and detoxification mechanisms in the root cell wall of *E. splendens*. A recently developed 6-plex Tandem Mass Tag (TMT) was used for relative and absolute quantitation methods to gain a comprehensive understanding of Cu tolerance/detoxification molecular mechanisms. LC–MS/MS approach was also conducted to analyze Cu-responsive CWPs and polysaccharides in this study. This proteomic study of the apoplast is the first in response to Cu. The results will significantly expand knowledge on the apoplast complexicity in metal-tolerant plants and develop a more comprehensive understanding of the involved cellular reactions in cell defense mechanisms.

Results

Isolation and purification of CWPs

Root CWPs from *E. splendens* were isolated with sucrose gradients and washed according to the method of Feiz et al. [30]. The activity of G'DH, a typical marker enzyme used to detect contamination of cytoplasmic proteins, was measured to validate the purification method [31]. The G6PDH activity in $CaCl_2$-extracted CWPs, NaCl-extracted CWPs, and cytosolic protein fractions were 8.36, 5.78, and 107.55 mU/mg protein, respectively (Figure 1). The G6PDH activity in $CaCl_2$- or NaCl-extracted CWPs was 10 times lower than that in cytosolic proteins ($p<0.01$). These results indicate that the contamination of intracellular proteins was minimal.

The identified proteins were classified into different gene ontology (GO) terms according to their cellular compartment(s), function(s), and biological processes (Table S1). Some of these identified proteins showed a compartment label (apoplast, cell wall, extracellular region, and plant-type cell wall) that is coherent with secreted proteins. The same proteins can also be attributed to various compartments in the GO classification. Three different computer programs (TargetP, Predotar, and WoLF PSORT) were used to predict the protein subcellular localization (Table S1). A protein was secreted when at least two out of the three algorithms predicted extracellular localization. Based on the results, 15 proteins were predicted to be secreted proteins in the fractions (Table S1).

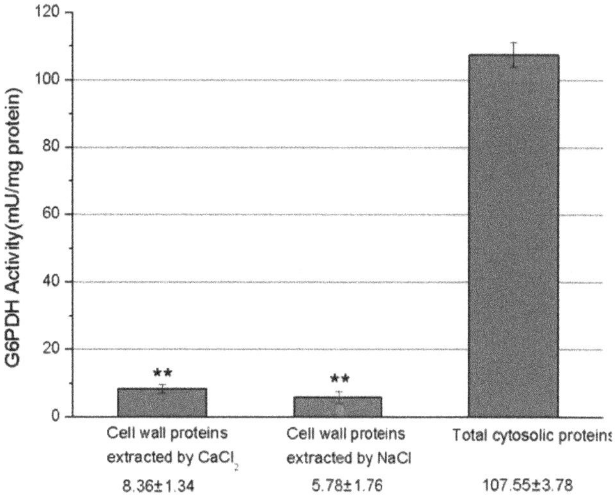

Figure 1. Evaluation of purity by the G6PDH activity in the cell wall proteins isolated from the roots of *Elsholtzia splendens*. The activity of G6PDH was assayed in $CaCl_2$-extracted cell wall proteins, NaCl-extracted cell wall proteins and total soluble proteins. One unit of G6PDH activity is defined as 1 μmol of NADPH turnover per min/mg protein. Results are presented as mean ±SE of G6PDH activity from three biological replicates. The asterisks indicate significant differences in the G6PDH activity of $CaCl_2$-extracted cell wall proteins, NaCl-extracted cell wall proteins compared with that of total soluble cytosolic proteins (** $p<0.01$).

Cu-induced changes in root CWP contents

Plants were grown on a medium supplemented with different $CuSO_4$ concentrations to determine changes in CWP abundance under Cu stress (Figure 2A). Cu toxicity led to a strong increase in CWP contents at low Cu concentrations. The CWP contents increased 1.39-fold at 25 μM Cu and 1.96-fold at 50 μM Cu. The highest content of the CWPs was observed at 50 μM Cu concentrations. The plant root began to decompose at 100 μM Cu (Figure 2C). The increase of Cu contents in the cell wall could proportionally increase with the Cu supply in the medium (slope = 0.3, $R^2 = 0.98$) at Cu concentrations below 75 μM. The Cu content in the root cell wall continued to increase significantly (slope = 0.45, $R^2 = 0.99$) above 75 μM Cu. The highest value of Cu content in the root cell wall reached 1.41 mg/g with the root cell wall dry weight at 100 μM Cu. These findings indicate an increase in Cu binding to cell walls with Cu supply in the medium (Figure 2A). Polypeptide separation by SDS-PAGE (Figure 2B) revealed about 30 polypeptide bands. Band intensities increased with the increase in Cu concentration below 50 μM (Figure S1). The staining intensity was the lowest among all the cell wall polypeptides at 100 μM Cu concentration.

Cu-induced remodeling in root cell wall polysaccharide

Evidence generally suggests that cell wall polysaccharides are crucial sites for Cu retention in plants. Therefore, changes in the contents of root cell wall polysaccharide of *E. splendens* were measured. Treatments with 25, 50, 75, and 100 μM Cu significantly increased the expression of CWPs and polysaccharides compared with controls (Figure 3). Cu supply enhanced the total sugar contents in various cell wall fractions. Sugar contents were higher in hemicellulose than in pectin and cellulose. Sugar contents in pectin increased 1.22-fold with 25 μM Cu. The major components of cell wall fractions are galacturonic acids. The content of galacturonic acids was considerably higher than that in

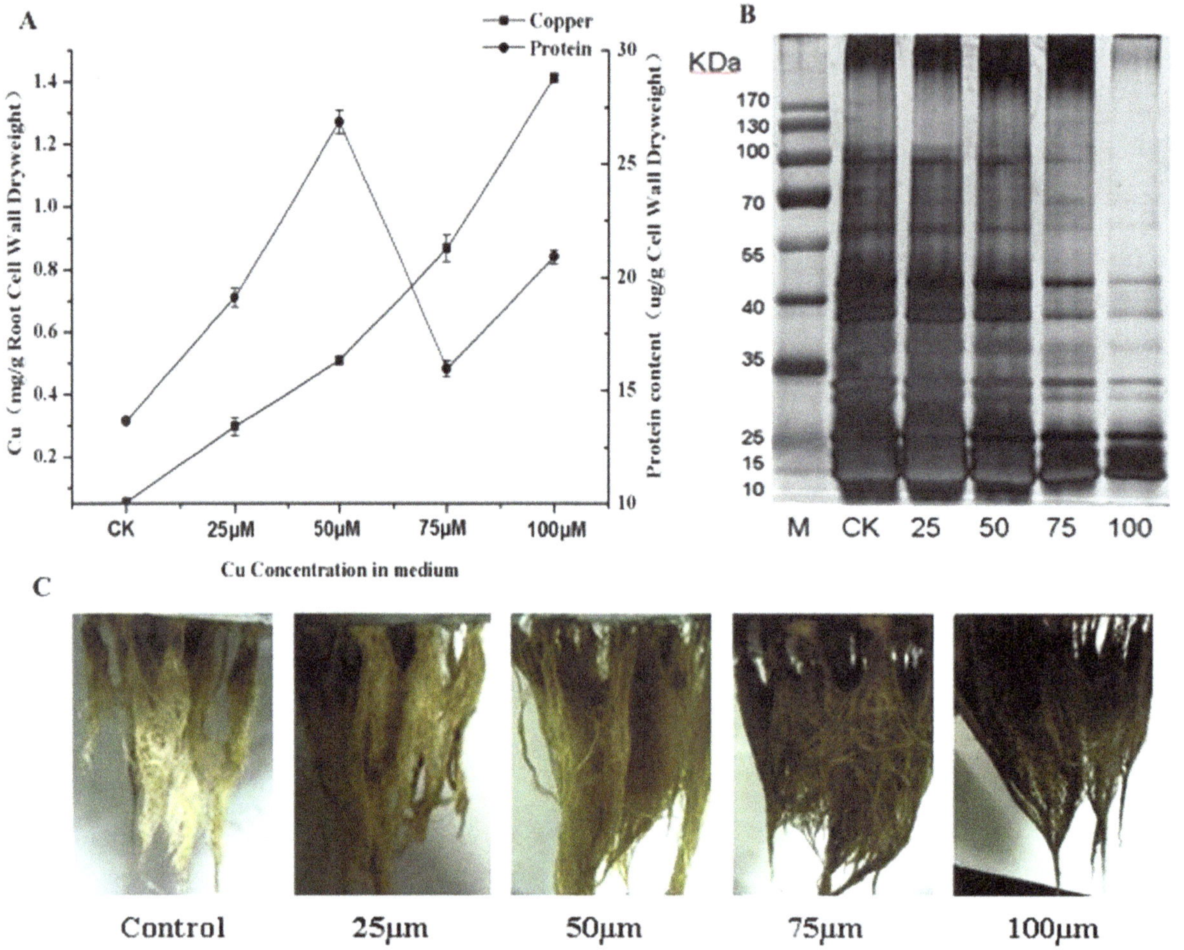

Figure 2. Effect of different copper concentration on the cell wall protein and copper ions in *Elsholtzia splendens*'s root cell. A: The content of copper and protein every root cell wall dry weight. B: SDS-PAGE of root cell wall protein under different copper stress. C: Root growth patterns of control and copper-stressed plants during different copper concentration. Data presented are mean ±SE (n = 20),*Significant mean differences from control at $p = 0.05$ in multiple comparison by LSD test.

the control during Cu treatment, particularly at higher Cu supply. The galacturonic acid values rapidly increased in pectin (1.28-fold), hemicellulose (1.22-fold), and cellulose (1.79-fold) at 100 μM Cu. A conserved disaccharide residue known as 2-keto-3-deoxyoctonic acid (KDO) is attached to C-3 of the backbone of rhamnogalacturonan II in pectin. The KDO concentration in pectin increased during Cu exposure. Thus, the highest concentration of KDO (1.49-fold) was found in the 100 μM Cu-treated group.

The results indicated that excess Cu affected the intracellular distribution of polysaccharides in the root cell wall. Given this reason, we focused on the relationship between wall polysaccharide composition and CWPs with increasing Cu intensity. The dendrogram proved the connection between the contents of polysaccharides and proteins as matched with different Cu concentrations (Figure 4). This finding suggests that Cu induced changes in the contents of root cell wall polysaccharides and proteins, which altered the intracellular distribution patterns in assigning effects on cellular activities.

Cu-responsive CWP identification using TMT LC-MS/MS

This study aimed to gain a better understanding of the metabolic processes and molecular mechanisms involved in Cu-responsive CWPs. A TMT-based shotgun quantitation approach was used to obtain an overall view of the proteome changes associated with Cu stress in the root cell wall of *E. splendens*. Unique proteins were successfully identified based on homology searching with a common protein confidence cutoff of 95% and relative quantitative information in control and Cu-treated samples, as indicated by the peak areas of the different TMT tags (Table 1, Table S2). A threshold of 1.5-fold change and a $p < 0.05$ were set as stringent criteria for significant differences between control and treatment. Fifty five proteins were classified as differentially expressed upon Cu stress using the criteria; 22 and 33 of these proteins showed increasing trends and decreasing abundance, respectively (Table 1, Table 2).

Functional classification of identified CWPs

Database search using Uni-Prot/NCBI accession numbers was conducted to obtain detailed information of differentially expressed proteins. These proteins were categorized into biological process and molecular function using an in-house Perl script according to the extracted GO terms from InterPro (www.ebi.ac.uk/interpro/) or Pfam (pfam.sanger.ac.uk/) (Figure 5). Functional classifications of the 55 differentially expressed proteins (Table 1, Table 2) were classified based on the GO terms that correspond to

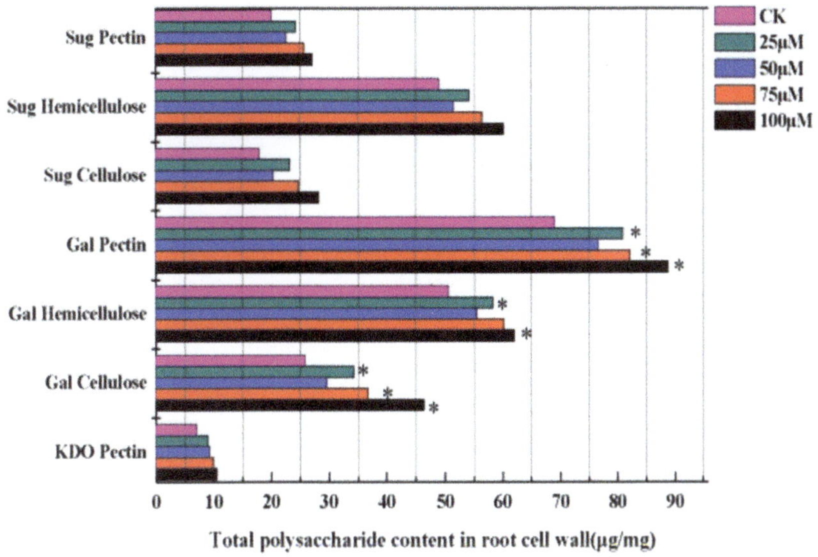

Figure 3. Effect of different copper concentrations on the contents of root cell wall polysacchride. Sug: sugar, Gal: galacturonic acids, KDO: 2-Keto-3-deoxyoctonic acid. Data presented are mean ±SE (n = 3). *Significant mean difference from control at $p = 0.05$ in multiple comparison by LSD test.

the biological process. The top three categories were those involved in metabolic processes (23.24%), cellular processes (16.20%), and response to stimuli (14.79%) (Figure 5A). These three categories were classified according to the GO molecular annotation into binding (41.38%), catalytic activity (31.03%), and transport activity (11.49%) (Figure 5B). The observed diversity in biological function and processes of the differentially expressed CWPs demonstrated that the response of the root cell wall of *E. splendens* to Cu stress was a complex process. Many physiological and biochemical changes were altered to counteract the adverse conditions.

Specific enzymes constituted by some small differentially expressed CWPs participated in various metabolic processes to counteract Cu stress. Classification pathways according to the

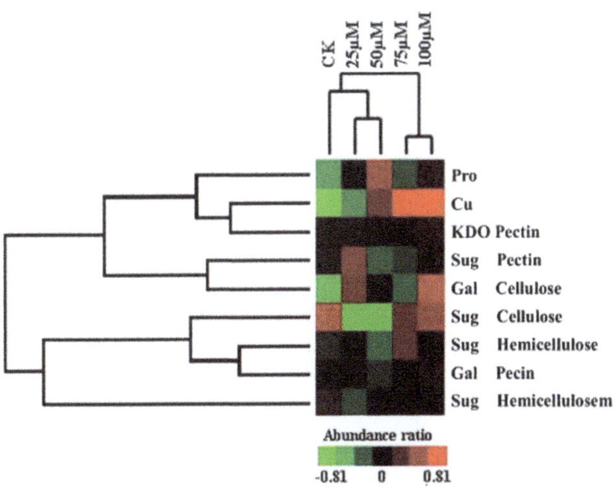

Figure 4. Hierarchical cluster result of cell wall polysaccharide abundance ratio using the average linkage distance between clusters is shown. The color weighting represents normalized levels of each variable from the high (red) to the low (green).

KEGG database searches categorized the cell wall enzymes into 12 classes (Table 3, Table S2). The more important category pathways were involved in oxidative phosphorylation, phenylalanine biosynthesis and metabolism, ascorbate and aldarate metabolism, and citrate cycle. All these pathways suggested the relatively altered abundance of the enzymes in various metabolic pathways that probably re-optimized the metabolic processes to combat against Cu stress.

Discussion

Cell wall polysaccharide remodeling in root under Cu stress

Cell walls provide a physical barrier to plants grown in Cu-contaminated soil, and cell wall polysaccharides have a crucial role in metal binding and accumulation [7]. The binding ability depends on the number of functional groups in polysaccharides, particularly for abundant carboxyl groups [7]. In the present study, the contents of the various cell wall polysaccharides significantly increased with the increase in Cu concentration. The expression of cell wall polysaccharides, which contain more carboxyl groups, provided more binding sites for Cu ions. Excess Cu also affected the intracellular distribution of polysaccharide in the root cell walls (Figure 3). Similar situations were also described in rice roots under Cd stress [45] and *Arabidopsis* roots under Al stress [46]. The starch and sucrose metabolism-related proteins (Table 3) – Chloroplast NAD-MDH (No. 1, Table 1) – were also markedly up-regulated in response to Cu stress based on our observations. These findings suggest that Cu or Cu-induced oxidative stress may activate some crucial enzymes in the glyconeogenesis pathway, which can alter the contents of cell wall polysaccharides.

Increasing physiological, biochemical, and molecular evidence also showed that the remodeling of the binding properties of root apoplasts was attributed to Cu resistance. Enhancement at the level of low-methylesterified pectins, where the polysaccharides bound more trivalent metal ions, is one of the most remarkable alterations. Pectin methylesterases (PMEs), known as pectinesterases, catalyzed

Table 1. Root cell wall proteins with significant increased in expression level under 50 μM Cu treatment identified by LC-ESI-MS/MS-based proteomics using SIEVE ($p<0.05$ and fold chang >1.5).

Serial No.	Protein name	gi number[a]	Percent coverage [b] (%)	Fold Change[c]	*p* value[d]	Species identified[e]	MW(kDa)	p*I*
1	Chloroplast NAD-MDH	3256066	8.68	1.50	0.00000	Arabidopsis thaliana	49.9	8.2
2	D-glyceraldehyde-3-phosphate Dehydrogenase(GAPDH)	51849658	17.26	1.52	0.00000	Periploca sepium	25.3	6.2
3	40S ribosomal protein S19	74231038	9.32	1.52	0.00271	Solanum peruvianum	15.7	5.5
4	Phosphoglycerate kinase	21272	5.08	1.53	0.00963	Spinacia oleracea	55.1	6.1
5	Putative malate dehydrogenase	50080249	15.29	1.55	0.00346	Oryza sativa Japonica Group	41.6	5.0
6	Cyc07	1856971	6.12	1.56	0.02389	Catharanthus roseus	35.6	6.0
7	2,3-bisphosphoglycerate-independent phosphoglycerate mutase	168588	1.61	1.57	0.00000	Zea mays	68.8	5.2
8	Non-symbiotic hemoglobin	11095158	13.75	1.58	0.01599	Medicago sativa	23.0	9.0
9	P0 ribosomal protein	1143507	6.83	1.59	0.00000	Lupinus luteus	41.0	9.8
10	Putative 40S ribosomal protein s12	643074	7.75	1.60	0.03853	Fragaria x ananassa	21.2	4.7
11	Unknown	255648032	11.18	1.62	0.02313	Glycine max	21.3	6.4
12	Mitochondrial chaperonin	415733	7.16	1.68	0.01830	Brassica napus	74.5	5.4
13	Maturase K	33332509	4.13	1.69	0.01270	Pitcairnia xanthocalyx	48.8	6.9
14	Pectin methylesterase	1222552	2.23	1.70	0.00633	Solanum lycopersicum	75.5	9.5
15	Porin-like protein	47847590	3.61	1.71	0.01348	Oryza sativa Japonica Group	36.1	6.4
16	Pollen allergen Que a 1 isoform	167472851	7.50	1.72	0.00062	Quercus alba	21.3	8.8
17	Putative 1,4-benzoquinone reductase	53749369	5.37	1. 73	0.00000	Oryza sativa Japonica Group	49.9	8.2
18	Thioredoxin h	13624884	9.32	1.89	0.02335	Pisum sativum	25.3	6.2
19	Predicted protein	162680354	5.16	1.91	0.00523	Physcomitrella patens subsp	24.8	8.9
20	Hypothetical protein OsI_31140	125563499	2.51	1.99	0.00035	Oryza sativa Indica Group	16.1	4.7
21	Hypothetical protein OsI_07053	218190658	1.47	2.70	0.00000	Oryza sativ a Indica Group	37.3	5.8
22	60s acidic ribosomal protein P2	551267	7.89	3.31	0.00000	Parthenium argentatum	56.7	5.4

[a] gi no., gene identification number as in GenBankTM.
[b] Coverage., sequence coverage.
[c] Fold change.
[d] p value, indicates the significance of up- or down- regulation of spots according to the F-test through analysis of variance ($p<0.05$).
[e] Species identified by Mascot search using NCBI.

Table 2. Root cell wall proteins with significant decreased in expression level under 50 μM Cu treatment identified by LC-ESI-MS/MS-based proteomics using SIEVE ($p<0.05$ and fold chang >1.5).

Serial No.	Protein name	gi number[a]	Percent coverage [b] (%)	Fold Change[c]	*p* value[d]	Species identified[e]	MW(kDa)	p*I*
23	Predicted protein	162671613	1.87	−2.68	0.00000	Physcomitrella patens subsp	71.6	4.9
24	Peroxidase	17066703	2.71	−2.59	0.00000	Capsicum annuum	39.5	6.9
25	Os07g0223100	113610742	18.75	−2.28	0.03704	Oryza sativa Japonica Group	6.1	6.0
26	Mitochondrial phosphate translocator	1842188	2.47	−2.25	0.00466	Betula pendula	45.3	9.4
27	Predicted protein	549049082	3.53	−2.23	0.01662	Populus trichocarpa	30.5	8.9
28	Actin	288300146	16.80	−2.17	0.00166	Cleistogenes songorica	46.2	5.5
29	Ubiquitin fusion protein	149391425	36.04	−2.08	0.02095	Oryza sativa Indica Group	16.6	10.0
30	Heat shock protein Hsp70	124360342	2.59	−1.98	0.02106	Medicago truncatula	80.9	6.4
31	Putative nascent polypeptide associated complex alpha chain	20160782	6.93	−1.97	0.00728	Oryza sativa Japonica Group	26.9	4.4
32	Hypothetical protein SORBIDRAFT_09g020360	241945088	8.50	−1.88	0.00003	Sorghum bicolor	49.1	5.2
33	Plasma membrane-type calcium ATPase	3335060	1.08	−1.86	0.03138	Arabidopsis thaliana	125.3	8.9
34	Cytochome b5	2695711	11.94	−1.83	0.01678	Olea europaea	18.3	5.5
35	Unknown	118482646	3.48	−1.82	0.00000	Populus trichocarpa	41.8	4.9
36	ATP synthase beta subunit	8452631	4.35	−1.80	0.04389	Cichorium intybus	18.1	6.2
37	Peptidyl-prolyl cis-trans isomerase	118104	7.56	−1.76	0.00000	Zea mays L	21.8	9.0
38	Ferredoxin–NADP+ reductase-like protein	7267299	8.61	−1.73	0.00081	Arabidopsis thaliana	48.3	8.9
39	Putative protein	5262759	0.82	−1.71	0.04231	Arabidopsis thaliana	124.3	8.5
40	Conserved hypothetical protein	223517971	3.65	−1.71	0.00000	Ricinus communis	24.2	4.2
41	Unnamed protein product	219742878	7.50	−1.70	0.01691	Glycine max	38.2	5.6
42	Unknown	255634488	7.50	−1.69	0.01879	Glycine max	38.0	5.4
43	Ferredoxin-NADP Reductase	6634773	7.59	−1.69	0.00157	Oryza sativa	50.9	9.2
44	Predicted protein	222844395	1.20	−1.68	0.00000	Populus trichocarpa	72.6	9.2
45	60S ribosomal protein L25	310935	14.94	−1.66	0.00002	Nicotiana tabacum	24.8	8.8
46	Vacuolar H(+)-ATPase subunit-like protein	7801655	2.27	−1.64	0.00000	Arabidopsis thaliana	58.0	5.9
47	Small G protein	974780	4.06	−1.61	0.01530	Beta vulgaris subsp. vulgaris	26.3	10.1
48	Elongation factor 1 subunit alpha	162036292	8.22	−1.58	0.02150	Cenchrus americanus	30.2	6.6
49	2-dehydro-3-deoxy phosphor heptonate aldolase	2546988	5.61	−1.52	0.01476	Morinda citrifolia	65.5	9.3
50	RAS-related GTP-binding protein	20756	5.34	−1.52	0.00000	Pisum sativum	26.7	8.3
51	Aspartic proteinase nepenthesin-1 precursor, putative	223535787	3.05	−1.52	0.00000	Ricinus communis	46.6	6.5
52	Hypothetical protein	147774267	0.73	−1.52	0.00000	Vitis vinifera	24.1	4.9
53	Hypothetical protein VITISV_025412	147822728	8.42	−1.52	0.00000	Vitis vinifera	21.9	7.3
54	Predicted protein	222855760	6.33	−1.51	0.00185	Populus trichocarpa	41.8	5.0
55	Hypothetical protein VITISV_023718	147801802	1.87	−1.51	0.00000	Vitis vinifera	71.1	7.8

[a] gi no., gene identification number as in GenBankTM.
[b] Coverage., sequence coverage.
[c] Fold change.
[d] p value, indicates the significance of up- or down- regulation of spots according to the F-test through analysis of variance ($p<0.05$).
[e] Species identified by Mascot search using NCBI.

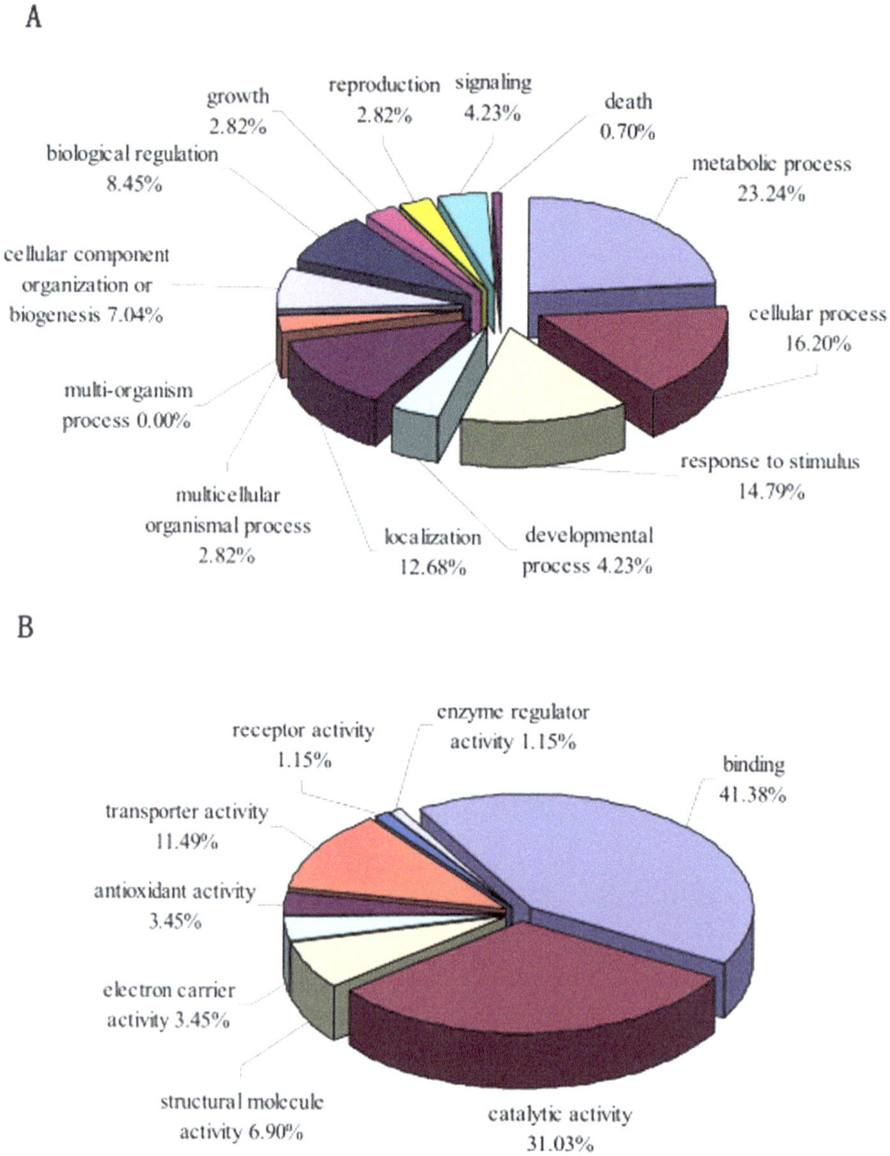

Figure 5. Functional cataloguing of 55 differentially expressed cell wall proteins in *Elsholtzia splendens*'s root based on GO annotation. The pie charts show the distribution of 55 differentially expressed cell wall proteins on of the Cu –responsive proteins into their functional classes in percentage. A: Biological Process Ontology, B: Molecular Function Ontology.

the demethylesterification of the homogalacturonan pectin domain in the cell wall. Demethylesterification of the pectin increases the abundance of free carboxylic acid groups on the galacturonic acid residue [47]. A flax PME has been recently implicated in wall remodeling following Cd treatment [48]. Interestingly, this study identified PMEs (No. 14, Table 1) as Cu-induced differentially expressed proteins; the protein had changes in abundance of 1.7-fold ($p \leq 0.5$). Catalyzed ROS by excess Cu^{2+} can also damage or cause degradation of essential complex molecules in cell wall polysaccharide. Generating hydroxyl radicals from H_2O_2 possesses a direct role in cell wall loosening through polysaccharide cleavage [49,50]. Hydroxyl radicals (OH) may cause non-enzymic scission of polysaccharides in vivo (e.g., in plant cell walls) [50]. Basing on these findings, we concluded that the chemical composition and distribution of plant cell wall polysaccharides are a factor in the outcome of plant–metal interaction, which elucidated a possible role in a novel Cu-resistance mechanism.

Cu-induced oxidative stress and antioxidant defense in cell wall

Cu is a redox active metal that catalyzes ROS production, such as superoxide ($O_2^{\cdot}$), hydrogen peroxide (H_2O_2), and hydroxyl radicals (OH·) through Habere Weiss and Fenton reactions [51]. ROS can act as signaling molecules for stress response. However, ROS can cause damage to many cellular components above a certain threshold. Most Cu-tolerance mechanisms are primarily involved in protecting the cellular structure. An important method is the control of the level of ROS or the limitation of damage caused by ROS. In the current study, significant changes in the abundance of some antioxidant and defense-related proteins suggested that ROS can be involved in a Cu-induced oxidative

Table 3. Category of differentially expressed cell wall proteins refers to the entry on the classification of pathways from KEGG database.

No.	Pathway	Enzyme	Enzyme ID	Change Folds	Protein Name
1	Phenylpropanoid biosynthesis??? phenylalanine metabolism	Lactoperoxidase	ec:1.11.1.7	−2.59	Peroxidase
				−1.69	Ferredoxin-NADP Reductase
				−1.66	Predicted protein
2	Phenylalanine, tyrosine and tryptophan biosynthesis	Synthase	ec:2.5.1.54	−1.52	Hypothetical protein VITISV_025412
				2.70	Hypothetical protein OsI_07053
3	Nitrogen metabolism	(6S)-tetrahydrofolate aminomethyltransferase dehydratas	ec:2.1.2.10	1.72	Pollen allergen Que a 1 isoform
			ec:4.2.1.1	−1.83	Cytochome b5
4	One carbon pool by folate	(6S)-tetrahydrofolate aminomethyltransferase	ec:2.1.2.10	1.72	Pollen allergen Que a 1 isoform
5	Starch and sucrose metabolism	Synthase	ec:2.4.1.34	1.50	Chloroplast NAD-MDH
6	Oxidative phosphorylation	ATPase	ec:3.6.3.6	−1.76	Peptidyl-prolyl cis-trans isomerase
		Dehydrogenase	ec:1.3.5.1	1.53	Phosphoglycerate kinase
7	Ascorbate and aldarate metabolism	Oxidase	ec:1.1.3.8	−1.64	Vacuolar H(+)-ATPase subunit-like protein
8	Glyoxylate and dicarboxylate metabolism	(Si)-synthase	ec:2.3.3.1	−1.86	Plasma membrane-type calcium ATPase
9	Sphingolipid metabolism	Phosphodiesterase	ec:3.1.4.12	−1.52	2-dehydro-3-deoxyphosphoheptonate aldolase
10	Citrate cycle (TCA cycle)	(Si)-synthase	ec:2.3.3.1	−1.86	Plasma membrane-type calcium ATPase
		dehydratas	ec:1.3.5.1	1.53	Phosphoglycerate kinase
11	Glycine, serine and threonine metabolism	(6S)-tetrahydrofolate aminomethyltransferase	ec:2.1.2.10	1.72	Pollen allergen Que a 1 isoform
12	Methane metbolisma	Lactoperoxidase	ec:1.11.1.7	−2.59	Peroxidase
				−1.69	Ferredoxin-NADP Reductase
				−1.66	Predicted protein

stress response. Analysis of *E. splendens* root CWPs confirmed the associated gene expression for enzymes involved in ROS scavenging and oxidative phosphorylation (Table 3), such as peroxidase (No. 24, Table 2), peptidyl-prolyl cis-trans isomerase (No. 37, Table 2), and phosphoglycerate kinase (No. 4, Table 1). Among these proteins, peptidyl-prolyl cis-trans isomerase, as a protein chaperone, possess complementary and sometimes overlapping roles in protecting the proteins [15].

Excess Cu generates oxidative stress, thereby hindering some important metabolic process, such as up-regulation of antioxidant and stress-related regulatory proteins, that help maintain cellular homeostasis [52]. Maturase K (No. 13, Table 1) changed in abundance of 1.69-fold. Maturase K has an important role in plant growth, cell division, and expansion as well as in protecting metabolic processes against H_2O_2 and other toxic derivatives of oxygen. Hydroquinone formation was accompanied by the oxidation of two moles of NADPH and the presence of an inducible 1,4-benzoquinone reductase. Thus, putative 1,4-benzoquinone reductase (No. 17, Table 1) plays an essential role in antioxidant defense pathway [53]. Studies also reported about wall-bound malate dehydrogenase (No. 5, Table 1) [54,55], which can regenerate NAD(P)H that is needed by cell wall peroxidases for free radical generation associated with lignin polymerization. Thioredoxin (No. 18, Table 1) is also involved in redox regulation by reducing disulfides on the target protein for detoxifying lipid hydroperoxides or repairing oxidized proteins and relaying the signal to mitogen-activated protein kinase pathway of stress signaling [56]. Phosphoglycerate kinase (No. 4, Table 1) can also interact with cytosolic catalase and has a role in relieving oxidative stress. Interestingly, down-regulation of key metabolic enzymes revealed that oxidant protection conferred by these proteins was also regulated during Cu treatment. Ferredoxin-NADP reductase (No. 43, Table 2) and ferredoxin-NADP reductase-like protein (No. 38, Table 2), known to sequester highly reactive Fe^{3+} and prevent formation of toxic$^{\cdot}$ OH species, were also identified [57]. Vacuolar H(+)-ATPase subunit-like protein (No. 46, Table 2) was involved in ascorbate and aldarate metabolism. L-ascorbic acid is characterized by plant tissues, and ascorbate is one of the most important antioxidant molecules [58]. Furthermore, 2-dehydro-3-deoxy-phosphorheptonate aldolase synthase (No. 49, Table 2) may be involved in aromatic pathways in the secondary metabolites, which are known to act as defense responses to abiotic stress [59].

Cu stress-activated signaling pathways

The communication between the cytoskeleton and the CWPs is one of the most characterized features of cellular mechanisms that enable cells to respond effectively to various extracellular signals. Several candidate components involved in signal transduction were identified in this study, including Hsp70, small G protein, and cytochrome. The biosynthesis and accumulation of HSP proteins (No. 30, Table 2) can generally contribute to the protection and repair of cells under stress [60]. These proteins may be involved in cell wall biogenesis [61]. Small G proteins (No. 47, Table 2) transduced signals from receptors to control a wide range of cellular functions, particularly for regulating Ca^{2+} channel expression at the cell surface [62]. These proteins are clustered into distinct families but all act as molecular switches, which are active in their GTP-bound form but inactive when GDP-bound [63]. RAS-related GTP-binding protein (No. 50, Table 2) also has a primarily role in modulating cellular functions that involve actin cytoskeleton (No. 28, Table 2), such as establishing cell polarity and morphology [64]. Cytochrome has been identified in the stem cell wall of *Medicago sativa* [65] and germinating embryos of *Oryza sativa* [66]. This first proteomic study showed that cytochromes (No. 6, Table 1; No. 34, Table 2) were Cu-responsive proteins in the root cell wall of *E. splendens*. More preliminary evidence suggested that porins can form aqueous transmembrane channels for transporting solutes and macromolecules across the extracellular surface [67] [68]. These pores are regulated by ATP and GTP with a gating mechanism that modulates the pore size and ion selectivity [54]. In this case, Cu resistance mechanism may be induced by active apoplastic permeability because of up-regulated porin-like proteins (No. 15, Table 1).

Cu stress-activated energy pathways and protein synthesis

ATP is an essential metabolite in cell walls associated with energy conversion [69]. The plasma membrane calcium ATPase (PMCA) (No. 33, Table 2) or Ca^{2+} pump transports Ca^{2+} ions out of the cells using the stored energy in ATP. Control of Ca^{2+} concentration is significant in the cytosol [70]. PMCA down-regulation at the protein levels was observed during the experiment, which may represent an adaptive mechanism to facilitate removal of Ca^{2+} in the maintenance of calcium homeostasis in abiotic stress. The vacuolar H(+)-ATPase (No. 46, Table 2) functions as a primary proton pump that generates electrochemical gradients of protons across the trans-membrane region [71], which provides the primary driving force for transporting numerous ions and metabolites against their electrochemical gradients [72]. Glyceraldehyde-3-phosphate de-hydrogenase (GAPDH) (No. 2, Table 1), as a classical cellular enzyme involved in glycolysis, is also differentially expressed in response to Cu exposure, where GAPDH is specifically targeted at the cell wall [73]. Mitochondrial phosphate translocator (No. 26, Table 2), involved in ATP/ADP transportation, is also identified in the cell wall. Amino acids and amino acid-derived molecules can be chelated with metal ions by high affinity ligands [74]. Pollen allergen Que a 1 isoform (No. 16, Table 1), which is involved in glycine, serine, and threonine metabolism (Table 3), was up-regulated in the root cell wall of *E. splendens*, whereas aspartic proteinase nepenthesin-1 precursor (No. 51, Table 2) was down-regulated by Cu stress. Different expression levels of amino acid-related proteins in the cell wall elucidated their possible role in Cu tolerance.

Other Cu stress-responsive proteins and their potential functions

The 2,3-bisphosphoglycerate independent phosphoglycerate mutase (No. 7, Table 1) was up-regulated in the study, which was also differentially expressed in rice anthers under cold stress [75]. Previous evidence indicated that the expression of non-symbiotic hemoglobins (nsHbs) from cotton is induced with Verticillium wilt fungus [76]. The proteins also had a role in the defense responses against pathogen invasions in *Arabidopsis* [76]. Our studies showed that the nsHbs (No. 8, Table 1) were up-regulated in the cell wall of *E. splendens* root under Cu stress, although their physiological function is yet to be determined. A significant amount of the unidentified proteins were classified as hypothetical or predictable. These proteins were annotated in databases as unknown, hypothetical, or putative proteins because of the theoretical translation of open reading frame sequences [77]. The roles of these proteins remain to be investigated.

Overview of regulated CWP functions

This study provides insights into the functional role of the cell wall of *E. splendens* under Cu stress, where several CWPs were identified with possible roles in Cu tolerance/detoxification. Cu stress conditions can alter the composition (polysaccharide and protein) of the cell wall both qualitatively and quantitatively. Approximately 40% of the differentially expressed CWPs showed higher abundance in response to Cu stress involved in antioxidant defense, cell wall polysaccharide remodeling, and metabolism process. Up to 60% of the CWPs were in low abundance in response to Cu stress that is involved in signal, energy, and protein synthesis. Polysaccharide analysis confirmed the cell wall remod-eling under Cu stress. The amount, composition, and distribution of the cell wall polysaccharides are consequential for plant adaptation to enhance Cu ion levels. Proteome analysis of cell wall confirmed that most proteins were associated with an antioxidant defense response. Hsp 70, small G protein, and RAS-related GTP-binding protein also have essential roles in signal transduction across the cell wall and through the cytoskeleton. Literature provides fundamental information about the role of polysaccharide composition of plant cell wall in metal tolerance and complementary evidence on continuous crosstalk between CWPs and the cytoskeleton. Therefore, knowledge has been expanded on plant stress-related signaling pathways in the cell wall. Cu regulation of these proteins may also not solely respond on abundance changes. Post-translational modification and dynamics are interesting subjects for future investigation. The present study is the first cell wall proteome and polysaccharide investigation of plants in response to Cu and important for the understanding of the plant cell wall response to environmental heavy metal stresses. We propose a possible protein interaction network (Figure 6) that provides new insights into Cu stress response in the root cell wall and facilitates further functional research of target proteins that are associated with Cu response based on the abundant changes in these proteins, as well as polysaccharides and their putative functions.

Materials and Methods

Ethics statement

No specific permissions were required for collecting *E. splendens* seeds from deposited Cu-mining soil in Zhuji County, Zhejiang Province, China. *E. splendens* is neither endangered nor protected. Authors maintained the population at sustainable levels. All plant work was conducted according to relevant national and international guidelines.

Plant material and Cu treatment

E. splendens seeds were collected from plants that grew on deposited Cu-mining soil in Zhuji County, Zhejiang Province, China. The seeds were washed with deionized water and soaked in distilled water to germinate in a controlled dark condition (25°C)

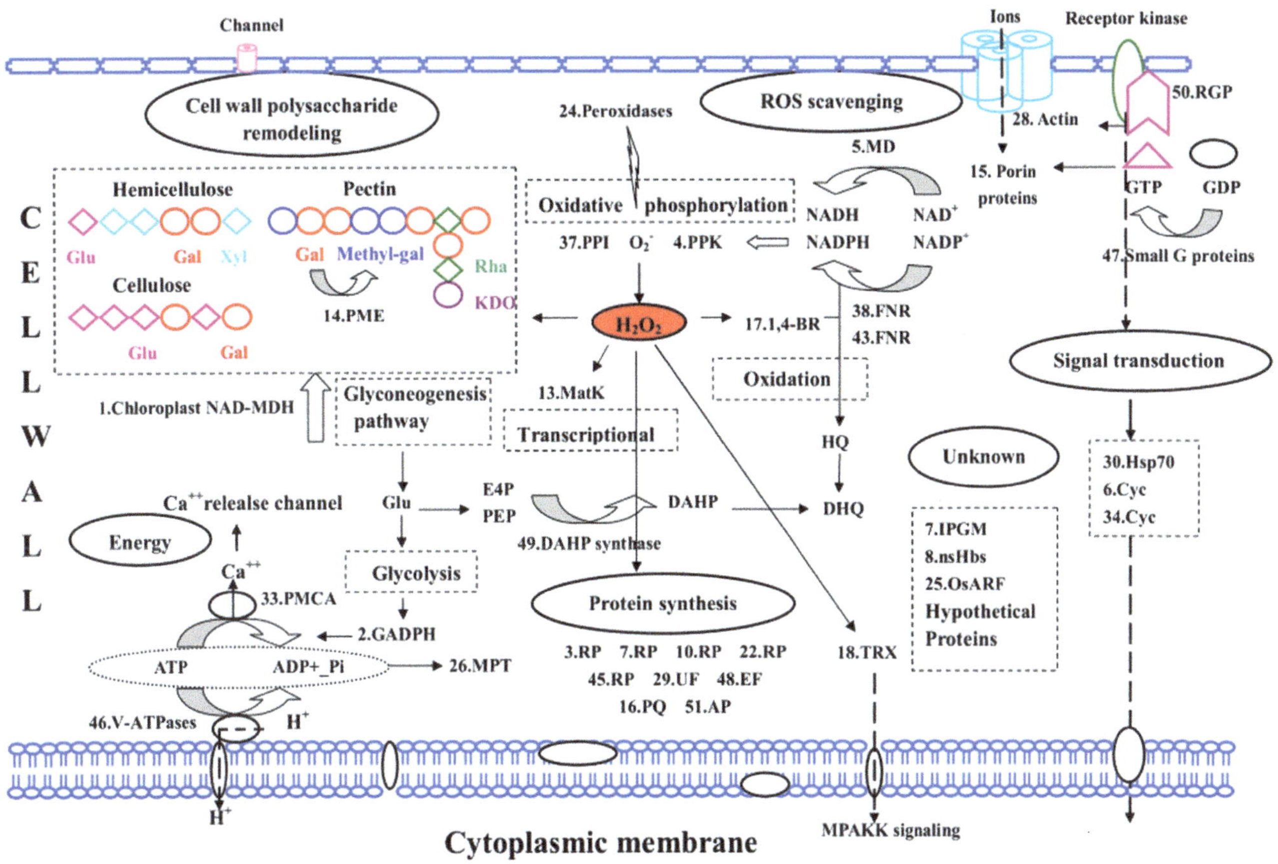

Figure 6. Pathways involved in cell defense, signaling, and cell wall remodeling under copper stress in the root cell wall of *Elsholtzia splendens*. Proteins identified in this study are displayed on the corresponding metabolic pathways and the number indicates the protein identification number. Gal, galacturonic acids; Methyl-gal, methylated galacturonic acids; KDO, 2-keto-3-deoxyoctonic acid; Rha, rhamnose; PME, Pectin methylesterase; Xyl, xylanase; Glu, glucose; MD, malate dehydrogenase; Trx, thioredoxin; MAPKK, mitogen-activated protein kinase kinase; PPI, peptidyl-prolyl cis-trans isomerase; PPK, phosphoglycerate kinase; MatK, maturase K; 1,4-BR, putative 1,4-benzoquinone reductase; oxidation; FNR, ferredoxin-NADP reductase; HQ, hydroquinone; DHQ, Durohydroquinone; DAHP, 3-deoxy-D-arabino-heptulosonic acid-7-phosphate; E4P, perythrose-4-phosphate; PEP, phosphoenolpyruvate; RGP, RAS-related GTP-binding protein; PLP, porin-like proteins; Cyc, cytochrome; PMCA, plasma membrane calcium ATPase; V-ATPases, vacuolar H(+)-ATPases; MPT, mitochondrial phosphate translocator; GADPH glyceraldehyde-3-phosphate dehydrogenase; RP, ribosomal protein; UF, ubiquitin fusion protein; EF, elongation factor 1 subunit alpha; PQ, Pollen allergen Que a 1 isoform; AP, aspartic proteinase nepenthesin-1 precursor; nsHbs. non-symbiotic hemoglobin; OsARF, Os070223100; iPGM, 2,3-bisphosphoglycerate-independent phosphoglycerate mutase.

for 48 h. The solution was changed to one-quarter-strength complete nutrient solution after germination. At the fourth leaf stage, uniform seedlings were transferred into vessels filled with full-strength nutrient solution that contained macronutrients (in mM): 1.0 $Ca(NO_3)_2$, 0.5 $MgSO_4$, and 0.5 K_2HPO_4; and micronutrients (in µmol/L): 27.0 Fe(III)–EDTA, 23.0 H_3BO_3, 0.8 $CuSO_4$, 0.5 Na_2MoO_4, 0.5 $ZnSO_4$, and 4.5 $MnSO_4$. The pH was adjusted to 5.8 with 0.1 M HCl and NaOH. Nutrient solutions were renewed every 3 d and aerated continuously through 0.2 µm filters. Plants were grown in a hydroponic nutrient solution chamber with a 16 h, 25°C day and an 8 h, 20°C night regimen at 60% to 70% relative humidity. Light conditions in the growth chamber were fixed at 5 µmol to 10 µmol photons m^{-2} s^{-1}. Plants were exposed to various concentrations of Cu (25, 50, 75, and 100 µM) after 28 d of growth, which was added as sulfate for 48 h. Each treatment (15 plants) was conducted in triplicate, and the control plants (CK) were grown in pure nutrient solution for comparison. Plant roots were rinsed with distilled water at harvest and then immersed in 5 mmol/L of $Ca(NO_3)_2$ for 20 min to remove the putative adsorbed Cu^{2+} [32]. Roots were separated, pooled, and rinsed with deionized water; plant parts were washed, immediately frozen in liquid nitrogen, and stored at −80°C for analysis [33].

Cell wall preparation

Cell wall purification and CWP extraction were prepared as described by Feiz et al. [34] with slight modifications. All procedures were conducted at 4°C unless mentioned otherwise. Fifteen independent root preparations were pooled to yield one biological replicate. Root tissues (4 g) were homogenized in pre-cold extraction buffer [5 mM acetate, 0.4 M sucrose, 1 mM phenylmethylsulfonyl fluoride (PMSF), pH 4.6, 125 mL] with a chilled mortar and pestle. After adding 0.4 g of polyvinylpoly-pyrrolidone (PVPP), the mixture was incubated at 4°C for 30 min with stirring. Cell walls were separated from soluble cytoplasmic fluid by centrifuging the homogenate at 1,000×*g* for 15 min. The pellet was washed by suspension in 125 mL of 5 mM acetate buffer at pH 4.6 with 0.6 M sucrose. The mixture was centrifuged at 1,000×*g* for 15 min. The pellet was further purified by resuspension in 125 mL of 5 mM acetate buffer at pH 4.6 with

1 M sucrose and centrifuged at 1,000×g for 15 min. The residue was washed thoroughly with 750 mL of 5 mM acetate buffer at pH 4.6 while filtered on a layer of Miracloth (Merck, Darmstadt, Germany). The supernatant was discarded, and the final pellet was freeze-dried overnight. The freeze-dried cell wall materials were then stored at −80°C for further use. Most intracellular proteins were removed from the cell walls given the advantage of sucrose gradients and extensive washing with low ionic strength acidic buffer. PVPP was treated with acid to increase polymerization and to remove metal ions and contaminants.

Cell wall composition extraction and analysis

Isolation of cell wall material. Cell wall materials were extracted according to Zhong and Lauchli (1993) [35]. The pectin fraction was extracted twice with 0.5% ammonium oxalate buffer that contained 0.1% $NaBH_4$ (pH 4) in boiling water bath for 1 h each and pooled the supernatants. Pellets were subsequently subjected to triple extractions with 4% KOH that contained 0.1% $NaBH_4$ at room temperature for 24 h, followed by similar extraction with 24% KOH that contained 0.1% $NaBH_4$. The supernatants from the 4% and 24% KOH extractions were collected and thus yielded the hemicellulose fractions. The remaining pellet from the 24% KOH extraction was then lyophilized, weighed, and considered to be the cellulose fraction [36].

Protein extraction and digestion. A portion (2 g) of the roots was homogenized with 1 mL of PBS (pH 7.6) that contained 65 mM K_2HPO_4, 2.6 mM KH_2PO_4, 400 mM NaCl, and 3 mM NaN_3 in a mortar and pestle. The homogenate was centrifuged twice at 15,000×g for 10 min, and the supernatant was collected as the total soluble protein. CWPs were extracted from the root cell wall fraction in two successive steps. The first step used $CaCl_2$ solution (5 mM sodium acetate buffer, pH 4.6, 0.2 M $CaCl_2$, and 1 mM PMSF; $CaCl_2$ can efficiently extract CWPs that exert weak electrostatic interactions with other cell wall components [37]). The second step involved two extractions with NaCl solution (5 mM sodium acetate buffer, pH 4.6, 1 M NaCl, and 1 mM PMSF; NaCl was also used to extract the strong ionically bound proteins) [34]. For each extraction, the sample was incubated with vortexing at 4°C, and the supernatant was collected after centrifugation at 4,000×g for 15 min. The supernatants from all the extraction steps were pooled and concentrated to 4 mL by centrifugation at 1,500×g at 10°C using a 3 kDa molecular weight cut-off Amicon Spin Tube (Millipore, MA). The protein sample was buffer-exchanged with ultrapure water.

CWP extracts were dissolved in 100 μL of lysis buffer (7 M urea, 2 M thiourea, 5 mM EDTA, 10 mM DTT, and 1 mM PMSF). The supernatant was transferred to a new tube, reduced with 10 mM DTT for 1 h at 56°C, and alkylated with 55 mM iodoacetamide for 45 min at room temperature in darkness. The protein was precipitated with four volumes of pre-chilled acetone for 30 min at −20°C. The pellet was dissolved in 0.5 M TEAB after centrifugation and sonicated for 5 min. The centrifugation step was repeated, and the supernatant was collected. Approximately 100 μg of proteins from each sample was digested with trypsin (Promega) overnight at 37°C in a 1:20 trypsin-to-protein mass ratio.

Analysis of Cu, protein, and cell wall polysaccharide content. Cell wall fraction was dried at 70°C to a constant weight and then digested at 145°C for 24 h in an acid mixture of HNO_3:$HClO_4$ (3:1, v:v). Cu concentrations were measured by Elmer flame atomic absorption spectrometry (AAS-3600).

Protein contents were determined by the Bradford method [38] with bovine serum albumin as the standard. CWPs were dissolved with 400 μL of SDS sample lysis buffer by boiling for 5 min and then loaded onto a 12% acrylamide mini-gel (5 cm to 8 cm) for 1D SDS-PAGE.

The contents of total sugars, galacturonic acids, and KDO were determined colorimetrically by sulfate-phenol [39], hydroxydiphenyl [40], and thiobarbituric acid [41], respectively.

Labeling TMT reagents

Peptide was desalted by Strata XC18 SPE column (Phenomenex) and vacuum-dried after trypsin digestion. Peptide was reconstituted in 0.5 M TEAB and processed according to the protocol for 6-plex TMT reagent kits. In brief, two units of TMT reagent (defined as the amount of reagent required to label 100 μg of protein) was thawed and reconstituted in 41 μL of acetonitrile. Peptide samples from 50 μM Cu were treated for 48 h, and CK samples were labeled with TMT tags 126, 128, and 130 as well as TMT tags 127, 129, and 131. Both samples were incubated at room temperature for 2 h. The peptide mixtures were then pooled, desalted, and dried by vacuum centrifugation.

LC–ESI–MS/MS analysis by Q Exactive

The labeled peptide was resuspended in buffer A (2% ACN, 0.1% FA) and centrifuged at 20,000×g for 2 min. The supernatant was transferred into a sample tube and loaded onto an Acclaim PepMap 100 C18 trap column (75 μm×2 cm; Dionex) by EASY nLC1000 nano UPLC (Thermo). The peptide was then eluted onto an Acclaim PepMap RSLC C18 analytical column (50 μm×15 cm; Dionex). An 85 min gradient program was run at 300 nL/min, which started from 3% to 35% B (80% ACN, 0.1% FA), followed by 5 min linear gradient to 90% B, and maintained at 90% B for 5 min.

The peptides were subjected to NSI source, followed by tandem mass spectrometry (MS/MS) in Q Exactive (Thermo) coupled online to the UPLC. Intact peptides were detected in the Orbitrap at a resolution of 70,000. Peptides were selected for MS/MS using 27% NCE with 12% stepped NCE; ion fragments were detected in the Orbitrap at a resolution of 17,500. A data-dependent procedure that alternated between one MS scan followed by 20 MS/MS scans was applied to the top 20 precursor ions above a threshold ion count of 3E4 in the MS survey scan with 5.0 s dynamic exclusion. The applied electrospray voltage was 1.8 kV. Automatic gain control was used to prevent overfilling of the ion trap; 1E5 ions were accumulated for generation of MS/MS spectra. The m/z scan range was 350 Da to 2000 Da for MS scans.

Proteomic database search

The instrument data file (.raw) was merged and transformed to an.mgf file by Proteome Discoverer (ver. 1.3.0.339; Thermo). Peptide and protein identifications were performed using the Mascot search engine (ver. 2.3.02; Matrix Science). Derived protein sequences from plants in the NCBI were collected, and a database containing 1,596,443 sequences was created. In the Mascot search engine version 2.3.02 software, all parameters were set as follows. Database searching was restricted to tryptic peptides. Carbamidomethyl (C), TMT 6-plex (N-term), and TMT 6-plex (K) were selected as fixed, and deamidated (NQ), Gln→pyro-Glu (N-term Q), and oxidation (M) were selected as variable modifications, where two missed cleavages were allowed with precursor error tolerance at 10 ppm and fragment deviation at 0.02 Da. The Mascot search results were quantified using Mascot 2.3.02 with the following criteria: protein ratio type = median, minimum unique peptides = 1, peptide threshold type = at least homolog. Peptides were not quantified if peptide score was too low or the deviation was too large. The final ratios of protein were then normalized by taking the median of all the proteins quantified. The complete list of identified peptides was encoded in an Excel (Microsoft) database for grouping of results into proteins and calculation of ratios and coefficient of variation.

To obtain the differentially expressed proteins, we selected the proteins according to the following principles. (a) The expression change of protein folds is more than 1.5. (b) The protein quantitative T-test p value is less than 0.05. A total of 55 differentially expressed proteins were obtained in our bioinformatics analysis data.

Bioinformatics studies and statistical analysis

Data were the average of at least three independent experimental replicates. One-way ANOVA (Duncan's test) and LSD test with $p<0.05$ as the significance level were performed to analyze the differences between the control and treatments. All data were presented as the mean value ± standard error and were analyzed using SPSS statistical software package (version 16.0).

The clustering of content abundance profiles was performed using Cluster 3.0. Euclidean metric was used for computing the distance between points and cluster centroids, which is a typical choice for K-means clustering analysis. The output was visualized using TreeView after hierarchical clustering.

GO information was used to categorize the biological processes of identified proteins from TMT data sets. GO annotations were extracted from the Uni-Prot database and were matched with corresponding gene locus identifiers embedded in the NCBI RefSeq database. Proteins were then classified based on their biological process using the Web Gene Ontology Annotation Plot [42] (http://wego.genomics.org.cn/cgi-bin/wego/index.pl). Functionally analyzed Cu stress-induced CWPs in biosynthesis pathways were identified through the Kyoto Encyclopedia of Genes and Genomes (KEGG) [43]. These pathways were classified into hierarchical categories according to the KEGG website (http://www.genome.jp/kegg/). Subcellular locations were predicted using TargetP (http://www.cbs.dtu.dk/services/TargetP/), Predotar (http://urgi.versailles.inra.fr/predotar/predotar.html), and WoLF PSORT (http://wolfpsort.org/) (http://www.genscript.com/psort/wolf_psort.html)software applications as previously described [44].

Supporting Information

Figure S1 SDS–PAGE of root cell wall proteins under different copper stress.

Table S1 Protein subcellular localization and go annotation of 55 significantly differentially expressed proteins.

Table S2 Details of 55 significantly differentially expressed proteins identified by TMT using ANOVA including their p –value.

Author Contributions

Conceived and designed the experiments: TTL JYS. Performed the experiments: TTL CFS. Analyzed the data: TTL CKH JYS. Contributed reagents/materials/analysis tools: YW. Wrote the paper: TTL.

References

1. Maksymiec W (1998) Effect of copper on cellular processes in higher plants. Photosynthetica 34: 321-342.
2. Teisseire H, Guy V (2000) Copper-induced changes in antioxidant enzymes activities in fronds of duckweed (*Lemna minor*)Plant science 153: 65-72.
3. Vinit-Dunand F, Epron D, Alaoui-Sossé B, Badot PM (2002) Effects of copper on growth and on photosynthesis of mature and expanding leaves in cucumber plants. Plant science 163: 53-58.
4. Bona E, Marsano F, Cavaletto M, Berta G (2007) Proteomic characterization of copper stress response in *Cannabis sativa* roots. Proteomics 7: 1121-1130.
5. Hall J (2002) Cellular mechanisms for heavy metal detoxification and tolerance. Journal of Experimental Botany 53: 1-11.
6. Cobbett C, Goldsbrough P (2002) Phytochelatins and metallothioneins: roles in heavy metal detoxification and homeostasis. Annual review of plant biology 53: 159-182.
7. Krzesłowska M (2011) The cell wall in plant cell response to trace metals: polysaccharide remodeling and its role in defense strategy. Acta Physiologiae Plantarum 33: 35-51.
8. Jarvis MC (2009) Plant cell walls: supramolecular assembly, signalling and stress. Structural Chemistry 20: 245-253.
9. Zhu J, Chen S, Alvarez S, Asirvatham VS, Schachtman DP, et al. (2006) Cell wall proteome in the maize primary root elongation zone. I. Extraction and identification of water-soluble and lightly ionically bound proteins. Plant physiology 140: 311-325.
10. Vorwerk S, Somerville S, Somerville C (2004) The role of plant cell wall polysaccharide composition in disease resistance. Trends in plant science 9: 203-209.
11. Dani V, Simon WJ, Duranti M, Croy RR (2005) Changes in the tobacco leaf apoplast proteome in response to salt stress. Proteomics 5: 737-745.
12. Soares NC, Francisco R, Vielba JM, Ricardo CP, Jackson PA (2009) Associating wound-related changes in the apoplast proteome of Medicago with early steps in the ROS signal-transduction pathway. Journal of Proteome Research 8: 2298-2309.
13. Jamet E, Canut H, Boudart G, Pont-Lezica RF (2006) Cell wall proteins: a new insight through proteomics. Trends in plant science 11: 33-39.
14. Bhushan D, Pandey A, Choudhary MK, Datta A, Chakraborty S, et al. (2007) Comparative proteomics analysis of differentially expressed proteins in chickpea extracellular matrix during dehydration stress. Molecular & Cellular Proteomics 6: 1868-1884.
15. Pandey A, Rajamani U, Verma J, Subba P, Chakraborty N, et al. (2010) Identification of extracellular matrix proteins of rice (*Oryza sativa* L.) involved in dehydration-responsive network: a proteomic approach. Journal of Proteome Research 9: 3443-3464.
16. Floerl S, Druebert C, Majcherczyk A, Karlovsky P, Kües U, et al. (2008) Defence reactions in the apoplastic proteome of oilseed rape (*Brassica napus var. napus*) attenuate Verticillium longisporum growth but not disease symptoms. Bmc Plant Biology 8: 129.
17. Goulet C, Goulet C, Goulet MC, Michaud D (2010) 2-DE proteome maps for the leaf apoplast of Nicotiana benthamiana. Proteomics 10: 2536-2544.
18. Brune A, Urbach W, Dietz K-J (1994) Zinc stress induces changes in apoplasmic protein content and polypeptide composition of barley primary leaves. Journal of Experimental Botany 45: 1189-1196.
19. Blinda A, Koch B, Ramanjulu S, Dietz K (1997) De novo synthesis and accumulation of apoplastic proteins in leaves of heavy metal-exposed barley seedlings. Plant, Cell & Environment 20: 969-981.
20. Fecht-Christoffers MM, Braun H-P, Lemaitre-Guillier C, VanDorsselaer A, Horst WJ (2003) Effect of manganese toxicity on the proteome of the leaf apoplast in cowpea. Plant physiology 133: 1935-1946.
21. Fecht-Christoffers MM, Führs H, Braun H-P, Horst WJ (2006) The role of hydrogen peroxide-producing and hydrogen peroxide-consuming peroxidases in the leaf apoplast of cowpea in manganese tolerance. Plant physiology 140: 1451-1463.
22. Alves M, Francisco R, Martins I, Ricardo C (2006) Analysis of *Lupinus albus* leaf apoplastic proteins in response to boron deficiency. Plant and Soil 279: 1-11.
23. Komatsu S, Kobayashi Y, Nishizawa K, Nanjo Y, Furukawa K (2010) Comparative proteomics analysis of differentially expressed proteins in soybean cell wall during flooding stress. Amino Acids 39: 1435-1449.
24. Kataoka T, Furukawa J, Nakanishi T (2003) The decrease of extracted apoplast protein in soybean root tip by aluminium treatment. Biologia plantarum 46: 445-449.
25. Kong F-J, Oyanagi A, Komatsu S (2010) Cell wall proteome of wheat roots under flooding stress using gel-based and LC MS/MS-based proteomics approaches. Biochimica et Biophysica Acta (BBA)-Proteins and Proteomics 1804: 124-136.
26. Zhou L, Bokhari SA, Dong C-J, Liu J-Y (2011) Comparative proteomics analysis of the root apoplasts of rice seedlings in response to hydrogen peroxide. Plos One 6: e16723.
27. Hossain Z, Nouri M-Z, Komatsu S (2011) Plant cell organelle proteomics in response to abiotic stress. Journal of Proteome Research 11: 37-48.

28. Yang MJ, Yang XE, Römheld V (2002) Growth and nutrient composition of *Elsholtzia splendens Nakai* under copper toxicity. Journal of plant nutrition 25: 1359-1375.
29. Lou L, Shen Z, Li X (2004) The copper tolerance mechanisms of *Elsholtzia haichowensis*, a plant from copper-enriched soils. Environmental and Experimental Botany 51: 111-120.
30. Feiz L, Irshad M, Pont-Lezica RF, Canut H, Jamet E (2006) Evaluation of cell wall preparations for proteomics: a new procedure for purifying cell walls from Arabidopsis hypocotyls. Plant Methods 2: 10.
31. Kwon H-K, Yokoyama R, Nishitani K (2005) A proteomic approach to apoplastic proteins involved in cell wall regeneration in protoplasts of Arabidopsis suspension-cultured cells. Plant and Cell Physiology 46: 843-857.
32. Harrison S, Lepp N, Phipps D (1979) Uptake of Copper by excised roots II. Copper desorption from the free space. Zeitschrift für Pflanzenphysiologie 94: 27-34.
33. Li F, Shi J, Shen C, Chen G, Hu S, et al. (2009) Proteomic characterization of copper stress response in *Elsholtzia splendens* roots and leaves. Plant Molecular Biology 71: 251-263.
34. Feiz L, Irshad M, Pont-Lezica RF, Canut H, Jamet E (2006) Evaluation of cell wall preparations for proteomics: a new procedure for purifying cell walls from Arabidopsis hypocotyls. Plant Methods 2.
35. Zhong H, Lauchli A (1993) Changes of cell wall composition and polymer size in primary roots of cotton seedlings under high salinity. Journal of Experimental Botany 44: 773-778.
36. Yu M, Shen R, Liu J, Chen R, Xu M, et al. (2009) The role of root border cells in aluminum resistance of pea (*Pisum sativum*) grown in mist culture. Journal of Plant Nutrition and Soil Science 172: 528-534.
37. Jamet E, Boudart G, Borderies G, Charmont S, Lafitte C, et al. (2008) Isolation of plant cell wall proteins. 2D PAGE: Sample Preparation and Fractionation: Springer. pp. 187-201.
38. Bradford MM (1976) A rapid and sensitive method for the quantitation of microgram quantities of protein utilizing the principle of protein-dye binding. Analytical biochemistry 72: 248-254.
39. Dubois M, Gilles KA, Hamilton JK, Rebers P, Smith F (1956) Colorimetric method for determination of sugars and related substances. Analytical chemistry 28: 350-356.
40. Dronnet V, Renard C, Axelos M, Thibault J-F (1996) Heavy metals binding by pectins: selectivity, quantification and characterisation. Progress in Biotechnology 14: 535-540.
41. Mohnen D (2008) Pectin structure and biosynthesis. Current opinion in plant biology 11: 266-277.
42. Ye J, Fang L, Zheng H, Zhang Y, Chen J, et al. (2006) WEGO: a web tool for plotting GO annotations. Nucleic acids research 34: W293-W297.
43. Kanehisa M, Goto S (2000) KEGG: kyoto encyclopedia of genes and genomes. Nucleic acids research 28: 27-30.
44. Day A, Fénart S, Neutelings G, Hawkins S, Rolando C, et al. (2013) Identification of cell wall proteins in the flax (*Linum usitatissimum*) stem. Proteomics.
45. Xiong J, An L, Lu H, Zhu C (2009) Exogenous nitric oxide enhances cadmium tolerance of rice by increasing pectin and hemicellulose contents in root cell wall. Planta 230: 755-765.
46. Yang JL, Zhu XF, Peng YX, Zheng C, Li GX, et al. (2011) Cell wall hemicellulose contributes significantly to aluminum adsorption and root growth in Arabidopsis. Plant physiology 155: 1885-1892.
47. Pelloux J, Rusterucci C, Mellerowicz EJ (2007) New insights into pectin methylesterase structure and function. Trends in plant science 12: 267-277.
48. Paynel F, Schaumann A, Arkoun M, Douchiche O, Morvan C (2009) Temporal regulation of cell-wall pectin methylesterase and peroxidase isoforms in cadmium-treated flax hypocotyl. Annals of Botany 104: 1363-1372.
49. Liszkay A, Kenk B, Schopfer P (2003) Evidence for the involvement of cell wall peroxidase in the generation of hydroxyl radicals mediating extension growth. Planta 217: 658-667.
50. Fry S, Dumville J, Miller J (2001) Fingerprinting of polysaccharides attacked by hydroxyl radicals in vitro and in the cell walls of ripening pear fruit. Biochem J 357: 729-737.
51. Thounaojam TC, Panda P, Mazumdar P, Kumar D, Sharma G, et al. (2012) Excess copper induced oxidative stress and response of antioxidants in rice. Plant Physiology and Biochemistry 53: 33-39.
52. Ahsan N, Lee D-G, Lee S-H, Kang KY, Lee JJ, et al. (2007) Excess copper induced physiological and proteomic changes in germinating rice seeds. Chemosphere 67: 1182-1193.
53. Spain JC, Wyss O, Gibson DT (1979) Enzymatic oxidation of p-nitrophenol. Biochemical and Biophysical Research Communications 88: 634-641.
54. Gross GG (1977) Cell wall-bound malate dehydrogenase from horseradish. Phytochemistry 16: 319-321.
55. Li Z-C, McClure JW, Hagerman AE (1989) Soluble and bound apoplastic activity for peroxidase, β-D-glucosidase, malate dehydrogenase, and nonspecific arylesterase, in barley (*Hordeum vulgare* L.) and oat (*Avena sativa* L.) primary leaves. Plant physiology 90: 185-190.
56. Apel K, Hirt H (2004) Reactive oxygen species: metabolism, oxidative stress, and signal transduction. Annu Rev Plant Biol 55: 373-399.
57. Jensen KA, Houtman CJ, Ryan ZC, Hammel KE (2001) Pathways for extracellular Fenton chemistry in the brown rot basidiomycete Gloeophyllum trabeum. Applied and Environmental Microbiology 67: 2705-2711.
58. Howard L, Talcott S, Brenes C, Villalon B (2000) Changes in phytochemical and antioxidant activity of selected pepper cultivars (Capsicum species) as influenced by maturity. Journal of Agricultural and Food Chemistry 48: 1713-1720.
59. Keith B, Dong X, Ausubel FM, Fink GR (1991) Differential induction of 3-deoxy-D-arabino-heptulosonate 7-phosphate synthase genes in Arabidopsis thaliana by wounding and pathogenic attack. Proceedings of the National Academy of Sciences 88: 8821-8825.
60. Timperio AM, Egidi MG, Zolla L (2008) Proteomics applied on plant abiotic stresses: role of heat shock proteins (HSP). Journal of proteomics 71: 391-411.
61. Wang SB, Hu Q, Sommerfeld M, Chen F (2004) Cell wall proteomics of the green alga *Haematococcus pluvialis* (Chlorophyceae). Proteomics 4: 692-708.
62. Béguin P, Nagashima K, Gonoi T, Shibasaki T, Takahashi K, et al. (2001) Regulation of Ca2+ channel expression at the cell surface by the small G-protein kir/Gem. Nature 411: 701-706.
63. Rittinger K, Walker PA, Eccleston JF, Nurmahomed K, Owen D, et al. (1997) Crystal structure of a small G protein in complex with the GTPase-activating protein rhoGAP. Nature 388: 693-697.
64. Hall A (1994) Small GTP-binding proteins and the regulation of the actin cytoskeleton. Annual review of cell biology 10: 31-54.
65. Watson BS, Lei Z, Dixon RA, Sumner LW (2004) Proteomics of *Medicago sativa* cell walls. Phytochemistry 65: 1709-1720.
66. Zhang H, Lian C, Shen Z (2009) Proteomic identification of small, copper-responsive proteins in germinating embryos of *Oryza sativa*. Annals of Botany 103: 923-930.
67. Odou M, Singer E, Romond M, Dubreuil L (1998) Isolation and characterization of a porin-like protein of 45 kilodaltons from Bacteroides fragilis. Fems Microbiology Letters 166: 347-354.
68. Massari P, Ho Y, Wetzler LM (2000) Neisseria meningitidis porin PorB interacts with mitochondria and protects cells from apoptosis. Proceedings of the National Academy of Sciences 97: 9070-9075.
69. Chivasa S, Ndimba BK, Simon WJ, Lindsey K, Slabas AR (2005) Extracellular ATP functions as an endogenous external metabolite regulating plant cell viability. The Plant Cell Online 17: 3019-3034.
70. Carafoli E, Lim D (2009) Plasma Membrane Calcium ATPase. Handbook of Neurochemistry and Molecular Neurobiology: Springer. pp. 581-596.
71. Nelson H, Nelson N (1990) Disruption of genes encoding subunits of yeast vacuolar H (+)-ATPase causes conditional lethality. Proceedings of the National Academy of Sciences 87: 3503-3507.
72. Sze H, Ward JM, Lai S (1992) Vacuolar H+-translocating ATPases from plants: structure, function, and isoforms. Journal of bioenergetics and biomembranes 24: 371-381.
73. Petersen J, Brinkmann H, Cerff R (2003) Origin, evolution, and metabolic role of a novel glycolytic GAPDH enzyme recruited by land plant plastids. Journal of molecular evolution 57: 16-26.
74. Sharma SS, Dietz K-J (2006) The significance of amino acids and amino acid-derived molecules in plant responses and adaptation to heavy metal stress. Journal of Experimental Botany 57: 711-726.
75. Imin N, Kerim T, Rolfe BG, Weinman JJ (2004) Effect of early cold stress on the maturation of rice anthers. Proteomics 4: 1873-1882.
76. Qu Z-L, Zhong N-Q, Wang H-Y, Chen A-P, Jian G-L, et al. (2006) Ectopic expression of the cotton non-symbiotic hemoglobin gene GhHbd1 triggers defense responses and increases disease tolerance in Arabidopsis. Plant and Cell Physiology 47: 1058-1068.
77. Zhou X, Li J, Cheng W, Liu H, Li M, et al. (2010) Gene structure analysis of rice ADP-ribosylation factors (OsARFs) and their mRNA expression in developing rice plants. Plant Molecular Biology Reporter 28: 692-703.

5

Cadmium-Induced Hydrogen Sulfide Synthesis Is Involved in Cadmium Tolerance in *Medicago sativa* by Reestablishment of Reduced (Homo)glutathione and Reactive Oxygen Species Homeostases

Weiti Cui[1], Huiping Chen[2], Kaikai Zhu[1], Qijiang Jin[1], Yanjie Xie[1], Jin Cui[1], Yan Xia[1], Jing Zhang[1], Wenbiao Shen[1]*

1 College of Life Sciences, Laboratory Center of Life Sciences, Nanjing Agricultural University, Jiangsu Province, Nanjing, China, **2** Key Laboratory of Protection and Development Utilization of Tropical Crop Germplasm Resources, Hainan University, Haikou, China

Abstract

Until now, physiological mechanisms and downstream targets responsible for the cadmium (Cd) tolerance mediated by endogenous hydrogen sulfide (H_2S) have been elusive. To address this gap, a combination of pharmacological, histochemical, biochemical and molecular approaches was applied. The perturbation of reduced (homo)glutathione homeostasis and increased H_2S production as well as the activation of two H_2S-synthetic enzymes activities, including L-cysteine desulfhydrase (LCD) and D-cysteine desulfhydrase (DCD), in alfalfa seedling roots were early responses to the exposure of Cd. The application of H_2S donor sodium hydrosulfide (NaHS), not only mimicked intracellular H_2S production triggered by Cd, but also alleviated Cd toxicity in a H_2S-dependent fashion. By contrast, the inhibition of H_2S production caused by the application of its synthetic inhibitor blocked NaHS-induced Cd tolerance, and destroyed reduced (homo)glutathione and reactive oxygen species (ROS) homeostases. Above mentioned inhibitory responses were further rescued by exogenously applied glutathione (GSH). Meanwhile, NaHS responses were sensitive to a (homo)glutathione synthetic inhibitor, but reversed by the cotreatment with GSH. The possible involvement of cyclic AMP (cAMP) signaling in NaHS responses was also suggested. In summary, LCD/DCD-mediated H_2S might be an important signaling molecule in the enhancement of Cd toxicity in alfalfa seedlings mainly by governing reduced (homo)glutathione and ROS homeostases.

Editor: Ji-Hong Liu, Key Laboratory of Horticultural Plant Biology (MOE), China

Funding: This research was supported by the Fundamental Research Funds for the Central Universities (KYZ201316), the National Natural Science Foundation of China (grants no. 30971711, J1210056, J1310015), and the Priority Academic Program Development of Jiangsu Higher Education Institutions. The funders had no role in study design, data collection and analysis, decision to publish, or preparation of the manuscript.

Competing Interests: The authors have declared that no competing interests exist.

* Email: wbshenh@njau.edu.cn

Introduction

Cadmium (Cd) contamination is a non-reversible accumulation process, with the estimated half-life and high plant-soil mobility, thus resulting in a serious threat to human health through food chains. Normally, Cd exposure leads to the inhibition of plant growth, decrease of crop yield, and even plant cell death [1,2]. Indirectly stimulated generation of reactive oxygen species (ROS) that modify the antioxidant defence and bring out oxidative stress is ascribed to one of the Cd toxicities in plants, and therefore lipid peroxidation is considered as a hallmark of Cd exposure [3].

In plants, there are a lot of antioxidant defence mechanisms, which could keep the normally formed ROS at a low level and prevent them from exceeding toxic thresholds [3,4]. The glutathione (GSH) and ascorbate were subsequently recognized as the heart of the redox hub [5]. In plants, GSH is synthesized by two ATP-dependent steps: γ-glutamylcysteine (γ-EC) is synthesized from L-glutamate and L-cysteine by γ-glutamyl cysteine synthetase (γ-ECS, also called as γ-GCS); and the second step, glycine is conjunct to γ-EC by glutathione synthetase (GS) [6,7]. In soybean and alfalfa plants, GSH homolog homoglutathione (hGSH) synthesized by homoglutathione synthetase (hGS) from β-alanine and γ-EC, is more abundant than GSH [8]. The rate of glutathione reductase (GR) reaction was the same with either oxidized glutathione (GSSG) or oxidized homoglutathione (hGSSGh) as the substrate [7]. Upon Cd exposure, it was confirmed that the rapid accumulation of peroxides and depletion of GSH and hGSH causes redox imbalance in *Medicago sativa* [9]. Subsequent experiments with comparing ten pea genotypes showing that, activities of ascorbate peroxidase (APX) decreased, but concentrations of GSH increased in the less Cd-sensitive genotypes [10].

Another sulphur-containing compound, hydrogen sulfide (H_2S), previously known as a toxic gas, has been progressively recognized as a gaseous signaling molecule with multiple functions in animals [11,12]. For example, H_2S has been revealed as a cytoprotectant and a regulator in various biological processes, such as oxidative stress suppression, smooth muscle relaxation, proliferation inhibition and apoptosis triggering [13–16]. Meanwhile, although previous reports observed that many plants can emit H_2S [17–19], there have been few studies on the physiological role of H_2S in *planta* during the last century.

In mammals, the majority of endogenous H_2S was produced by two enzymes, cystathionine β-synthase (CBS, EC 4.2.1.22) and cystathionine γ-lyase (CSE, EC 4.4.1.1), from L-cysteine [20]. Cysteine-degrading enzymes such as cysteine desulfhydrases are hypothesized to be involved in H_2S release in plants [21]. Previously, two specific desulfhydrases, L-cysteine desulfhydrase (LCD, EC 4.4.1.1; also called L-CDes or L-DES) and D-cysteine desulfhydrase (DCD, EC 4.4.1.15; also called D-CDes or D-DES), have been isolated and partially analyzed from *Arabidopsis thaliana* [22–24]. The LCD, which is considered as the most important enzyme with H_2S production in plants, shares a 100% sequence homolog with CSE in mammals [25]. By using sodium hydrosulfide (NaHS) as a H_2S donor, ample evidence further suggested that H_2S can protect plants against various stress-induced damage, such as salinity stress [26], drought [27–29], heavy metal exposure [30,31], and heat shock [32]. Additionally, H_2S can act as an inducer in several developmental processes, including adventitious root formation [33] and flower senescence [34]. However, exogenously applied H_2S donor without checking the kinetics of H_2S synthesis including corresponding metabolic enzyme activities or transcripts, may not fully replicate the function of endogenous H_2S in plants.

Cyclic AMP (adenosine 3′, 5′-cyclic monophosphate, cAMP) is a well-known second messenger playing important roles in many physiological processes. The cAMP is synthesized by adenylyl cyclase and broken down by cNMP phosphodiesterase. Dedioxyadenosine (DDA) and 1,3-diazinane-2,4,5,6-tetrone (alloxan) are well characterized as the inhibitors of adenylyl cyclase. Likewise, cNMP phosphodiesterase is sensitive to the inhibitor 1-methyl-3-(2-methylpropyl)-7*H*-purine-2,6-dione (IBMX) [35,36]. In animals, there is ample evidences to show H_2S-activited cAMP level or H_2S-regulated cAMP homeostasis [37,38]. It was found that H_2S acted via cAMP-mediated PI3K/Akt/p70S6K signal pathways to inhibit hippocampal neuronal apoptosis and protect neurons from OGD/R-induced injury [39]. However, the functions of cAMP signaling in H_2S-alleviated Cd stress in plants are still poorly understood.

Thus, the aim of this study was to investigate the signaling role of endogenous H_2S in the tolerance of *Medicago sativa* seedlings to Cd stress. For this purpose, we preliminarily investigated the synthesis of endogenous H_2S under Cd stress, which has not been fully performed. Furthermore, the effects of H_2S on GSH and hGSH metabolism, as well as ROS homeostasis were checked. Our results further indicated that Cd stress triggered endogenous H_2S production catalyzed by LCD/DCD pathways, and the elevated H_2S acts as a signal improving the homeostasis of GSH pool and keeping ROS under control, both of which finally contributed to Cd tolerance. Finally, the possible involvement of cAMP signaling in NaHS responses was also suggested.

Materials and Methods

Plant material, growth condition

Commercially available alfalfa (*Medicago sativa* L. Victoria) seeds were surface-sterilized with 5% NaClO for 10 min, and rinsed extensively in distilled water before being germinated for 1 d at 25°C in the darkness. Uniform seedlings were then selected and transferred to the plastic chambers and cultured with nutrient medium (quarter-strength Hoagland's solution) in the illuminating incubator (14 h light with a light intensity of 200 $\mu mol \cdot m^{-2} \cdot s^{-1}$, 25±1°C, and 10 h dark, 23±1°C). Five-day-old seedlings were then incubated in quarter-strength Hoagland's solution with or without varying concentrations of NaHS (Sigma-Aldrich; St Louis, MO, USA) or the other indicated chemicals (2 mM DL-propargylglycine (PAG), 1 mM GSH, 1 mM L-buthionine-sulfoximine (BSO), 50 μM 8-Br-cAMP (8Br), 200 μM alloxan (All), 1 mM DDA, and 500 μM IBMX) alone, or the combination of treatments for 6 h followed by the indicated time points of incubation in 200 μM $CdCl_2$. Seedlings without chemicals were used as the control (Con). The pH for both nutrient medium and treatment solutions was adjusted to 6.0.

After various treatments, above-ground parts and root tissues of seedlings were sampled immediately or flash-frozen in liquid nitrogen, and stored at −80°C for further analysis. Among these, above-ground parts and root tissues of 240 seedlings were respectively used for the determination of Cd contents. Seedling root tissues were also used for fresh weight determination (10 seedlings), thiobarbituric acid reactive substances (TBARS) content determination (120 seedlings), and other indicated tests (30 seedlings).

Determination of H_2S content, LCD and DCD activity

Hydrogen sulfide content was determined according to the method previously reported [19,34]. 100 mg of alfalfa seedling roots from 30 seedlings were ground under liquid nitrogen and extracted by 1 ml phosphate buffered saline (50 mM, pH 6.8) containing 0.1 M EDTA and 0.2 M ascorbic acid. After centrifugation at 13000 g for 15 min at 4°C, 400 μl of the supernatant was injected to 200 μl 1% zinc acetate and 200 μl 1 N HCl. After 30 min reaction, 100 μl 5 mM dimethyl-*p*-phenylenediamine dissolved in 7 mM HCl was added to the trap followed by the injection of 100 μl 50 mM ferric ammonium sulfate in 200 mM HCl. After 15 min incubation at room temperature, the amount of H_2S was determined at 667 nm. Solutions with different concentrations of Na_2S were used in a calibration curve.

100 mg of alfalfa seedling roots from 30 seedlings were used for activity determination. The activities of LCD and DCD were determined as described by the methods previously reported [23,40]. L-cysteine desulfhydrase (LCD) activity was measured by the release of H_2S from L-cysteine in the presence of dithiothreitol (DTT). The formation of methylene blue was determined at 670 nm. To removal of the background, content of H_2S in the extracted protein solution was measured by same way with 50% trichloroacetic acid (TCA) instead of L-cysteine. The final LCD activity was calculated from the difference between the measured LCD activity and the background. D-cysteine desulfhydrase (DCD) activity was measured by the same method with following modifications: D-cysteine instead of L-cysteine, the pH of Tris-HCl was 8.0 rather than 9.0. Solutions with different concentrations of Na_2S were prepared, treated in the same way as the assay samples and were used for the quantification of enzymatically formed H_2S.

Determination of thiobarbituric acid reactive substances (TBARS), (h)GSH and (h)GSSG(h) contents

Lipid peroxidation was estimated by measuring the amount of TBARS as previously described [41]. About 400 mg of root tissues from 120 seedlings was ground in 0.25% 2-thiobarbituric acid (TBA) in 10% TCA using a mortar and pestle. After heating at 95°C for 30 min, the mixture was quickly cooled in an ice bath and centrifuged at 10,000×*g* for 10 min. The absorbance of the supernatant was read at 532 nm and corrected for unspecific turbidity by subtracting the absorbance at 600 nm. The concentration of lipid peroxides together with oxidatively modified proteins of plants were thus quantified in terms of TBARS amount using an extinction coefficient of 155 mM^{-1} cm^{-1} and expressed as nmol g^{-1} fresh weight (FW).

(h)GSH (GSH + hGSH) and (h)GSSG(h) (GSSG + hGSSGh) contents were measured by the 5,5′dithio-bis-(2-nitrobenzoic acid) (DTNB)-glutathione reductase (GR) recycling assay [41,42]. Frozen root tissues from 30 seedlings were homogenized in cold 5% 5-sulfosalicylic acid. The homogenate was centrifuged at 12,000×*g* for 20 min at 4°C and the supernatant was collected. Total glutathione ((h)GSH plus (h)GSSG(h)) was determined in the homogenates spectrophotometrically at 412 nm, using GR, DTNB, and NADPH. (h)GSSG(h) contents were determined by the same method in the presence of 2-vinylpyridine and (h)GSH contents were calculated from the difference between total glutathione and (h)GSSG(h).

Thiol analysis by reversed-phase HPLC

Low-molecular-weight thiols and their corresponding disulfides contents in root tissues from 30 seedlings were measured according to the methods previously reported [43–45], through derivatization with monobromobimane (mBBr) after reduction with DTT with or without previously blocked with *N*-ethylmaleimide (NEM), and separation by reversed-phase HPLC (Agilent Technologies, 1200 series Quaternary, Foster city, USA).

Histochemical analyses

Histochemical detection of lipid peroxidation and loss of plasma membrane integrity was performed with Schiff's reagent and with Evans blue described by previous reports [41,45].

Real-time quantitative RT-PCR analysis

Total RNA from root tissues of 30 seedlings was extracted using Trizol reagent (Invitrogen) according to the manufacturer's instructions. DNA-free total RNA (2 μg) from different treatments was used for first-strand cDNA synthesis in a 20-μL reaction volume containing 2.5 units of avian myeloblastosis virus reverse transcriptase XL (TaKaRa) and oligo dT primer.

Real-time quantitative RT-PCR reactions were performed with Mastercycler realplex^2 real-time PCR system (Eppendorf, Hamburg, Germany) using the SYBR *Premix Ex Taq* (TaKaRa) according to the user manual. The cDNA was amplified using primers (Table S1). The expression levels of the genes are presented as values relative to the corresponding control samples under the indicated conditions, with normalization of data to the geometic average of two internal control genes *MSC27* and *Actin2* [46].

Visualization of endogenous ROS by LSCM

Endogenous ROS was imaged using the fluorescent probe H_2DCFDA, and then scanned described by [45,47].

Statistical analysis

Values are means ± SD of three different experiments with three replicated measurements. Differences among treatments were analysed by one-way ANOVA, taking $P<0.05$ as significant according to Duncan's multiple range test.

Results

(h)GSH depletion and increased endogenous H_2S synthesis triggered by Cd stress

Considering alfalfa plants contain a thiol tripeptide homolog, hGSH, instead of or in addition to GSH [8,9], we detected the concentrations of GSH and hGSH. As shown in Table 1, the content of hGSH in alfalfa seedling roots under the control conditions, was about 8-fold higher than that of GSH. Similarly, hGSSGh is the main component of (h)GSSG(h) (total of hGSSGh and GSSG), because the GSSG content was almost negligible.

To further elucidate the correlation among GSH pool, H_2S and Cd tolerance, the time course of (homo)glutathione ((h)GSH; total of hGSH and GSH, and (h)GSSG(h)) contents, and H_2S synthesis were investigated in alfalfa seedling roots upon Cd stress. As expected, a decrease of (h)GSH content (especially hGSH) and an increase of (h)GSSH(h) (especially hGSSGh) level were progressively triggered by Cd stress within 12 h, thus leading to a decreased (h)GSH/(h)GSSH(h) ratio (12 h; Figure 1A-C), an important parameter for the intracellular redox status in *planta* upon Cd stress [3,45]. The ratio of hGSH/hGSSGh exhibited the similar tendency (Table 1). These results were consistent with the observed Cd toxicity, confirmed by the histochemical staining detecting the aggravated loss of plasma membrane integrity and lipid peroxidation with Evans blue and Schiff's reagent, increased TBARS content and growth stunt of seedling roots (Figure S1).

Because H_2S synthesis could be induced by oxidative stress and depletion of GSH both in animals and plants [48–50], we simultaneously investigated the production of H_2S in seedling roots after the exposure to Cd. Similar to the recent report [51], the production of H_2S was continuously increased after the exposure to Cd alone for 12 h (Figure 1D). The changes in activities of two H_2S synthetic enzymes LCD and DCD displayed similar tendencies (Figure 1E and F). Apparently, the reduced (homo)glutathione depletion and increased endogenous H_2S synthesis preceded Cd toxicity in alfalfa seedlings.

NaHS not only mimics intracellular H_2S content, but also alleviates Cd toxicity

Previous results revealed that the exogenously applied NaHS, a H_2S donor, alleviates Cd toxicity in bermudagrass seedlings [51]. Therefore, a preliminary work was carried out to compare the oxidative damage and growth performance of alfalfa seedlings upon Cd exposure with or without the indicated concentrations of NaHS pretreatment. Firstly, the results of histochemical staining and TBARS contents revealed that NaHS at 100 (in particular) and 500 μM was able to significantly decreased Cd-induced lipid peroxidation (Figure S1A and B). These beneficial roles were also supported by the changes of fresh weight of ten alfalfa seedling roots, showing that NaHS at 100 and 500 μM had the greatest effects on the alleviation of the inhibition of root growth caused by Cd stress (Figure S1C). The beneficial roles of 100 μM NaHS alone were also observed. Subsequent work confirmed that H_2S rather than other sulphur-containing derivatives and sodium exhibited the cytoprotective role in the improvement of Cd toxicity by using a series of sulphur- and sodium-containing chemicals including Na_2S, Na_2SO_4, Na_2SO_3, $NaHSO_4$, $NaHSO_3$, and NaAc, in comparison with the positive roles of NaHS (Figure S2).

Accordingly, we observed that the treatment with 100 μM NaHS for 3 h resulted in the enhancement of endogenous H_2S level in alfalfa seedling roots, which also mimicked a physiological response elicited by Cd alone for 12 h (Figure 2A). The addition of Cd to the NaHS-pretreated plants further strengthened the increased H_2S content. Therefore, 100 μM NaHS was used to mimic the physiological role of intracellular H_2S in the subsequent experiments.

Table 1. Concentrations of low molecular weight thiols and their disulfides, and hGSH/hGSSGh ratio in root tissues.

Treatment	cysteine (nmol g^{-1} FW)	cysteine disulfide (nmol g^{-1} FW)	γ-EC (nmol g^{-1} FW)	γ-EC disulfide (nmol g^{-1} FW)	GSH (nmol g^{-1} FW)	GSSG (nmol g^{-1} FW)	hGSH (nmol g^{-1} FW)	hGSSGh (nmol g^{-1} FW)	hGSH/hGSSGh
Con→Con	30±1 d	3.8±0.8 c	10±1 e	1.5±0.1	27±2 bc	0.2±1.9	252±16 b	28±2 c	8.86
Con→Cd	33±1 cd	5.7±0.8 b	14±2 d	1.7±0.1	21±2 c	0.2±1.4	112±13 f	33±1 bc	3.41
NaHS→Cd	40±2 c	4.0±0.6 c	18±2 bc	1.4±0.5	26±4 c	0.1±1.4	163±14 de	33±4 bc	4.89
NaHS→Con	34±2 cd	3.4±0.7 c	8±0 e	1.2±0.5	36±1 b	0.2±1.1	309±14 a	30±2 bc	10.23
NaHS + PAG→Cd	54±7 b	4.3±0.4 bc	21±3 b	1.2±0.5	29±5 bc	0.3±0.7	144±8 e	41±6 a	3.55
NaHS + PAG + GSH→Cd	65±8 a	4.7±1.6 bc	27±1 a	1.4±0.5	46±11 a	0.7±0.6	179±7 d	36±5 ab	4.91
PAG→Cd	52±6 b	7.4±0.7 a	21±1 b	1.6±0.2	29±1 bc	0.9±0.9	82±12 g	33±3 bc	2.41
PAG→Con	56±4 b	4.0±0.3 c	17±3 cd	1.2±0.5	29±2 bc	0.7±0.7	206±28 c	36±2 ab	5.67

Seedlings were pretreated with or without 100 μM NaHS, 2 mM PAG, 1 mM GSH, individual or combination for 6 h, and then exposed to 200 μM $CdCl_2$ for another 12 h. Values are means ± SD of three independent experiments with three replicates for each. Different letters within columns indicate significant differences ($P<0.05$) according to Duncan's multiple range test.

Changes of low molecular weight thiols and their disulfides as well as representative transcripts in response to NaHS

To determine the influence of H_2S at physiologically concentrations on (h)GSH depletion, GSH pool and corresponding metabolism associated genes were investigated. As shown in Figure 2B, the time-course analysis revealed that (h)GSH contents in seedling roots were significantly enhanced by the pretreatment with NaHS for 6 h, and remained high through 24 h of further incubation in the control solution. Meanwhile, NaHS pretreatment was able to slow down the decreased (h)GSH levels caused by Cd exposure. Changes of the (h)GSH/(h)GSSG(h) ratio also exhibited the similar tendencies (Figure 2C). Comparatively, Cd-induced cysteine and γ-EC (in particular), and cysteine disulfide contents were differentially strengthened or blocked by NaHS pretreatment, respectively (Table 1).

These results arises the question that, whether this increases in metabolites are, at least in part, duo to changes in the expression of genes involved in (h)GSH metabolism. Therefore, the expression of *ECS*, *GS*, and *GR1* genes, were analyzed by real-time RT-PCR. Results of Figure 3A and B revealed that the transcripts of *ECS*, *GS* and *GR1* (especially) in seedling roots approximately displayed a time-dependent increase during Cd stress for 24 h, while the transcriptional profiles of these genes in the control samples were relatively constant during the same period. The pretreatment with NaHS for 6 h in culture solution increased above transcripts, which were differentially strengthened by thereafter Cd stress.

NaHS-induced Cd tolerance, (h)GSH and ROS homeostases were sensitive to PAG, but rescued by GSH

To further verify the involvement of endogenous H_2S in Cd tolerance, DL-propargylglycine (PAG), an effective H_2S synthetic inhibitor [27], and GSH, applied individually and in combination, were used in the subsequent experiment. After 72 h exposure to Cd, the alfalfa seedlings displays severe growth inhibition both in roots and above ground parts, compared to control samples, both of which were improved by NaHS pretreatment (Figure 4A). By contrast, the improvement of seedling growth inhibition as well as the reestablishment of (h)GSH homeostasis triggered by NaHS were sensitive to PAG, but blocked by exogenously applied GSH (Figure 4, Figure S3A). An aggravated Cd toxicity in seedling growth inhibition was also observed when PAG was pretreated.

In an attempt to assess the potential role of endogenous H_2S in ROS homeostasis in Cd-stressed seedlings, ROS production was visualized by staining with H_2DCFDA. As expected, ROS in root tips with Cd alone were produced considerably, suggesting a perturbation in ROS homeostasis (Figure 5). However, the pretreatment with NaHS reduced the ROS abundance. Further results revealed that PAG pretreatment increased the H_2DCFDA fluorescence in Cd-stressed seedling roots, which was further blocked by the addition of GSH. The changes of TBARS content, an indictor of lipid peroxidation, exhibited the similar tendencies (Figure S3B).

Cd treatment caused the accumulation of Cd contents both in shoot and root (particularly) tissues (Figure S4). Similar to the previous reports [31], NaHS decreased Cd accumulation, which was significantly reversed by PAG, but was further blocked by the cotreatment with GSH.

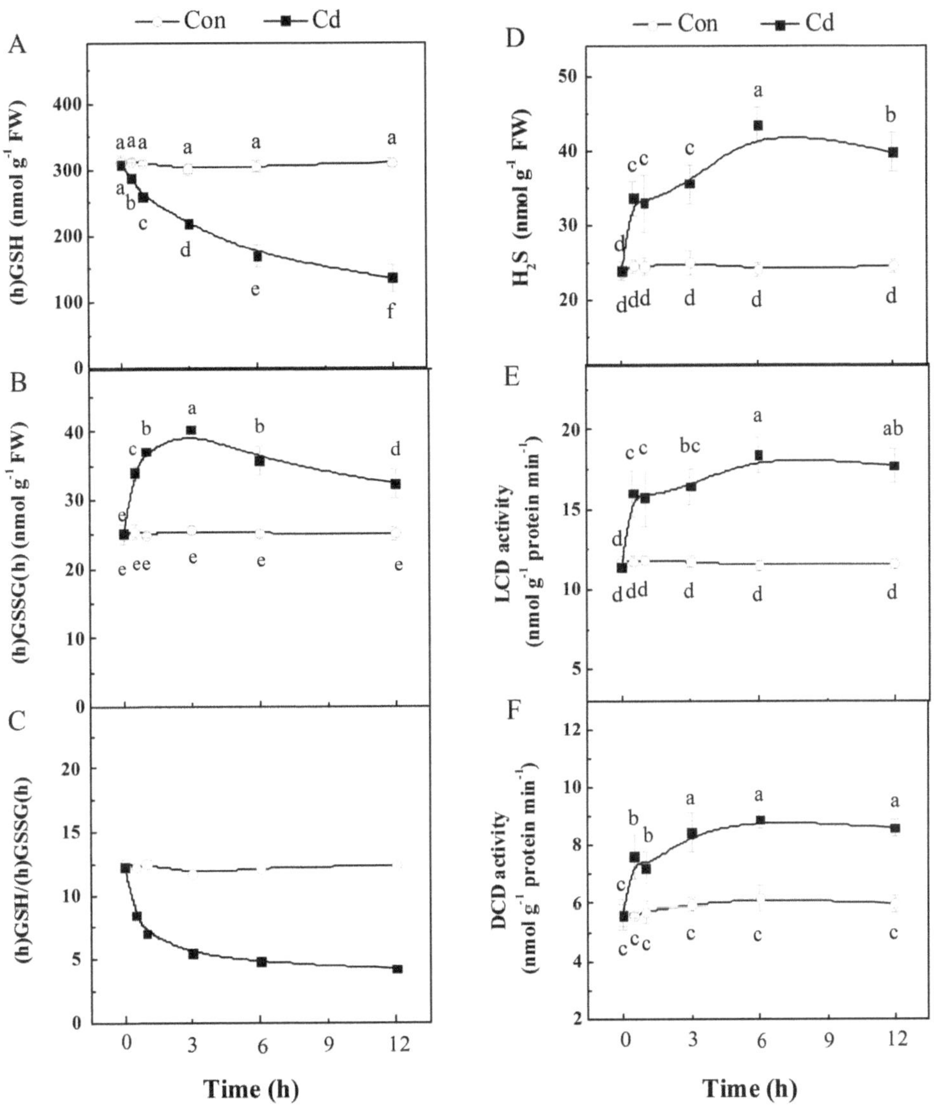

Figure 1. Time course changes of GSH pool and H_2S synthesis upon Cd stress. Upon 200 μM $CdCl_2$ treatment for 12 h, contents of (h)GSH (A), (h)GSSG(h) (B) and H_2S (D), the ratio of (h)GSH/(h)GSSG(h) (C), and the activities of LCD (E) and DCD (F) in root tissues were analyzed. Values are means ± SD of three independent experiments with three replicates for each. Bars denoted by the same letter did not differ significantly at $P<0.05$ according to Duncan's multiple range test.

Transcripts of representative antioxidant defense genes were sensitive to PAG, but rescued by GSH

Since ROS homeostasis was reestablished by NaHS in stressed conditions, the real-time RT-PCR test of corresponding genes involved in their metabolism, i.e. *Cu, Zn-SOD*, *APX1*, and *GPX* [3,5], were analysed. The results of Figure 6 revealed that in comparison with Cd alone samples, NaHS pretreatment followed by Cd exposure resulted in the enhancement in the transcript levels of *Cu, Zn-SOD*, *APX1*, and *GPX* in alfalfa seedling roots. The addition of PAG, however, significantly blocked the increases in the transcripts levels of these representative antioxidant enzymes induced by NaHS, all of which were reversed when GSH was added together with PAG.

NaHS responses were sensitive to a (h)GSH synthetic inhibitor, but reversed by the added GSH

The involvement of (h)GSH homeostasis in NaHS-induced cytoprotective against Cd stress were further investigated using a (h)GSH synthetic inhibitor and GSH applied exogenously. Pretreatment with NaHS, and L-buthionine-sulfoximine (BSO) at 1 mM, a concentration expected to be effective [52], exhibited an aggravated Cd toxicity, which was confirmed by the severe growth stunt and TBARS overproduction, in comparison with Cd plus NaHS (Figure 7A and B). Similarly, NaHS-mediated reestablishment of (h)GSH homeostasis in Cd stressed alfalfa seedling roots was also perturbed by BSO (Figure 7C and D), which was confirmed by the significant decreased (h)GSH content and the ratio of (h)GSH/(h)GSSG(h), respect to Cd alone. By contrast, above BSO responses were sensitive to the addition of GSH when

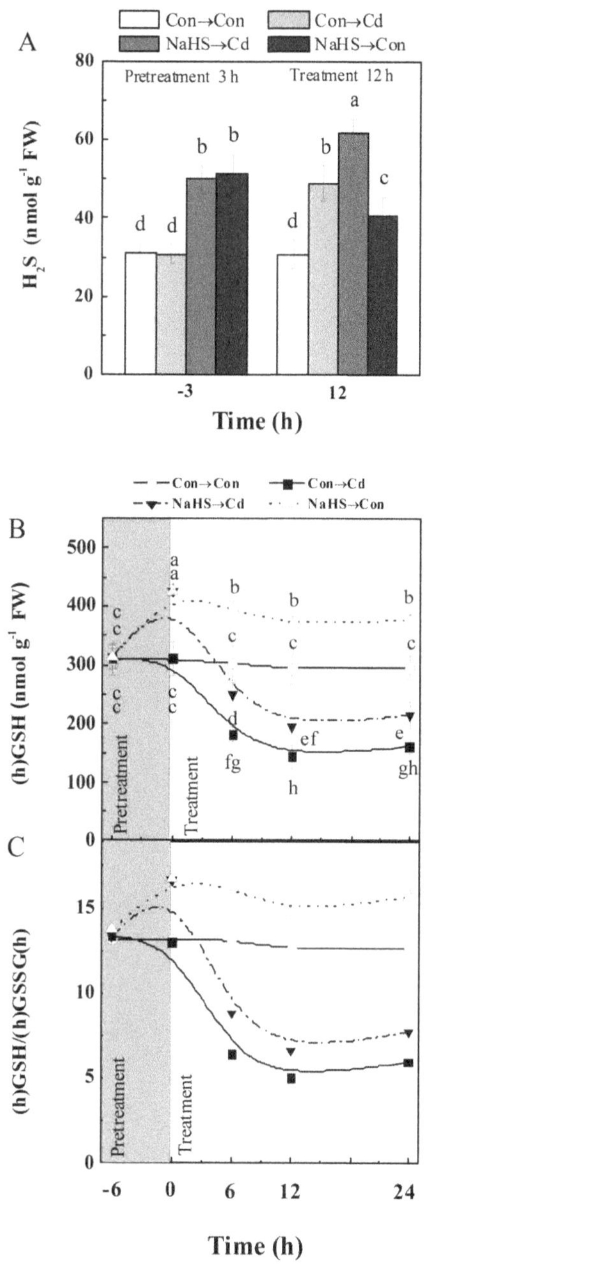

Figure 2. NaHS increased endogenous H_2S and (h)GSH contents, and the ratio of (h)GSH/(h)GSSG(h) upon Cd stress. Endogenous H_2S concentration in root tissues (A) was detected at 3 h after the beginning of 100 μM NaHS pretreatment (−3 h), and 200 μM $CdCl_2$ or chemical-free control treatments for 12 h (12 h). Meanwhile, contents of (h)GSH (B) and the ratio of (h)GSH/(h)GSSG(h) (C) in root tissues were detected at the indicated time points of treatments. Values are means ± SD of three independent experiments with three replicates for each. Bars denoted by the same letter did not differ significantly at $P<0.05$ according to Duncan's multiple range test.

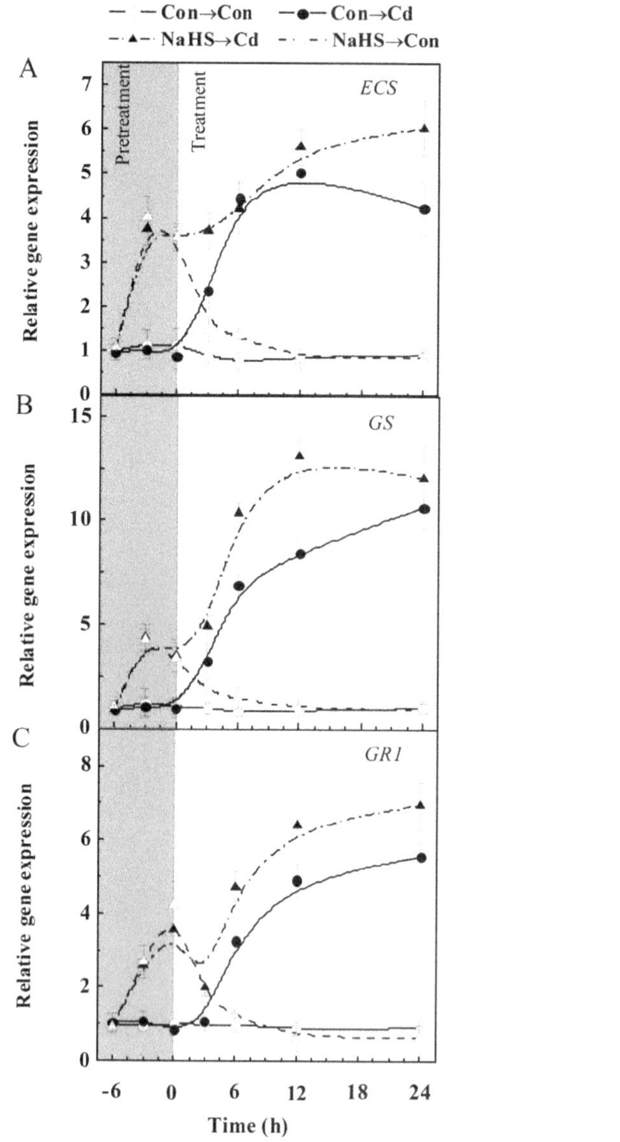

Figure 3. Time course of transcripts responsible for (h)GSH metabolism regulated by NaHS and Cd. Seedlings were pretreated with or without 100 μM NaHS for 6 h and then exposed to 200 μM $CdCl_2$ for another 24 h. The expression levels of *ECS* (A), *GS* (B) and *GR1* (C) in root tissues analyzed by real-time RT-PCR are presented as values relative to the control at the beginning of pretreatment, normalized against expression of two internal reference genes in each sample. Values are means ± SD of three independent experiments with three replicates for each.

applied together. Above results clearly indicated a requirement for (h)GSH homeostasis in NaHS-mediated alleviation of Cd toxicity.

cAMP signaling might be involved in NaHS responses

To testify the hypothesis that H_2S response is associated with cAMP signaling pathway, a pharmacological approach was used to manipulate endogenous cAMP. Results presented in Figure 8A and B indicated that the pretreatment with 8-Br-cAMP, a membrane-permeable analogue of cAMP, alleviated Cd-induced decrease of fresh weight and increase of TBARS content in alfalfa seedling roots. Both of two adenylyl cyclase inhibitors, alloxan and DDA, blocked NaHS-alleviated Cd stress. Moreover, similar to the beneficial actions of 8-Br-cAMP (when was cotreated with PAG followed by Cd stress), a cNMP phosphodiesterase inhibitor IBMX also reversed the PAG responses in the aggravation of fresh weight loss and lipid peroxidation caused by Cd stress. Results from the real-time RT-PCR showed that 8-Br-cAMP and IBMX pretreatments followed by Cd stress, mimicked the effect of NaHS on *GR1* up-regulation, regardless of whether PAG was added or

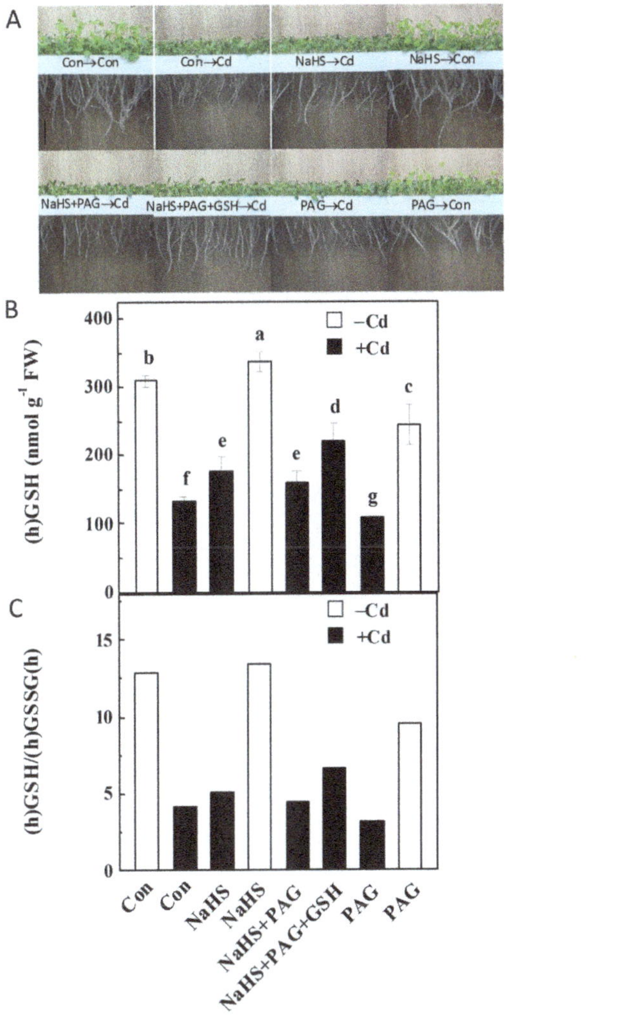

Figure 4. NaHS, PAG and GSH pretreatments differentially regulated seedling growth, (h)GSH content, and (h)GSH/(h)GSSG(h) ratio. Corresponding phenotypes were photographed after 200 μM $CdCl_2$ treatment for 72 h, with or without 100 μM NaHS, 2 mM PAG, 1 mM GSH, individual or combination pretreatments for 6 h (A). Scale bar, 2 cm. Contents of (h)GSH (B), and the ratio of (h)GSH/(h)GSSG(h) (C) in root tissues were also analyzed after 200 μM $CdCl_2$ treatment for 12 h, with or without 100 μM NaHS, 2 mM PAG, 1 mM GSH, individual or combination pretreatment for 6 h. Values are means ± SD of three independent experiments with three replicates for each. Bars denoted by the same letter did not differ significantly at $P<0.05$ according to Duncan's multiple range test.

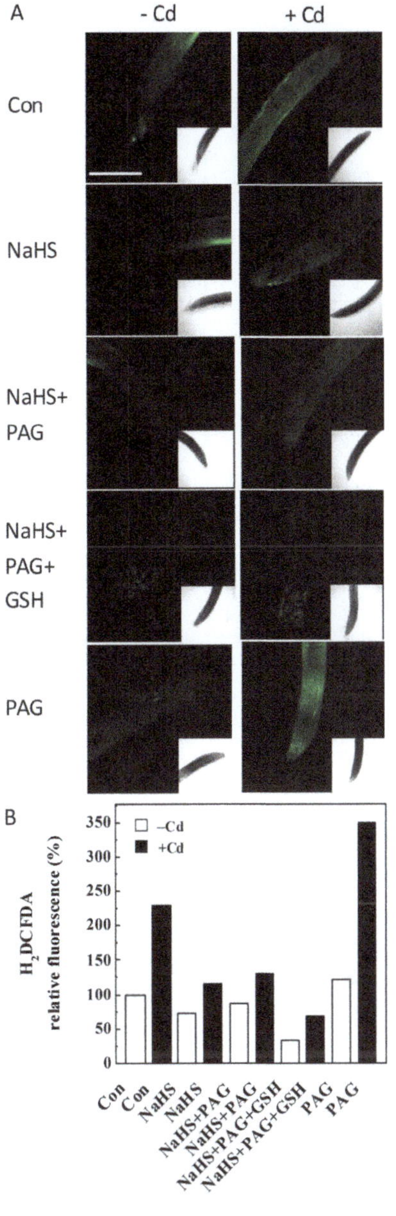

Figure 5. NaHS and GSH pretreatments alleviated Cd-induced ROS production, but blocked by PAG. LSCM results (A). Seedlings were pretreated with or without 100 μM NaHS, 2 mM PAG, 1 mM GSH, individual or combination for 6 h, and then exposed to 200 μM $CdCl_2$ for another 6 h. After various treatments, the roots were respectively stained with H_2DCFDA, then washed thoroughly to removal extra dye and immediately photographed by LSCM. Scale bar, 0.5 mm. The relative DCF fluorescence intensity in the corresponding roots (B).

not (Figure 8C). Two inhibitors alloxan and DDA partially blocked NaHS plus Cd-induced *GR1* transcripts. A similar tendency was found in the changes in *GPX* transcripts (Figure 8F). Results presented in Figure 8D and E further revealed the negative effects of adenylyl cyclase inhibitors on the transcripts of *Cu, Zn-SOD* and *APX1* in NaHS-pretreated seedling roots upon Cd, in comparison with the positive responses of 8-Br-cAMP and IBMX in the presence or absence of PAG.

Discussion

Although H_2S is a hazardous gaseous molecule with a strong odor of rotten eggs, it has been described as an important regulator with a variety of biological roles in animals and recently in plants [11–16,25–34,53–56]. Moreover, recent works on *Populus euphratica* cells [57] and bermudagrass seedlings [51], demonstrated that exogenously applied NaHS, a H_2S donor, resulted in an enhanced Cd tolerance in these species. However, possible physiological mechanisms and downstream targets responsible for the observed Cd tolerance triggered by intracellular H_2S remain elusive. In this report, we discovered endogenous H_2S production in response to Cd stress, and further provided evidence demonstrating a requirement of (h)GSH and ROS homeostases, at least partially, in the intracellular H_2S-medaited plant adaptation

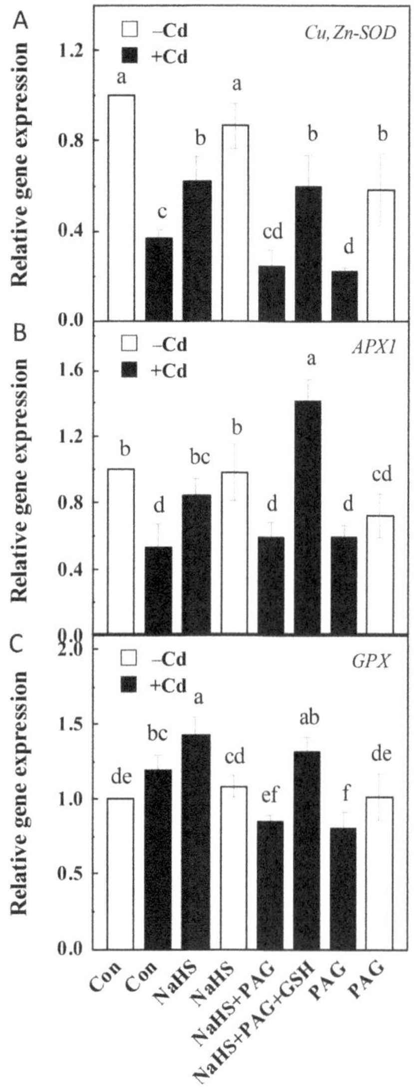

Figure 6. Transcripts of *Cu, Zn-SOD*, *APX1*, and *GPX* regulated by NaHS, PAG, GSH and Cd. Seedlings were pretreated with or without 100 μM NaHS, 2 mM PAG, 1 mM GSH, individual or combination for 6 h, and then exposed to 200 μM $CdCl_2$ for another 12 h. The expression levels of *Cu,Zn-SOD* (A), *APX1* (B), and *GPX* (C) transcripts in root tissues analyzed by real-time RT-PCR are presented as values relative to the control, normalized against expression of two internal reference genes in each sample. Values are means ± SD of three independent experiments with three replicates for each. Bars denoted by the same letter did not differ significantly at $P<0.05$ according to Duncan's multiple range test.

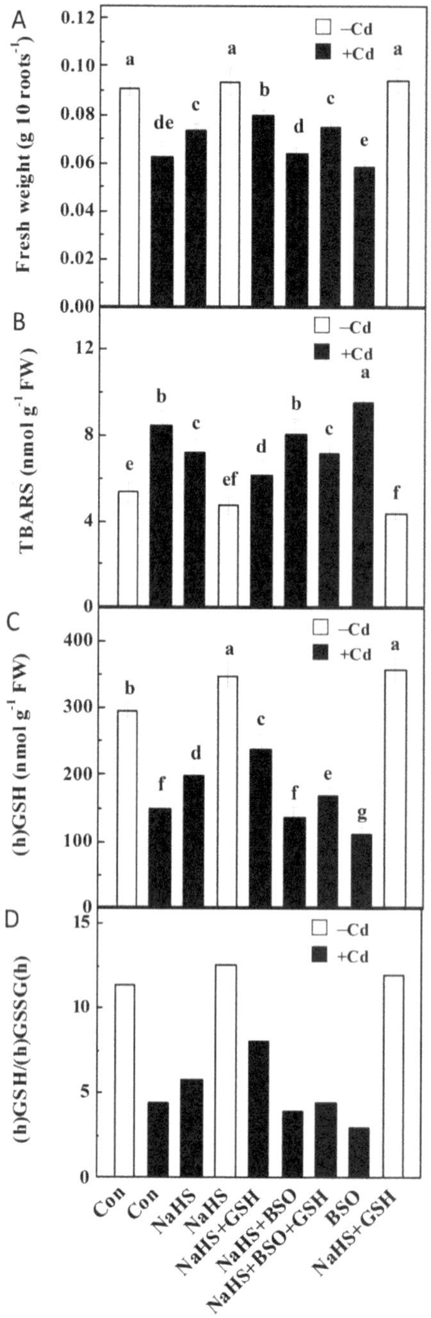

Figure 7. NaHS, GSH and BSO pretreatments differentially regulated seedling growth, TBARS accumulation, (h)GSH content, and (h)GSH/(h)GSSG(h). Fresh weight of 10 roots (A), TBARS accumulation (B), (h)GSH contents (C), and (h)GSH/(h)GSSG(h) ratio (D) in root tissues were determined after seedlings were pretreated with or without 100 μM NaHS, 1 mM GSH, 1 mM BSO, individual or combination for 6 h, and then exposed to 200 μM $CdCl_2$ for 72 h (A), 24 h (B) and 12 h (C and D). Values are means ± SD of three independent experiments with three replicates for each. Bars denoted by the same letter did not differ significantly at $P<0.05$ according to Duncan's multiple range test.

against Cd toxicity. Therefore, our results presented in this work are vital for both fundamental and applied plant biology.

Endogenous H_2S production in response to Cd stress: the possible involvement of LCD/DCD

In animals, it was previously reported that diverse stress-inducing stimuli could result in the production of H_2S, including oxidative stress [49], depletion of cysteine (or its derivatives) [58] and glutathione [50]. Recent work in Arabidopsis [25] and bermudagrass seedlings [51] reported drought- and Cd-induced H_2S production. Because the signal compound H_2S is very reactive [53], the rapid regulation of the activity of H_2S biosynthetic enzymes seems essential to fulfill H_2S-depenent functions. In this work, we further showed that Cd-triggered endogenous H_2S production might be related to LCD/DCD pathways (Figure 1D-F), since the similar increasing changes in the levels of intracellular

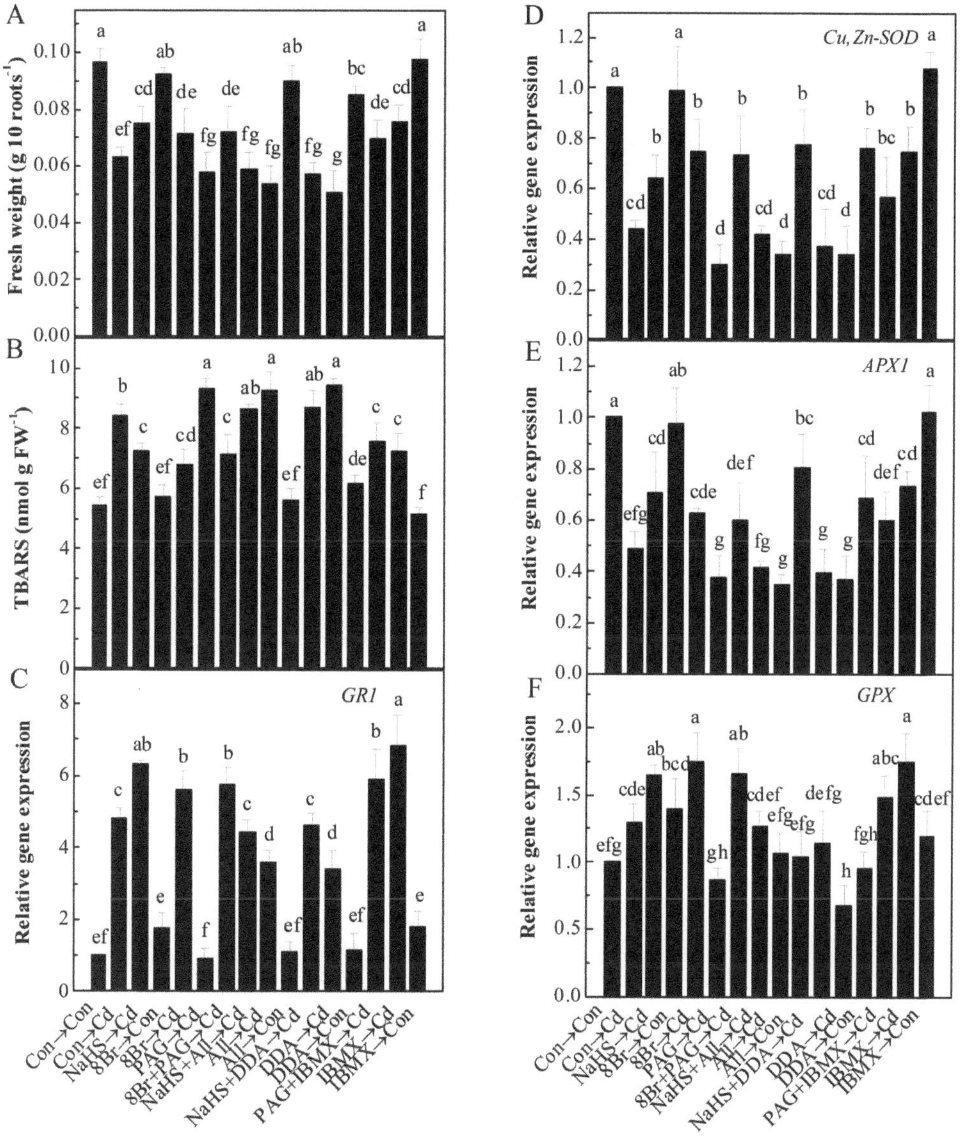

Figure 8. cAMP pathway might be involved in H_2S-alleviated Cd toxicity. Fresh weight of 10 roots (A), TBARS accumulation (B), *GR1* (C), *Cu,Zn-SOD* (D), *APX1* (E), and *GPX* (F) gene expression in alfalfa seedling roots upon Cd stress. Seedlings were pretreated with or without 100 μM NaHS, 50 μM 8-Br-cAMP (8Br), 2 mM PAG, 200 μM alloxan (All), 500 μM IBMX, 1 mM DDA alone, or the combination of treatments for 6 h, and then exposed to 200 μM $CdCl_2$ for 72 h (A), 24 h (B) and 12 h (C–F). The expression levels of corresponding genes analyzed by real-time RT-PCR are presented as values relative to the control, normalized against expression of two internal reference genes in each sample. Values are means ± SD of three independent experiments with three replicates for each. Bars denoted by the same letter did not differ significantly at $P<0.05$ according to Duncan's multiple range test.

H_2S as well as LCD/DCD activities were observed in the seedling roots of alfalfa challenged with Cd for 12 h. Meanwhile, similar to previous reports in wheat [30], bermudagrass [51], *Spinacia oleracea* seedlings [59], and strawberry plants [60], NaHS-induced H_2S production in alfalfa plants was also observed (Figure 2A).

In plants, both LCD and DCD are hypothesized to be involved in intracellular H_2S synthesis [21,27]. Several LCD/DCD candidates have been cloned and partially analyzed from the model plant *Arabidopsis* to *Brassica napus* [24,61]. Our above findings are consistent with those reported by Bloem et al. [40], in which they found that *Brassica napus* was able to react to *Pyrenopeziza brassicae* infection with a greater potential to release H_2S, which was reflected by an increasing LCD activity with fungal infection. More recently, auxin-induced DES-mediated H_2S generation was also found to be involved in lateral root formation in tomato seedlings [62]. In view of the fact that all H_2S synthetic enzymes are not fully elucidated, our results suggested that LCD/DCD pathways might be, at least partially, related to Cd-induced H_2S production in alfalfa seedlings. In a future study, the role of other enzymatic and non-enzymatic pathways-mediated induction of H_2S synthesis in alfalfa seedlings upon Cd stress need be further elucidated.

The mechanism underlying the role of intracellular H_2S in the alleviation of Cd toxicity: reestablishment of reduced (homo)glutathione and ROS homeostases

Ample evidence revealed a clear relationship between metal stress and redox homeostasis and antioxidant capacity

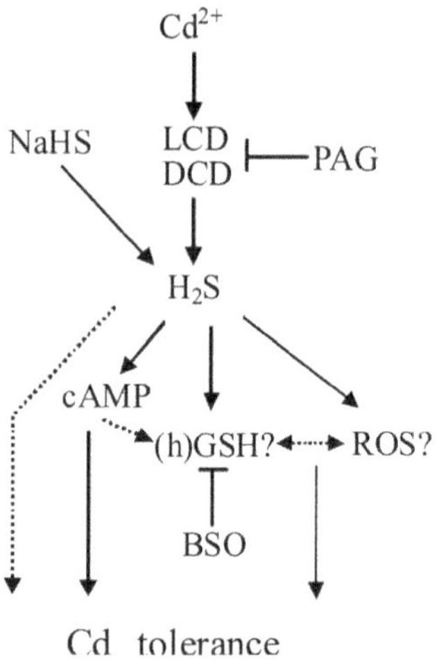

Figure 9. Simplified scheme of mechanisms involved in Cd tolerance by LCD/DCD-produced H_2S-modulated (h)GSH and ROS homeostases. Abbreviations: NaHS, sodium hydrosulfide; PAG, DL-propargylglycine; LCD, L-cysteine desulfhydrase; DCD, D-cysteine desulfhydrase; H_2S, hydrogen sulfide; ROS, reactive oxygen species; (h)GSH, reduced (homo)glutathione; BSO, L-buthionine-sulfoximine; cAMP, cyclic AMP. The dashed line denotes possible signaling cascade. T bars, inhibition.

[3,9,63–66]. Also, GSH could function as a heavy metal-ligand and an antioxidant [5,67]. In plants, H_2S serves as a signal as well as a novel antioxidant in hormonal and defense responses against abiotic stress [53,60]. Genetic evidence further revealed that the GSH deficiency mutant *pad2-1* shows the more oxidized redox state in contrast to wild type [68]. Arabidopsis mutants deficient in phytochelatins (PCs) and GSH biosynthesis respectively, *cad1* and *cad2*, are consequently more sensitive to Cd [6,69,70], that showed the crucial role of PCs, especially their precursor GSH in responding to Cd challenge. In the assays described here, as expected, when alfalfa seedling plants were upon Cd exposure, (h)GSH homeostasis is altered, which is reflected by the fact that the concentrations of reduced GSH and hGSH dropped (Table 1, Figure 1A–C), possible as a consequence of initiated PCs biosynthesis [71,72]. Similarly, a low ratio of (h)GSH/(h)GSSH(h), an important redox index related to Cd tolerance in alfalfa plants [45], was also observed. These changes thereafter cause redox imbalance and in turn Cd toxicity (Figures S1A, S3 and S4; Figures 4A and 5).

Our further experiments provide strong evidence to support the existence of a causal relationship between the endogenous H_2S signal and the alleviation of Cd toxicity in alfalfa seedlings partly by reestablishment of (h)GSH and ROS homeostases, which might be associated with the cAMP pathway. This conclusion is based on several pieces of evidence: (i) increased H_2S metabolism as well as the perturbation of (h)GSH homeostasis in alfalfa seedling roots are two early responses to the exposure of Cd (Figure 1, Table 1). These changes were consistent with the phenotypes of Cd toxicity (Figure 4A, Figure S1C). (ii) Application of a H_2S-releasing compound NaHS (also called as H_2S donor), not only mimics intracellular H_2S content triggered by Cd, but also alleviates Cd toxicity (Figures 2 and 4). Consistently, we also detected reestablishment of (h)GSH homeostasis, which was reflected by a higher (h)GSH content and ratio of (h)GSH/(h)GSSG(h) upon Cd stress. The observed Cd tolerance might be due to the available (h)GSH by the up-regulation of (h)GSH synthesis related genes, *ECS* and *GS* (Figure 3A and B), as well as *GR1* (Figure 3C), because besides the synthesis of PCs, availability of GSH and concerted activity of GR seem to play a important role for plants to combat oxidative stress and Cd toxicity [7,72,73]. While, the inhibition of H_2S production caused by its synthetic inhibitor PAG blocked NaHS-induced Cd tolerance and reestablishment of (h)GSH and ROS homeostases, the latter of which was confirmed by the histochemical staining detecting the alleviation of plasma membrane integrity and lipid peroxidation, decreased ROS content and up-regulation of *Cu,Zn-SOD*, *APX1* and *GPX* transcripts, as well as declined TBARS level (Table 1, Figures 2–6, and Figures S1, S3 and S4). (iii) Above mentioned PAG responses were further rescued by exogenously applied GSH (Table 1, Figures 4–6). (iv) NaHS responses were sensitive to a (h)GSH synthetic inhibitor, but reversed by the added GSH (Figure 7), both of which suggesting a requirement of (h)GSH homeostasis for NaHS cytoprotective roles; and (v) Previous reports in animals showed H_2S-activited cAMP level or H_2S-regulated cAMP homeostasis [37,38]. Here, we found that two adenylyl cyclase inhibitors, alloxan and DDA, blocked the beneficial responses conferred by NaHS in alfalfa seedlings subjected to Cd stress (Figure 8). On the contrary, an analogue of cAMP 8-Br-cAMP and a cNMP phosphodiesterase inhibitor IBMX mimicked the effects of NaHS on the alleviation of Cd toxicity as well as the regulation of (h)GSH homeostasis and ROS metabolism (*GR1*, *Cu,Zn-SOD*, *APX1*, and *GPX*, etc). Above pharmacological evidence indicated the involvement of cAMP signaling in NaHS responses. Additionally, NaHS-triggered cytoprotective roles were confirmed to act as a H_2S-dependent fashion (Figure S2). Above results clearly established a casual link between intracellular H_2S in the alleviation of Cd toxicity and reestablishment of (h)GSH and ROS homeostases.

Conclusions

In summary, our pharmacological, histochemical, biochemical and molecular evidence suggested that the intracellular H_2S was able to ameliorate Cd toxicity in alfalfa seedlings at least partly by reestablishment of (h)GSH and ROS homeostases. Figure 9 illustrates a simplified scheme of mechanisms involved in Cd tolerance by LCD/DCD-produced H_2S-modulated (h)GSH and ROS homeostases, since 1) LCD/DCD-produced H_2S acts as a signal triggered by Cd to regulated (h)GSH metabolisms; 2) both (h)GSH and ROS homeostases could be reestablished by H_2S and further linked to Cd tolerance; 3) cAMP signaling pathway might be related to NaHS-triggered Cd tolerance, partially through the regulation of GSH homeostasis and ROS metabolism. Taking into account that H_2S participates in stressful responses and developmental process, our study therefore may extend our understanding of the complex system integrating environmental and developmental signals.

Supporting Information

Figure S1 NaHS pretreatment alleviates Cd toxicity.

Figure S2 H_2S or HS^-, but not other compounds derived from NaHS contribute to NaHS responses.

Figure S3 Effects of NaHS, PAG and GSH pretreatments on the fresh weight (A) and TBARS concentrations (B) in alfalfa seedling roots upon Cd stress.

Figure S4 Effects of NaHS, PAG and GSH pretreatments on Cd concentrations in alfalfa seedlings upon Cd stress.

Table S1 The sequences of primers for real-time RT-PCR.

Author Contributions

Conceived and designed the experiments: WS. Performed the experiments: WC KZ QJ. Analyzed the data: WC Y. Xie WS. Contributed reagents/materials/analysis tools: HC JC Y. Xia JZ. Wrote the paper: WC HC WS.

References

1. Gao J, Sun L, Yang X, Liu JX (2013) Transcriptomic analysis of cadmium stress response in the heavy metal hyperaccumulator *Sedum alfredii* hance. PLoS ONE 8(6): e64643.
2. Ye Y, Li Z, Xing D (2013) Nitric oxide promotes MPK6-mediated caspase-3-like activation in cadmium-induced *Arabidopsis thaliana* programmed cell death. Plant Cell Environ 36: 1–15.
3. Sharma SS, Dietz KJ (2009) The relationship between metal toxicity and cellular redox imbalance. Trends Plant Sci 14: 43–50.
4. Tkalec M, Štefanić PP, Cvjetko P, Šikić S, Pavlica M, et al. (2012) The effects of cadmium-zinc interactions on biochemical responses in tobacco seedlings and adult plants. PLoS ONE 9(1): e87582.
5. Foyer CH, Noctor G (2011) Ascorbate and glutathione: the heart of the redox hub. Plant Physiol 155: 2–18.
6. Cobbett CS, May MJ, Howden R, Rolls B (1998) The glutathione-deficient, cadmium-sensitive mutant, *cad2-1*, of *Arabidopsis thaliana* is deficient in γ-glutamylcysteine synthetase. Plant J 16: 73–78.
7. Cruz de Carvalho MH, Brunet J, Bazin J, Kranner I, d' Arcy-Lameta A, et al. (2010) Homoglutathione synthetase and glutathione synthetase in drought-stressed cowpea leaves: expression patterns and accumulation of low-molecular-weight thiols. J Plant Physiol 167: 480–487.
8. Matamoros MA, Moran JF, Iturbe-Ormaetxe I, Rubio MC, Becana M (1999) Glutathione and homoglutathione synthesis in legume root nodules. Plant Physiol 121: 879–888.
9. Ortega-Villasante C, Rellán-Álvarez R, Del Campo FF, Carpena-Ruiz RO, Hernández LE (2005) Cellular damage induced by cadmium and mercury in *Medicago sativa*. J Exp Bot 56: 2239–2251.
10. Metwally A, Safronova VI, Belimov AA, Dietz KJ (2005) Genotypic variation of the response to cadmium toxicity in *Pisum sativum* L. J Exp Bot 56: 167–178.
11. Abe K, Kimura H (1996) The possible role of hydrogen sulfide as an endogenous neuromodulator. J Neurosci 16: 1066–1071.
12. Li L, Rose P, Moore PK (2011) Hydrogen sulfide and cell signaling. Annu Rev Pharmacol Toxicol 51: 169–187.
13. Hosoki R, Matsuki N, Kimura H (1997) The possible role of hydrogen sulfide as an endogenous smooth muscle relaxant in synergy with nitric oxide. Biochem Biophys Res Commun 237: 527–531.
14. Du J, Hui Y, Cheung Y, Bin G, Jiang H, et al. (2004) The possible role of hydrogen sulfide as a smooth muscle cell proliferation inhibitor in rat cultured cells. Heart Vessels 19: 75–80.
15. Yang G, Sun X, Wang R (2004) Hydrogen sulfide-induced apoptosis of human aorta smooth muscle cells via the activation of mitogen-activated protein kinases and caspase-3. FASEB J 18: 1782–1784.
16. Kimura Y, Goto Y, Kimura H (2010) Hydrogen sulfide increases glutathione production and suppresses oxidative stress in mitochondria. Antioxid Redox Signal 12: 1–13.
17. Wilson LG, Bressan RA, Filner P (1978) Light-dependent emission of hydrogen sulfide from plants. Plant Physiol 61: 184–189.
18. Hällgren JE, Fredriksson SÅ (1982) Emission of hydrogen sulfide from sulfur dioxide-fumigated pine trees. Plant Physiol 70: 456–459.
19. Sekiya J, Schmidt A, Wilson LG, Filner P (1982) Emission of hydrogen sulfide by leaf tissue in response to L-cysteine. Plant Physiol 70: 430–436.
20. Wang R (2002) Two's company, three's a crowd: can H_2S be the third endogenous gaseous transmitter? FASEB J 16: 1792–1798.
21. Papenbrock J, Riemenschneider A, Kamp A, Schulz-Vogt HN, Schmidt A (2007) Characterization of cysteine-degrading and H_2S-releasing enzymes of higher plants – from the field to the test tube and back. Plant Biol 9: 582–588.
22. Léon S, Touraine B, Briat JF, Lobréaux S (2002) The *AtNFS2* gene from *Arabidopsis thaliana* encodes a NifS-like plastidial cysteine desulphurase. Biochem J 366: 557–564.
23. Riemenschneider A, Wegele R, Schmidt A, Papenbrock J (2005) Isolation and characterization of a D-cysteine desulfhydrase protein from *Arabidopsis thaliana*. FEBS J 272: 1291–1304.
24. Álvarez C, Calo L, Romero LC, García I, Gotor C (2010) An *O*-acetylserine(thiol)lyase homolog with L-cysteine desulfhydrase activity regulates cysteine homeostasis in Arabidopsis. Plant Physiol 152: 656–669.
25. Jin Z, Shen J, Qiao Z, Yang G, Wang R, et al. (2011) Hydrogen sulfide improves drought resistance in *Arabidopsis thaliana*. Biochem Biophys Res Commun 414: 481–486.
26. Wang Y, Li L, Cui W, Xu S, Shen W, et al. (2012) Hydrogen sulfide enhances alfalfa (*Medicago sativa*) tolerance against salinity during seed germination by nitric oxide pathway. Plant Soil 351: 107–119.
27. García-Mata C, Lamattina L (2010) Hydrogen sulphide, a novel gasotransmitter involved in guard cell signalling. New Phytol 188: 977–984.
28. Lisjak M, Srivastava N, Teklic T, Civale L, Lewandowski K, et al. (2010) A novel hydrogen sulfide donor causes stomatal opening and reduces nitric oxide accumulation. Plant Physiol Biochem 48: 931–935.
29. Zhang H, Jiao H, Jiang CX, Wang SH, Wei ZJ, et al. (2010) Hydrogen sulfide protects soybean seedlings against drought-induced oxidative stress. Acta Physiol Plant 32: 849–857.
30. Zhang H, Hu LY, Hu KD, He YD, Wang SH, et al. (2008) Hydrogen sulfide promotes wheat seed germination and alleviates oxidative damage against copper stress. J Integr Plant Biol 50: 1518–1529.
31. Li L, Wang Y, Shen W (2012) Roles of hydrogen sulfide and nitric oxide in the alleviation of cadmium-induced oxidative damage in alfalfa seedling roots. Biometals 25: 617–631.
32. Li ZG, Gong M, Xie H, Yang L, Li J (2012) Hydrogen sulfide donor sodium hydrosulfide-induced heat tolerance in tobacco (*Nicotiana tabacum* L) suspension cultured cells and involvement of Ca^{2+} and calmodulin. Plant Sci 185–186: 185–189.
33. Lin YT, Li MY, Cui WT, Lu W, Shen WB (2012) Haem oxygenase-1 is involved in hydrogen sulfide-induced cucumber adventitious root formation. J Plant Growth Regul 31: 519–528.
34. Zhang H, Hu SL, Zhang ZJ, Hu LY, Jiang CX, et al. (2011) Hydrogen sulfide acts as a regulator of flower senescence in plants. Postharvest Biol Tec 60: 251–257.
35. Ma W, Qi Z, Smigel A, Walker RK, Verma R, et al. (2009) Ca^{2+}, cAMP, and transduction of non-self perception during plant immune responses. Proc Natl Acad Sci U S A 106: 20995–21000.
36. Jin XC, Wu WH (1998) Involvement of cyclic AMP in ABA- and Ca^{2+}-mediated signal transduction of stomatal regulation in *Vicia faba*. Plant Cell Physiol 40: 1127–1133.
37. Kimura H (2000) Hydrogen sulfide induces cyclic AMP and modulates the NMDA receptor. Biochem Biophys Res Commun 267: 129–133.
38. Lu M, Liu YH, Ho CY, Tiong CX, Bian JS (2012) Hydrogen sulfide regulates cAMP homeostasis and renin degranulation in As4.1 and rat renin-rich kidney cells. Am J Physiol Cell Physiol 302: C59–C66.
39. Shao JL, Wan XH, Chen Y, Bi C, Chen HM, et al. (2011) H_2S protects hippocampal neurons from anoxia-reoxygenation through cAMP-mediated PI3K/Akt/p70S6K cell-survival signaling pathways. J Mol Neurosci 43: 453–460.
40. Bloem E, Riemenschneider A, Volker J, Papenbrock J, Schmidt A, et al. (2004) Sulphur supply and infection with *Pyrenopeziza brassicae* influence L-cysteine desulphydrase activity in *Brassica napus* L. J Exp Bot 55: 2305–2312.
41. Cui W, Gao C, Fang P, Lin G, Shen W (2008) Alleviation of cadmium toxicity in *Medicago sativa* by hydrogen-rich water. J Hazard Mater 260: 715–724.
42. Smith IK (1985) Stimulation of glutathione synthesis in photorespiring plants by catalase inhibitors. Plant Physiol 79: 1044–1047.
43. Herschbach C, Pilch B, Tausz M, Rennenberg H, Grill D (2002) Metabolism of reduced and inorganic sulphur in pea cotyledons and distribution into developing seedlings. New Phytol 153: 73–80.
44. Meyer AJ, Brach T, Marty L, Kreye S, Rouhier N, et al. (2007) Redox-sensitive GFP in *Arabidopsis thaliana* is a quantitative biosensor for the redox potential of the cellular glutathione redox buffer. Plant J 52: 973–986.
45. Cui W, Li L, Gao Z, Wu H, Xie Y, et al. (2012) Haem oxygenase-1 is involved in salicylic acid-induced alleviation of oxidative stress duo to cadmium stress in *Medicago sativa*. J Exp Bot 63: 5521–5534.
46. Vandesompele J, De Preter K, Pattyn F, Poppe B, Van Roy N, et al. (2002) Accurate normalization of real-time quantitative RT-PCR data by geometric averaging of multiple internal control genes. Genome Biol 3: research0034.
47. Kováčik J, Babula P, Klejdus B, Hedbavny J, Jarošová M (2014) Unexpected behavior of some nitric oxide modulators under cadmium excess in plant tissue. PLos ONE 9(3): e91685.
48. Rennenberg H, Filner P (1982) Stimulation of H_2S emission from pumpkin leaves by inhibition of glutathione synthesis. Plant Physiol 69: 766–770.
49. Kwak WJ, Kwon GS, Jin I, Kuriyama H, Sohn HY (2003) Involvement of oxidative stress in the regulation of H_2S production during ultradian metabolic oscillation of *Saccharomyces cerevisiae*. FEMS Microbiol Lett 219: 99–104.
50. Sohn HY, Kum EJ, Kwon GS, Jin I, Adams CA, et al. (2005) *GLR1* plays an essential role in the homeodynamics of glutathione and the regulation of H_2S production during respiratory oscillation of *Saccharomyces cerevisiae*. Biosci Biotechnol Biochem 69: 2450–2454.
51. Shi H, Ye T, Chan Z (2014) Nitric oxide-activated hydrogen sulfide is essential for cadmium stress response in bermudagrass (*Cynodon dactylon* (L). Pers.). Plant Physiol Biochem 74: 99–107.
52. Rüegsegger A, Schmutz D, Brunold C (1990) Regulation of glutathione synthesis by cadmium in *Pisum sativum* L. Plant Physiol 93: 1579–1584.
53. Lisjak M, Teklic T, Wilson ID, Whiteman M, Hancock JT (2013) Hydrogen sulfide: environmental factor or signalling molecule? Plant Cell Environ 36: 1607–1616.

54. García-Mata C, Lamattina L (2013) Gasotransmitters are emerging as new guard cell signaling molecules and regulators of leaf gas exchange. Plant Sci 201–202: 66–73.
55. Kimura Y, Kimura H (2004) Hydrogen sulfide protects neurons from oxidative stress. FASEB J 18: 1165–1167.
56. Li ZG, Gong M, Liu P (2012) Hydrogen sulfide is a mediator in H_2O_2-induced seed germination in *Jatropha Curcas*. Acta Physiol Plant 34: 2207–2213.
57. Sun J, Wang R, Zhang X, Yu Y, Zhao R, et al. (2013) Hydrogen sulfide alleviates cadmium toxicity through regulations of cadmium transport across the plasma and vacuolar membranes in *Populus euphratica* cells. Plant Physiol Biochem 65: 67–74.
58. Sohn HY, Kuriyama H (2001) The role of amino acids in the regulation of hydrogen sulfide production during ultradian respiratory oscillation of *Saccharomyces cerevisiae*. Arch Microbiol 176: 69–78.
59. Chen J, Wu FH, Wang WH, Zheng CJ, Lin GH, et al. (2011) Hydrogen sulphide enhances photosynthesis through promoting chloroplast biogenesis, photosynthetic enzyme expression, and thiol redox modification in *Spinacia oleracea* seedlings. J Exp Bot 62: 4481–4493.
60. Christou A, Manganaris GA, Papadopoulos I, Fotopoulos V (2013) Hydrogen sulfide induces systemic tolerance to salinity and non-ionic osmotic stress in strawberry plants through modification of reactive species biosynthesis and transcriptional regulation of multiple defence pathways. J Exp Bot 64: 1953–1966.
61. Xie Y, Lai D, Mao Y, Zhang W, Shen W, et al. (2013) Molecular cloning, characterization, and expression analysis of a novel gene encoding L-cysteine desulfhydrase from *Brassica napus*. Mol Biotechnol 54: 737–746.
62. Fang T, Cao Z, Li J, Shen W, Huang L (2014) Auxin-induced hydrogen sulfide generation is involved in lateral root formation in tomato. Plant Physiol Biochem 76: 44–51.
63. Dawood M, Cao F, Jahangir MM, Zhang G, Wu F (2012) Alleviation of aluminum toxicity by hydrogen sulfide is related to elevated ATPase, and suppressed aluminum uptake and oxidative stress in barley. J Hazard Mater 209–210: 121–128.
64. Jin CW, Mao QQ, Luo BF, Lin XY, Du ST (2013) Mutation of *mpk6* enhances cadmium tolerance in *Arabidopsis* plants by alleviating oxidative stress. Plant Soil 371: 387–396.
65. Lagorce A, Fourçans A, Dutertre M, Bouyssiere B, Zivanovic Y, et al. (2012) Genome-wide transcriptional response of the archaeon *Thermococcus gammatolerans* to cadmium. PLoS ONE 7(7): e41935.
66. Thapa G, Sadhukhan A, Panda SK, Sahoo L (2012) Molecular mechanistic model of plant heavy metal tolerance. Biometals 25: 489–505.
67. Dixit P, Mukherjee PK, Ramachandran V, Eapen S (2013) Glutathione transferase from *Trichoderma virens* enhances cadmium tolerance without enhancing its accumulation in transgenic *Nicotiana tabacum*. PLoS ONE 6(1): e16360.
68. Dubreuil-Maurizi C, Vitecek J, Marty L, Branciard L, Frettinger P, et al. (2011) Glutathione deficiency of the Arabidopsis mutant *pad2-1* affects oxidative stress-related events, defense gene expression, and the hypersensitive response. Plant Physiol 157: 2000–2012.
69. Howden R, Andersen CR, Goldsbrough PB, Cobbett CS (1995) A cadmium-sensitive, glutathione-deficient mutant of *Arabidopsis thaliana*. Plant Physiol 107: 1067–1073.
70. Howden R, Goldsbrough PB, Andersen CR, Cobbett CS (1995) Cadmium-sensitive, *cad1* mutants of *Arabidopsis thaliana* are phytochelatin deficient. Plant Physiol 107: 1059–1066.
71. Grill E, Löffler S, Winnacker EL, Zenk MH (1989) Phytochelatins, the heavy-metal-binding peptides of plants, are synthesized from glutathione by a specific γ-glutamylcysteine dipeptidyl transpeptidase (phytochelatin synthase). Proc Natl Acad Sci USA 86: 6838–6842.
72. Mishra S, Srivastava S, Tripathi RD, Govindarajan R, Kuriakose SV, et al. (2006) Phytochelatin synthesis and response of antioxidants during cadmium stress in *Bacopa monnieri* L. Plant Physiol Biochem 44: 25–37.
73. Verbruggen N, Juraniec M, Baliardini C, Meyer CL (2013) Tolerance to cadmium in plants: the special case of hyperaccumulators. Biometals 26: 633–638.

6

Community Profiling of Culturable Fluorescent Pseudomonads in the Rhizosphere of Green Gram (*Vigna radiata* L.)

Rupak K. Sarma[1], Animesh Gogoi[1], Budheswar Dehury[2], Rajal Debnath[1], Tarun C. Bora[1], Ratul Saikia[1]*

1 Biotechnology Division, CSIR-North East Institute of Science and Technology, Jorhat, Assam, India, **2** Department of Agricultural Biotechnology, Assam Agricultural University, Jorhat, Assam, India

Abstract

Study on microbial diversity in the unexplored rhizosphere is important to understand their community structure, biology and ecological interaction with the host plant. This research assessed the genetic and functional diversity of fluorescent pseudomonads [FP] in the green gram rhizophere. One hundred and twenty types of morphologically distinct fluorescent pseudomonads were isolated during vegetative as well as reproductive growth phase of green gram. Rep PCR, ARDRA and RISA revealed two distinct clusters in each case at 75, 61 and 70% similarity coefficient index respectively. *16S rRNA* partial sequencing analysis of 85 distantly related fluorescent pseudomonads depicted *Pseudomonas aeruginosa* as the dominant group. Out of 120 isolates, 23 (19%) showed antagonistic activity towards phytopathogenic fungi. These bacterial isolates showed varied production of salicylic acid, HCN and chitinase, 2, 4-diacetylphloroglucinol (DAPG), phenazine-1-carboxylic acid (PCA) and pyoluteorin (PLT). Production efficiency of inherent level of plant growth promoting (PGP) traits among the 120 isolates demonstrated that 10 (8%) solubilised inorganic phosphates, 25 (20%) produced indoles and 5 (4%) retained ACC deaminase activity. *Pseudomonas aeruginosa* GGRJ21 showed the highest production of all antagonistic and plant growth promoting (PGP) traits. In a greenhouse experiment, GGRJ21 suppressed root rot disease of green gram by 28–93% ($p=0.05$). Consistent up regulation of three important stress responsive genes, i.e., *acdS*, *KatA* and *gbsA* and elevated production efficiency of different PGP traits could promote GGRJ21 as a potent plant growth regulator.

Editor: Tom Coenye, Ghent University, Belgium

Funding: The work is supported by a Network Project (AMAAS), sponsored by Indian Council of Agricultural Research (ICAR), Government of India, New Delhi. The funder had no role in study design, data collection and analysis, decision to publish, or preparation of the manuscript.

Competing Interests: The authors have declared that no competing interests exist.

* Email: rsaikia19@gmail.com

Introduction

Fluorescent pseudomonads (FP) are one of the most diverse and ecologically significant groups under γ-proteobacteria that has been well studied in relation to their beneficial interactions with plants [1]. This ubiquitous bacterial group is widely accepted as most prominent plant growth promoting rhizobacteria (PGPR) [2], biocontrol agent [3] and a potential agent that may stimulate plant growth and development under varied abiotic stress conditions [4–6]. In the recent years, a wide attention was paid to decipher the diversity of fluorescent pseudomonads with keen reference to their biocontrol and biofertilizing abilities. Despite of other PGPR and non fluorescent pseudomonad isolates, wide recognition of fluorescent pseudomonads as potent plant growth promoter as well as biocontrol agent are mainly due to: 1) higher rhizosphere competence, i.e. extensive colonization in the ecto- and endorhizosphere when introduced through seed inoculation [7,8] and 2) production efficiency of different secondary metabolites that can inhibit other microorganisms [1,3]. Therefore, exploration of genetic and functional diversity of FPs from crop rhizosphere has great practical importance, with relevance to their application as effective biofertilizing and biocontrol agents.

The biocontrol activity of FP against different phytopathogens is mainly due to the production of diverse types of extracellular metabolites and antibiotic compounds [3]. Different phenazines, phenolics, polyketides, pyrrole-type compounds and siderophore from fluorescent pseudomonads render synergistic effect against the pathogenic microorganisms [9,10]. Voisard et al. [11] and Keel et al. [12] reported the detrimental effect of fluorescent pseudomonads generated HCN and 2, 4-diacetylphloroglucinol (DAPG) against different soil borne phytopathogenic fungi. Similarly inherent production efficiency of indoles, 1-aminocyclo-propane-1-carboxylate (ACC) deaminase, as well as phosphate solubilizing capability helps to place most of the γ-proteobacteria under PGPR class [4,13,14]. During the last decade several workers have reported induced drought tolerance in plants using *Pseudomonas* spp. [15–17]. Although the role of *Pseudomonas* spp. on water stress tolerance is not a new area to excavate; however to the best of our knowledge, very scanty amount of work is yet available on the amelioration of water stress through the use fluorescent pseudomonad isolates in acidic soils of North East India [4]. The genetic diversity and functional characterization of this large group in rhizosphere soils of different plants, *viz.*, rice

[18,19], cotton [20], banana [13], wheat [21], and canola [14] have been already reported in different parts of the world. However, there is no report on total species richness of the ubiquitous group from rhizosphere of green gram. Green gram rhizosphere may sustain a rich repository of FP isolates that can be beneficial for plant health promotion. Previously we had reported the diversity of alkaline proteinase producing fluorescent pseudomonads from rhizosphere of green gram [22]. This was only a concise report on genetic diversity based on alkaline proteinase production. Thus, with continuation of the previous work, the present research was undertaken to investigate: (i) the total genetic diversity of green gram rhizosphere adhered fluorescent pseudomonads through PCR based molecular tools, *i.e.*, rep PCR (BOX-PCR and ERIC-PCR), amplified ribosomal DNA restriction analysis (ARDRA), ribosomal intergenic space analysis (RISA), and *16S rRNA* sequence analysis, (ii) functional diversity with relation to biocontrol and PGP traits along with their nature in water stress tolerance and (iii) mRNA expression level of three important drought responsive genes, *acdS*, *KatA* and *gbsA* in the stress tolerant isolate by real time quantitative polymerase chain reaction (qPCR).

Materials and Methods

Soil sampling and isolation of bacteria

Rhizosphere adhering soil samples were collected from ten different locations of green gram cultivating areas of Jorhat district of Assam, located in 26.75°N and 94.22°E of North East India. Sampling sites were selected based on minimal annual precipitation, *i.e* mainly drought prone areas. Sampling was carried out during the month of October (vegetative growth phase) and February (reproductive growth phase), 2011–2012. Soils were clay loam in texture with pH of 3.5 to 4. The soil samples from each location were combined and passed through 0.2 cm sieve and preserved at 4°C until use. A total of 120 fluorescent pseudomonad colonies were obtained upon growth on King's B agar (KB) and *Pseudomonas* isolation agar (Hi Media, Mumbai, India) medium by incubating at 30±2°C for 24 hours The isolates were stored in 20% glycerol stock at −80°C until use.

Ethics statement

Since the fields were public agricultural land; therefore, no further specific permission was required for obtaining samples from these locations.

Microbial strains

Fungal pathogens *Fusarium oxysporum* f. sp. *raphani* (FoRN5), *Fusarium oxysporum* f. sp. *ciceri* (FocRs9), *Fusarium semitectum* (FsNJ9) and *Rhizoctonia solani* (RsNJ10) were obtained from the Culture Bank of Biotechnology Division, North-East Institute of Science Technology, Jorhat, Assam, India.

Morphological and biochemical characterization

Isolates were gram stained and examined under light microscope. Biochemical characterization, *viz* fluorescent pigments, motility, nitrate reduction, catalase, oxidase, methyl red, starch hydrolysis, nitrate reduction and gelatin liquification tests were carried out with five replications as described in Bergey's manual of determinative bacteriology [23].

Genotypic Analysis

Genomic DNA of 120 fluorescent pseudomonads were extracted by GenElute Bacterial Genomic DNA Extraction kit (Sigma, USA), following the Manufacturer's protocol. The DNA purity and quantity were checked by spectrophotometer at 260 and 280 nm. The genotypic analysis of 120 *Pseudomonas* strains were carried out by rep PCR using BOX-AIR1 primer (5′CTACGG-CAAGGCGACGCTGACG3′) as described by Louws et al. 1994 [24] as well as ERIC F (5′AAGTAAGTGACTGGGGT-GAGCG3′) and ERIC R (5′TGTAAGCTCCTGGGGATT-CAC) as mentioned earlier by de Bruijn, 1992 [25] with three repetitions. A 10 μl PCR product together with 500 bp DNA marker (Bangalore Genei, Banglore, India) was separated on a 1.5% agarose gel stained with ethidium bromide in 1x TAE. A snapshot of the gel was taken by gel documentation system (UVP BioImaging system, Upland, California, USA) and stored as TIFF file for further analysis.

Amplified ribosomal DNA restriction analysis (ARDRA)

Amplification of *16S rRNA* region was performed by using bacterial universal primers P^A (5′AGAGTTTGATCCTGGC-TAG3′) and rP2 (5′ACGGCTACCTTGTTACGACTT3′) as described earlier by Edwards et al. 1989 [26]. Twenty microliters (100 ng) of purified *16S rRNA* PCR products were digested for 2 h with 1.5 U of *Hae*III, *Alu*I and *Msp*I restriction endonucleases respectively, as recommended by the manufacturer (Banglore Genei, Banglore, India). The restriction fragments were analyzed on a 2.5% agarose gel in 1X TAE electrophoresis buffer containing 10 μg ml^{-1} ethidium bromide and run at 40 V for 3 h. Whole experiment was repeated thrice to avoid any experimental error.

Ribosomal intergenic space analysis (RISA)

The 16S–23S rDNA intergenic spacer region was amplified with universal primers G1 (5′GAAGTCGTAACAAGG3′) and L1 (5′CAAGGCATCCACCGT3′), as reported by Jensen et al. 1993 [27] with three repetitions. Purified amplicons (100 ng) were digested with single tetra cutter, *Msp*I. The products were separated on a (1.5%) agarose gel and run for 3 h at 40 V. The Gel was documented and stored as TIFF file for further analysis.

Molecular Phylogenetic Analysis

Out of 120 bacterial isolates, 85 distantly related isolates were selected on the basis of clustering by rep PCR, RISA and ARDRA analysis. The clustering methods are described later. Furthermore, PCR amplified *16S rRNA* gene from the bacterial isolates was purified using Wizard PCR Preps (Promega, Madison, WI, USA) and then sequenced with an Applied Biosystems 310 automatic sequencer (Foster City, CA, USA). The ABI Prism dye terminator sequencing kits were used with the same primers used in *16S rRNA* amplification. Edited sequences were submitted in NCBI GenBank and accession numbers were obtained for the same. The reference *16S rRNA* gene sequences (*P. aeruginosa* NR 026078, *P. otitidis* NR 043289, *P. fulva* NR 040859, *P. monteilii* NR 024910, *P geniculata* NR024708) of NCBI GenBank along with our own sequenced bacterial isolates were subjected for phylogenetic inference.

In vitro Screening for Antimicrobial Activity

Fluorescent pseudomonad isolates were tested for *in vitro* antagonism towards phyto pathogenic fungi, i.e. *F. oxysporum* f. sp. *raphani*, *F. oxysporum* f. sp. *ciceri*, *F. semitectum* and *R. solani* through agar well diffusion assay on potato dextrose agar [28] with three replicates for each bacterium. The minimum inhibitory concentration (MIC), *i.e.* the lowest concentration of the bacterial secondary metabolite was determined through batch cultures containing different volumes of 48 h old crude bacterial supernatant (1×10^9 cfu ml^{-1}) against all the phytopathogens (1×10^6 conidia ml^{-1}). Thus 60 µl was calculated to be the average MIC and loaded on to wells (5 mm diameter) of PDA plates pre inoculated with fungal spore suspension (1×10^6 conidia ml^{-1}). Assay plates were incubated at 28°C for 3 days and inhibition zone were recorded.

Screening for Antimicrobial Traits

Assay for HCN. Quantitative assay for HCN production was carried out as described earlier by Kremer and Souissi, 2001 [29] with slight modification. Whatman No. 1 filter paper strips of 9 cm long and 0.5 cm wide, pre-equilibrated with alkaline picrate solution (0.5% picric acid in 2% sodium carbonate) were placed inside the conical flasks with bacterial culture in overhang position and incubated in a rotary shaker at 30°C for 48 h. Filter papers with color change from orange to yellow were removed, extracted by 5 ml 1.0 M NaOH and titrated with 4.25 ml acetic acid. Extracted cyanide into NaOH was further allowed to react with barbituric acid - pyridine reagent and absorbance was read at 575 nm. HCN was quantified as nmoles mg^{-1} cellular protein. The experiment was repeated thrice for each bacterium.

Chitinase Assay. The fluorescent pseudomonad isolates were grown in 100 ml of colloidal chitin supplemented peptone medium (colloidal chitin 0.2%, glucose 0.5%, peptone 0.2%, K_2HPO_4 0.1%, $MgSO_4 7H_2O$ 0.05% and NaCl 0.05%, pH 6.8) at 28±1°C. Bacterial cultures were centrifuged at 12000×g for 15 minutes. Optical density of the supernatants was read at 280 nm (Specord 200, Analytik Jena, Germany) taking N-acetylglucosamine (GlcNac) as a standard. The enzyme activity was expressed as nmolGlcNac produced min^{-1} ml^{-1} considering uninoculated medium as blank with three replications. A previously reported pronounced chitinase producer *Streptomyces roseochromogenus* TSR12 [30] was used as positive control (Table 1).

Siderophore Production. Production of siderophore was screened through the chrome azurol S agar (CAS) assay [31]. The hydroxamate nature of siderophore was further detected by Neilands spectrophotometric assay [32]. Quantification was done as described earlier by Reeves et al. 1983 [33]. The absorbance for dihydroxyl phenols was read in a spectrophotometer (Specord 200, Analytik Jena, Germany) at 700 nm. A standard curve was drawn with dihydroxy benzoic acid, and the quantity of siderophore synthesized was expressed as µmol of benzoic acid per ml of culture filtrate. The whole experiments, i.e. CAS assay as well as quantitative estimation was repeated individually for three times to avoid experimental errors.

Assay for in vitro salicylic acid (SA) production. Apart from the above antimicrobial traits, SA was estimated to study the indirect mechanism for disease suppression by the *Pseudomonas* isolates. SA triggers plant defense responses by stimulating induced systemic resistance (ISR) in the host plant. Forty eight hours old bacterial culture grown in King's B broth (KB) was centrifuged at 2800×*g* for 20 minutes at 4°C. Supernatants were acidified to pH 2.0 using 1 N HCl and filtered through nylon membrane under vacuum. Filtrates were partitioned twice with chloroform and ultimately dried under nitrogen steam at 40°C. Samples were analyzed (µg SA ml^{-1} bacterial culture) using HPLC (Waters, Milford, USA) after dissolving them in 23% methanol in 20 mM sodium acetate buffer of pH 5.0 [34]. The whole experiment was carried out with three replications.

Detection of Pseudomonas Antibiotic genes. Since, the antibiotics 2, 4-diacetylphloroglucinol (DAPG), pyoluteorin (PLT) and different phenazine (Phz) derivatives have been described in biocontrol *Pseudomonas* spp. as the main cause of their antagonistic activity [35,36], we further screened for their responsible genes within my antagonistically potential isolates. PCR detection for *DAPG*, phenazine-1-carboxylic acid (*PCA*) and *PLT* was performed as mentioned by Mavrodi et al. 2001 and Raaijmakers et al. 1997 [37,38]. Further, extraction and phenotypic assays of 2, 4-DAPG and phenazine were carried out as mentioned by earlier workers [10,39,40]. PLT was extracted as described by Sarniguet et al. 1995 [41]. TLC was carried out on silica gel G60 to purify the extracted antibiotics. Activated plates were developed at 110°C for 20 minutes were spotted with an ethanolic solution of standard antibiotic (0.5 µg) and 20 µl of the extract running the samples using different mobile phases. Chloroform-methanol 9:1 (v/v) solvent system was used for DAPG and PCA; however for PLT, chloroform-acetone 9:1 (v/v) solvent system was used. The corresponding spots by PCA, and DAPG were detected by UV irradiation at 254 nm [21]. PLT spots were detected by spraying with an aqueous 0.5% (w/v) Fast Blue RR salt solution (Sigma Aldrich, USA). Further, antifungal activity of the purified compounds was performed by agar well diffusion assay [28].

Quantitative Assay of Plant Growth Promoting Traits

Estimation of indoles. Pure bacterial cultures were inoculated in DF salts minimal medium [42] with L-tryptophan of different concentration (0, 50, 100, 200 and 500 µg ml^{-1}). Bacterial cultures were grown for 48 hours and harvested by centrifugation (4,000×*g* for 20 min at 4°C). Screening of indole production was carried out by mixing the supernatant with Salkowski's reagent (50 ml, 35% perchloric acid and 1 ml 0.5 M $FeCl_3$) in the ratio of 1:4 (supernatant: reagent) at room temperature (28°C) for 20 minutes. Development of pink color indicated production of indoles. Indole production was quantified by spectrophotometric absorption (Specord 200, Analytik Jena, Germany) at 535 nm with three replications [43]. Standard curve was prepared by using pure IAA (Sigma Aldrich, USA).

ACC deaminase Production. Fluorescent pseudomonads were grown in DF salts minimal medium [42] supplemented with 10 µg of ACC (Sigma-Aldrich, USA). ACC deaminase activity was quantified colorimetrically (Specord 200, Analytik Jena, Germany) as discussed earlier and expressed as α-ketobutyrate produced mg of $protein^{-1}$ h^{-1} [44]. The whole experiment was carried out in three replicates.

Quantitative Estimation of Soluble Phosphate. Efficiency of the fluorescent pseudomonads as phosphate solubilizer (PSB) was screened by standard protocol using modified Pikovskaya agar (PKA) medium (gl^{-1} 0.5, yeast extract; 10, dextrose; 5, aluminium phosphate; 0.5, ammonium sulphate; 0.2, potassium chloride; 0.1, magnesium sulphate; 0.0001, manganese sulphate; 0.0001, ferrous sulphate; final pH 3±2) as growth medium. Since use of tricalcium phosphate (TCP) was reported as relatively weak and unreliable factor for isolation of phosphate solubilizing bacteria; for isolation of PSBs from acidic soil of Assam, aluminium phosphate ($AlPO_4$) was used replacing TCP in PKA [45]. Clear zone was observed around the bacterial colony after 5 days of incubation at 30°C and portrayed phosphate solubilization by the organism. Quantification of phosphatase activity was carried out as mentioned by Fiske and Subbarow, 1925 with three replications [46].

Green house experiment for biocontrol activity. In vivo biocontrol activity of *P. aeruginosa* GGRJ21 againt *Rhizoctonia solani* was performed as described earlier by Saikia et al. 2011 [47]. *Rhizoctonia solani* is the causal organisms of root rot disease of green gram. Soil from green gram cultivated area was collected and autoclaved. Earthen pots were filled up with 5 kg of sterilized

Table 1. Quantitative estimation of Antimicrobial and plant growth promoting traits of fluorescent pseudomonads.

Fluorescent pseudomonads isolates	salicylic acid ($\mu g\ ml^{-1}$)	HCN (nmoles mg cellular protein^{-1})	Chitinase (nmol GlcNac $min^{-1}\ ml^{-1}$)	2,4 DAPG ($ng\ ml^{-1}$)	PCA (absorbance at 367 nm)	siderophore (µmol benzoic acid ml^{-1})	Indole ($\mu g\ ml^{-1}$) At 100 µg ml^{-1} tryptophan	ACC deaminase (µmol α ketobutyrate mg $protein^{-1}\ h^{-1}$)	Phosphate solubilization ($\mu g\ ml^{-1}$)
GGRJ1	4.093 ± 0.21^{f}	22.9 ± 0.81^{b}	46.09 ± 1.23^{b}	-	-	4.51 ± 0.55^{e}	263.33 ± 1.52^{e}	-	-
GGRJ2	-	-	-	-	-	-	-	-	4.20 ± 0.11^{d}
GGRJ5	-	-	-	-	-	-	328.63 ± 1.49^{d}	8.32 ± 1.09^{b}	58.84 ± 0.74^{b}
GGRJ7	-	-	-	-	-	-	192 ± 2.51^{f}	-	-
GGRJ9	-	-	-	-	-	-	158.43 ± 2.37^{ab}	-	-
GGRJ12	-	-	-	-	-	-	282.46 ± 2.37^{e}	9.51 ± 0.88^{b}	-
GGRJ14	11.21 ± 1.07^{d}	19.21 ± 0.21^{b}	22.77 ± 2.20^{c}	0.64 ± 0.09^{b}	6.03 ± 0.35^{b}	2.62 ± 0.57^{e}	466.9 ± 1.41^{b}	-	-
GGRJ15	-	-	-	-	-	-	$86.8\pm1.36b^{c}$	12.28 ± 1.22^{a}	-
GGRJ18	-	-	-	-	-	-	$97.13\pm1.01b^{c}$	5.51 ± 0.95^{c}	-
GGRJ19	-	-	-	-	-	-	574.54 ± 1.67^{a}	-	-
GGRJ20	9.19 ± 0.67^{d}	18.37 ± 0.95^{b}	42.70 ± 1.54^{b}	-	-	9.48 ± 0.57^{c}	-	-	-
GGRJ21	20.50 ± 0.62^{a}	30.76 ± 1.07^{a}	69.33 ± 1.52^{a}	0.863 ± 0.06^{a}	6.8 ± 0.49^{a}	17.22 ± 0.33^{a}	591.14 ± 1.06^{a}	14.21 ± 0.41^{a}	88.4 ± 1.50^{a}
GGRJ22	10.43 ± 1.56^{d}	7.27 ± 0.13^{c}	-	0.256 ± 0.05^{d}	1.43 ± 0.4^{d}	5.35 ± 0.25^{d}	-	-	-
GGRJ23	4.16 ± 0.80^{f}	3.80 ± 0.96^{d}	6.56 ± 1.07^{e}	0.616 ± 0.05^{b}	5.63 ± 0.4^{b}	3.23 ± 0.75^{e}	$75.34\pm1.02b^{c}$	-	-
GGRJ24	-	-	-	-	-	-	336.5 ± 1.34^{d}	-	-
GGRJ25	7.09 ± 0.58^{e}	15.58 ± 0.73^{b}	50.44 ± 1.99^{b}	0.816 ± 0.06^{a}	-	6.53 ± 0.21^{d}	437.2 ± 2.15^{c}	-	-
GGRJ27	12.14 ± 1.06^{d}	17.43 ± 0.51^{b}	15.94 ± 1.20^{d}	0.5 ± 0.01^{c}	-	3.25 ± 0.25^{e}	483.9 ± 1.13^{b}	-	-
GGRJ29	-	-	-	-	-	-	585.13 ± 1.8^{a}	-	-
GGRJ30	14.48 ± 1.63^{c}	7.98 ± 1.12^{c}	-	0.34 ± 0.04^{d}	2.56 ± 0.55^{c}	13.81 ± 0.70^{b}	-	-	66.66 ± 1.52^{b}
GGRJ31	-	-	-	-	-	-	262.3 ± 1.67^{e}	-	-
GGRJ33	16.94 ± 0.65^{b}	14.61 ± 0.85^{b}	-	0.27 ± 0.07^{d}	-	14.77 ± 1.15^{b}	213.9 ± 2.56^{f}	-	-
GGRJ34	3.18 ± 0.23^{f}	16.35 ± 1.34^{b}	3.40 ± 0.61^{e}	-	-	7.10 ± 0.18^{d}	-	-	-
GGRJ35	5.27 ± 1.62^{e}	9.52 ± 0.56^{c}	-	0.16 ± 0.05^{e}	1.56 ± 0.4^{d}	5.54 ± 0.89^{d}	145.6 ± 2.25^{ab}	-	46.13 ± 0.96^{c}
GGRJ36	15.97 ± 0.36^{b}	26.39 ± 0.82^{a}	41.27 ± 1.12^{b}	0.47 ± 0.06^{c}	3.37 ± 0.41^{c}	2.39 ± 0.16^{e}	183.77 ± 1.63^{f}	-	-
GGRJ39	-	-	-	-	-	-	312.94 ± 2.10^{d}	-	-
GGRJ42	-	-	-	-	-	-	436.9 ± 1.65^{c}	-	-
GGRJ43	-	-	-	-	-	-	326.83 ± 1.68^{d}	-	-
GGRJ46	2.58 ± 0.73^{f}	5.66 ± 1.07^{d}	-	0.21 ± 0.06^{e}	-	3.86 ± 0.22^{e}		-	-
GGRJ51	-	-	-	-	-	-	455.56 ± 2.65^{b}		-
GGRJ52	-	-	-	-	-	-	-	-	34 ± 2^{c}
GGRJ62	13.4 ± 0.69^{c}	3.58 ± 0.52^{d}	9.43 ± 0.94^{e}	0.2 ± 0.04^{e}	-	4.43 ± 0.28^{e}	-	-	-
GGRJ63	16.94 ± 1.42^{b}	6.47 ± 0.76^{d}	7.78 ± 0.61^{e}	-	-	11.61 ± 0.81^{b}	-	-	43.76 ± 1.56^{c}
GGRJ66	3.88 ± 0.39^{f}	12.9 ± 0.64^{c}	35.48 ± 1.07^{c}	0.35 ± 0.08^{d}	-	8.89 ± 0.11^{c}	-	-	-

Table 1. Cont.

Fluorescent pseudomonads isolates	salicylic acid (μg ml^{-1})	HCN (nmoles mg cellular protein^{-1})	Chitinase (nmol GlcNac min^{-1} ml^{-1})	2,4 DAPG (ng ml^{-1})	PCA (absorbance at 367 nm)	siderophore (μmol benzoic acid ml^{-1})	Indole (μg ml^{-1}) At 100 μg ml^{-1} tryptophan	ACC deaminase (μmol α ketobutyrate mg protein^{-1} h^{-1})	Phosphate solubilization (μg ml^{-1})
GGRJ67	16.87±0.50^{b}	10.48±1.15^{c}	-	-	-	13.29±0.51^{b}	-	-	-
GGRJ68	3.89±0.40^{f}	2.16±0.35^{e}	-	-	-	6.05±0.75^{d}	-	-	14±1.73^{d}
GGRJ70	17.7±1.05^{b}	26.28±1.14^{a}	30.80±1.40c	-	-	4.21±0.94^{e}	-	-	-
GGRJ71	-	-	-	-	-	-	-	-	12.11±0.40^{d}
GGRJ76	-	-	-	-	-	-	-	-	35.68±0.35^{c}
KFP1	6.25±0.96^{e}	0.63±0.09^{e}	-	0.56±0.12^{b}	-	5.73±0.24^{d}	429.5±3.1^{c}	-	-
KFP2	4.88±0.79^{e}	4.69±0.56^{d}	-	-	-	11.31±0.85^{b}	423.86±2.4^{c}	-	-
KFP3	13.13±1.10^{c}	1.91±0.23^{e}	-	-	-	12.88±1.24^{b}	-	-	-
KFP7	16.24±0.85^{b}	5.90±1.1^{d}	-	-	-	14.35±0.78^{b}	-	-	-
Streptomyces roseochromogenus TSR12			72.13±1.28^{a}						

Means within a column sharing same superscript are not significantly different according to Turkey's test at p = 0.05; ± means standard deviation (SD); - means no activity.

soil. Green gram (*var K851*) seeds were surface sterilized with 2% sodium hypochlorite for 30 s, subsequently rinsed with sterile distilled water (SDW) and dried with sterilized air stream. Ten seeds per pot were sown in the previously prepared pots. Twenty days old plants were considered for in vivo experiments. The whole experiment was carried out with six different experimental conditions i.e. (i) pathogen alone (2×10^{2} spore ml^{-1}, 5 ml pot^{-1}), (ii) GGRJ21 alone (1×10^{8} CFU ml^{-1}, 5 ml pot^{-1}), (iii) simultaneous inoculation of pathogen plus GGRJ21, (iv) pre-inoculation of GGRJ21 and then pathogen inoculatation after 2 days (v) post inoculation of GGRJ21 after 2 days of pathogen inoculation and (vi) control (sterile distilled water treatment). Bacterial as well as the pathogen inoculums were prepared as mentioned by Saikia et al. 2011 [47]. Experiment was carried out as complete randomized block design (CRD) with 10 replications. Five microlitres of fungal and bacterial inoculums were applied to each pot. Disease severity was assessed 21 days post inoculation (either pathogen or GGRJ21). The severity of root rot was visually scored by assessing necrotic lesions on the roots and hypocotyls using a rating scale of 0–5 described earlier by Filion et al. 2003 [48].

Bacterial growth under water stress condition. Bacterial growth was monitored in 250 ml of nutrient broth (NB) (gl^{-1} 5, peptic digest of animal tissue; 5, sodium chloride; 1.5, beef extract; 1.5, yeast extract; final pH 7.4±2) medium with different water potentials (−0.05, −0.15, −0.30, −0.49, and −0.73 M Pa). The desired osmotic stress condition was developed in the growth medium by adding appropriate amounts of polyethylene glycol (PEG 6000) [49]. One millilitre of overnight grown culture (1×10^{9} CFU ml^{-1}) was inoculated to the prepared NB and incubated at 30°C for 24 hours with continuous agitation of 120 rpm. Bacterial growth kinetics was further recorded colorimetrically (Specord 200, Analytik Jena, Germany) by measuring absorption (600 nm) as a function of time with three replications for each bacterium.

RNA Isolation and Two Steps Real Time PCR

Fluorescent *Pseudomonas* isolate GGRJ21 was grown under different osmotic conditions (−0.30, −0.49, and −0.73 MPa) [49]. Bacterial cells were harvested in log phase (OD approx. 2.10 at 600 nm). Bacteria grown in normal condition in NB medium were taken as control. Total RNA was isolated from $\sim1\times10^{9}$ cells using Geneipure^{TM} Bacterial total RNA isolation kit (Genei, Bangalore, India) as per the manufacturer instruction. One microgram of each RNA sample was reverse transcribed to cDNA with 2X Verso cDNA synthesis kit (Thermo Scientific, USA) using random hexamers. Quantitative amplification reactions of cDNAs from reference genes and target genes were carried out on StepOnePlus^{TM} Real Time PCR System (Applied Biosystems, USA) using Thermo DYNAMO^{TM} 4C SYBR Green qPCR Kit (Thermo scientific, USA). The reaction conditions were set as follows: 10 min at 42°C; 10 min at 95°C; 40 cycles of cDNA amplification for 15 s at 95°C, 30 s at 60°C, 30 s at 72°C with fluorescent signal recording. At the end, a final step of 15 s at 95°C, 1 min at 60°C and fluorescence measured at each 0.7°C variation (from 60°C to 95°C) was included to obtain the melting curve. Four reference genes; *gyrA* (DNA gyrase subunit A), *gmk* (guanylate kinase), sigma factor RpoD (*rpoD*) and *16S rRNA* was selected for normalization of real time PCR reaction. Bacterial osmotic stress responsive genes, *i.e., acdS* (encoding ACC deaminase), *katA* (encoding for catalase) and *gbsA* (encoding for glycine betain) were selected as target genes. The sequences of the genes studied were obtained from NCBI GenBank and the primers were designed with the aid of the OLIGO software (version 5.0; Molecular Biology Insights).

Table 2. Primers for real time PCR: sequences, final concentration, product size, T_m and Efficiency of PCR amplification.

Target	Forward primer (5′-3′)	Reverse primer (5′-3′)	Final concentration (pmol μl^{-1})	Product size (bp)	T_m	PCR efficiency value (E±SD)
gyrA	GACATGATCCCGGAAGAAGA	CAGGTGGGCAATGTAGTCCT	10	150	59	1.963±0.042
gmk	CACGACCTGATCCTGGAAAT	TCAATGATCTCGTCGCTGTC	10	152	59	1.857±0.052
rpoD	CGAAACGATCAACAAGCTCA	TCACCCAGATGGGAGTCTTC	10	192	60	1.923±0.038
16S rRNA	CTCTAAGGAGACTGCCGGTG	CGGACTACGATCGGTTTTGT	10	158	60	2.211±0.056
acdS	GCCTCTTCTTCGATACCGTG	AATCATCTCGGCGGTGTTAC	10	179	59	2.012±0.049
katA	AAGACCGATATGTTCCTGCG	GTTGAGATCGGGGAACTTGA	10	180	60	1.977±0.044
gbsA	GTGCTGACCCTGCAGGTATT	CCTCGAATTCGTCGTAGACC	10	184	60	1.983±0.051

T_m means melting temperature; ±SD means standard deviation of the mean value.

The sequences and other properties of the primers are shown in Table 2. Triplicate reaction was maintained for each gene.

Data Analysis

ARDRA as well as RISA results were analyzed by considering, the character state "1" for clearly detected bands in the gel track and assigned "0" if it was absent or impossible to determine. The data matrix was generated by Jaccard's similarity coefficient algorithm. Each pair-wise comparison was constructed from the similarity matrix by the un-weighted pair group method with arithmetic mean (UPGMA). For rep-PCR (BOX-PCR and ERIC-PCR) fingerprinting analysis, the photographs were imported into the software package BioNumerics version 2.5 (Applied Maths, Belgium). Similarity matrices from densitometric curves of the gel tracks were calculated using the Pearson's product moment correlation coefficient followed by dendrogram construction using UPGMA algorithm. The *16S rRNA* gene sequences of our isolates along with their closest homolog were aligned in Clustal W, where a total ninety (*16S rRNA* of 85 bacterial isolates from the present study and 5 reference sequences from GenBank) sequences were considered. Both distance-based and character-based method was used for inferring the phylogenetic relationship among various FPs. For distance based method Neighbor-Joining (NJ) algorithm was employed whereas for character-based method Unweighted Pair Group with Arithmetic Mean (UPGMA) algorithm was employed. In both the cases Kimura-2-parameter substitution model [50] was employed. The robustness of the inferred phylogenetic trees was tested by bootstrap analysis [51] with 1000 iterations of the original dataset. The triplicate data generated during quantitative evaluation of antimicrobial, and PGPR traits were analyzed by means of one-way ANOVA and means were compared by the Tukey's test, using the SPSS software (ver. 10.1, SPSS Inc., www.spss.com) at the significance level $p = 0.05$. The sequences of isolates in the present study were submitted to GenBank of NCBI and obtained accession numbers. During quantitative reverse transcription PCR, tenfold serial dilution of cDNA curves were produced to calculate the amplification efficiency for all genes through the equation $E = 10^{(-1/\text{slope})}$ [52]. Threshold cycle (C_T) was compared with $\log_{10}$ relative copy number of the sample from a dilution series. Normalization of PCR reaction with four different reference genes was performed. Relative expression level obtained for target genes were compared when three candidate normalizer genes were used individually. Then, the best combination was obtained by geNorm software [53]. Expression levels were determined as the number of cycles needed for the amplification, to reach a threshold fixed in the exponential phase of PCR reaction (C_T) [54]. C_T values from the ABI Step one plus System (Applied Biosystems, Foster City, California, USA) were analyzed through $2^{-\Delta\Delta C_T}$ method, where $\Delta\Delta C_T = \Delta C_T\text{sample} - \Delta C_T\text{control}$ [55]. The green house experimental data were analyzed by LSD test at $P<0.05$ using Duncan's multiple range tests.

Results

Isolation and Phenotypic Characterization of the pseudomonad isolates

Aerobic incubation of soil suspension from lower dilution series in different *Pseudomonas* specific media for 24 hrs at 30±2°C resulted in 437 colonies of distinct morphotypes. The initial screening through UV light ($\lambda = 356$ nm) revealed that only 120 were fluorescent *Pseudomonas* out of total 437 bacterial isolates. Microscopic observation showed that all the isolates were rod shaped, motile and gram negative. All the isolates showed positive reaction for catalase, oxidase, and gelatine liquefaction. The biochemical characteristics of the isolates are summarized in Table S1.

Rep PCR Fingerprinting

The BOX PCR fingerprinting revealed a banding pattern from 200 to 5000 bp, similarly for ERIC-PCR, the molecular weights of the amplified products were estimated at approximately 100–5500 bp among the 120 pseudomonad isolates. A total of 25 unique rep-PCR fingerprints were detected among the strains isolated during the vegetative-growth as well as reproductive phase of the crop (Fig. S1). The dendrogram showed two major clusters; one large (Cluster-I) and one small (Cluster-II) (Fig. S2). Both the clusters attained a similarity coefficient value of 75%. Further, cluster I is sub-divided into nine distinct sub-clusters. The sub-cluster-Ia is formed of 3 isolates with a coefficient value of 88%, whereas its closest sub-cluster-Ib is comprised of 54 isolates shared approximately 95% similarity coefficient. Sub-cluster-If with 48 members constituted the next large group after Ib. Most of the members of sub cluster Ib were isolated during the reproductive phase of the pulse crop. All the members of sub-cluster-Id and If were obtained while the crop was in vegetative stage. The isolates from GGRJ86-GGRJ120 (subcluster If) were 100% identical to each other.

Amplified Ribosomal DNA Restriction Analysis (ARDRA)

Restriction digestion profiles of *16S rRNA* amplicons with *Hae*III, *Alu*I and *Msp*I on 2% agarose gels were compared to

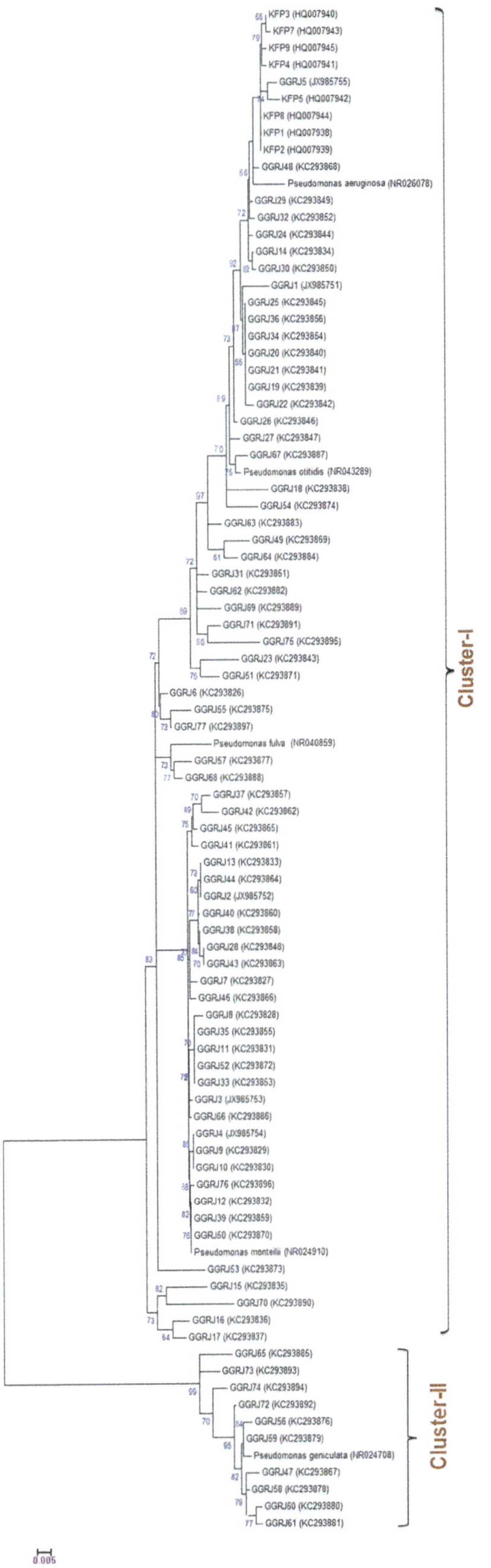

Figure 1. Phylogenetic analyses of fluorescent pseudomonads based on the nucleotide sequence of *16S rRNA*. The multiple sequence alignment was done in CLUSTALW program embedded in MEGA version 5.10. The pair-wise evolutionary distances were calculated using Kimura-2 parameter model. The phylogenetic tree was constructed by Neighbor-Joining (NJ) method with 1000 replicates using bootstrap. A total of 5 reference fluorescent pseudomonad strains were used for the tree construction. Bar, .0.005 shows the substitutions per nucleotide position.

avoid redundancy among the strains. According to the type of restriction enzyme, the *16S rRNA* fragment patterns were different, and showed notable genetic differences within the microbial communities. *Hae*III showed 17 restriction patterns; whereas *Alu*I and *Msp*I revealed 9 and 20 restriction patterns, respectively (Fig. S3a). Based on ARDRA fingerprinting with three endonucleases, dendrogram was constructed. The resulted dendrogram can be divided into two major clusters (Fig. S4). The large cluster-I formed three distinct sub-clusters (Ia, Ib and Ic). The sub-cluster Ic comprises of maximum number of isolates sharing the similarity coefficient of 76%. All the members from Ic were isolated from reproductive growth phase of green gram, representing high bacterial diversity. Interestingly, two pseudomonad strains GGRJ1 and GGRJ2 isolated during the reproductive phase of the pulse were out-clustered from the other and grouped in a separate sub-cluster Ia. The sub-cluster Ib with 24 numbers of FP shared 78% similarity with rest of the bacterial strains. Most of the isolates of sub-cluster Ib with were associated with reproductive phase of the pulse crop. Most of the members from sub-cluster Ic representing poor genetic diversity were related to vegetative phase of the crop. Pseudomonad isolates from GGRJ86 to GGRJ120 were under Cluster II and shared closest genetic similarity that correlates with the findings from rep PCR genotyping. Majority of isolates from II were associated with vegetative phase of the pulse crop.

Ribosomal Intergenic Space Analysis (RISA)

Comparative analysis of the 16S–23S rRNA intergenic space indicated a higher degree of genetic diversity among the isolates (Fig. S3b and S5). The dendrogram showed the formation of two large clusters at 70% similarity coefficient. Cluster I can be divided to two sub-clusters (Ia and Ib). Both the sub-clusters Ia and Ib included almost equal number of isolates. All the members of sub-cluster Ib were associated with vegetative growth phase of green gram showed close affinity among the isolates from GGRJ86 to GGRJ120. Cluster II seemed to be somehow different from cluster I comprising 17 isolates with three distinct sub-clusters depicted a significant genetic diversity among them.

Molecular Phylogenetic Analysis

Based on the genotypic analysis, 35 closest fluorescent pseudomonads (GGRJ86 to GGRJ120) were excluded from further molecular phylogenetic study. All of them showed close affinity to *P. aeruginosa*. *16S rRNA* sequence of rest 85 isolates was compared against NCBI nucleotide database and confirmed the presence of 23 different species of pseudomonads (Table S1). Moreover, it was found that most of the isolates shared highest percentage of sequence identity with that of *P. aeruginosa*. Phylogenetic relatedness among the 85 FP isolates was established through NJ and UPGMA method. For proper comparison, five reference sequences from NCBI genebank database were used during the study. Both NJ and UPGMA methods yielded same type of tree topology with two major clusters (Fig. 1 and Fig. S6). As evidenced from the Fig. 1 that the larger cluster *i.e.*, cluster-I consist of 75 numbers of isolates with three reference bacteria (*i.e.*, *P. aeruginosa* strain DSM 50071 (NR026078); *P. monteilii* strain

Table 3. Antagonostic activity of fluorescent pseudomonads against phytopathogenic fungi. Activity was monitored on the basis of inhibition zone.

Sl. No.	*Pseudomonas* isolates	Zone of inhibition (ZOI) (mm)			
		R. solani	*F. oxysporum f. sp. raphani*	*F. oxysporum f. sp. ciceri*	*F. semitectum*
1	GGRJ1	17 ± 1^{a}	12 ± 1^{a}	10 ± 0.76^{a}	-
2	GGRJ14	12 ± 1.21^{b}	13 ± 1.56^{a}	-	10 ± 0.77^{a}
3	GGRJ20	-	17 ± 0.65^{b}	14 ± 1.2^{b}	12 ± 1^{a}
4	GGRJ21	23 ± 0.83^{c}	19 ± 1^{b}	21 ± 1.11^{c}	16 ± 1.42^{b}
5	GGRJ22	-	-	-	12 ± 0.43^{a}
6	GGRJ23	17 ± 1^{a}	-	14 ± 0.95^{b}	-
7	GGRJ25	14 ± 1.54^{b}	9 ± 1.41^{c}	-	15 ± 1.3^{b}
8	GGRJ27	16 ± 1^{a}	-	14 ± 1^{b}	8 ± 0.2^{c}
9	GGRJ30	10 ± 1.1^{d}	-	-	7 ± 0.76^{c}
10	GGRJ33	-	14 ± 1^{a}	-	7 ± 0.28^{c}
11	GGRJ 34	-	7 ± 0.86^{c}	13 ± 1.13^{b}	-
12	GGRJ35	-	-	-	7 ± 0.65^{c}
13	GGRJ36	-	8 ± 0.78^{c}	20 ± 1.24^{c}	6 ± 0.73^{c}
14	GGRJ46	-	8 ± 1^{c}	-	-
15	GGRJ62	-	-	9 ± 0.76^{a}	-
16	GGRJ63	-	12 ± 0.64^{a}	8 ± 1.21^{d}	-
17	GGRJ66	-	-	18 ± 1.54^{c}	12 ± 1.42^{a}
18	GGRJ67	-	8 ± 1.12^{c}	-	-
19	GGRJ68	-	-	9 ± 0.54^{a}	-
20	GGRJ70	-	7 ± 0.76^{c}	20 ± 1.32^{c}	6 ± 0.76^{c}
21	KFP1	8 ± 1^{d}	14 ± 1^{a}	-	13 ± 1^{a}
22	KFP2	-	8 ± 0.65^{c}	7 ± 1^{d}	15 ± 1.3^{b}
23	KFP3	-	-	6 ± 0.67^{d}	7 ± 0.43^{c}

Means within a column sharing same superscript are not significantly different according to Turkey's test at $p = 0.05$; $\pm$ means standard deviation (SD) and - means no activity.

CIP 104883 (NR024910) and *P. fulva* strain AJ 2129 (NR040859)) at 70% cut off value. The reference strain, *P. aeruginosa* formed largest clusters with its closest relatives representing the major group,. All the *Pseudomonas* isolates of cluster-II conferred their placement in Gamma-β proteobacteria group along with the reference strain (*P. geniculata* strain ATCC 19374 (NR024708). Further, the placement of various clades in distinc phylogenetic position from NJ clustering was compared with UPGMA clustering (Fig. S6) which conferred correct phylogenetic position of each isolate.

Antimicrobial Activity

Bacterial secondary metabolites were tested against the plant pathogenic fungi *Rhizoctonia solani*, *Fusarium oxysporum* f. sp. *raphani*, *F. oxysporum* f. sp. *ciceri* and *F. semitectum*. Out of 120 bacterial isolates, 23 isolates exhibited mixed antagonistic activity against the pathogens. The strain *P. aeruginosa* GGRJ 21 was found to be the most prominent, showing activity against all the fungal pathogens (Table 3). After GGRJ21, antifungal activity of GGRJ1, GGRJ14, GGRJ20, GGRJ25, GGRJ27, GGRJ36, GGRJ70 and KFP2 were found to be promising. GGRJ21 exhibited pronounced antagonistic activity against *R. solani*, one of the important fungal pathogen of green gram with 23 mm of inhibition zone.

Antimicrobial Traits

Variable production of HCN, chitinase, siderophore and salicylic acid by the 23 isolates were determined by standard protocols. GGRJ21 was found to be excellent in production of antimicrobial traits amongst the 23 isolates (Table 1). The intrinsic ability to produce HCN was varied greatly among the antagonistically potential isolates. Values were expressed as nmoles mg^{-1} cellular protein, varied greatly from 0.72 to 30. Extracellular chitinase production was recorded in 13 isolates. Highest chitinase production was monitored in strains GGRJ21, followed by GGRJ25, GGRJ1, GGRJ20 and GGRJ36. Similarly, a distinct variation in siderophore production was monitored among the isolates; where GGRJ21 again proved its superior nature with a production efficacy of 17.2 μmol benzoic acid ml^{-1}. Highest production of salicylic acid by GGRJ21 was calculated as 20.5 μg ml^{-1}, followed by GGRJ70, GGRJ33, GGRJ67 and KFP7 having the production efficiency of 17.7, 16.9, 16.8 and 16.2 μg ml^{-1} respectively.

Detection of the antibiotic coding genes for 2, 4-diacetylphloroglucinol (*DAPG*), phenazine-1-carboxylic acid (*PCA*) and pyoluteorin (*PLT*) from bacterial genomic DNA were conducted by using gene-specific primers within the 23 antagonistically potential *Pseudomonas* isolates. PCR amplification reaction with the nucleotide primers Phl2a and Phl2b revealed amplification of 745-bp fragment of the 1,001-bp *phlD (DAPG)* gene in the 14

Table 4. Effect of *Pseudomonas aeruginosa* GGRJ21 on root rot disease suppression of green gram during infection with *Rhizoctonia solani.*

Treatments	Disease severity		
	Trial 1	Trial 2	Trial 3
Control (sterile distilled water treatment)	0	0	0
Pathogen alone	5.17±0.351[a]	4.93±0.4[a]	5.13±0.55[a]
GGRJ21 alone	0	0	0
Simultaneous inoculation of pathogen plus GGRJ21	2.99±0.11[b]	2.87±0.28[b]	2.81±0.17[b]
Pre-inoculation of GGRJ21 and then pathogen was inoculated after 2 days	1.14±0.15[c]	1±0.2	1.11±0.22[c]
Post inoculation of GGRJ21 after 2 days of pathogen inoculation	3.69±0.16[b]	3.67±0.67[b]	3.51±0.09[b]

Means within a column sharing same superscript are not significantly different according to Turkey's test at p = 0.05; ± means standard deviation (SD).

isolates (GGRJ14, GGRJ21, GGRJ22, GGRJ23, GGRJ25, GGRJ27, GGRJ30, GGRJ33, GGRJ35, GGRJ36, GGRJ46, GGRJ62, GGRJ66 and KFP1). The phenazine biosynthetic gene cluster contains 7 genes, of which five (*phzC-G*) are essential and two others (*phzA* and *B*) substantially enhancing the level of synthesis of phenazine-1-carboxylic acid (PCA). The Primers PCA2a and PCA3b amplified a region of 1150 bp (within *phzC* and *phzD*) in 7 bacterial isolates (GGRJ14, GGRJ21, GGRJ22, GGRJ23, GGRJ30, GGRJ35, and GGRJ36). Pyoluteorin (PLT) biosynthesis linked with the gene cluster of 10 genes (*pltL, pltB, pltC, pltE, pltF, pltG, pltA, pltD, pltM* and *pltR*). PCR screening revealed the presence of *pltB* gene (type I polyketide synthases) in the single pseudomonad isolate, GGRJ21 with a product size of 779 bp. The phenotypical observation showed varied efficiency of 2, 4 - DAPG production among the 14 *phlD* positive isolates. *Pseudomonas* isolate GGRJ21 showed maximum production efficiency (0.863 $ngml^{-1}$) of 2, 4 - DAPG (Table 1). Similarly, varied degree of phenazine production was observed. *Pseudomonas* isolate GGRJ21 was again marked as dominant antibiotic producer than rest of the isolates (Table 1). The purified antibiotic extracts, *i.e. PLT* and *PCA* revealed a significant antibiosis against the phytopathogens except DAPG (Figure S7).

Plant Growth Promoting Traits

Out of 120 numbers of isolates, production of indoles, ACC deaminase and phosphatase activity were found to occur in 25 (20.8%), 5 (4.1%), and 10 (8.3%) numbers of fluorescent isolates respectively (Table 1). GGRJ21 showed higher level of indole production (591.14 μg ml^{-1} at 100 μg ml^{-1} tryptophan concentration) among the isolates. The indole production was recorded maximum at 100 μg/ml tryptophan concentration; while a gradual decrease was recorded at higher tryptophan concentration (Table S2). GGRJ21 was again dominant producer of ACC deaminase and phosphatase with quantitative estimation values of 14.2 μmole α-KBh^{-1} mg protein^{-1} h^{-1} and 88.4 μg ml^{-1} phosphate solubilized.

Green House Experiment

Red lesions on hypocotyls and tap root system of the green gram seedlings clearly indicated the deleterious attack of *R. solani* on the host plant. However, the rate of disease severity was less in the GGRJ21 treated plants as compared to the plant samples with the pathogen alone with significantly (p = 0.05) less number of lesions (Table 4). GGRJ21 suppressed root rot disease of green gram by 28–93%. Control healthy plants without any inoculation and seedlings with GGRJ21 alone did not show any disease symptoms. Disease severity in GGRJ21 pre inoculated seedlings was significantly different (p = 0.05) with low disease rate as compared to other treatment condition.

Bacterial Growth Kinetics under Osmotic Stress

Though ten numbers of isolates from drought prone rhizosphere microhabitat showed growth under osmotic stressed condition; however at high stressed condition (−0.30, −0.49, and −0.73 M Pa) only *Pseudomonas* isolate GGRJ21 showed vigorous growth among them (Table S3). Twenty hours observation under osmotic stressed condition clearly emphasized the capability of GGRJ21 to survive under extreme stressed condition.

Relative Quantification of Stress Responsive Genes

Among the four housekeeping genes; *gyrA*, *gmk*, *rpoD* and *16S rRNA; 16S rRNA* was verified to have the lowest average expression stability (M) when samples experiencing osmotic stress were analyzed (data not shown). The relative expression level of *acdS*, *katA* and *gbsA* gene transcripts in GGRJ21 were analyzed over four treatment condition; A, B, C, D (where, A-GGRJ 21 grown in normal NB medium, B-GGRJ21 grown in -0.3 mPA, C-GGRJ21 grown in 0.49 mPA and D-GGRJ21 grown in -0.73 mPA). Sample A was used as calibrator for all the three experiments. Thus the C_T value obtained in sample A was taken as the control value in order to calculate the fold change in gene expression over each of the three samples.

The transcripts of the *acdS*, *katA* and *gbsA* gene from bacterial mass growing under different conditions were detected. Transcript copy numbers of each gene were found to be in increasing order for bacterial samples growing in lower to higher osmotic stressed condition (Fig. 2a, b, c). Thus *acdS*, *katA* and *gbsA* showed significant up-regulation in the bacterial cells after a gradual increase of osmotic stress from −0.3 mPA to −0.73 mPA. Almost 3 fold increase in expression level of *acdS* was noticed, which was the best in comparison to the other two target genes, *i.e.*, *katA* and *gbsA*. Moreover, highest transcript copy numbers for each case was recorded for cell cultures growing under −0.73 mPA. Overall, the transcriptional activity of the three target genes up regulated over a function of increasing osmotic stress condition. A linear relationship was obtained between threshold cycles (C_T) and the log copy number of cDNA for all genes with an correlation coefficient (R^2) ranging from 0.96 to 0.99, indicating that C_T values changed proportionally to the serial dilution of the samples. The E value, within the range 1.857 to 2.211, indicated the efficient amplification near the theoretical optimum level of 2.

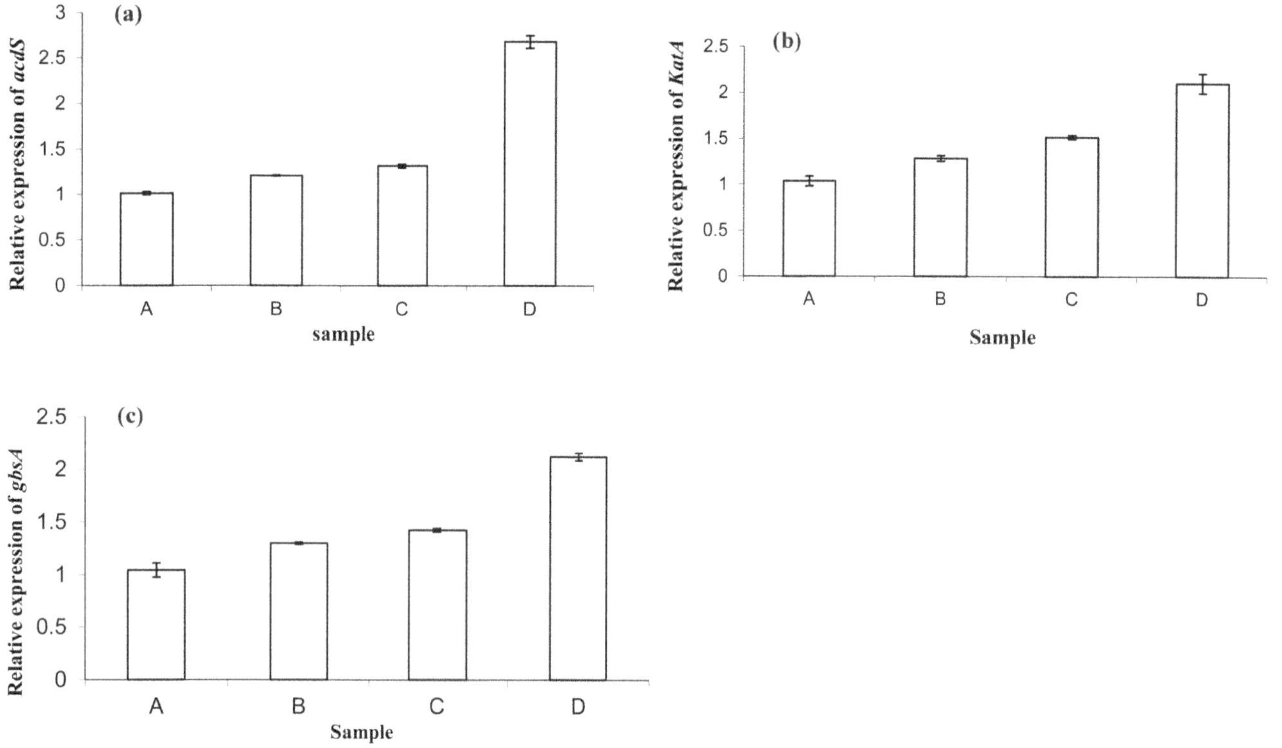

Figure 2. Relative gene expression level of, a) ***acdS*****, b)** ***katA*** **and c)** ***gbsA*** **in** ***P. aeruginosa*** **(GGRJ21) growing in different osmotic stress condition.** A-GGRJ 21 grown in normal NB medium, B-GGRJ21 grown in –0.3 mPA, C-GGRJ21 grown in 0.49 mPA and D-GGRJ21 grown in –0.73 mPA.

Discussion

Among the plant growth promoting rhizobacteria (PGPR), fuorescent pseudomonads, account for a significant proportion of the culturable rhizosphere population [56]. Since this ubiquitous group possess PGPR as well as biocontrol activities, their occurrence in the crop rhizosphere might be very essential for substantial crop development.

In this present investigation, we examined the genetic diversity and functional characterization of culturable fluorescent pseudo monads isolated from green gram rhizosphere. A comparative analysis was performed between the pseudomonad strains isolated during vegetative as well as reproductive growth phase of the pulse crop. Different PCR-based techniques *i.e.*, Rep PCR ribotyping, ARDRA and RISA, were employed to investigate the genetic variability of fluorescent pseudomonad isolates. Although the different PCR based genotypic analysis tools have their own level of genetic resolution to differentiate the organisms, they were not sufficient to resolve the fine genetic variation of all the isolates in this study. Genetic distance among the isolates within a same cluster was varied significantly depending upon the molecular markers used. For instances, isolates like GGRJ40, GGRJ41, GGRJ43, GGRJ44, GGRJ45, GGRJ46, GGRJ47, GGRJ55, GGRJ63, GGRJ75 and KFP8 were placed in sub-cluster If, Ic and Ia of rep PCR, ARDRA and RISA respectively, but their distant location in the clusters denoted intrageneric diversity within the isolates. In addition, genotypic fingerprints revealed close evolutionary relationships within the FPs isolated from vegetative growth phase of green gram with monophyletic clustering. However, significant diversity was observed in case of isolates linked with reproductive growth phase of the pulse crop, forming para/poly phyletic group among them. The lack of genotypic correlation within the FPs linked with reproductive growth phase of the pulse crop was observed because of the existence of high degree of genetic diversity within the isolates as reported by earlier workers [57,58].

16S rRNA sequence analysis of isolates delineated 85 isolates representing 23 distinct species of FPs. Reconstruction of molecular phylogeny grouped 85 isolates into two major clusters. All the isolates of cluster II were linked to reproductive growth phase of the host plant and showed closest homology to the Gamma-β proteobacteria group. This exception colonization suggests the variation in root microenvironment pattern during reproductive growth phase of the pulse crop. However, cluster-I with 75 isolates depicted predominant occurrence of *P. aeruginosa*, *P. otitidis* and *P. plecoglossicida*. Among the green gram associated FPs, *P. aeruginosa* clearly signifies the dominant chronicle with19 bacterial isolates followed by *P. otitidis*, *P. plecoglossicida, P. monteilii* and *P. mosselii*. Earlier studies conducted by Weisburg et al. 1991 and Widmer et al. 1998 [59,60] have shown that *P. fluorescens* and *P. putida* are the presumed to be dominant species in rice and wheat rhizosphere. Similarly, studies conducted in rice rhizosphere revealed dominant population of *P. fluorescens*, *P. aeruginosa*, and *P. putida* [61]. Nayak et al. 2008 reported *P. monteilii* as the dominant species in the banana rhizosphere along with other species of fluorescent pseudomonads [13]. Moreover, as compared to other PCR-based genotyping, *16S rRNA* analysis depicted a clear resolution of genetic distance and diversity among the diverse taxa of FPs; as evidenced from phylogenetic trees inferred through NJ and UPGMA method.

Variation in genetic composition between the vegetative and reproductive phase associated FP communities clearly suggests the role of ecological conditions on microbial species distribution;

directly correlated with the previous findings [62–64]. Earlier studies by McArthur et al.1988 [65] revealed that habitat variability due to change in soil properties could influence the genetic diversity in natural populations of a soil bacterium. Again structural diversities among the microorganisms in the rhizosphere soil may be also due to differences in root exudation during different growth phase, as well as for soil physicochemical properties and agronomic practices [66,67].

Based on metabolic characteristics, bacterial functional traits were studied extensively. In natural condition different plant growth promoting (PGP) as well as biocontrol traits of root associated microflora plays a crucial role in better plant growth and development [66]. In recent years, much attention has been given to the antagonistic activities and bio fertilizing abilities of fluorescent pseudomonads associated with agricultural crop plants [7]. Antagonistic activity of HCN against different phytopathogens has been already reported [68]. In the present investigation, all the antagonically active FP showed a varied range of HCN production. However, eight strains (GGRJ21, GGRJ70, GGRJ36, GGRJ1, GGRJ14, GGRJ20, GGRJ27 and GGRJ34) with high HCN production capability showed pronounced antagonistic activity against almost all the tested fungal pathogens. Kloepper et al.1980 [69] demonstrated the role of siderophore production as one of the significant characteristics of biocontrol agents. Our observation shortlisted ninety six percent of the *Pseudomonas* isolates with biocontrol activity (23 out of 24) produced siderophore (hydroxamate type) with varying degrees of production efficiency that correlates with the previous report [70]. Establishment of iron delimiting condition in the rhizosphere is the main mechanism of siderophore based inhibition the root colonizing pathogenic fungi [71]., Chitin amendment significantly increased the chitinase activity of *Pseudomonas* strains. Among the thirteen strains, *P. aeruginisa* GGRJ21 responded well to the addition of chitin and produced 69.33 nmolGlcNac/min/ml chitinase in chitin- amended medium. Moderate response to the addition of chitin was noticed in other strains. However, the hydrolytic nature of chitinase is only potential to inhibit or degrade the chitin containing pathogens. Thus, chitin production may not be equally relevant in protecting all types of phytopathogens. It have been established that, majority of PGPR activate induced systemic resisrance (ISR) via a SA-independent pathway involving jasmonate and ethylene signals [72,73]. Most of the fluorescent pseudomonads with antagonistic activity revealed significant levels of SA production. SA production triggers ISR development in the host plant by controlling the major pathogen related (PR) gene expression, which might be the one proficient path for disease management.

Moreover, wide range of antibiotic production efficiency i.e. DAPG, PCA and PLT by FP isolates may play a significant role in suppressing various soil borne pathogens around the green gram rhizosphere. DAPG are an important group of polyketide antibiotics that may have the capability to suppress root and seedling diseases on a variety of crops [13]. DAPG retains a wide range of antifungal, antibacterial as well as antinematocidal activity. The antagonistic activity of DAPG was recorded against a wide array of fungal pathogens, i.e, *Thielaviopsis basicola* [12], *Gaeumannomyces graminis* [12], *Fusarium oxysporum* f. sp. *lycopersici* [74], *Fusarium oxysporum f. sp. radicis-lycopersici* [75] etc. Local antagonism in the plant root system and induction of plant defense mechanism may be the main reason for DAPG antagonism against *Thielaviopsis basicola* and *Gaeumannomyces graminis* [12]. Duffy and Defago 1997 [75] stated the role of zinc amendments on enhncement of DAPG production and supression of *Fusarium oxysporum f. sp. radicis-lycopersici* by reducing the fusaric acid production. Saomeya et al. 2007 [74] observed a synergistic effect of DAPG and chitinolytic enzymes against *Fusarium oxysporum* f. sp. *lycopersici*. However, during the study purified DAPG was found to be ineffective against the fungal pathogen tested. The observation was found to be corroborated with the previous findings of Reddi and Borovkov 1970 [76], where they had enumerated the poor activity of DAPG against fungi and many gram negative bacteria. Thus the result concluded the narrow spectrum activity of DAPG extracted from the fluorescent pseudomonad isolates. Similarly, a number of broad spectrum antifungal colored phenazines of fluorescent pseudomonads origin have been reported in previous studies [77–80]. Our present investigation was also reflected the similar predisposition with broad range of antagonistic activity of purified phenazine-1-carboxylic acid against all phytopathogenic fungi tested. During the study, PLT activity was retained by only one isolate, *i.e.* GGRJ21, which showed broad range of antibiotic activity against all the phytopathogens. Broad range of anti microbial activity of purified PLT from fluorescent pseudomonads has been reflected from the earlier investigations [12,81,82].

Screening for the production of different biofertlizing attributes from green gram rhizosphere associated FP isolates, depicted a significant variation in the production of a wide array of metabolites and enzymes. A total of 25 numbers of isolates showed indole production. The spectrophotometric assay clearly distinguished gradual increase in IAA production with increase in L-tryptophan concentration from 0 to 100 $\mu g\ ml^{-1}$. The result, *i.e* enhancement of L-tryptophan-derived auxin biosynthesis can be directly correlated with the previous findings [83,84]; *i.e* further amendment of L-tryptophan (beyond 100 $\mu g\ ml^{-1}$) negatively stimulated the in vitro level of auxins production by the FP isolates. Thus suggested a negative feedback loop [85]. Although ACC deaminase activity is very prominent in pseudomonads group; during our study, we screened out only five pseudomonad isolates with this particular enzymatic activity. Moreover, the enzyme production efficiency was found to be better than the previous reports [86,87]. These specific enzyme positive isolates may play a vital role in root elongation and seed germination as reported earlier [88]. During the study, we had screened out all the *Pseudomonas* isolates for their phosphate solubilising activity. However only ten numbers of isolates showed phosphate solubilization with less phosphatase activity than some of the previous reports [18,89]. Lower proportion of phosphate solubilization might be due to release of lesser concentration of organic acids from the PSBs required to mobilize major quantities of P into the soil solution [90,91]. However, organic acid production does not seem to be the only mechanism of inorganic phosphate solubilization; as in some cases pH reduction did not correlate with the mineral phosphate solubilization [92]. Release of protons accompanying respiration or ammonium assimilation was reported to be another mechanism for phosphate solubilization [93].

Among the 120 bacterial isolates, *Pseudomonas aeruginosa* GGRJ21 was selected further for in vivo experiments mainly due its wide range of antagonistic, plant growth promotion as well as effective range of stress tolerant propensity. The *in vivo* pot experiment with *P. aeruginosa* (GGRJ21) and *R. solani*, presented a practical and proficient green approach to control root rot disease of green gram. *In vivo* trial in green house condition strongly supported the *in vitro* antifungal activity of GGRJ21. Prior inoculation of GGRJ21, suppressed the rate of disease severity much more efficiently than the control. This may be due to systemic resistance induced by the biocontrol agent in the host plant [94,95]. However, other possible mechanisms could be through antibiosis [94], mycoparasitism [96] or for competition for nutrients and/or space [97]. The relative biocontrol activity of GGRJ21 could be corroborated with the earlier reports [2,98–

100]. However, greater disease suppressive propensity could be enough to promote *P. aeruginosa* GGRJ21 as a most efficient biocontrol agent than the previously reported microbial strains.

The osmotic strength of the environment is one of the major physical parameters that determines survival rate of the organism in their own habitat. Variation in osmotic stress in the environment is an indisputable reality for microbes colonizing any environment [101]. Being an opportunistic organism, pseudomonads acquire up adaptive cellular machinery, which assist them for survival in diverse and often stressful environmental conditions [101]. Several genes differentially expressed during stressed condition, which may play a key role in the prevalence and persistence of the bacterium in osmotically stressful infection sites [102]. Accumulation of *N*-acetylglutaminylglutamine amide and glycinebetaine as cytoplasmic osmoprotectants in response to osmotic stress is one of the important distinctiveness of *P. aeruginosa* [103]. According to the earlier workers, transcriptional product of two genes *gbsA* and *gbsB* was essential for glycine betaine biosynthesis in bacteria. Since *P. aeruginosa* (GGRJ21), showed stable growth kinetics under optimum osmotic stress condition, we further analyzed the relative expression study of *gbsA*. As would be expected, *gbsA* showed enhanced levels of expression during the early stages of osmotic up-shock. Interestingly, the expression of gene *katA* associated with the catalase was substantially over-expressed in osmotically shocked cells, thus protect the *Pseudomonas* cell from the toxicity of hydrogen peroxide (H_2O_2). Further we have analyzed the relative expression study of *acdS* (gene encoding for ACC deaminase) regulation and found almost three fold up-regulation during osmotic shock at −0.73 mPA condition. Plants under water stress experiences high level of ethylene biosynthesis, which thereby causes fatal effect on plant cell. However, the application of PGPR containing ACC deaminase is very crucial agent to regulate the plant ethylene by converting, ACC into α-ketobutyrate and ammonia, thus helps the plant to live under adverse climatic condition. Since *P. aeruginosa* (GGRJ21) showed a remarkable synthesis of ACC deaminase due to up-regulation of *acdS* gene under osmotic stress condition, application of the *Pseudomonas* strain may interact with green gram and confer a resistance support to the pulse crop under water deficit condition. Moreover, we have already examined the positive effect of *P. aeruginosa* (GGRJ21) towards the alleviation of drought stress in green gram plant in normal environmental condition [4].

Conclusion

Knowledge of the PGPR diversity and their bio-control, biofertilizing activity is not only essential to understand their ecological role in the rhizosphere, but also for utilization in sustainable agriculture. The present study showed a high degree of functional and genotypic diversity among fluorescent pseudomonads in the green gram rhizosphere for the first time. *Pseudomonas aeruginosa* GGRJ21 with innate biocontrol, osmotolerant and biofertilizing potential could provide a vital bio-resource for plant growth and development, promotion, disease control, and subsequent enhancement of crop yield in the host plant.

Supporting Information

Figure S1 Rep-PCR genomic fingerprints of 120 dominant strains generated with BOX AIR1 (A) and ERIC1 (B) primer with 500 bp DNA marker.

Figure S2 Dendrogram showing the genetic diversity of fluorescent pseudomonads of green gram rhizosphere. Clustering analysis based on the combined fingerprints of ERIC and BOX-PCR was performed using the UPGMA method followed by Jaccard's coefficient.

Figure S3 Restriction patterns of PCR amplified fragment of (a) 16S rDNA digested with *HaeIII*, *AluI* and *MspI* and (b) 16S–23S rDNA intergenic spacer region digestion with *MspI*.

Figure S4 Dendrogram showing the genetic diversity of fluorescent pseudomonads of green gram rhizosphere. Clustering analysis of ARDRA fingerprints was performed using the UPGMA method followed by Jaccard's coefficient.

Figure S5 Dendrogram based on RISA analysis (restriction digestion of 16S–23S rDNA intergenic spacer region sequences by *MspI*) showing the intra specific relationships among 120 members of the fluorescent pseudomonads.

Figure S6 Phylogenetic analyses of fluorescent pseudomonads based on the nucleotide sequence of *16S rRNA* using UPGMA method in MEGA 5.2. The Bar, 0.005, shows the substitutions per nucleotide position.

Figure S7 Antifungal activity of purified phenazine-1-carboxylic acid (PCA) and pyoluteorin (PLT) against phytopathogenic fungi.

Table S1 Eighty-five *Pseudomonas* isolates with their taxonomic, phenotypic and biochemical characteristics.

Table S2 Quantitative estimation of indoles with tryptophan concentrations of 100, 200 and 500 μg ml^{-1}. Values are mean of three replicates.

Table S3 Optical density (OD) of 10 *Pseudomonas* isolates at 600 nm after 24 hours of growth under different osmotic stress condition considering normal growth in nutrient broth (NB) as control. Values are mean of three replicates.

Acknowledgments

The authors are also thankful to the Director, CSIR-NEIST for providing necessary facilities to carry out the work.

Author Contributions

Conceived and designed the experiments: RKS RS. Performed the experiments: RKS AG. Analyzed the data: RKS RS BD RD. Contributed reagents/materials/analysis tools: RKS RS AG TCB. Contributed to the writing of the manuscript: RKS RS BD.

References

1. Haas D, Defago G (2005) Biological control of soil-borne pathogens by fluorescent pseudomonads. Nat Rev Microbiol 3: 307–319.
2. Saikia R, Singh T, Kumar R, Srivastava J, Srivastava AK, et al. (2003) Role of salicylic acid in systemic resistance induced by *Pseudomonas fluorescens* against *Fusarium oxysporum* f. sp. *ciceri* in chickpea. Microbiol Res 158: 871–881.
3. Salman M (2010) Determination of antibiotic activity on plasmids from fluorescent pseudomonad isolates CW2, WB15 and WB52 against pre-emergence damping-off caused by *Pythium ultimum* and *Rhizoctonia solani* in cucumber. Biol Control 53: 161–167.
4. Sarma RK, Saikia R (2014) Alleviation of drought stress in mung bean by strain *Pseudomonas aeruginosa* GGRJ21. Plant Soil 377: 111–126.
5. Saravanakumar D, Samiyappan R (2007) ACC deaminase from *Pseudomonas fluorescens* mediated saline resistance in groundnut (*Arachis hypogea*) plants. J Appl Microbiol 102: 1283–1292.
6. Mayak S, Tirosh T, Glick BR (2004) Plant growth promoting bacteria that confer resistance to water stress in tomato and pepper. Plant Sci 166: 525–530.
7. Hofte M, Altier N (2010) Fluorescent pseudomonads as biocontrol agents for sustainable agricultural systems. Res Microbiol 61: 464–471.
8. Weller DM (1988) Biological control of soil-borne plant pathogens in the rhizosphere with bacteria. Annu Rev Phytopathol 26: 379–407.
9. Pfender WF, Kraus J, Loper JE (1993) A genomic region from *Pseudomonas fluorescens* Pf-5 required for pyrrolnitrin production and inhibition of *Pyrenophora tritici-repentis* in wheat straw. Phytopathology 83: 1223–1228.
10. Pierson LS, Thomashow LS (1992) Cloning and heterologous expression of the phenazine biosynthetic locus from *Pseudomonas aureofaciens* 30–84. Mol Plant- Microbe Interact 5: 330–339.
11. Voisard C, Keel C, Haas D, Defago G (1989) Cyanide production by *Pseudomonas fluorescens* helps suppress black root rot of tobacco under gnotobiotic conditions. The EMBO J 8: 351–358.
12. Keel C, Schnider U, Maurhofer M, Voisard C, Laville J, et al. (1992) Suppression of root diseases by *Pseudomonas fluorescens* CHA0: Importance of the bacterial secondary metabolite 2,4-diacetylphloroglucinol. Mol Plant-Microbe Interact 5: 4–13.
13. Naik PR, Sahoo N, Goswami D, Ayyadurai N, Sakthivel N (2008) Genetic and functional diversity among fluorescent pseudomonads isolated from the rhizosphere of banana. Microb Ecol 56: 492–504.
14. Patten CL, Glick BR (2002) Role of *Pseudomonas putida* indole acetic acid in development of the host plant root system. Appl Environ Microbiol 68: 3795–3801.
15. Saravanakumar D, Kavino M, Raguchander T, Subbian P, Samiyappan R (2011) Plant growth promoting bacteria enhances water stress resistance in green gram plants. Acta Physiol Plant 33: 203–209.
16. Sandhya V, Ali SZ, Grover M, Reddy G, Venkateswarlu B (2009) Alleviation of drought stress effects in sunflower seedlings by the exopolysaccharides producing *Pseudomonas putida* strain GAP-P45. Biol Fert Soils 46: 17–26.
17. Wright CA, Beattie GA (2003) *Pseudomonas syringae* pv. *tomato* cells encounter inhibitory levels of water stress during the hypersensitive response of *Arabidopsis thaliana*. Proc Natl Acad Sci USA 101: 3269–3274.
18. Naik PR, Raman G, Narayanan KB, Sakthivel N (2008) Assessment of genetic and functional diversity of phosphate solubilizing fluorescent pseudomonads isolated from rhizosphere soil. BMC Microbiol 8: 230. doi 10.1186/1471-2180-8-230.
19. Ayyadurai N, Naik PR, Sakthivel N (2007) Functional characterization of antagonistic fluorescent pseudomonads associated with rhizosphere soil of rice (*Oryza sativa* L.). J Microbiol Biotech 17: 919–927.
20. Hu HB, Xu YQ, Chen F, Zhang XH, Hur BK (2005) Isolation and characterization of a new fluorescent *Pseudomonas* strain that produces both phenazine-1-carboxylic acid and pyoluteorin. J Microbiol Biotechnol 115: 86–90.
21. Thomashow LS, Weller DM, Bonsall RF, Pierson LS (1990) Production of the antibiotic phenazine-1-carboxylic acid of fluorescent *Pseudomonas* species in the rhizosphere of wheat. Appl Environ Microbiol 56: 908–912.
22. Sarma RK, Debnath R, Saikia R, Handique PJ, Bora TC (2012) Phylogenetic analysis of alkaline proteinase producing fluorescent pseudomonads associated with green gram (*Vigna radiata* L.) rhizosphere. Folia Microbiol 57: 129–137.
23. Bergey DH, Holt JG, Krieg NR, Sneath PHA (1994) Bergey's Manual of Determinative Bacteriology. Lippincott Williams & Wilkins.
24. Louws FJ, Fulbright DW, Stephens CT, de Bruijn FJ (1994) Specific genomic fingerprints of phytopathogenic *Xanthomonas* and *Pseudomonas* pathovars and strains generated with repetitive sequences and PCR. Appl Environ Microbiol 60: 2286–2295.
25. de Bruijn FJ (1992) Use of repetitive (repetitive extragenic palindromic and enterobacterial repetitive intergenic consensus) sequences and the polymerase chain reaction to fingerprint the genomes of *Rhizobium meliloti* isolates and other soil bacteria. Appl Environ Microbiol 58: 2180–2187.
26. Edwards U, Rogall TH, Blocker H, Emde M, Bottger EC (1989) Isolation and direct complete nucleotide determination of entire genes. Characterization of a gene coding for 16S ribosomal RNA. Nuc Acids Res 17: 7843–7853.
27. Jensen MA, Webster JA, Straus N (1993) Rapid identification of bacteria on the basis of polymerase chain reaction-amplified ribosomal DNA spacer polymorphisms. Appl Environ Microbiol 59: 945–952.
28. Schillinger U, Lucke FK (1989) Antibacterial activity of *Lac-tobacillus* sake isolated from meat. Appl Environ Microbiol 55: 1901–1906.
29. Kremer RJ, Souissi T (2001) Cyanide production by rhizobacteria and potential for suppression of weed seedling growth. Curr Microbiol 43: 182–186.
30. Debnath R, Saikia R, Sarma RK, Yadav A, Bora TC, et al. (2013) Psychrotolerant antifungal Streptomyces isolated from Tawang, India and the shift in chitinase gene family. Extremophiles 17: 1045–1059.
31. Alexander DB, Zuberer DA (1991) Use of chrome azurol S reagents to evaluate siderophore production by rhizospherebacteria. Biol Fert Soils 12: 39–45.
32. Jalal MAF, van der Helm D (1991) Isolation and spectroscopic identification of fungal siderophores. In: Winkelmann G, editor. CRC Handbook of Microbial Iron Chelates. CRC Press, Boca Raton. 235–269.
33. Reeves M, Pine L, Neilands JB, Bullows A (1983) Absence of siderophore activity in *Legionella* sp. grown in iron deficient media. J Bacteriol 154: 324–329.
34. Yalpani N, Silverman P, Wilson TMA, Kleier DA, Raskin I (1991) Salicylic acid is a systemic signal and an inducer of pathogenesis-related proteins in virus-infected tobacco. The Plant Cell 3: 809–818.
35. Thomashow LS, Weller DM (1996) Current concepts in the use of introduced bacteria for biological disease control: Mechanisms and antifungal metabolites. In: Stacey G, Keen NT editors. *Plant-Microbe Interactions* Vol 1. Chapman and Hall, New York. 187–235.
36. Ligon JM, Hill DS, Hammer PE, Torkewitz NR, Hofmann D, et al. (2000) Natural products with antifungal activity from *Pseudomonas* biocontrol bacteria. Pest Manag Sci 56: 688–695.
37. Mavrodi OV, Gardener BBM, Mavrodi DV, Bonsall RF, Weller DM, et al. (2001) Genetic diversity of PhlD from 2,4-diacetylphloroglucinol-producing fluorescent *Pseudomonas* spp. Phytopathology 91: 35–43.
38. Raaijmakers J, Weller DM, Thomashow LS (1997) Frequency of antibiotic producing *Pseudomonas* spp. in natural environments. Appl Environ Microbiol 63: 881–887.
39. Raaijmakers JM, Bonsall RF, Weller DM (1999) Effect of population density of *Pseudomonas fluorescens* on production of 2,4-diacetylphloroglucinol in the rhizosphere of wheat. Phytopathology 89: 470–475.
40. Chancey ST, Wood DW, Pierson LS (1999) Two-component transcriptional regulation of N-acyl-homoserine lactone production in *Pseudomonas aureofaciens*. Appl Environ Microbiol 65: 2294–2299.
41. Sarniguet A, Kraus J, Henkels MD, Muehlchen AM, Loper JE (1995) The sigma factor sigma s affects antibiotic production and biological control activity of *Pseudomonas fluorescens* Pf-5. Proc Natl Acad Sci USA 92: 12255–12259.
42. Dworkin M, Foster J (1958) Experiments with some microorganisms which utilize ethane and hydrogen. J Bacteriol 75: 592–601.
43. Gordon SA, Weber RP (1951) Colorimetric estimation of indole acetic acid. Plant Physiology 26: 192–195.
44. Honma M, Shimomura T (1978) Metabolism of 1-aminocyclopropane- 1-carboxylic acid. Agric Biol Chem 42: 1825–1831.
45. Bashan Y, Kamnev AK, de-Bashan LE (2013) Tricalcium phosphate is inappropriate as a universal selection factor for isolating and testing phosphate-solubilizing bacteria that enhance plant growth: a proposal for an alternative procedure. Biol Fertil Soils 49: 465–479.
46. Fiske CH, Subbarow Y (1925) A colorimetric determination of phosphorus. J Biol Chem 66: 375–400.
47. Saikia R, Gogoi DK, Mazumder S, Yadav A, Sarma RK, et al. (2011) *Brevibacillus laterosporus* strainBPM3, a potential biocontrol agent isolated from a natural hot water spring of Assam, India. Microbiol Res 166: 216–25.
48. Filion M, St-Arnaud M, Jabaji-Hare SH (2003) Quantification of *Fusarium solani* f. sp. *phaseoli* in mycorrhizal bean plants and surrounding mycorrhizosphere soil using realtime polymerase chain reaction and direct isolations on selective media. Phytopathology 93: 229–235.
49. Michel BE, Kaufmann MR (1973) The osmotic potential of polyethylene glycol 6000. Plant Physiol 51: 914–916.
50. Kimura M (1980) A simple method for estimating evolutionary rate of base substitutions through comparative studies of nucleotide sequences. J Mol Evol 16: 111–120.
51. Felsenstein J (1985) Confidence limits on phylogenies: an approach using the bootstrap. Evolution 39: 783–791.
52. Pfaffl MW (2001) A new mathematical model for relative quantification in real-time RT-PCR. Nuc Acids Res 29(9): e45.
53. Vandesompele J, De Preter K, Pattyn F, Poppe B, Van Roy N, et al. (2002) Accurate normalization of real-timequantitative RT-PCR data by geometric averaging of multiple internal control genes. Genome Biol. doi:10.1186/gb-2002-3-7-research0034.
54. Walker NJ (2002) A technique whose time has come. Science 296: 557–559.
55. Winer J, Jung CK, Shackel I, Williams PM (1999) Development and validation of real-time quantitative reverse transcriptase-polymerase chain reaction for monitoring gene expression in cardiac myocytes in vitro. Anal Biochem 270: 41–49.
56. Lambert B, Leyns F, Rooyen LV, Gossele F, Papon Y, et al. (1987) Rhizobacteria of maize and their atifungal activities. Appl Environ Microbiol 53: 1866–1871.

57. Cordero P, Cavigliasso A, Príncipe A, Godino A, Jofré E, et al. (2012) Genetic diversity and antifungal activity of native *Pseudomonas* isolated from maize plants grown in a central region of Argentina. Syst Appl Microbiol 35: 342–351.
58. Misko AL, Germida JJ (2002) Taxonomic and functional diversity of pseudomonads isolated from the roots of field grown canola. FEMS Microbiol Ecol 42: 399–407.
59. Weisburg WG, Barns SM, Pelletier DA, Lane DJ (1991) 16S Ribosomal DNA amplification for phylogenetic study. J Bacteriol. 173: 697–703.
60. Widmer F, Seidler RJ, Gillevet PM, Watrud LS, Di Giovanni GD (1998) A highly selective PCR protocol for detecting *16S rRNA* genes of the genus *Pseudomonas* (sensustricto) in environmental samples. Appl Environ Microbiol 64: 2545–2553.
61. Rangarajan S, Saleena LM, Nair S (2002) Diversity of *Pseudomonas* spp. isolated from rice rhizosphere populations grown along a salinity gradient. Microb Ecol 43: 280–289.
62. Lottmann J, Berg G (2001) Phenotypic and genotypic characterization of antagonistic bacteria associated with roots of transgenic and non-transgenic potato plants. Microbiol Res 156: 75–82.
63. McSpadden GBB, Schroeder KL, Kalloger SE, Raaijmakers JM, Thomashow LS, et al. (2000) Genotypic and phenotypic diversity of phlD-containing *Pseudomonas* strains isolated from the rhizosphere of wheat. Appl Environ Microbiol 66: 1939–1946.
64. Lemanceau P, Corberand T, Gardan L, Latour X, Laguerre G, et al. (1995) Effect of two plant species, flax (*Linum usitatissinum* L.) and tomato (*Lycopersicon esculentum* Mill.), on the diversity of soilborne populations of fluorescent pseudomonads. Appl Environ Microbiol 61: 1004–1012.
65. Mc Arthur JV, Kovacic DA, Smith MH (1988) Genetic diversity in natural populations of a soil bacterium across a landscape gradient. Proc Natl Acad Sci USA 85: 9621–9624.
66. Brimecombe MJ, Lelj FAD, Lynch JM (2001) The Rhizosphere. The Effect of Root Exudates on Rhizosphere Microbial Populations. In: Pinton R, Varanini Z, Nannipieri P, editors. The Rhizosphere. Biochemistry and Organic Substances at the Soil-Plant Interface. Marcel Dekker, New York. 95–140.
67. Marschner P, Yang CH, Lieberei R, Crowley DE (2001) Soil and plant specific effects on bacterial community composition on the rhizosphere. Soil Biol Biochem 33: 1437–1445.
68. Ramette A, Moenne-Loccoz Y, Defago G (2006) Genetic diversity and biocontrol potential fluorescent pseudomonads producing phloroglucinols and hydrogen cyanide from swiss soils naturally suppressive or conducive to *Thielaviopsis basicola* mediated black root rot of tobacco. FEMS Microbiol Ecol 55: 369–381.
69. Kloepper JW, Leong J, Teintze M, Schroth M (1980) *Pseudomonas* siderophores: a mechanism expla- ining disease suppressive soils. Curr Microbiol 4: 317–320.
70. Maksimov I, Abizgil'dina R, Pusenkova L (2011) Plant growth promoting rhizobacteria as alternative to chemical crop protectors from pathogens (review). Appl Biochem Microbiol 47: 333–345.
71. Podile AR, Kishore GK (2006) *Plant growth promoting rhizobacteria*. Springer.
72. Pieterse CMJ, van Wees SCM, van Pelt JA, Knoester M, Laan R, et al. (1998) A novel signaling pathway controlling induced systemic resistance in *Arabidopsis*. Plant Cell 10: 1571–1580.
73. Pettersson M, Baath E (2004) Effects of the properties of the bacterial community on pH adaptation during recolonization of a humus soil. Soil Biol Biochem 36: 1383–1388.
74. Someya N, Tsuchiya K, Yoshida T, Noguchi MT, Akutsu K, et al. (2007) Co-inoculation of an antibiotic-producing bacterium and a lytic enzyme-producing bacterium for the biocontrol of tomato wilt caused by Fusarium oxysporum f. sp. lycopersici. Biocontrol Sci 12: 1–6.
75. Duffy BK, Défago G (1997) Zinc improves biocontrol of Fusarium crown and root rot of tomato by *Pseudomonas fluorescens* and represses the production of pathogen metabolites inhibitory to bacterial antibiotic biosynthesis. Phytopathology 87: 1250–1257.
76. Reddi TK, Borovkov AV (1970) Antibiotic properties of 2,4-diacetylphloroglucinol produced by *Pseudomonas fluorescens* strain 26-0. Antibiotiki (Moscow). 19–21.
77. Gurusiddaiah S, Weller DM, Sarkar A, Cook RJ (1986) Characterization of an antibiotic produced by a strain of *Pseudomonas fluorescens* inhibitory to *Gaeumannomyces graminis* var. *tritici* and *Pythium* spp. Antimicrob Agents Chemother 29: 488–95.
78. Kerr JR (2000) Phenazine pigments: antibiotics and virulence factors. Rev Infect Dis 2: 184–94.
79. Price-Whelan A, Dietrich LEP, Newman DK (2006) Rethinking 'secondary' metabolism: physiological roles for phenazine antibiotics. Nat Chem Biol 2: 71–78.
80. Puopolo G, Masi M, Raio A, Andolfi A, Zoina A, et al. (2013) Insights onthe susceptibility of plant pathogenic fungi to phenazine1carboxylic acid and its chemical derivatives. Nat Prod Res 27: 956–966.
81. Dutrecq A, Debras P, Stevaux J, Marlier M (1991) Activity of 2,4-diacetylphloroglucinol isolated from a strain of *Pseudomonas fluorescens* to *Gaeumannomyces graminis* var. *Tritici*. In: Beemster ABR, Bollen GJ, Gerlagh M, Ruissen MA, Schippers B et al. Biotic interactions and soil-borne diseases. Elsevier Science Publishers, Amsterdam, The Netherlands. 252–257.
82. Garagulya AD, Kirprianova EA, Boiko OI (1974) Antibiotic effect of bacteria from the genus *Pseudomonas* on phytopathogenic fungi. Mikrobiol Zh 36: 197–202.
83. Khalid A, Arshad M, Zahir ZA (2004) Screening plant growthpromoting rhizobacteria for improving growth and yield of wheat. J Appl Microbiol 96: 473–480.
84. Bharucha U, Patel K, Trivedi UB (2013) Optimization of Indole Acetic Acid Production by *Pseudomonas* putida UB1 and its Effect as Plant Growth-Promoting Rhizobacteria on Mustard (*Brassica nigra*). Agric Res 2: 215–221.
85. Gollnick P, Babitzke P, Antson A, Yanofsky C (2005) Complexity in regulation of tryptophan biosynthesis in Bacillus subtilis. Annu Rev Genet 39: 47–68.
86. Wang C, Ramette A, Punjasamarnwong P, Zala M, Natsch A, et al. (2001) Cosmopolitan distribution of phlD-containing dicotyledonous crop-associated pseudomonads of worldwide origin. FEMS Microbiol Ecol 37: 105–116.
87. Campbell BG, Thomson JA (1996) 1-Aminocyclopropane-1-carboxylate deaminase genes from *Pseudomonas* strains. FEMS Microbiol Lett138: 207–210.
88. Belimov AA, Safronova VI, Sergeyeva TA, Egorova TN, Matveyeva VA, et al. (2001) Characterization of plant growth promoting rhizobacteria isolated from polluted soils and containing 1-aminocyclopropane-1-carboxylate deaminase. Can J Microbiol 47: 642–652.
89. Meyer JB, Frapolli M, Keel C, Maurhofer M (2011) Pyrroloquinoline quinone biosynthesis gene pqqC, a novel molecular marker for studying the phylogeny anddiversity of phosphate-solubilizing pseudomonads. Appl Environ Microbiol 77: 7345–54.
90. Castagno LN, Estrella MJ, Sannazzaro AI, Grassano AE, Ruiz OA (2011) Phosphate solubilization mechanism andin vitro plant growth promotion activity mediated by Pantoea eucalyptiisolated from Lotus tenuis rhizosphere inthe Salado River Basin (Argentina). J Appl Microbiol 110: 1151–65.
91. Nahas E (1996) Factors determining rock phosphate solubilization by microorganism isolated from soil. World J Microb Biotech 12: 18–23.
92. Subba Rao NS (1982) Advances in agricultural microbiology. India: Oxford and IBH Publications Company. 229–305.
93. Park KH, Lee CY, Son HJ (2009) Mechanism of insoluble phosphate solubilization by *Pseudomonas fluorescens* RAF15 isolated from ginseng rhizosphere and its plant growth-promoting activities. Lett Appl Microbiol 49: 222–428.
94. Kloepper JW, Ryu CM, Zhang S (2004) Induced systemic resistance and promotion of plant growth by *Bacillus* spp. Phytopathology 94: 1259–1266.
95. Akram A, Ongena M, Duby F, Dommes J, Thonart P (2008) Systemic resistance and lipoxygenaserelated defence response induced in tomato by *Pseudomonas putida* strain BTP1. BMC Plant Biol doi:10.1186/1471-2229-8-113.
96. Diby P, Saju KA, Jisha PJ, Sarma YR, Kumar A, et al. (2005) Mycolytic enzymes produced by *Pseudomonas fluorescens* and *Trichoderma* spp. against *Phytophthora capsici*, the foot rot pathogen of black pepper (*Piper nigram* L.). Ann Microbiol 55: 129–133.
97. Riley MA, Gordon DM (1999) The ecological role of bacteriocin in bacterial competition. Trends Microbiol 7: 129–133.
98. Thahir BS, Radhaiah A, Nagalakshmi DM, Eswara RNP (2012). Biocontrol potential of indigenous *Pseudomonas spp.* against *Sclerotium rolfsii* causing stem rot of groundnut. Int J Food Agri Vet sci 2: 134–141.
99. Gardener M (2007) Diversity and ecology of biocontrol *Pseudomonas* spp. in agricultural systems. Phytopathology 97: 221–226.
100. Weller DM (2007) *Pseudomonas* biocontrol agents of soilborne pathogens: looking back over 30 years. Phytopathology 97: 250–256.
101. Wood JM (1999) Osmosensing by bacteria: signals and membrane-based sensors. Microbiol Mol Biol Rev 63: 230–262.
102. Aspedon A, Palmer K, Whiteley M (2006) Microarray Analysis of the Osmotic Stress Response in *Pseudomonas aeruginosa*. J Bacteriol 188: 2721–2725.
103. D'Souza-Ault MR, Smith LT, Smith GM (1993) Roles of *N* acetylglutaminylglutamine amide and glycine betaine in adaptation of *Pseudomonas aeruginosa* to osmotic stress. Appl Environ Microbiol 59: 473–478.

7

Herbivore Impacts on Marsh Production Depend upon a Compensatory Continuum Mediated by Salinity Stress

Jeremy D. Long*, Laura D. Porturas

Biology Department and Coastal & Marine Institute Laboratory, San Diego State University, San Diego, California, United States of America

Abstract

Plant communities are disturbed by several stressors and they are expected to be further impacted by increasing anthropogenic stress. The consequences of these stressors will depend, in part, upon the ability of plants to compensate for herbivory. Previous studies found that herbivore impacts on plants can vary from negative to positive because of environmental control of plant compensatory responses, a.k.a. the Compensatory Continuum Hypothesis. While these influential studies enhanced our appreciation of the dynamic nature of plant-herbivore interactions, they largely focused on the impact of resource limitation. This bias limits our ability to predict how other environmental factors will shape the impact of herbivory. We examined the role of salinity stress on herbivory of salt marsh cordgrass, *Spartina foliosa*, by an herbivore previously hypothesized to influence the success of restoration projects (the scale insect, *Haliaspis spartinae*). Using a combination of field and mesocosm manipulations of scales and salinity, we measured how these factors affected *Spartina* growth and timing of senescence. In mesocosm studies, *Spartina* overcompensated for herbivory by growing taller shoots at low salinities but the impact of scales on plants switched from positive to neutral with increasing salinity stress. In field studies of intermediate salinities, scales reduced *Spartina* growth and increased the rate of senescence. Experimental salinity additions at this field site returned the impact of scales to neutral. Because salinity decreased scale densities, the switch in impact of scales on *Spartina* with increasing salinity was not simply a linear function of scale abundance. Thus, the impact of scales on primary production depended strongly upon environmental context because intermediate salinity stress prevented plant compensatory responses to herbivory. Understanding this context-dependency will be required if we are going to successfully predict the success of restoration efforts and the ecological consequences of anthropogenic disturbances.

Editor: Cristina Armas, Institute of Ecology and Biodiversity, Chile

Funding: This research was supported by grants from the United States National Science Foundation Biological Oceanography program (OCE-0825846 awarded to JL and Geoff Trussell), California Sea Grant (NA10OAR4170060), and CSU COAST (awarded to LP). The funders had no role in study design, data collection and analysis, decision to publish, or preparation of the manuscript.

Competing Interests: The authors have declared that no competing interests exist.

* Email: jlong@mail.sdsu.edu

Introduction

Environmental stress can disturb communities directly by reducing survivorship or forcing migrations, or indirectly by altering the traits responsible for determining how species interact [1–5]. An important plant trait that can be affected by stress is the ability to compensate for herbivory [6]. In a foundational study, Maschinski and Whitham (1989) proposed the Compensatory Continuum Hypothesis whereby plant compensatory responses can vary in sign and strength, largely via environmental factors. The discovery that compensatory responses to herbivory can be shaped by the environment profoundly influenced our understanding of plant-herbivore interactions [7,8]. However, most studies focused on the effect of resource limitation on compensatory responses [8]. Indeed, a well-cited revision of the Compensatory Continuum Hypothesis, the Limiting Resource Model, focused on the primacy of the level and identity of the limiting resource [9]. The roles of other environmental stressors on herbivory remain unclear. If compensatory responses generally fall along a continuum determined by environmental stress, then understanding this complexity will become important to successfully predicting the consequences of anthropogenic stressors like pollution and climate change.

Although there is emerging evidence that compensatory responses vary with multiple environmental contexts, we have an incomplete appreciation for the role of environmental stress other than resource limitation. For example, overcompensation appears more commonly for monocot herbs when nutrients are not limiting but overcompensation is more common for dicots under nutrient limitation [8]. Water-stress can also determine compensatory responses and the consequences of herbivory ([10–13] but see [14–16] for exceptions). In most of these cases, water stress leads to reduced compensatory responses. Unfortunately, the consequences of other common stressors, like salinity, are poorly known. Although the immediate consequences of salinity stress may largely result from changes in water uptake, salinity stress can create unique problems for plants that influence their ability to compensate for herbivory [6,17]. For example, water and salinity stress may have very different impacts on natural communities because the former increases soluble nitrogen and nutritional quality of plants [14,15,18], whereas the latter can decrease nutritional quality [19,20]. Also, salinity stress may influence compensatory responses because of negative, salt-specific effects on hormonal regulation, photosynthesis, and cellular function [17]. Furthermore, salinity stress might directly suppress herbivore performance via increases in plant tissue salt concentrations [21]. Thus, previous studies may fail to predict the impact of salinity stress on plant compensatory responses and their consequences because of stress-specific effects on plant physiology.

Because of their distinct gradients in environmental stress and plant distributions [22,23] and their susceptibility to anthropogenic stressors like climate change, salt marshes represent important habitats where salinity stress may interact with herbivory to influence primary production. Three observations support this hypothesis. First, experimental salinity stress increased the abundance of marsh planthoppers ([24,25] but see [20] for an exception). Second, palatability of marsh plants depended upon whether plants were collected from high salinity versus low salinity sites [26]. Third, salt addition stress magnified the negative impact of snail herbivory on *Spartina* biomass [27], perhaps because salt impaired the ability of *Spartina* to tolerate herbivory or increase resistance after attack [28]. Taken together, these studies suggest that salinity stress may shape the impact of herbivores on marsh plants and that the synergy of these factors may contribute to marsh dieback [27]. Given that climate change is expected to alter soil salinities via changes in sea level and precipitation [29,30], there is a pressing need to understand the impact of salinity stress on marsh herbivory.

We examined the impact of salinity stress on top-down control of Pacific cordgrass, *Spartina foliosa*, by the specialist scale insect, *Haliaspis spartinae*. Scale insects are common pests that can reduce the growth and survival of plants [31–33]. Although no study has experimentally tested the impact of *H. spartinae* on cordgrass, it has been implicated in the failure of some restoration projects to achieve desired outcomes [34,35]. These authors noted that, "In 1992, where the highest densities of *Haliaspis* occurred, cordgrass was shorter than in previous years and senesced unusually early in the season" [34]. Because cordgrass height is one measure of restoration success (in part because endangered birds require tall plants to make nests [35]), it was hypothesized that the scales may prevent successful restoration of these sensitive habitats [34,35]. Thus, an improved understanding of scale impacts and how these change with environmental conditions could lead to improved restoration outcomes.

We tested for a causative relationship between scales and plant performance, and the role of salinity stress on scale herbivory, by manipulating scale presence and salinity in a series of mesocosm and field studies. We predicted that 1) scales would exert the strongest negative impacts on *Spartina* at intermediate salinities where compensatory responses would be impaired, and 2) further stressing *Spartina* with high salinities would weaken scale impacts.

Methods

We conducted three major experiments to examine the impact of scale insects (*Haliaspis spartinae*, hereafter scales) on *Spartina foliosa* stems (hereafter *Spartina*). First, we used a field experiment to demonstrate that scales can negatively influence *Spartina* performance. Second, we used a mesocosm experiment to examine the effect of reduced salinities on *Spartina*-scale interactions. Finally, we conducted a field experiment to determine the effect of elevated soil salinities on these interactions. The latter two experiments were conducted because 1) the heaviest infestations of scales on *Spartina* (# per stem) occur in the high marsh and 2) high soil salinities in this zone suppress *Spartina* growth [36]. Thus, we hypothesized that top-down control of *Spartina* by scales may be influenced by soil salinity.

Infestations of *Haliaspis*, a *Spartina* specialist, appear highest in Southern California marshes. We have observed heavy infestations of *Haliaspis* from South San Diego Bay to Upper Newport Bay, California. Throughout this range, there is variation within and between marshes in *Haliaspis* densities. Although *Haliaspis* is believed to be indigenous to these marshes, there is some evidence that the distribution and abundance of this scale is expanding. For example, in the late 1980s and early 1990s, *Haliaspis* was abundant in human-constructed marshes but rare or absent in more natural marshes [34,37–39]. Our recent surveys suggest that *Haliaspis* has now expanded into some of these natural marshes at extremely high densities.

The two field experiments were conducted in the high zone of Sweetwater Marsh in South San Diego Bay. Sweetwater is a tidal, hypersaline marsh with soil salinities that frequently exceed 50 ppt. This marsh also has the highest maximum densities of scales per stem that we have encountered during our surveys of southern California marshes (maximum number of scales per stem >3000). Within Sweetwater Marsh, the highest scale densities per stem are observed in the high marsh zone. Salt marsh perennials including *Sarcorcornia virginica*, *Frankenia grandifolia*, *Jaumea carnosa*, *Batis maritima*, and *Suaeda californica* dominate this habitat. This research was permitted by the U.S. Fish and Wildlife Service (SUP 81680-12002) – the organization that manages Sweetwater Marsh.

Effects of scale removal (2011 Field experiment)

To examine the impact of scales on cordgrass performance, we manipulated scale presence on *Spartina* at Sweetwater Marsh (32.6412°N, 117.1142°W). On 20 May 2011, we randomly selected 20 *Spartina* stems separated by at least 1 m and that contained high densities of mature scales relative to nearby stems (70±19 scales per stem, mean ± SE). Stems were randomly assigned to Scale or No Scale treatments, and scales were removed from half of the stems (No Scale stems, N = 10) by brushing the adaxial surface of leaves with a soft toothbrush. This procedure effectively removes scales because they cluster on the adaxial surface of leaves. As a procedural control, we brushed the abaxial surface of Scale stems. For 20 weeks, we maintained treatments by manually removing scales from No Scale stems with a toothbrush every 1–2 weeks. Because scales have a short larval dispersal period followed by a sessile juvenile and adult period, manual removals effectively create Scale/No Scale treatments [40].

Every 1–2 weeks, we recorded Scale density (# per stem), and *Spartina* senescence, stem height, and seed presence. *Spartina* senescence was determined when the stem's meristem was completely brown. As a conservative estimate, all plants that had not senesced by our final sampling date (7 October 2011) were assumed to have senesced one week later. Stem height was determined by recording the maximum vertical distance from shoot base to the tip of the longest leaf. Seed presence was recorded as a binomial variable (present/absent). After 20 weeks (7 October 2011), we collected shoots, recorded seed presence, dried the shoots, and recorded final dry shoot biomass. We observed early senescence of many plants in the Scale treatment. These were included in all performance measurements. On 19 August 2011, we observed leaves that had been grazed entirely across their width. Although we did not identify this grazer, this type of damage is consistent with large, vertebrate grazers (e.g. birds). One Scale stem and two No Scale stems contained such damage. To exclude further vertebrate grazing, we placed tomato cages lined with chicken wire (opening ~3 cm) around all stems.

We compared two measures of growth of Scale and No Scale plants using two-tailed, two-sample t-tests. First, we calculated the proportional change in shoot height during the 20-week experiment. Second, we compared final dry shoot biomass. We used a two-tailed Fisher's Exact Test to compare the proportion of starting plants that ever produced seeds because 9 of 10 Scale plants did not produce seeds. We compared days to senescence using a two-tailed, two-sample t-test.

Effects of scale removal and salinity reduction (2011 Mesocosm experiment)

We examined the impact of reduced salinities on *Spartina*-scale interactions using mesocosms because logistical hurdles prevented us from reducing salinities in the field. A limitation of this design is that our mesocosm experiment may have been confounded by other factors. We sought to minimize differences between the field and our mesocosm experiment by conducting mesocosm research outdoors at SDSU's Coastal & Marine Institute Laboratory that is located 14 km from Sweetwater Marsh. This lab is located adjacent to San Diego Bay (i.e. the same water body from which all organisms were collected). Thus, ambient environmental conditions during our mesocosm studies were likely similar to field conditions.

We collected mud cores (~15 cm diameter × 15 cm deep) from Sweetwater Marsh that each contained a single *Spartina* stem with scales. We transferred these to the marine lab, and planted each stem-containing mud core into a 2.6 L plastic pot (15 cm height) with holes in the bottom for drainage. Stems were randomly assigned to each of four different treatments created by manipulating Scales (Scale, No Scale) and Salinity (Freshwater, Saltwater). While we recognize that freshwater exposure of southern California *Spartina* is extremely rare, this treatment was included because previous studies showed that salt marsh plants grow optimally in reduced salinity [36]. Most studies of plant stress define stress as any condition that reduces the optimal performance of plants (e.g. [41]). Additionally, ambient soil salinities as low as 15 ppt have been reported for *Spartina foliosa*, especially during high rainfall years [42].

All treatments had 8 replicates, with the exception of the Scale-Saltwater treatment, which had 7. Scales were removed from No Scale stems with a toothbrush and scale treatments were maintained weekly. Stems from each treatment were randomly assigned to shallow pools of a specified salinity treatment ($N = 2$). Salinity was manipulated by filling the pools to the soil-air interface with either freshwater or saltwater. The saltwater pools were filled with water from the flow-through seawater system that takes in water from San Diego Bay (~32 ppt). Freshwater pools were filled with water from a tap. Twice per week, we measured the pool salinity with a refractometer (Model RHS-10ATC, Aquatic Eco-Systems Inc.), drained and refilled each pool, and then re-measured pool salinity. We chose to measure the salinity of the pool water rather than the porewater (soil salinity) so that we would minimize disturbance to the soil in the pots. Thus, the magnitude by which our freshwater mesocosm reduced soil salinity is currently unclear. During the water changes, each pot was allowed to drain outside of the pools for ~30 minutes. All pools were kept outside in full sunlight.

We maintained these treatments for six months (April–October 2011). Every week, we recorded Scale density (# per stem starting 3 May 2011), and *Spartina* senescence and stem height (see above for details). Because the replicates of each treatment were divided amongst two replicate pools, we calculated an average for all replicates within each pool at each time point. During measurements, each pot was allowed to drain outside of the pools for 1–4 hours. We used stem height to calculate the average proportional growth at each time point. These data were analyzed with a Repeated Measures ANOVA. We visually inspected our data and the resulting covariance matrix to confirm that they met the assumptions of this test. However, we were concerned that our data violated the sphericity assumption, perhaps because of the high number of repeated measurements, so we applied a Greenhouse-Geisser correction to this analysis ($\varepsilon = 0.061$). We compared the number of days to senescence using ANOVA. Both ANOVA analyses examined the effect of the fixed factors of Scales (Scale, No Scale) and Salinity (Freshwater, Saltwater) on the response variable.

Effects of scale removal and salinity addition (2012 Field experiment)

To examine the influence of elevated salinity on the top-down control of *Spartina* by scales, we conducted a manipulative experiment at Sweetwater Marsh during 2012. We selected two sites in March 2012 [hereafter North (32.6412°N, 117.1142°W) and South (32.6388°N, 117.1099°W)]. These sites were selected because of their different elevations (+1.6 m and +2.1 m above Mean Lower Low Water, respectively). Also, we observed negative effects of scales on *Spartina* at the North site during the 2011 experiment. At each site, we randomly selected 1×1 m plots that contained *Spartina* and were separated by at least 1 m. Plots were randomly assigned to a Salinity treatment (Ambient or Elevated) and a Scale treatment (Scale or No Scale). We replicated each of four treatments at both sites ($N = 7$ or 8; North or South, respectively).

For two months, we pre-treated Elevated Salinity plots with 750 g of solar evaporated sea salt (produced by South Bay Salt Works, Chula Vista, CA) every 2 weeks during low tide. On 26 May 2012, we increased salt additions in Elevated Salinity plots to 1500 g every 2 weeks to create salinities near the maximum levels we have recorded in the field. We collected a single soil core (~5 cm deep) from each plot for soil salinity analysis on seven dates (31 March, 14 April, 28 April, 12 May, 8 June, 22 June, 6 July, 9 July). Soil salinity was not sampled after 9 July so as to minimize further disturbance to plots. On each of the first seven dates, we sampled plots prior to that day's salt addition so they reflected salinity differences persisting two weeks after the previous salt addition. In contrast, soil samples from the last date that soil salinities were measured (9 July) were collected 3 d after salt addition. This measurement provided better insight into soil salinity dynamics because the greatest impact of salt additions on soil salinity was observed 3 d after salt additions. We measured soil salinity using a modification of an existing protocol [43]. Briefly, soil cores were weighed, dried for 2 days at 60°C, and then reweighed. Then, we rehydrated cores with the original water volume plus 15 ml deionized water and measured the salinity of the slurry supernatant with a refractometer (Model RHS-10ATC, Aquatic Eco-Systems Inc.). We used these measurements to determine the soil salinity of the original samples.

We marked a single target *Spartina* stem near the center of each Scale and No Scale plot with flagging tape. Scales were removed from No Scale stems using toothbrushes. All stems were initially measured on 26 May 2012, and stems were re-measured and treatments were maintained every 2 weeks thereafter. We measured scale density and recorded the date of senescence. Because previous experiments suggested that plant growth did not extend past September, we stopped this experiment on 14 September 2012.

Salinities were compared using Repeated Measures ANOVA. We excluded the soil salinity data from 6 July 2012 from this analysis because we lost 6 samples during processing and these were not distributed evenly across treatments. We excluded one Enhanced Salinity, Scale plot from the North site because a sample was lost on 6/8/12. We also conducted a three factor ANOVA to examine the effects of Site (North, South), Scales (Scale, No Scale), and Salt (Ambient, Elevated) on the number of days to senescence. We excluded one plant assigned to the North, Scale, Ambient plot because scales never colonized it.

Results

Effects of scale removal (2011 Field experiment)

We observed dispersal of juvenile scales beginning in late June and ending in late July 2011 (Fig. S1). During this time, scale densities on Scale plants increased from 154±62 to 771±184 scales per stem (mean ± SE). Although we occasionally observed recruitment onto No Scale plants, the maximum density of scales we observed on No Scale stems was only 124±15 scales per stem. In contrast, we observed an order of magnitude greater maximum scale densities on Scale stems 1531±358 scales per stem. Thus, we effectively maintained differences in scale densities between our treatments.

Scales suppressed *Spartina* shoot height growth by 69% (Fig. 1a; t = 3.381, p = 0.003) and final dry shoot biomass by 37% (Fig. 1b; t = 2.531, p = 0.021). By 7 October 2011, 5 No Scale stems, but only 1 Scale stem, produced seeds. However, the proportion of stems producing seeds was not significantly different at $\alpha = 0.05$ (Fig. 1c; Fisher's Exact, p = 0.141).

No plants senesced during the first three months of the 2011 field experiment (Fig. 2). However, 70% of Scale stems senesced between mid-August and the end of the experiment. None of the No Scale stems senesced by the end of the experiment. If we conservatively assume that all No Scale plants senesced a week after the experiment ended, then No Scale stems senesced at least 27 days later than Scale stems (Fig. 2 inset; t = 3.452, p = 0.003). Importantly, the earlier senescence in Scale stems occurred during seed production.

Effects of scale removal and salinity reduction (2011 Mesocosm experiment)

Salinity of the pools in Freshwater treatments ranged between 0–3.5 ppt, with a mean of 0.3 ppt. In contrast, the salinity in Saltwater treatments was typically an order of magnitude larger (range = 27–59 ppt, mean = 39.8 ppt). *Spartina* in the high marsh can encounter salinities in excess of 80 ppt. Salinity tended to increase in pools after water changes because of evaporation from the pools. Similar to the 2011 field experiment, we saw a major dispersal event of scales in late June/early July (Fig. S2). However, we also saw a pronounced secondary dispersal event in mesocosms that started late in August 2011. Scale densities on Scale stems were higher in Freshwater than Saltwater treatments after the initial dispersal event (Fig. S2). This trend was not evident after the second dispersal event.

Plants senesced earlier in Saltwater treatments (days to senescence = 102±18 and 150±11 for Scale and No Scale stems, respectively) relative to Freshwater treatments (188±0 for both Scale treatments because no Freshwater plant had senesced by the end of the experiment; $F_{1,4} = 35.7$, p = 0.004). Although there was a trend for scales to cause earlier senescence in Saltwater treatments, neither Scales nor the interaction between Scales and Salinity influenced days to senescence ($F_{1,4} = 5.196$, p = 0.085).

Importantly, Salinity and Scales interacted to influence *Spartina* growth (Fig. 3, Table S1). Scales stimulated *Spartina* growth in Freshwater treatments but had no effect on growth in Saltwater treatments. As expected, *Spartina* growth increased during the experiment.

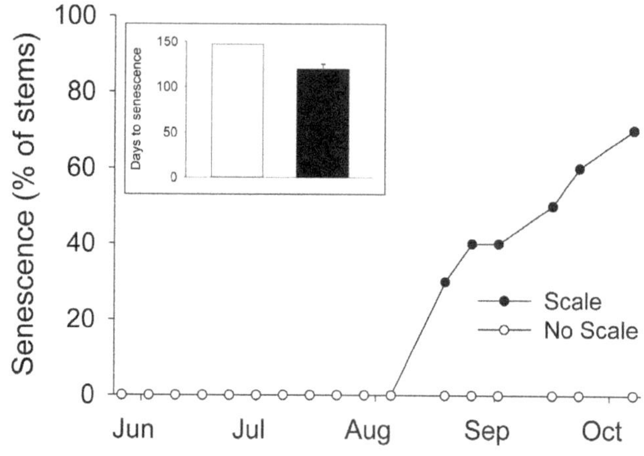

Figure 2. Scale effects on the timing of *Spartina* senescence during the 2011 field experiment. Senescence is reported as the cumulative percentage of stems senesced at each time point. Inset shows the number of days until *Spartina* senescence for No Scale (white bar) and Scale stems (black bar). Stems not senescing at the termination of the experiment were assigned a senescence date of 10/14/11. Because all 10 No Scale stems fell into this category but only 3 Scale stems did, the difference in days to senescence is conservative. This also explains the lack of error bars associated with the No Scale inset. Values in inset are means ± SE.

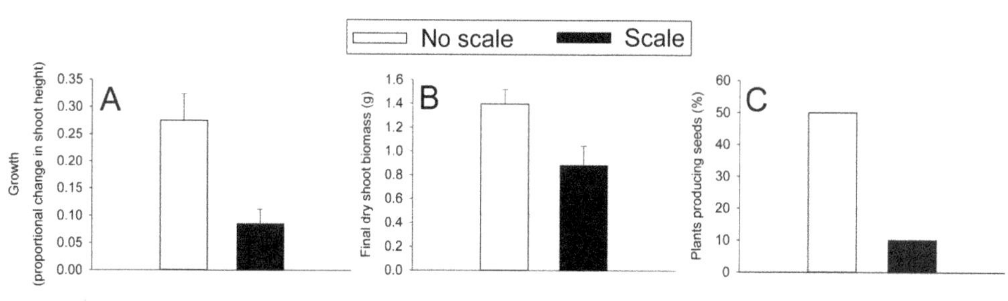

Figure 1. Scale effects on *Spartina* performance during the 2011 field experiment. The three performance measured included A) growth, B) final dry shoot biomass, and C) percentage of plants producing seeds. N = 10. Values are means ± SE.

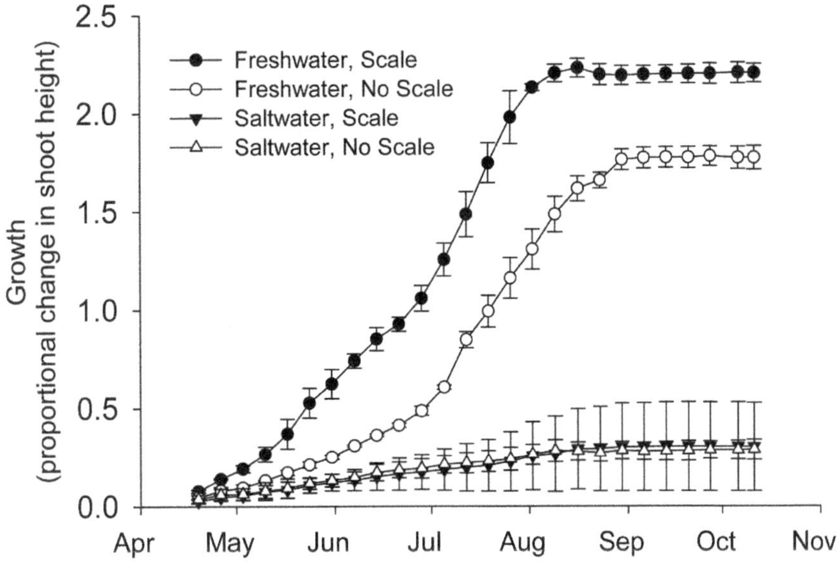

Figure 3. Effects of scales and reduced salinity on *Spartina* growth during the 2011 mesocosm experiment. Growth is reported as the proportional change in shoot height. Overcompensation to scale herbivory in freshwater treatments was absent in saltwater treatments. N = 7–8. Values are means ± SE.

Effects of scale removal and salinity addition (2012 Field experiment)

Plots with Elevated Salinity had higher porewater salinities than Ambient plots ($F_{1,54} = 21.2$, $p<0.001$; Fig. S3). The increase in soil salinity in Elevated plots was most pronounced three days after adding salt (9 July 2012) relative to two weeks after adding salt (all other dates, Fig. S3). Although overall soil salinities were similar at the North and South sites ($F_{1,54} = 0.663$, $p = 0.419$), Elevated Salinity plots contained 22% higher salinity on 9 July 2012 at the North versus the South site (84 vs. 69 ppt, respectively; $F_{1,28} = 4.20$, $p = 0.019$). Salinity did not differ with Scale treatment ($F_{1,54} = 0.117$, $p = 0.734$). All soil salinity measurements were within the range of soil salinities that can be currently experienced by *Spartina* at Sweetwater (we have measured a maximum soil salinity of 86 ppt).

Compared to the 2011 experiments (Figs. S1–S2), mean scale densities at the beginning of June were lower in 2012 (Fig. S4). For example, although our stems had ~100 scales per stem at the start of June 2011, mean scale densities in 2012 did not reach 100 scales per stem in any treatment until 6 July 2012. Thus, stems in the 2011 experiments may have experienced greater herbivore pressure than the 2012 experiment. The within-treatment variation of scale abundance increased with time, in part because of early plant senescence within this experiment. Because we removed scales from No Scale plants every two weeks, scale densities were always low on these treatments (Fig. S4).

For days to senescence, we observed a significant three-way interaction between Salt, Scale, and Site (Figure 4, Table S2). To better understand this interaction, we examined the interaction between Salt x Scale at each site separately using the appropriate error term. At the North site, the effect of Scales on senescence depended upon Salinity treatment ($F_{1,51} = 7.046$, $p = 0.011$). In contrast, we observed no interaction at the South site ($F_{1,51} = 0.437$, $p = 0.512$).

Discussion

The impact of scales on *Spartina* performance depended strongly upon environmental context. When grown under the lowest salinities (freshwater mesocosms), *Spartina* overcompensated for herbivory by growing taller shoots. However, increasing salinity stress switched the impact of scales from positive to neutral (2011 Mesocosm experiment). Additionally, scales negatively impacted *Spartina* by reducing growth and expediting senescence in the field. These data provide experimental evidence that *Haliaspis* can negatively impact *Spartina* and may have influenced the success of previous restoration efforts [33]. At the hypersaline field site where we observed negative effects of scales in 2011 and 2012, increasing soil salinity further returned the impact of scales to neutral (2012 Field experiment). Interestingly, scale densities and salinity were negatively related suggesting that scale impacts were not simply a linear function of scale abundance.

Previous studies examining the impact of environmental stress on plant compensatory responses largely focused on resource limitation as a stressor [8]. There, low nutrient conditions tended to weaken compensatory responses of monocots. Our study suggests that salinity stress acts similarly to weaken compensatory responses of *Spartina foliosa*. Reduced compensatory abilities with increasing salt stress might be particularly common in dominant salt marsh monocots like *Spartina* and *Juncus*. Salt stress decreases foliar nitrogen in *Spartina* [19], perhaps because of reduced uptake abilities [44,45]. Taken together, these observations suggest that salinity may alter the impact of marsh herbivores on marsh plants via reductions in compensatory abilities resulting from nutrient limitations. However, this hypothesis should be considered tentative, especially given the numerous exceptions to the general effect of resource limitation on monocots seen previously [8] and because salinity stress can directly impact other aspects of plant physiology [17].

While the interactive effects of environmental stress and herbivores can be complex and difficult to predict [7–9,46], there is some evidence that the impact of marsh herbivory might generally be more intense during intermediate salt stress because

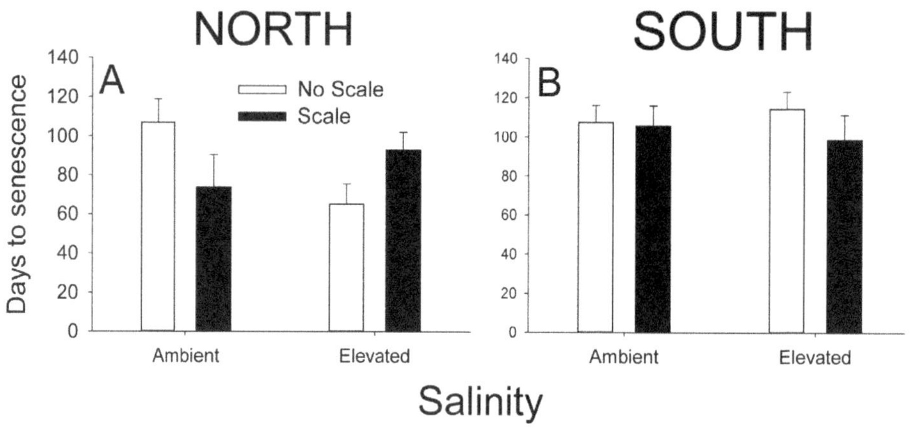

Figure 4. Effects of scales, elevated salinity, and site on *Spartina* senescence during the 2012 field experiment. Senescence is reported as the number of days after the start of the experiment that plants senesced at the North (A) and South Sites (B). N = 7–8. Values are means ± SE.

of reduced compensatory abilities. A study of a Northwest Atlantic Ocean salt marsh found that snail herbivory on *Spartina alterniflora* was much stronger in plots with experimental salt additions that increased salinity from ~35 to 57 ppt [27]. We observed positive or neutral effects of scale herbivory at low salinities (<4 to 40 ppt), negative effects at intermediate salinities (~50–55 ppt), and neutral effects at extremely high salinities (> 80 ppt). In contrast, previous research in a Southwest Atlantic Ocean salt marsh found that salt additions attempting to elevate ambient salinity to 60 ppt removed the negative effects of crab herbivores on marsh expansion seen at lower salinities [47]). In that case, crab impacts may have been low in salt addition plots because crabs avoided these areas [47]. Thus, salinity effects on marsh production may also depend upon how salinity directly affects herbivores.

Scales negatively influenced *Spartina* during the 2012 field experiment at only one of two sites (North site), suggesting that scale effects were site-specific. It is unlikely that salinity differences caused this specificity because soil salinity in Ambient Salinity plots showed no clear pattern with site. However, inundation can influence *Spartina* performance and two observations suggest that inundation differed between these sites. First, the North site was ~0.5 m lower in elevation. Second, soil salinities in Elevated Salinity plots three days after salt additions were much higher at the North site.

In addition to abiotic differences, at least two biotic factors may have influenced the site-specificity of scale effects. First, plant-plant competition may influence plant compensatory responses to herbivores [7,48]. Both sites have plant communities characteristic of high marsh but the makeup of these communities differed between the sites. *Jaumea carnosa* was more abundant and *Frankenia salina* and *Batis maritima* were less abundant at the site with negative scale effects. High marsh plants can suppress *Spartina* growth [49], and may influence *Spartina*'s ability to tolerate herbivory. Second, scale recruitment happened earlier at the North site, suggesting that scale pressure was greater at this site. Regardless of the mechanism, scales exerted a negative impact on *Spartina* at the site where overall plant senescence occurred earlier further supporting our contention that scale impacts depend upon plant stress.

Because we found the greatest impact of herbivores at intermediate salinities with reduced herbivore densities, our data suggest that the per capita effect of scales on *Spartina* was most negative at intermediate salinities. The realized impact of herbivory under stressful environmental conditions will also be determined by how the stressor impacts higher trophic levels, both directly and indirectly [50]. Unfortunately, the role of predators in this system is poorly known. Two observations suggest that predators may be important. First, potential predators such as beetles and parasitoids are present in these marshes [51,52]. Second, we observed a distinct secondary dispersal event in our mesocosms that was not present in the field in 2011. The lack of obvious predators in our mesocosms, but not the field, may have allowed for this dispersal event.

Understanding the complex interactions of salt stress and herbivory may help improve restoration strategies. In southern California, where loss of historic salt marshes has been extensive [53], a common restoration approach is to transplant *Spartina* into restoration sites. Perhaps because of the stress associated with transplantation, such approaches have been met with variable success [34,54,55]. One possible factor associated with restoration success has been heavy infestations of scale insects [34]. Our results suggest that scales may limit the success of restoration projects at intermediate soil salinities. Understanding this context-dependence may be particularly important given the anticipated changes in salinity with climate change that may alter the top-down control of marshes by herbivores.

Supporting Information

Figure S1 Densities of scales during the 2011 field experiment. Scales were removed from No Scale stems every 1–2 weeks. N = 10. Values are means ± SE.

Figure S2 Densities of scales during the 2011 mesocosm experiment. Scales were removed from No Scale stems every week. N = 7–8. Values are means ± SE.

Figure S3 Soil salinities of plots during the 2012 field experiment. Because salinity did not depend upon Scale treatment, data were combined for Scale and no Scale stems to facilitate comparisons of the North and South sites. Soil salinity was always measured two weeks after salt additions, except on July 9 when sampling occurred three days after salt additions. Values are means ± SE.

Figure S4 Densities of scales on stems during the 2012 field experiment. To highlight site and salt effects, data are shown separately for Scale (A) and No Scale (B) stems. Values are means ± SE.

Table S1 ANOVA table examining the effects of factors on *Spartina* growth during the 2011 Mesocosm experiment.

Table S2 ANOVA table examining the effects of factors on days to *Spartina* senescence during the 2012 Field experiment.

Acknowledgments

We thank Jeff Crooks and Brian Collins for providing access to Sweetwater Marsh. Steve Schroeter, Mark Page, and Luke Miller contributed thoughtful discussion on these topics. Heidi Long, Shelby Rinehart, and two anonymouns reviewers provided comments that improved the manuscript. Most importantly, data collection benefited from the significant efforts of the *Haliaspis* Chaps and *Spartina* Spartans (H. Ewertz, J. Hayward, P. Hitchcock, N. Nourn, and A. Warneke). This is Contribution No. 36 of the Coastal and Marine Institute Laboratory, San Diego State University.

Author Contributions

Conceived and designed the experiments: JL. Performed the experiments: JL LP. Analyzed the data: JL. Contributed reagents/materials/analysis tools: JL. Contributed to the writing of the manuscript: JL LP.

References

1. Menge BA, Sutherland JP (1987) Community regulation: variation in disturbance, competition, and predation in relation to environmental stress and recruitment. American Naturalist 130: 730–757.
2. Larsson S (1989) Stressful times for the plant stress-insect performance hypothesis. Oikos 56: 277–283.
3. White TCR (1969) An index to measure weather-induced stress of trees associated with outbreaks of psyllids in Australia. Ecology 50: 905–909.
4. Price PW (1991) Plant vigor and herbivore attack. Oikos 62: 244–251.
5. Huberty AF, Denno RF (2004) Plant water stress and its consequences for herbivorous insects: a new synthesis. Ecology 85: 1383–1398.
6. Rosenthal JP, Kotanen PM (1994) Terrestrial plant tolerance to herbivory. Trends in Ecology & Evolution 9: 145–148.
7. Maschinski J, Whitham TG (1989) The continuum of plant responses to herbivory: the influence of plant association, nutrient availability, and timing. American Naturalist 134: 1–19.
8. Hawkes CV, Sullivan JJ (2001) The impact of herbivory on plants in different resource conditions: a meta-analysis. Ecology 82: 2045–2058.
9. Wise MJ, Abrahamson WG (2005) Beyond the compensatory continuum: environmental resource levels and tolerance of herbivory. Oikos 109: 417–428.
10. Gonzales WL, Suarez LH, Molina-Montenegro MA, Gianoli E (2008) Water availability limits tolerance of apical damage in the Chilean tarweed *Madia sativa*. Acta Oecologica 34: 104–110.
11. Arthur AL, Hoffman AA, Umina PA (2013) Impact of *Halotydeus destructor* on crop seedlings at different plant developmental stages and levels of moisture stress. Environmental Entomology 42: 998–1012.
12. Louthan AM, Doak DF, Goheen JR, Palmer TM, Pringle RM (2013) Climatic stress mediates the impacts of herbivory on plant population structure and components of individual fitness. Journal of Ecology 101: 1074–1083.
13. Chase JM, Leibold MA, Downing AL, Shurin JB (2000) The effects of productivity, herbivory, and plant species turnover in grassland food webs. Ecology 81: 2485–2497.
14. Willis AJ, Ash JE, Groves RH (1993) Combined effects of two arthropod herbivores and water stress on growth of Hypericum species. Oecologia 96: 517–525.
15. Joern A, Mole S (2005) The plant stress hypothesis and variable responses by blue grama grass (*Bouteloua gracilis*) to water, mineral nitrogen, and insect herbivory. Journal of Chemical Ecology 31: 2069–2090.
16. Grinnan R, Carter TE Jr, Johnson MTJ (2013) The effects of drought and herbivory on plant-herbivore interactions across 16 soybean genotypes in a field experiment. Ecological Entomology 38: 290–302.
17. Munns R (2002) Comparative physiology of salt and water stress. Plant Cell and Environment 25: 239–250.
18. Stewart GR, Lahrer F (1980) Accumulation of amino acids and related compounds in relation to environmental stress. In: Miflin BJ, editor. Biochemistry of plants: a comprehensive treatise on amino acids and derivatives. New York: Academic Press. 609–635.
19. Bowdish TI, Stiling P (1998) The influence of salt and nitrogen on herbivore abundance: direct and indirect effects. Oecologia 113: 400–405.
20. Stiling P, Moon D (2005) Are trophodynamic models worth their salt? Top-down and bottom-up effects along a salinity gradient. Ecology 86: 1730–1736.
21. Bradley PM, Morris JT (1991) Relative importance of ion exclusion, secretion and accumulation in *Spartina alterniflora* Loisel. Journal of Experimental Botany 42: 1525–1532.
22. Pennings SC, Callaway RM (1992) Salt marsh plant zonation: the relative importance of competition and physical factors. Ecology 73: 681–690.
23. Zedler JB, Nordby CS, Kus BE (1992) The ecology of Tijuana Estuary: a National Estuarine Research Reserve. Washington, D.C.: NOAA Office of Coastal Resource Management, Sanctuaries and Research Division.
24. Moon DC, Stiling P (2002) Top-down, bottom-up, or side to side? Within-trophic-level interactions modify trophic dynamics of a salt marsh herbivore. Oikos 98: 480–490.
25. Jimenez JM, Wieski K, Marczak LB, Ho C-K (2012) Effects of an omnivorous katydid, salinity, and nutrients on a planthopper-*Spartina* food web. Estuaries and Coasts 35: 475–485.
26. Goranson CE, Ho CK, Pennings SC (2004) Environmental gradients and herbivore feeding preferences in coastal salt marshes. Oecologia 140: 591–600.
27. Silliman BR, van de Koppel J, Bertness MD, Stanton LE, Mendelssohn IA (2005) Drought, snails, and large-scale die-off of southern U.S. salt marshes. Science 310: 1803–1806.
28. Long JD, Mitchell JL, Sotka EE (2011) Local consumers induce resistance differentially between *Spartina alterniflora* populations in the field. Ecology 92: 180–188.
29. Adam P (2002) Saltmarshes in a time of change. Environmental Conservation 29: 39–61.
30. Hopkinson CS, Lugo AE, Alber M, Covich AP, Van Bloem SJ (2008) Forecasting effects of sea-level rise and windstorms on coastal and inland ecosystems. Frontiers in Ecology and the Environment 6: 255–263.
31. Cockfield SD, Potter DA (1986) Interaction of Euonymus scale (Homoptera: Diaspididae) feeding damage and severe water stress on leaf abscission and growth of *Euonymus fortunei*. Oecologia 71: 41–46.
32. Vranjic JA, Ash JE (1997) Scale insects consistently affect roots more than shoots: the impact of infestation size on growth of eucalypt seedlings. Journal of Ecology 85: 143–149.
33. Howard FW, Hamon A, McLaughlin M, Weissling T, Yang S-L (1999) *Aulacaspis yasumatsui* (Hemiptera: Sternorrhyncha: Diaspididae), a scale insect pest of cycads recently introduced into Florida. Florida Entomologist 82: 14–27.
34. Boyer KE, Zedler JB (1996) Damage to cordgrass by scale insects in a constructed salt marsh: effects of nitrogen additions. Estuaries 19: 1–12.
35. Zedler JB (1993) Canopy architecture of natural and planted cordgrass marshes: selecting habitat evaluation criteria. Ecological Applications 3: 123–138.
36. Crain CM, Silliman BR, Bertness SL, Bertness MD (2004) Physical and biotic drivers of plant distribution across estuarine salinity gradients. Ecology 85: 2539–2549.
37. Boyer KE (1994) Scale insect damage in constructed salt marshes: nitrogen and other factors. San Diego, CA: San Diego State University. 81 p.
38. Covin JD, Zedler JB (1988) Nitrogen effects on *Spartina foliosa* and *Salicornia virginica* in the salt marsh at Tijuana Estuary, California. Wetlands 8: 51–65.
39. Johnson KM (1991) The effects of host quality on a phytophagous insect (Homoptera: Delphacidae) and its predators in a California salt marsh system. San Diego, CA: San Diego State University. 85 p.
40. Spitzer BW (2004) Maternal effects in the soft scale insect Saissetia coffeae (Hemiptera: Coccidae). Evolution 58: 2452–2461.
41. Louda SM, Collinge SK (1992) Plant resistance to insect herbivores: a field test of the environmental stress hypothesis. Ecology 73: 153–169.
42. Trnka S, Zedler JB (2000) Site conditions, not parental phenotype, determine the height of *Spartina foliosa*. Estuaries 23: 572–582.
43. Pennings SC, Moore DJ (2001) Zonation of shrubs in western Atlantic salt marshes. Oecologia 126: 587–594.
44. Bandyopadhyay BK, Pezeshki SR, DeLaune RD, Lindau CW (1993) Influence of soil oxidation-reduction potential and salinity on nutrition, N^{15} uptake, and growth of *Spartina patens*. Wetlands 13: 10–15.
45. Brown CE, Pezeshki SR, DeLaune RD (2006) The effects of salinity and soil drying on nutrient uptake and growth of *Spartina alterniflora* in a simulated tidal system. Environmental and Experimental Botany 58: 140–148.
46. Wise MJ, Abrahamson WG (2007) Effects of resource availability on tolerance of herbivory: a review and assessment of three opposing models. American Naturalist 169: 443–454.
47. Canepuccia AD, Alberti J, Daleo P, Pascual J, Farina JL, et al. (2010) Ecosystem engineering by burrowing crabs increases cordgrass mortality caused by stem-boring insects. Marine Ecology Progress Series 404: 151–159.

48. Hjalten J, Danell K, Ericson L (1993) Effects of simulated herbivory and intraspecific competition on the compensatory ability of birches. Ecology 74: 1136–1142.
49. Pennings SC, Grant M-B, Bertness MD (2005) Zonation in low-latitude salt marshes: disentangling the roles of flooding, salinity, and competition. Journal of Ecology 93: 159–167.
50. Denno RF, Gratton C, Peterson MA, Langellotto GA, Finke DL, et al. (2002) Bottom-up forces mediate natural-enemy impact in a phytophagous insect community. Ecology 83: 1443–1458.
51. Williams K (1990) Control of scale insect infestation at Chula Vista Wildlife Reserve. San Diego, California.
52. Japoshvili G, Russell E (2012) A new parasitization record of *Haliaspis spartinae* (Diaspididae) and *Encarsia ellisvillensis* sp. nov. (Chalcidoidea: Aphelinidae) from the United States. Annals of the Entomological Society of America 105: 493–497.
53. Macdonald K (1990) South San Diego Bay Enhancement Plan. San Diego, California.
54. Langis R, Zalejko M, Zedler JB (1991) Nitrogen assessments in a consructed and a natural salt marsh of San Diego Bay. Ecological Applications 1: 40–51.
55. Zedler JB, Callaway JC (1999) Tracking wetland restoration: do mitigation sites follow desired trajectories. Restoration Ecology 7: 69–73.

8

QTLs for Tolerance of Drought and Breeding for Tolerance of Abiotic and Biotic Stress: An Integrated Approach

Shalabh Dixit[1], B. Emma Huang[2], Ma. Teresa Sta Cruz[1], Paul T. Maturan[1], Jhon Christian E. Ontoy[1], Arvind Kumar[1]*

1 International Rice Research Institute (IRRI), Los Baños, Laguna, Philippines, **2** Computational Informatics, CSIRO, Dutton Park, Queensland, Australia

Abstract

Background: The coupling of biotic and abiotic stresses leads to high yield losses in rainfed rice (*Oryza sativa* L.) growing areas. While several studies target these stresses independently, breeding strategies to combat multiple stresses seldom exist. This study reports an integrated strategy that combines QTL mapping and phenotypic selection to develop rice lines with high grain yield (GY) under drought stress and non-stress conditions, and tolerance of rice blast.

Methodology: A blast-tolerant BC_2F_3-derived population was developed from the cross of tropical *japonica* cultivar Moroberekan (blast- and drought-tolerant) and high-yielding *indica* variety Swarna (blast- and drought-susceptible) through phenotypic selection for blast tolerance at the BC_2F_2 generation. The population was studied for segregation distortion patterns and QTLs for GY under drought were identified along with study of epistatic interactions for the trait.

Results: Segregation distortion, in favour of Moroberekan, was observed at 50 of the 59 loci. Majority of these marker loci co-localized with known QTLs for blast tolerance or NBS-LRR disease resistance genes. Despite the presence of segregation distortion, high variation for DTF, PH and GY was observed and several QTLs were identified under drought stress and non-stress conditions for the three traits. Epistatic interactions were also detected for GY which explained a large proportion of phenotypic variance observed in the population.

Conclusions: This strategy allowed us to identify QTLs for GY along with rapid development of high-yielding purelines tolerant to blast and drought with considerably reduced efforts. Apart from this, it also allowed us to study the effects of the selection cycle for blast tolerance. The developed lines were screened at IRRI and in the target environment, and drought and blast tolerant lines with high yield were identified. With tolerance to two major stresses and high yield potential, these lines may provide yield stability in rainfed rice areas.

Editor: Keqiang Wu, National Taiwan University, Taiwan

Funding: The work was funded by Global Rice Science Partnership (GRiSP), Generation Challenge Program (GCP), Federal ministry for economic cooperation and development (BMZ), Germany, and the Bill and Melinda Gates Foundation (BMGF). The funders had no role in study design, data collection and analysis, decision to publish, or preparation of the manuscript.

Competing Interests: The authors have declared that no competing interests exist.

* Email: a.kumar@irri.org

Introduction

Developing crop varieties with higher yield potential is a challenging task in modern-day plant breeding. The Green Revolution led to a historic breakthrough in increasing the yield potential of crops, such as rice and wheat, through the development of semi-dwarf varieties that has fed the increasing population till now. However, it is predicted that a 60% increase in agricultural production needs to be achieved by 2050 to feed the ever-growing population [1].

Unlike other cereals, rice is cultivated across a much wider range of ecosystems, including irrigated, rainfed upland, rainfed lowland, flooded, and deepwater ecosystems across tropical and temperate climate conditions. Despite this diversity of cultivation systems and environments, the major focus of research in increasing rice yield potential has been for the irrigated ecosystem in the post-Green Revolution period. However, no major breakthrough in breaking the yield potential barrier has been achieved until now.

The low-yielding rainfed lowland rice ecosystem has great potential for yield improvement. This ecosystem occupies about 38% of the total rice area, but contributes only 21% to total rice production [2]. In Asia alone, 34 million ha of rainfed lowland rice and 8 million ha of rainfed upland rice [3] experience drought stress of varying intensity almost every year. These drought events are often coupled with increased incidence of other biotic stresses, such as blast, brown spot, and bacterial blight, leading to a further

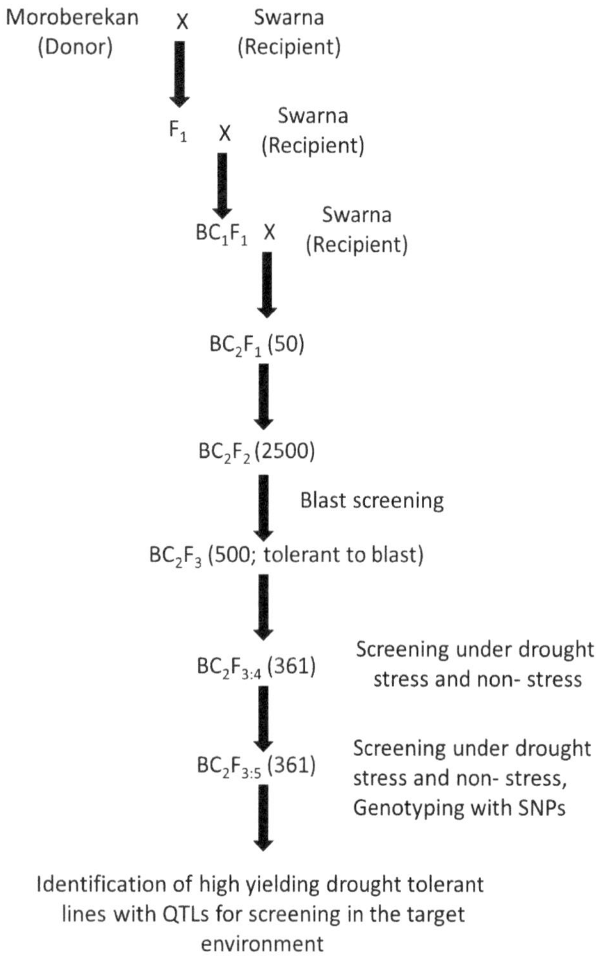

Figure 1. Population development and phenotypic screening strategy used to develop a blast-tolerant BC_2F_3-derived mapping population for mapping QTLs for high grain yield under drought stress and non-stress conditions.

decline in rice yield. Rice yields in these ecosystems remain low at 1.0–2.5 t ha^{-1} [4].

Despite the importance of drought as a major factor in yield reduction in rainfed ecosystems, few efforts have been made to develop high-yielding drought-tolerant cultivars. As a result, a large proportion of these rainfed areas are covered with high-yielding varieties such as Swarna, Samba Mahsuri, IR64, and MTU1010 that were specifically developed for irrigated ecosystems and suffer heavy yield losses under drought. On the other hand, several drought-tolerant cultivars, such as N22 and Moroberekan, have evolved in these rainfed ecosystems through natural or artificial selection over the years. These cultivars carry genes for tolerance of a wide range of biotic and abiotic stresses, but lack farmers' and consumers' preference because of their low yield potential and poor grain quality. In many cases, a single cultivar shows tolerance of a combination of biotic and abiotic stresses. This provides a unique opportunity for rice breeders to develop high-yielding varieties with tolerance of multiple abiotic and biotic stresses.

The lack of effective selection criteria and low heritability of grain yield (GY) under drought have been reported as the two major reasons for the slow progress in breeding drought-tolerant

Table 1. Details of experiments conducted for identification of QTLs and screening of drought-tolerant lines identified from the mapping population at IRRI and in IRRI-SA.

Experiment	Season	Ecosystem	Population size	Location	Experimental design	Replications	Plot size
1	DS2011	LSS	361	IRRI	38×10 AL	2	1.0
2	DS2012	LMS	361	IRRI	38×10 AL	2	2.0
3	DS2011	LNS	361	IRRI	38×10 AL	2	1.0
4	DS2012	LNS	361	IRRI	38×10 AL	2	2.0
5	WS2012	USS	99	IRRI	12×10 AL	2	1.0
6	DS2013	LSS	77	IRRI	10×10 AL	2	2.0
7	DS2013	LNS	77	IRRI	10×10 AL	2	2.0
8	WS2013	LMiS	95	IRRI-SA	8×13 AL	2	2.4
9	WS2013	LNS	95	IRRI-SA	8×13 AL	2	3.2

DS = dry season, WS = wet season, LSS = lowland severe stress, LMS = lowland moderate stress, LNS = lowland non-stress, UMS = upland moderate stress.

varieties [5]. With increased precision in large-scale phenotyping, many studies have shown moderate to high heritability for GY under drought [6–8]. These studies used direct selection for GY in progenies derived from high-yielding drought-susceptible varieties and drought-tolerant donors as a successful strategy to improve yield under drought. This strategy has also proven beneficial to combining high yield potential with drought tolerance and has resulted in the development and release of several high-yielding rice varieties with good yield under drought in countries such as India, Nepal, Bangladesh, Indonesia, and the Philippines [9]. Along with the release of varieties through conventional breeding approaches, several large-effect QTLs for GY under drought have also been identified [2,10–15]. Similarly, several studies have also reported independent and epistatic QTLs for GY and other traits of agronomic importance [16–19]. However, only a few studies have reported QTLs leading to a yield advantage under both drought stress and non-stress environments, taking grain yield as the primary selection criterion. Occurrence of drought is normally coupled with increased incidence of diseases such as rice blast, brown spot, and bacterial blight. However, very few studies have been undertaken to understand the genetics of these abiotic and biotic stresses simultaneously in a mapping population.

In our study, an advanced backcross (AB) QTL approach was used on a large backcross population developed from a cross between tropical japonica, drought- and rice blast-tolerant donor Moroberakan with highly popular recipient *indica* rice variety Swarna. The objectives were to: (1) develop a strategy to combine breeding approaches for tolerance of blast and drought with QTL mapping, (2) understand the effect of a selection cycle for blast tolerance during population development on segregation distortion and population structure, (3) identify QTLs for GY under drought stress and non-stress conditions, and (4) develop high-yielding blast- and drought-tolerant breeding lines from the mapping population and evaluate them in the target environment.

Materials and Methods

Experiments for this study were conducted at the Experiment Station of the International Rice Research Institute (IRRI), Los Baños, Laguna, Philippines, during the dry seasons (DS) of 2011, 2012, and 2013 and the wet season (WS) of 2012. IRRI is located at 14°13′ N and 121°15′ E, at an elevation of 21 m above mean sea level. The study also presents results from two experiments conducted at the IRRI South Asia breeding hub (IRRI-SA) located at the International Crop Research Institute for Semi-Arid Tropics (ICRISAT), Patancheru, Hyderabad (AP), India in WS2013. ICRISAT is located at 17°31′ N and 78°15′ E at an elevation of 516 m above sea level.

Plant material

A BC_2F_3-derived population developed from the cross of Moroberekan and Swarna was used in this study. Moroberekan, the tolerant donor, is an upland adapted tropical japonica [20] cultivar from New Guinea. It is a long-duration line with sturdy plant type, deep roots, and tolerance of drought and rice blast. However, this variety has poor yield potential due to its low tillering ability and lower number of grains per panicle. On the other hand, Swarna (MTU7029), the drought-susceptible recipient parent used in this study, is a lowland-adapted high-yielding indica [2,21] variety derived from the cross Vashishtha × Mahsuri. It is a long-duration semi-dwarf variety with high tillering ability and grain yield but is highly susceptible to diseases such as rice blast and bacterial blight. This variety is grown on a large area in rainfed and irrigated ecologies across India, Nepal, and Bangladesh and is regarded as a mega-variety of rice.

Figure 1 presents the strategy used to develop the mapping population. The BC_2F_1 plants developed through the cross of the two parents were advanced through the single seed descent (SSD) method to develop a large BC_2F_2 population. This population was screened for tolerance of rice blast. Blast-tolerant plants identified through this screening were advanced to the BC_2F_3 generation through SSD. Seeds from BC_2F_3 single plants were used to develop a $BC_2F_{3:4}$ mapping population to screen for GY under drought stress and non-stress conditions. Furthermore, single plant selections were made from the mapping population under drought stress. These selections were advanced, purified, and screened at IRRI, Philippines and IRRI-SA, India to identify lines that perform the best in both environments.

Stress ecosystems

Field experiments for mapping QTLs were conducted in upland and lowland ecosystems. Throughout the study, the term "upland" refers to field experiments conducted under dry direct-seeded, unpuddled, aerobic conditions, whereas "lowland" refers to field experiments conducted under flooded, puddled, transplanted, and anaerobic conditions. Experiments were conducted under both drought stress and non-stress conditions. In this study, fully irrigated experiments with no drought stress imposed throughout the season are referred to as non-stress experiments; whereas, those experiments on which drought stress was imposed during the reproductive stage (starting at 40 and 50 days after seeding, respectively, under upland and lowland conditions) are referred to as drought stress experiments. In order to further classify the stress experiments into mild, moderate, or severe stress, the percentage reduction in yield as compared to non-stress conditions was used as a criterion [6].

Phenotypic evaluation and data collection

Details of all the experiments, including population size, experimental design, number of replications, and plot size are presented in Table 1. A population of 361 BC_2F_3-derived lines was screened under lowland stress and non-stress conditions in DS2011 and DS2012. A subset of 100 lines from this population was screened under upland stress conditions in WS2012, along with parents Moroberekan and Swarna. Breeding lines developed through plant selection from this population were screened under lowland stress and non-stress conditions in DS2013 at IRRI and WS2013 at IRRI-SA. All experiments were conducted as replicated field trials in α lattice design in continuous plots. The experimental setup has proved effective for screening large rice mapping populations for grain yield under drought using smaller plot sizes. The method has been used in several QTL identification studies, which have led to the identification of large and consistent QTLs now being used in marker-assisted breeding (MAB) programs [2,10–15]. Field and crop management were carried out using the protocol presented in File S1.

In all experiments, data on days to flowering (DTF), plant height (PH) at maturity, and grain yield (GY) were recorded. DTF was recorded as the number of days from seeding to the day that 50% of the plants had flowering tillers. The PH of three plants from each plot were measured at maturity from the ground to the tip of the tallest tiller, and then averaged to get the mean PH for analysis. GY from each plot was harvested at physiological maturity, dried to a moisture content of 14%, and then weighed [11]. This data set was then used to calculate the quantity used in QTL analysis, namely GY of the genotypes in kg ha^{-1}.

Generation of genotypic data

Fresh leaves for all lines were collected and freeze-dried. DNA was extracted from freeze-dried leaf samples by a modified CTAB method in deep-well plates. The DNA was then quantified and purified and lines were genotyped using KASPar SNP assays. These SNPs were selected as subsets from the set of 1,536 and 44K SNP chips [22,23] converted to SNP assays and made available through the integrated breeding platform (https://www.integratedbreeding.net/snp-marker-conversion). A total of 2,015 SNP markers were screened for polymorphism between the two parents. Out of the 2,015 SNPs screened for polymorphism, 591 polymorphic SNP loci were identified. A total of 200 polymorphic markers spaced uniformly across the genome were used to generate the genotypic data of the population. To ensure data quality, lines having percentages higher than expected of heterozygote alleles (>40%), Moroberekan alleles (>40%), or missing data (>30%) were filtered before analysis. Furthermore, markers were removed if the frequency of genotypes deriving from Moroberekan (homozygote and heterozygote) was <2%, as they were essentially monomorphic.

Statistical analysis

ANOVA and means. Data from all experiments for computation of means and standard error of difference (SED) were analyzed using CROPSTAT version 7.2.3. Mixed model analysis of data from individual years was carried out using the model

$$y_{ijk} = \mu + g_i + r_j + b_k(r_j) + e_{ijk}$$

where y_{ijk} is the measurement recorded in a plot, μ is the overall mean, g_i is the effect of the ith genotype, r_j is the effect of the jth replicate, $b_k(r_j)$ is the effect of the kth block within the jth replicate, and e_{ijk} is the error. Genotypic effects were considered fixed and the replicates and block effects were random.

A combined analysis was conducted for lowland experiments to obtain line means across years (2011 and 2012) under stress and non-stress conditions. The model used was

$$y_{ijkl} = \mu + g_i + m_l + r_j(m_l) + b_k[r_j(m_l)] + g_i m_l + e_{ijkl}$$

where y_{ijkl} is the measurement recorded in a plot, μ is the overall mean, g_i is the effect of the ith genotype, m_l is the effect of the lth year, $r_j(m_l)$ is the effect of the jth replicate within the lth year, $b_k[r_j(m_l)]$ is the effect of the kth block within the jth replicate of the lth year, $g_i m_l$ is the effect of the interaction between the ith genotype and lth year, and e_{ijkl} is the error. The effects of replicate within year and block within replicate of year were considered random, whereas the other effects were considered fixed. Due to the high variability for DTF in the population, adjusted mean GY were also calculated by taking DTF as covariate using the above models for QTL mapping. This was done to estimate the mean GY of lines free from the variation caused due to their difference in flowering time (File S3).

Heritability. Broad-sense heritability (H) of the traits for single years was calculated as shown below:

$$H = \frac{\sigma^2_{G'}}{\sigma^2_{G'} + \left[\frac{\sigma^2_e}{r}\right]}$$

where $\sigma^2_{G'}$ is the genetic variance, σ^2_E is the plot residual variance and r is the number of replications. The formula used to calculate H across years (2011 and 2012) was

$$H = \frac{\sigma^2_G}{\sigma^2_G + \left[\frac{\sigma^2_{GM}}{M} + \frac{\sigma^2_e}{rM}\right]}$$

[24], where σ^2_G is the genetic variance, σ^2_{GY} is the genotype × year variance, σ^2_E is the plot residual variance, and r and M are the number of replicates and years, respectively. Variance components were calculated using the REML algorithm of PROC VARCOMP of SAS V.9.1 [25]. A class analysis was conducted to understand the effect of DTF and PH on GY. Lines were divided into five DTF and GY classes and heat maps were constructed showing GY response in all possible combinations of these classes under drought stress and non-stress conditions. 'Levelplot' in R was used to construct the heat maps.

QTL mapping

Genetic analysis of the population included estimation of segregation distortion, composite interval mapping (CIM), and estimation of epistatic effects. The level of segregation distortion was assessed using the function 'geno.table' in the R package qtl [26]. In order to understand the co-location of markers showing significant distortion to blast QTLs and known genes for disease resistance in rice, a physical map of SNP markers (used in this study), QTLs for blast resistance (reported in Moroberekan-derived mapping populations), and known disease resistance genes in rice was constructed. Information about disease resistance genes as well as those within the identified QTLs and their locations was obtained from the MSU Rice Genome Annotation Project, whereas QTL information was obtained from www.gramene.org and used along with the SNP markers from this study to construct a map using GGT 2 [27]. To map QTLs for GY, DTF, and PH, the genome was scanned for QTLs at a step size of 5 cM, using composite interval mapping ('cim') in R/qtl [26] with 5 cofactors. The genome-wide significance level was determined based on 1,000 permutations. A two-dimensional scan was also performed to identify epistatic QTLs using 'scantwo' in R/qtl with 1,000 permutations.

Results

Population development and segregation patterns

Our study reports a population development strategy using donors tolerant of more than one biotic and/or abiotic stress for simultaneous QTL mapping and pureline development in elite genetic background (Fig. 1). In this study, tropical japonica cultivar Moroberekan was used as the donor parent for tolerance of drought and rice blast, whereas *indica* variety Swarna was used as the recipient. Our strategy involved developing a large BC_2F_2 population and screening this population for blast tolerance. Only blast-tolerant plants were advanced further to develop the BC_2F_3-derived mapping population, which was used to detect QTLs for grain yield under drought and to develop purelines by plant selection in the mapping population. Since selection pressure was applied for blast tolerance, segregation distortion was expected within the mapping population. The genotypic data generated from BC_2F_3-derived lines were used to conduct the test for segregation distortion. A total of 50 and 9 marker loci showed segregation distortion in favour of the donor parent Moroberekan and recipient parent Swarna, respectively (Table S1). The largest segregation distortions were seen on chromosomes 3, 4, and 7 (Table S1). Segregation distortion in favour of Swarna was seen at

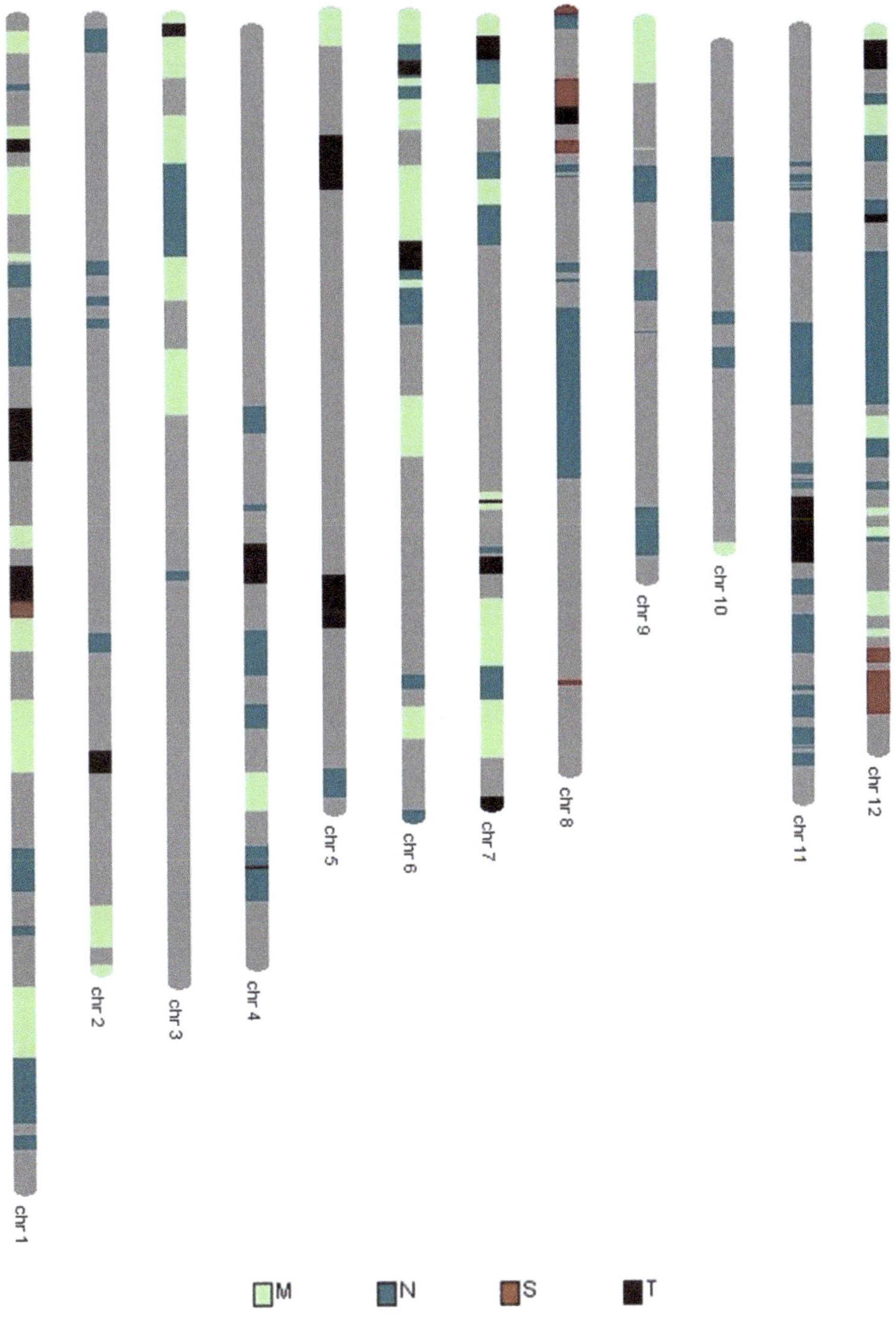

Figure 2. Physical map of markers showing segregation distortion, known genes for disease resistance in rice, and QTLs reported for blast resistance in the Moroberekan/CO39 population. M = markers showing distortion in favour of Moroberekan, S = markers showing distortion in favour of Swarna, N = genes coding for NBS-LRR proteins, and T = QTLs reported for blast resistance in the Moroberekan/CO39 population.

chromosomes 1, 8, and 12. The physical map of SNP markers from this study, known disease-resistance genes, and meta QTLs for blast resistance along the rice genome, highlighted NBS-LRR disease resistance genes close to the markers showing significant segregation distortion in favour of Moroberekan on chromosomes 3, 6, and 7 (Fig. 2). Since the majority of blast resistance genes code for NBS-LRR type proteins, the presence of these genes close to the markers showing distortion may be expected as a result of phenotypic selection for blast tolerance.

Phenotypic variation

Table 2 presents the results of ANOVA, means, heritability, and percentage yield reduction (under drought stress) in the experiments conducted at IRRI and IRRI-South Asia (IRRI-SA)

Table 2. Means ± SED (M), broad-sense heritability (H) and P values for grain yield, days to flowering, and plant height and percentage yield reduction (YR) of trial mean grain yield under stress and non-stress conditions.

	Grain yield (kg ha^{-1})					Days to flowering					Plant height (cm)					
Experiment	**P1**	**P2**	**M**	**H**	***P***	**P1**	**P2**	**M**	**H**	***P***	**P1**	**P2**	**M**	**H**	***P***	**YR**
1	88	87	677±563	0.60	****	109	110	99±4	0.85	****	72	56	68±9	0.8	****	93
2	954	2326	2115±688	0.61	****	100	102	92±5	0.83	****	102	76	82±8	0.87	****	45
Combined	384	1211	1390±626	0.55	****	104	106	95±5	0.87	****	87	66	75±9	0.70	****	66
3	2234	4552	4370±868	0.78	****	97	98	91±5	0.72	****	126	91	102±8	0.84	****	
4	1248	4012	3859±1049	0.54	****	93	93	88±7	0.78	****	118	92	99±10	0.69	****	
Combined	1701	4282	4118±676	0.71	****	95	95	90±5	0.72	****	122	90	101±6	0.80	****	
5	691	89	827±357	0.92	****	95	112	100±5	0.88	****	107	75	89±7	0.85	****	80
6	0	0	1274±433	0.93	****	105	105	87±2	0.98	****	84	58	74±6	0.86	****	83
7	2893	6857	7380±1628	0.67	****	94	95	86±2	0.95	****	128	91	100±7	0.82	****	
8	NA	NA	4757±856	0.83	****	NA	NA	94±2	0.85	****	NA	NA	97±6	0.59	***	40
9	NA	NA	7895±1027	0.31	****	NA	NA	93±3	0.92	****	NA	NA	116±7	0.61	****	

[a]P1 = Moroberekan, P2 = Swarna, M = population mean, H = heritability, **** = significant at 0.01% *P* level.

Table 3. Genetic correlation between grain yield (GY), days to flowering (DTF), and plant height (PH) under lowland drought stress and non-stress conditions.

	DTF(LSS)	**PH(LSS)**	**GY(LSS)**	**DTF(LMS)**	**PH(LMS)**	**GY(LMS)**	**DTF (LNS I)**	**PH (LNS I)**	**GY(LNS I)**	**DTF (LNS II)**	**PH(LNS II)**	**GY (LNS II)**
DTF(LSS)	1											
PH(LSS)	−0.597***	1										
GY(LSS)	−0.813***	0.551***	1									
DTF(LMS)	0.936***	−0.495***	−0.716***	1								
PH(LMS)	−0.042	0.727***	−0.018	−0.040	1							
GY(LMS)	−0.519***	0.168***	0.711***	−0.679***	−0.145**	1						
DTF (LNS I)	0.998***	−0.699***	−0.794***	0.869***	−0.166***	−0.568***	1					
PH (LNS I)	−0.238***	0.831***	0.185***	−0.170***	0.969***	−0.063	−0.297***	1				
GY(LNS I)	−0.297***	0.348***	0.308***	−0.318***	0.264***	0.579***	−0.403***	0.317***	1			
DTF (LNS II)	0.877***	−0.426***	−0.672***	0.845***	−0.137*	−0.664***	0.810***	−0.166**	−0.226***	1		
PH(LNS II)	0.114*	0.668***	−0.204***	0.095	0.801***	−0.238***	−0.086	0.894***	0.123*	0.037	1	
GY (LNS II)	−0.401***	0.245***	0.471***	−0.457***	0.084	0.949****	−0.482***	0.058	0.861***	−0.368***	−0.095	1

LSS = lowland severe stress, LMS = lowland moderate stress, LNS = lowland non-stress. *, **, ***, **** = significant at 5%, 1%, 0.1% 0.01% *P* levels, respectively.

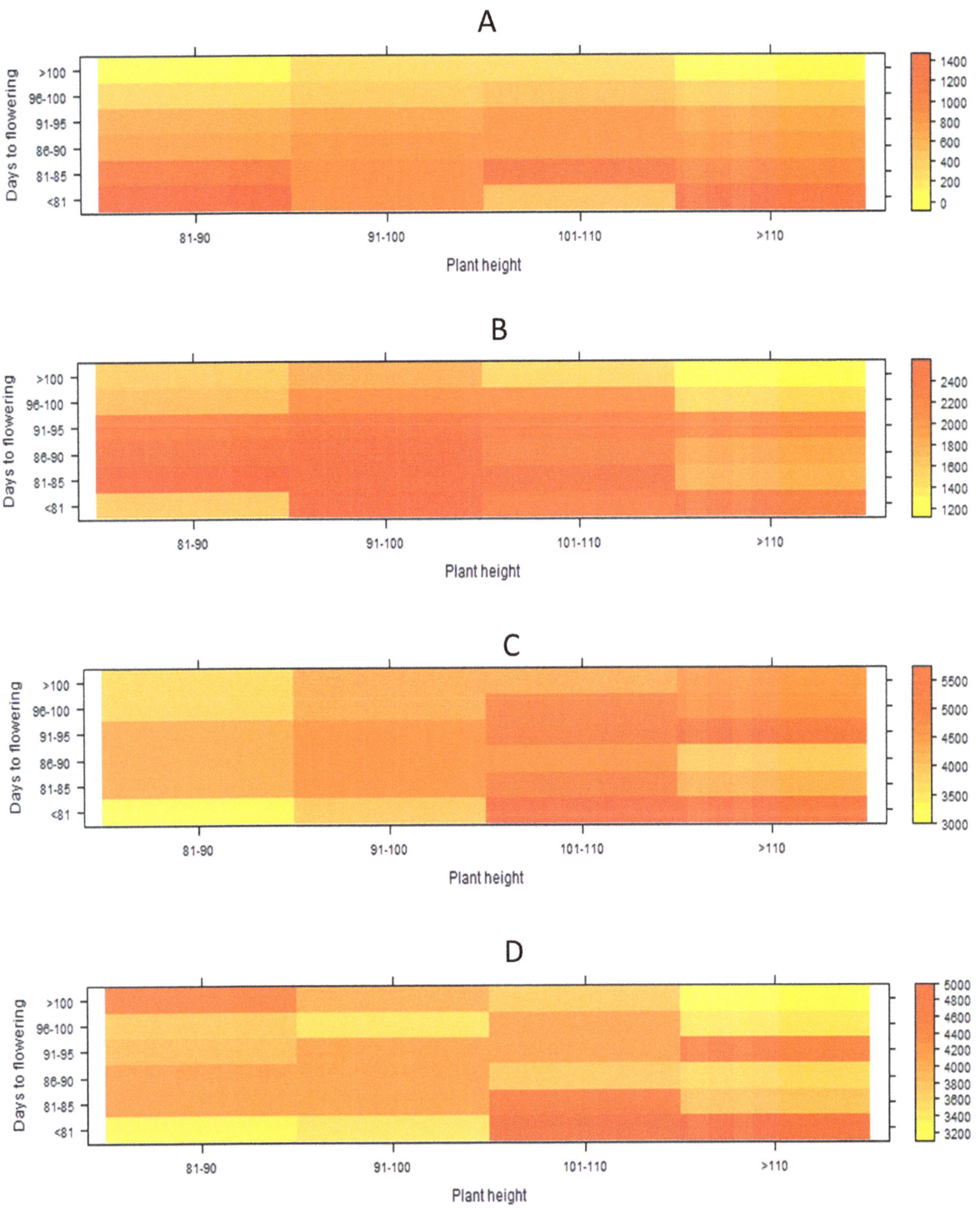

Figure 3. Heat maps showing the effect of DTF and PH on yield performance of lines under (A) lowland severe stress (LSS), (B) lowland moderate stress (LMS), (C) lowland non-stress (LNS I) and (D) LNS II conditions. Increasing intensity of colors from yellow to orange indicates increasing yield levels while white color shows the unavailable PH class.

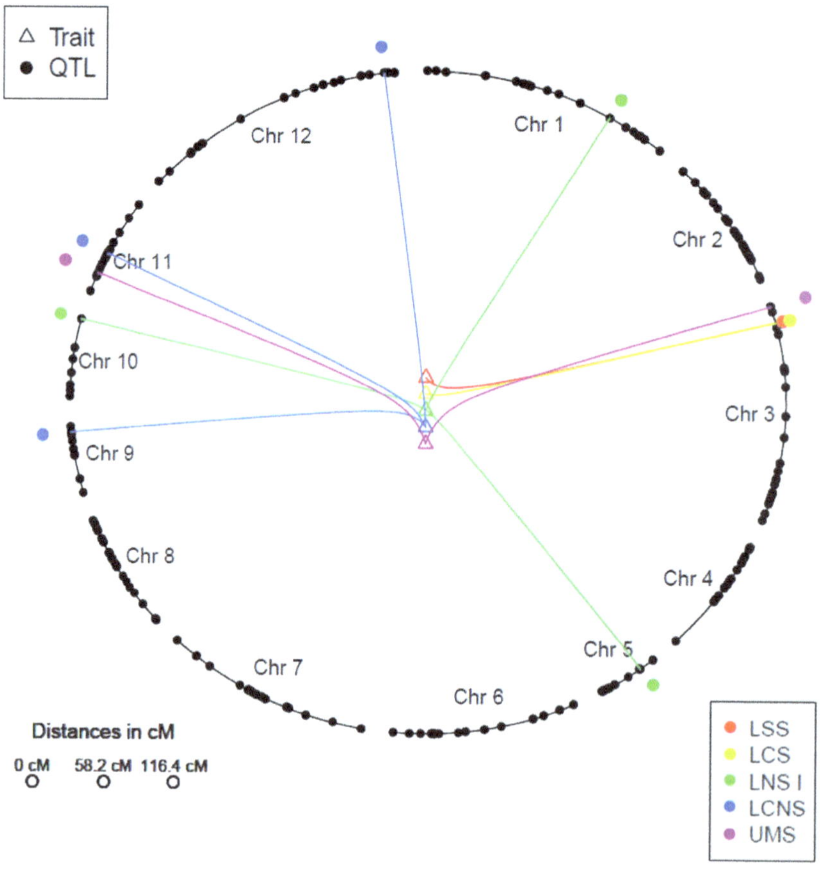

Figure 4. Circular genome plot showing the positions of QTLs identified for grain yield (adjusted for flowering time) under drought stress and non-stress conditions. LSS = lowland severe stress, LNS = lowland non-stress, LCS = lowland combined stress, LCNS = lowland combined non-stress, UMS = upland moderate stress.

under upland and lowland conditions. The full population was screened in Experiments 1-4 under stress and non-stress conditions. Yield reduction in Experiments 1 and 2 varied with stress levels in the respective seasons from severe for Experiment 1 to moderate in Experiment 2. Experiments 6 and 7 followed similar trends as Experiments 1 and 2 had in terms of yield. Both parents had similar yield under severe stress, however, Swarna (P2), being the high-yielding parent, had higher yield under moderate stress conditions in the lowland. The high yield of Swarna under moderate stress was also due to the poor adaptation of Moroberekan, an upland cultivar to lowland conditions. The population means remained higher than those for parents in both stress experiments, suggesting the role of high yield and drought tolerance in deciding the yield under drought.

The advantage of Moroberekan (P1) over Swarna was seen more clearly under upland conditions (Experiment 5), in which a large difference was recorded in the yield of both parents. Mild stress levels were recorded in Experiment 8 as this experiment was conducted in the wet season under naturally occurring drought stress. Moderate to high heritability estimates were recorded for yield in all experiments.

In all DS experiments, both parents had similar DTF under stress and non-stress conditions, but their progeny showed a wide range of DTF values. While Swarna is a well-known long-duration variety, there was a possibility of Moroberekan contributing to the earliness of the progeny. While, this earliness was not seen in the cultivar itself in DS experiments, this difference became clear in the WS experiment (5) when Moroberekan flowered much earlier than Swarna. Similar observations were also recorded in seed multiplication trials under lowland conditions in WS2013 (data not presented).

Trait correlation and effect

Genetic correlations between DTF, PH, and GY under lowland stress and non-stress conditions (DS2011 and DS2012) are presented in Table 3. These four experiments were conducted using the whole population, and hence, were used to understand trait correlations. Clear correlation patterns between DTF and GY were seen in all experiments. The two traits were negatively correlated in both stress and non-stress experiments. Regardless of stress severity levels, earliness of lines related to higher yield under stress environments. Also, under non-stress conditions, a similar pattern was observed, indicating that the reduced growth period did not lead to reduced yield potential. The degree of correlation however was much higher in stress experiments than in non-stress experiments. In contrast to DTF, no clear correlation patterns between PH and GY were observed in both stress and non-stress conditions.

In order to further understand the relationships between DTF, PH, and GY, a class analysis was conducted to measure GY for varying combinations of DTF and PH levels (Fig. 3). At any level of PH, greater differences were seen in GY for varying levels of DTF than vice versa. This agrees with the higher correlation observed between DTF and GY than between PH and GY. In

Table 4. QTLs for grain yield adjusted for flowering time (GY), plant height (PH), and flowering time (DTF) identified under severe and/or moderate reproductive-stage drought stress and non-stress conditions in DS2011 and DS2012.

Trait	QTL	Ecosystem	Peak marker	Chromosome	cM	R^2	A	*P*
GY	$qDTY_{3.2}$	LSS	Id3000757	3	15.0	16.0	281.6	<0.001
	$qDTY_{3.2}$	LCS	Id3000757	3	15.0	14.0	211.3	<0.001
	$qDTY_{1.1}$	LNS I	id1021344	1	154.0	5.5	552.5	0.009
	$qDTY_{5.1}$	LNS I	wd5002636	5	13.5	17.0	1743.0	0.009
	$qDTY_{10.2}$	LNS I	id10006250	10	69.3	4.9	−350.0	0.047
	$qDTY_{9.2}$	LCNS	id9007259	9	54.1	4.7	−353.4	0.012
	$qDTY_{11.2}$	LCNS	id11009456	11	36.7	10.0	1223.0	0.025
	$qDTY_{12.2}$	LCNS	ud12000118	12	212.0	4.8	620.7	0.050
	$qDTY_{3.2}$	UMS	id3000019	3	0.0	19.0	417.6	0.014
	$qDTY_{11.1}$	UMS	id11002801	11	17.4	25.0	1967.0	0.001
PH	$qDTH_{1.1}$	LSS	Id1021344	1	154.0	4.2	5.1	0.003
	$qDTH_{2.1}$	LMS	Id2006643	2	70.9	6.3	5.6	<0.001
	$qDTH_{8.1}$	LMS	Id8007896	8	95.0	4.9	4.2	<0.001
	$qDTH_{10.1}$	LMS	Id10004477	10	43.0	4.4	2.0	<0.001
	$qDTH_{1.1}$	LCS	Id1021344	1	150.0	31.0	9.5	<0.001
	$qDTH_{1.1}$	LNS I	Id1021344	1	154.0	9.1	9.1	<0.001
	$qDTH_{2.1}$	LNS I	Id2007213	2	75.0	4.5	6.5	0.008
	$qDTH_{10.1}$	LNS I	Id10004477	10	43.0	3.6	4.4	0.002
	$qDTH_{2.1}$	LNS II	Id2007213	2	74.3	9.1	6.6	0.008
	$qDTH_{3.1}$	LNS II	Id3002556	3	58.2	4.9	−3.3	0.031
	$qDTH_{1.1}$	LCNS	Id1021344	1	154.0	9.1	7.6	<0.001
	$qDTH_{2.1}$	LCNS	Id2007213	2	74.3	7.4	6.5	<0.001
	$qDTH_{7.1}$	LCNS	Id7005631	7	107.0	2.8	4.6	0.024
DTF	$qDTF_{3.1}$	LSS	Id3000757	3	20.0	30	−5.0	<0.001
	$qDTF_{7.1}$	LSS	Id7003271	7	91.0	8.4	−5.6	<0.001
	$qDTF_{10.1}$	LSS	Id10006250	10	69.3	3.4	1.9	0.040
	$qDTF_{11.1}$	LSS	Id11000855	11	5.0	4.9	−2.5	0.028
	$qDTF_{3.1}$	LMS	Id3000757	3	20.0	29.0	−4.9	<0.001
	$qDTF_{8.1}$	LMS	Id8004692	8	60.4	13.0	−6.7	<0.001
	$qDTF_{3.1}$	LCS	Id3000757	3	20.0	34.0	−5.2	<0.001
	$qDTF_{8.1}$	LCS	Id8004692	8	60.4	13.0	−6.6	<0.001
	$qDTF_{1.1}$	LNS I	Id1014783	1	130.0	7.9	−2.7	0.031
	$qDTF_{3.1}$	LNS I	Id3000946	3	25.0	17.0	−3.4	<0.001
	$qDTF_{3.1}$	LNS II	Id3000019	3	0.0	14.0	−2.6	<0.001

Table 4. Cont.

Trait	QTL	Ecosystem	Peak marker	Chromosome	cM	R^2	A	P
	$qDTF_{1.1}$	LCNS	Id1014783	1	125.0	3.9	−1.5	0.023
	$qDTF_{3.1}$	LCNS	Id3000946	3	25.0	21.0	−3.3	<0.001
	$qDTF_{6.1}$	LCNS	Id6010576	6	118.0	7.8	−4.0	0.044
	$qDTF_{2.2}$	UMS	Id2008112	2	90.0	11.0	−7.0	<0.001
	$qDTF_{3.1}$	UMS	Id3000019	3	0.0	19.0	−4.2	<0.001

qDTY, *qDTH*, and *qDTF* are QTLs for GY, PH, and DTF, respectively. Numbers before decimals denote the chromosome number, whereas those after are the order of their detection. A = additive effect of the peak marker, R^2 = percentage of phenotypic variance explained by the QTL, LSS = lowland severe stress, LMS = lowland moderate stress, LCS = lowland combined stress, LNS = lowland non-stress, LCNS = lowland combined non-stress, USS = upland moderate stress.

general, lines with medium maturity (DTF = 81–95 days) were the most consistent at all levels of plant height across all four experiments. However, lines with plant height from 101–110 cm showed the most consistent performance across the four experiments as compared to those beyond these levels.

Genetic variation

Analysis of GY identified several QTLs for each trait significant within and across environments (Fig 4.,Table 4). The most consistent QTL identified for GY was $qDTY_{3.2}$, which had an effect under drought stress across both upland and lowland conditions. The QTL explained a phenotypic variance effect of 16.0% and 19.0% under lowland severe stress (LSS) and upland moderate stress (UMS), respectively (Table 4). Apart from this, $qDTY_{11.1}$ was identified was identified under UMS condition. The QTL explained a phenotypic variance of 25.0%. To the best of our knowledge, this is the first time $qDTY_{11.1}$ is being reported for GY under upland drought stress.

In this study, large QTLs for DTF were identified, which reflect the negative correlations between DTF and GY under drought. A QTL for DTF ($qDTF_{3.1}$) was also identified at the $qDTY_{3.2}$ locus in this study. Several other QTLs for DTF were also observed at chromosomes 2, 6, 7, 8, 10 and 11 although $qDTF_{3.1}$ remained to be the most consistent QTL across drought stress and non-stress conditions (Table 4). Similarly, for PH, several QTLs were identified across the genome with $qDTH_{1.1}$ as the largest and most consistent QTL for the trait (Table 4). Majority of the GY QTLs under stress and non-stress were contributed by Moroberekan; however, two QTLs ($qDTY_{9.2}$ and $qDTY_{10.2}$) for yield under non-stress were also contributed by Swarna. Similarly, for PH and DTF, majority of the QTLs were contributed by the donor parent Moroberekan, but one QTL for each trait ($qDTH_{3.1}$ for PH and $qDTF_{10.1}$ for DTF) was also contributed by the recipient Swarna.

While trangressive segregants were observed in the population for all three traits, the contribution of both parents to higher performance of progeny lines and the presence of epistatic effect have become clear. In order to further explain the phenotypic variance present for grain yield under drought stress and non-stress conditions, tests for epistatic interactions were done. Epistatic interactions for GY under non-stress were observed between several loci under non-stress conditions. However, $eQTL_{10.1}$ consistently showed epistatic interactions with other loci across the genome (Table 5). Other consistant loci include $eQTL_{1.1}$ and $eQTL_{8.1}$ at chromosomes 1 and 8 respectively (Table 5). The high and consistent epistatic interactions of these loci suggest their importance in explaining the genetic basis of GY in lines under non-stress conditions. In this study, lines with high yield under both drought stress and non-stress conditions were identified (Table S2). Despite being more early-maturing than the recipient parent Swarna, these lines were able to maintain their yield potential equal to or more than Swarna under non-stress conditions.

Lines identified from the mapping population were evaluated in the target environment (IRRI-SA) for GY under naturally occurring drought stress and non-stress conditions to understand the consistency of performance of the lines. Figure 5 presents the scatter plots between the GY of selected lines at IRRI and IRRI-SA. Although the stress levels were very mild, the high-yielding drought-tolerant lines selected in DS2013 at IRRI were able to hold similar yield patterns at IRRI-SA. The R^2 values were non-significant, but no negative correlations were observed (Fig. 5). More significant correlations may have been apparent under higher severity of stress in the target environment. Lines finally

Table 5. Epistatic QTLs affecting grain yield (kg ha^{-1}) adjusted for flowering time under drought stress and non-stress conditions in DS2011 and DS2012.

Experiment	*QTL 1*	Chr (1)	cM (1)	Marker (1)	*QTL 2*	Chr 2	cM (2)	Marker (2)	*P*	R^2
LNS I	$eQTL_{8.1}$	8	75	id8006997	$eQTL_{10.1}$	10	55	id10004477	0.001	11.9
LNS II	$eQTL_{1.1}$	1	200	id1010697	$eQTL_{10.1}$	10	45	id10004477	0.015	13.6
LNS II	$eQTL_{1.1}$	1	200	id1010697	$eQTL_{12.1}$	12	175	id12008700	0.037	14.7
LCNS	$eQTL_{4.1}$	4	55	id4001882	$eQTL_{9.1}$	9	40	id9002357	0.039	10.8
LCNS	$eQTL_{7.1}$	7	80	id7002907	$eQTL_{10.1}$	10	50	id10004477	0.021	8.8
LCNS	$eQTL_{8.1}$	8	75	id8006997	$eQTL_{10.1}$	10	55	id10004477	<0.001	12.1
LCNS	$eQTL_{10.1}$	10	50	id10004477	$eQTL_{8.1}$	11	25	id11006588	0.024	8.9

LNS = lowland non-stress, LCNS= lowland combined non-stress, A = additive effect, R^2 = phenotypic variance explained.

identified from these experiments will undergo another testing for tolerance to blast to check for any loss of tolerance and grain quality testing to shortlist a final set of lines for deployment in the target environment.

Gene content of $qDTY_{3.2}$ and $qDTY_{11.1}$

Gene content of $qDTY_{3.2}$ **and** $qDTY_{11.1}$ was classified into 7 classes and percentage of genes falling into each class was calculated (Fig 6.). A large proportion of the genes in bot QTLs were genes coding for known proteins other than those known to be directly related to stress response. Apart from this, large proportion of genes were also related to other expressed proteins in both QTLs. $qDTY_{3.2}$ however showed a much higher percentage of genes related to enzyme synthesis and function while $qDTY_{11.1}$ showed a much higher percentage of retrotransposon proteins as compared to $qDTY_{3.2}$. 8% of the genes within $qDTY_{3.2}$ 6% of genes for $qDTY_{11.1}$ were genes known to be directly associated to stress response or those which could be crucial for maintenance of plant function under stress. These included genes coding for heat shock proteins, no apical meristem proteins, membrane, cell cycle related proteins and zinc finger proteins (File S2). $qDTY_{3.2}$ also showed the presence of flowering related genes such as AP2 domain containing proteins and MADS box family genes while $qDTY_{11.1}$ showed the presence of a large number of nucleotide-binding site leucine- rich repeats (NBS-LRR) genes which are known to be major candidates for blast resistance.

Discussion

Our study reports an integrated strategy to develop drought- and blast-tolerant lines along with simultaneous identification of QTLs for GY under drought. The study also attempts to explain the genetics behind the selection cycle for blast tolerance and its effect on drought tolerance. Several previous studies have used different kinds of mapping populations and strategies to identify QTLs for tolerance of rice blast and grain yield under drought. However, to the best of our knowledge, there has been no attempt to identify QTLs for high grain yield under drought stress on a mapping population tolerant of rice blast while simultaneously developing high-yielding, drought-tolerant breeding lines. Ballini et al. [28] conducted a meta-analysis to pinpoint the location of major genes reported for blast tolerance in rice. Similarly, Swamy et al. [29] conducted a meta-analysis on QTLs for grain yield under drought and reported several meta-QTLs controlling the trait across the rice genome. Both studies provide an extensive coverage of literature, indicating the large number of studies targeting these traits independently. However, reports of the two traits targeted in the same study seldom exist. In our study, the test for segregation distortion in the population showed most of the distortion to be toward the blast-tolerant parent Moroberekan, showing that it was the major contributor for blast tolerance in the population. Interestingly, distortion in favour of Swarna was also observed at some loci. While some of these marker loci co-locate with NBS-LRR genes, it is likely that these regions play a role in imparting tolerance to blast. The breeding strategy not only allowed us to select for but also to detect loci with small and large effects contributed by both parents. With observations of increased incidence of diseases such as blast in drought conditions, it becomes important to develop lines tolerant to both traits to have more stable yields in rainfed environments. Identifying QTLs independently affecting two different traits and pyramiding them together is a two-step process that is normally used to accomplish this task. The alternative mapping strategy presented in Figure 1 allows the simultaneous achievement of both these goals,

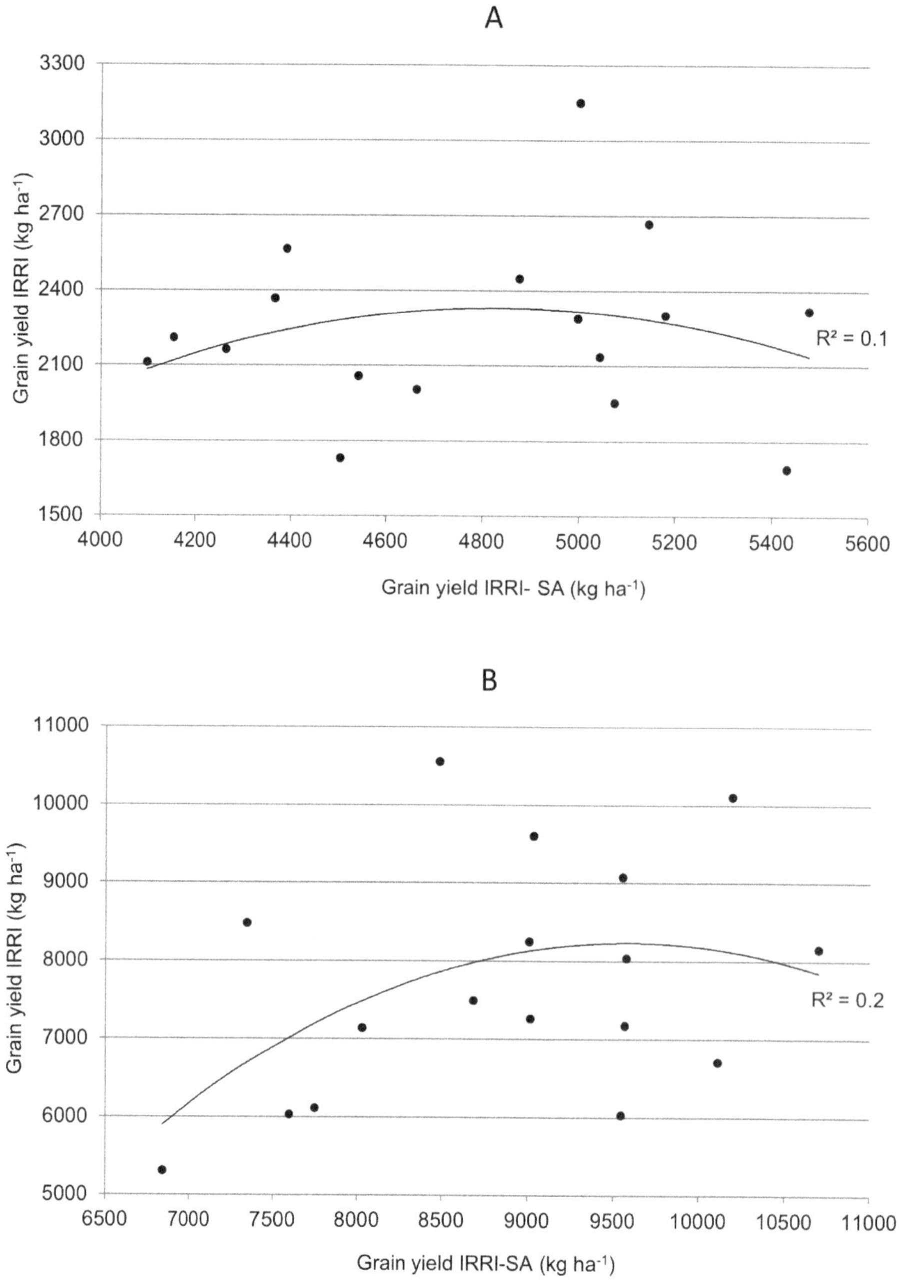

Figure 5. Correlation of GY performance of selected lines under drought stress and non-stress conditions in IRRI and IRRI-SA.

particularly for traits where QTLs/genes conferring tolerance are unknown. We observed that, through systematic early-generation screening of large populations, lines with higher tolerance of one of the two targeted traits (blast in this study) may be obtained. These lines can then be used as mapping populations for QTL identification for the second target trait (high GY under drought stress and non-stress in this study), thus enabling the identification of QTLs and development of high-yielding lines with tolerance of multiple stresses in a reduced time span. Despite the segregation distortion due to screening for blast tolerance, high genetic variation could be seen for DTF, PH, and GY under drought and non-stress conditions. Clear pattern of genetic correlations between the three traits were observed and specific DTF and PH classes showed GY advantage under different stress and non-stress conditions (Table 3, Fig. 3).

Advanced backcross (AB) lines were particularly suitable in such cases where a traditional cultivar is used as a donor for an elite recipient background. Moroberekan has long been known as an important donor for tolerance of drought. In addition, being a tropical japonica accession, it is likely to carry genes that may provide opportunity to increase yield potential under normal situations. Despite these advantages, it has not been successfully used in the breeding program because of its poor combining ability and probable adverse linkage drag. In this study, the earlier segregating generations, such as F_1- and BC_1-derived populations, were not found suitable for the study because of the high percentage of lines with poor performance and poor plant type.

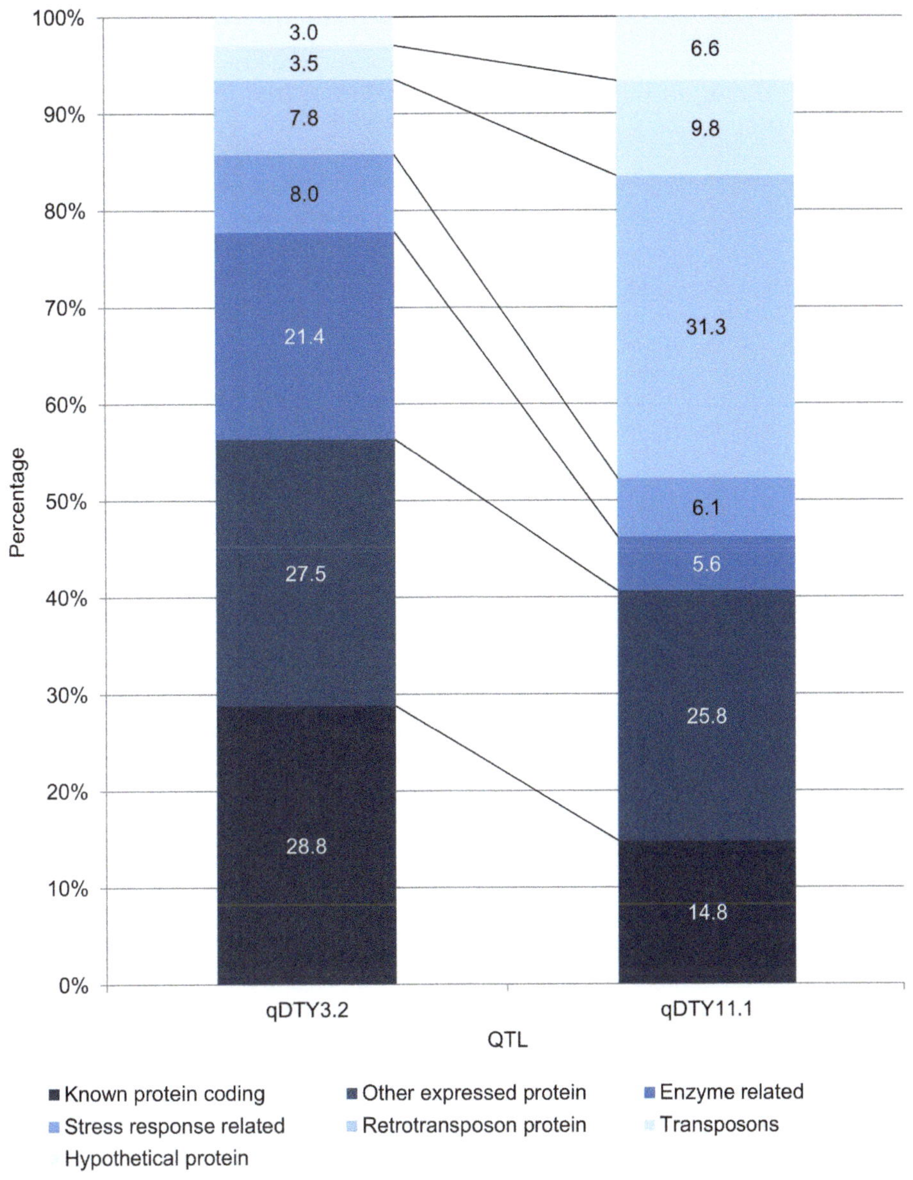

Figure 6. Comparison of gene content of *qDTY*$_{3.2}$ and *qDTY*$_{11.1}$ based on percentage of genes belonging to seven gene classes within the two QTLs.

However, the BC_2-derived lines showed a very high recovery of recipient parent Swarna. In our evaluation, several lines with 10–20 days earlier duration than Swarna, but that had higher yield than Swarna under both irrigated control and drought conditions, were identified (Table S2). DS screening at IRRI has proven highly beneficial in uniform screening of large segregating populations to identify high-yielding drought-tolerant lines. Moderate to high correlation has been reported previously between yield in DS screening at IRRI and WS screening in India [30]. Similar patterns were also seen for the lines selected in this study (Figure 5).

qDTY$_{3.2}$ was the most consistent QTL identified for GY under drought in this study (Table 4, Figure 4). This QTL has been reported in several previous studies for showing independent or epistatic effects on GY under upland and lowland conditions [11,13,31,32]. The QTL region coincides with the *HD9* locus, which controls DTF in rice. While *qDTF*$_{3.1}$ was found to collocate with this locus in the present study, it is likely that the Moroberekan allele at this locus causes earliness, leading to yield advantage under drought.

The role of drought escape mechanism coming into play due to this locus cannot be denied. However, the large variety of genes directly or indirectly related to stress response at this locus (File S2) shows that this locus could affect other traits related to stress response apart from flowering. Several existing reports also suggest the effect of this locus on a wide range of traits. Yadav et al. [13] reported the effect of this locus on grain yield under drought while Bernier et al. [10] reported its effect on plant height and biomass. Similarly, the region has been reported to affect traits including chlorophyll content, root volume under drought, low-temperature germination, and spikelet fertility [32–34]. The wide range of QTLs detected at this locus for traits related to yield and adaptation to abiotic stresses indicates the existence of higher genetic diversity on this chromosomal region. These results are suggestive of either multiple-linked QTLs or pleiotropic effects of one or more than one QTL on a wide range of traits.

The high genic diversity along with the effect of this region on flowering time may be the reason of its effect on yield under drought under wider range of environments as compared to $qDTY_{11.1}$ which showed effect under upland conditions only.

Similar to $qDTY_{3.2}$, $qDTY_{11.1}$ also showed the presence of a wide range of stress response related genes (File S2). However, NBS-LRR genes were by far the most dominating group within this region, while these genes are well known candidates for major blast resistance genes [28]; it is likely that the QTL may affect tolerance to blast along with affecting yield under upland drought.

Epistatic interactions are one of the major factors that determine the phenotypic expression of a mapping population [35]. It has also been seen in a large number of QTL mapping studies conducted in the past that, despite nearly complete coverage of the chromosomes, the identified QTLs do not explain the entire phenotypic variance observed for a trait [35]. This potentially indicates the presence of epistatic interactions contributing to phenotypic variation in a segregating population along with other factors contributing the same. The presence of transgressive segregants in mapping populations is also suggestive of epistasis. In this study, transgressive segregants for all three traits were observed. The study of epistatic interactions showed the presence of loci having significant interactions with several other loci across the genome. These loci could be very important in understanding the genetic basis of yield under drought and non-stress conditions. Understanding the genetic composition of these loci and their effect on plant function may allow us to use these loci for maximum advantage in breeding programs.

Our study involved a two-way approach to develop lines tolerant of blast and drought with high yield potential. Some of these lines show plant type and grain type similar to Swarna with added tolerance of drought and blast. These lines may act as a replacement of the variety where Swarna is highly preferred. On the other hand, lines with different plant and grain types can be tested as prospective new high-yielding cultivars. In addition, lines with the identified QTLs are also being used in other marker-assisted breeding programs as improved donors. The study has allowed us to demonstrate the combination of conventional breeding approaches with QTL identification and to understand the effect of selection pressure on QTL mapping. This strategy can be very effective in cases where traits are likely to be controlled by multiple genes/QTLs across the genome. In such cases, phenotypic selection for one trait is less likely to affect the other, leaving sufficient variation in the population for QTL mapping.

Supporting Information

Table S1 Markers showing significant segregation distortion across the genome with the direction of distortion.

Table S2 Grain yield performance of lines selected under severe stress and non-stress conditions in dry-season screening at IRRI, and under naturally occurring wet-season drought and non-stress conditions at IRRI-SA.

File S1 Standard evaluation system used for large scale screening of rice lines to determine their tolerance reaction rice blast (A), field management of lowland experiments with details of crop cultivation, fertilizer management, and management of insect pests and weeds (B).

File S2 List of genes with possible function related to biotic and abiotic stress response within $qDTY_{3.1}$, $qDTY_{11.1}$.

File S3 Phenotypic and genotypic data used for QTL mapping including trait values, marker names, positions and genotypes for individual lines in the mapping population.

Acknowledgments

The authors are thankful for the support received from Dr. H. Leung, Dr. M. Thomson, during the process of genotyping and development of the manuscript, Dr Nickolai Alexandrov, Dr Casiana Vera Cruz, Dr Bo Zhou and Ms Rose Imee Zhella Morantte for the support received during the process of manuscript development and Teodoro Cuevas, Marcial Panting, Jocelyn O. Guevarra, Danilo De Ocampo, Carmela Malabanan, Ariel Formarejo, and Ruth Erica Carpio for the technical support received in the completion of this study.

Author Contributions

Conceived and designed the experiments: SD AK. Performed the experiments: SD MTSC PTM. Analyzed the data: SD BEH. Contributed reagents/materials/analysis tools: MTSC JCEO. Wrote the paper: SD BEH AK.

References

1. Alexandratos N, Bruinsma J (2012) World agriculture towards 2030/2050. ESA Working Paper No. 12–03.
2. Vikram P, Mallikarjuna Swamy BP, Dixit S, Ahmed HU, Sta Cruz MT, et al. (2011) $qDTY_{1.1}$, a major QTL for rice grain yield under reproductive-stage drought stress with a consistent effect in multiple elite genetic backgrounds. BMC Genet 12: 89.
3. Huke RE, Huke EH (1997) Rice area by type of culture: South, Southeast, and East Asia. IRRI, Los Baños, Philippines.
4. Serraj R, Bennett J, Hardy B (2008) Drought frontiers in rice: crop improvement for increased rainfed production. Singapore: World Scientific Publishing and Los Baños (Philippines): International Rice Research Institute.
5. Ouk M, Basnayake J, Tsubo M, Fukai S, Fischer KS, et al. (2006) Use of drought response index for identification of drought tolerant genotypes in rainfed lowland rice. Field Crops Res 99: 48–58.
6. Kumar A, Bernier J, Verulkar S, Lafitte HR, Atlin GN (2008) Breeding for drought tolerance: direct selection for yield, response to selection and use of drought-tolerant donors in upland and lowland-adapted populations. Field Crops Res 107: 221–231.
7. Venuprasad R, Lafitte HR, Atlin GN (2007) Response to direct selection for grain yield under drought stress in rice. Crop Sci 47: 285–293.
8. Venuprasad R, Sta Cruz MT, Amante M, Magbanua R, Kumar A, et al. (2008) Response to two cycles of divergent selection for grain yield under drought stress in four rice breeding populations. Field Crops Res 107: 232–244.
9. Kumar A, Verulkar SB, Mandal NP, Variar M, Shukla VD, et al. (2012) High-yielding, drought-tolerant, stable rice genotypes for the shallow rainfed lowland drought-prone ecosystem. Field Crops Res 133: 37–47.
10. Bernier J, Kumar A, Venuprasad R, Spaner D, Atlin GN (2007) A large-effect QTL for grain yield under reproductive-stage drought stress in upland rice. Crop Sci 47: 507–516.
11. Venuprasad R, Dalid CO, Del Valle M, Zhao D, Espiritu M, et al. 2009 Identification and characterization of large-effect quantitative trait loci for grain yield under lowland drought stress in rice using bulk-segregant analysis. Theor Appl Genet 120: 177–190.
12. Ghimire KH, Quiatchon LA, Vikram P, Mallikarjuna Swamy BP, Dixit S, et al. (2012) Identification and mapping of a QTL ($qDTY_{1.1}$) with a consistent effect on grain yield under drought. Field Crops Res 131: 88–96.
13. Yadav RB, Dixit S, Raman A, Mishra KK, Vikram P, et al. (2013) A QTL for high grain yield under lowland drought in the background of popular rice variety Sabitri from Nepal. Field Crops Res 144: 281–287.

14. Mishra KK, Vikram P, Yadaw RB, Swamy BPM, Dixit S, et al. (2013) *$qDTY_{12.1}$*: a locus with a consistent effect on grain yield under drought in rice. BMC Genet 14: 12.
15. Dixit S, Singh A, Sta Cruz MT, Maturan PT, Amante M, et al. (2014) Multiple major QTLs lead to stable yield performance of rice cultivars across variable drought intensities. BMC Genet 15: 16.
16. You A, Lu X, Jin H, Ren X, Liu K, et al. (2006) Identification of quantitative trait loci across recombinant inbred lines and testcross populations for traits of agronomic importance in rice. Genetics 172: 1287–1300.
17. Yamagishi J, Miyamoto N, Hirotsu S, Laza RC, Nemoto K (2004) QTLs for branching, floret formation, and pre-flowering floret abortion of rice panicle in a temperate *japonica* × tropical *japonica* cross. Theor Appl Genet 109: 1555–1561.
18. Xing YZ, Tan YF, Hua JP, Sun XL, Xu CG, et al. (2002) Characterization of the main effects, epistatic effects and their environmental interactions of QTLs on the genetic basis of yield traits in rice. Theor Appl Genet 105: 248–257.
19. Thomson MJ, Tai TH, McClung AM, Lai X-H, Hinga ME, et al. (2003) Mapping quantitative trait loci for yield, yield components and morphological traits in an advanced backcross population between *Oryza rufipogon* and the *Oryza sativa* cultivar Jefferson. Theor Appl Genet 107: 479–493.
20. McNally KL, Childs KL, Bohnert R, Davidson RM, Zhao K, et al. 2009 Genomewide SNP variation reveals relationships among landraces and modern varieties of rice. Proc Nat Acad Sci USA 106(30): 12273–12278.
21. Neeraja CN, Maghirang-Rodriguez R, Pamplona A, Heuer S, Collard B, et al. (2007) A marker-assisted backcross approach for developing submergence-tolerant rice cultivars. Theor Appl Genet 115: 767–776.
22. Zhao K, Wright M, Kimball J, Eizenga G, McClung A, et al. (2010) Genomic diversity and introgression in *O. sativa* reveal the impact of domestication and breeding on the rice genome. PLOS One 5 (5): e10780.
23. Zhao K, Tung C-W, Eizenga GC, Wright MH, Ali ML, et al. (2011) Genome-wide association mapping reveals a rich genetic architecture of complex traits in *Oryza sativa*. Nat Comm 2: 467/DOI: 10.1038/ncomms1467.
24. Cooper M, DeLacy IH, Basford KE (1996) Relationship among analytical methods used to analyse genotypic adaptation in multi-environment trials, pp-193–224. In M. Cooper and G.L. Hammer editors Plant adaptation and crop improvement University Press, Cambridge, UK.
25. SAS Institute Inc (2004) SAS online doc 9.1.3. SAS Institute Inc., Cary, NC, USA.
26. Broman KW, Wu H, Sen S. Churchill GA (2003) R/qtl: QTL mapping in experimental crosses. Bioinformatics 19: 889–890.
27. Berloo RV (2008) GGT 2.0: versatile software for visualization and analysis of genetic data. J Heredity 99(2): 232–236.
28. Ballini E, Morel J-B, Droc G, Price A, Courtois B, et al. (2008) A genome-wide meta-analysis of rice blast resistance genes and quantitative trait loci provides new insights into partial and complete resistance. Mol. Plant Microbe Interact 21(7): 859–868.
29. Swamy BPM, Vikram P, Dixit S, Ahmed HU, Kumar A (2011) Meta-analysis of grain yield QTL identified during agricultural drought in grasses showed consensus. BMC Genomics 12: 319.
30. Verulkar SB, Mandal NP, Dwivedi JL, Singh BN, Sinha PK, et al. (2010) "Breeding resilient and productive rice genotypes adapted to drought-prone rainfed ecosystems of India", Field Crops Res 117: 197–208.
31. Dixit S, Mallikarjuna Swamy BP, Vikram P, Bernier J, Sta Cruz MT, et al. (2012) Increased drought tolerance and wider adaptability of *qDTY12.1* conferred by its interaction with *qDTY2.3* and *qDTY3.2*". Mol Breed 30: 1767–1779.
32. Ding X, Li X, Xiong L (2011) Evaluation of near-isogenic lines for drought resistance QTL and fine mapping of a locus affecting flag leaf width, spikelet number, and root volume in rice Theor Appl Genet 123: 815–826.
33. Ishimaru K, Yano M, Aoki N, Ono K, Hirose T, et al. (2001) Toward the mapping of physiological and agronomic characters on a rice function map: QTL analysis and comparison between QTLs and expressed sequence tags. Theor Appl Genet 102: 793–800.
34. Fujino K, Sekiguchi H, Sato T, Kiuchi H, Nonoue Y, et al. (2004) Mapping of quantitative trait loci controlling low-temperature germinability in rice (*Oryza sativa* L.). Theor Appl Genet 108: 794–799.
35. Li Z, Pinson SRM, Park WD, Paterson AH, Stansel JW (1997) Epistasis for three grain yield components in rice (*Oryza sativa* L.). Genetics 145: 453–465.

9

Nutrients Can Enhance the Abundance and Expression of Alkane Hydroxylase CYP153 Gene in the Rhizosphere of Ryegrass Planted in Hydrocarbon-Polluted Soil

Muhammad Arslan[1,3]*, Muhammad Afzal[1], Imran Amin[2], Samina Iqbal[1], Qaiser M. Khan[1]

1 Soil and Environmental Biotechnology Division, National Institute for Biotechnology and Genetic Engineering, Faisalabad, Pakistan, **2** Agricultural Biotechnology Division, National Institute for Biotechnology and Genetic Engineering, Faisalabad, Pakistan, **3** Earth Sciences Department, King Fahd University of Petroleum and Minerals, Dhahran, Saudi Arabia

Abstract

Plant-bacteria partnership is a promising strategy for the remediation of soil and water polluted with hydrocarbons. However, the limitation of major nutrients (N, P and K) in soil affects the survival and metabolic activity of plant associated bacteria. The objective of this study was to explore the effects of nutrients on survival and metabolic activity of an alkane degrading rhizo-bacterium. Annual ryegrass (*Lolium multiflorum*) was grown in diesel-contaminated soil and inoculated with an alkane degrading bacterium, *Pantoea* sp. strain BTRH79, in greenhouse experiments. Two levels of nutrients were applied and plant growth, hydrocarbon removal, and gene abundance and expression were determined after 100 days of sowing of ryegrass. Results obtained from these experiments showed that the bacterial inoculation improved plant growth and hydrocarbon degradation and these were further enhanced by nutrients application. Maximum plant biomass production and hydrocarbon mineralization was observed by the combined use of inoculum and higher level of nutrients. The presence of nutrients in soil enhanced the colonization and metabolic activity of the inoculated bacterium in the rhizosphere. The abundance and expression of CYP153 gene in the rhizosphere of ryegrass was found to be directly associated with the level of applied nutrients. Enhanced hydrocarbon degradation was associated with the population of the inoculum bacterium, the abundance and expression of CYP153 gene in the rhizosphere of ryegrass. It is thus concluded that the combination between vegetation, inoculation with pollutant-degrading bacteria and nutrients amendment was an efficient approach to reduce hydrocarbon contamination.

Editor: Dwayne Elias, Oak Ridge National Laboratory, United States of America

Funding: The work was supported by the Higher Education Commission of Pakistan (HEC-No. 20-1997/R&D/11/9210). The funder had no role in study design, data collection and analysis, decision to publish, or preparation of the manuscript.

Competing Interests: The authors have declared that no competing interests exist.

* Email: arsilan324@gmail.com

Introduction

Various human activities, such as exploration, extraction, refining, storage and transportation of petroleum oil and its products, are polluting the environment with hydrocarbons. The hydrocarbons in soil and water adversely affect plant health and microbial population. Furthermore, hydrocarbons exert toxic effects to animals and humans [1]. Therefore, the removal of hydrocarbons from soil and water is one the main issues in field of environmental sciences [2].

Phytoremediation, i.e., use of plants for the remediation of contaminated soil and water, is one of the most suitable technology for the cleanup of contaminated environment [3]. However, the presence of hydrocarbons in soil reduces plant growth and consequently phytoremediation efficiency [4,5]. To improve plant growth and phytoremediation of soil contaminated with hydrocarbons, the combined use of suitable plants and hydrocarbon-degrading bacteria has been recently proposed [6,7]. In the root zone, plants and their associated rhizobacteria form synergistic relationships with each other. Plant roots releases nutrients that enhance the proliferation, colonization and activity of microorganisms in the rhizosphere [8,9]. Cébron et al. (2009) also proposed the similar effects of vegetation on bacterial community structures especially about the polycyclic aromatic hydrocarbon degraders [10]. In return, rhizobacteria reduce phytotoxicity of hydrocarbons by the virtue of their pollutant-degrading pathways and metabolic activities [2]. In addition to this, rhizobacteria also produce plant growth promoting hormones which promote the plant growth and development [11].

Degradation of hydrocarbon depends upon the abundance and expression of hydrocarbon-degrading genes in soil [6,12,13]. Therefore, it is important to analyze the levels of abundance and expression of pollutant-degrading genes in the contaminated soil during phytoremediation [14].

During phytoremediation of hydrocarbon contaminated soil, competition for nutrients between plants and their associated microorganisms enhances. The hydrophobic nature of hydrocarbons reduces the availability of nutrients to the plants and microorganisms [15,16]. Hence, nutrients such as nitrogen, phosphorus and potassium, act as limiting factors for plant growth and microbial proliferation [17]. Furthermore, nutrients level in soil affects mineralization of organic pollutants [18,19]. However, the effects of nutrients on the abundance and expression of catabolic genes in hydrocarbon contaminated soil have been rarely evaluated. The objective of the current investigation was to evaluate the effects of two different nutrient levels on the abundance and expression of alkane-degrading gene (CYP153) in the rhizospheric soil of ryegrass grown in hydrocarbon-contaminated soil.

Materials and Methods

1. Bacterial strain

Pantoea sp. strain BTRH79, previously isolated and characterized for hydrocarbon degradation [20] and ACC deaminase activity [21], was used in this study. This strain possessed alkane hydroxylase CYP153 gene with a substrate range between C_4–C_{16}. *Pantoea* sp. strain BTRH79 was cultured in LB broth for overnight and harvested by centrifugation, further washed and suspended in autoclaved 0.9% (w/v) NaCl solution. This bacterial suspension was used as inoculum.

2. Experiment setup

Agricultural loamy soil, after passing through 2 mm sieve and spiking with diesel (15 g kg^{-1} soil), was filled into plastic pots (1.5 kg soil pot^{-1}) and subsequently pots were placed in the greenhouse. Spiking was done by thoroughly mixing the appropriate amount of diesel fuel (15 g) with fresh soil designated for one pot. The water holding capacity of the soil was 45.3 g 100 g^{-1}. Different treatments were:

1. Ryegrass vegetated in uncontaminated soil without bacterial inoculation (control);
2. Ryegrass vegetated in uncontaminated soil with bacterial inoculation (control);
3. Soil contaminated with diesel without vegetation but with inoculation;
4. Ryegrass vegetated in diesel contaminated soil without bacterial inoculation;
5. Ryegrass vegetated in diesel contaminated soil with bacterial inoculation;
6. Ryegrass vegetated in diesel contaminated and inoculated soil with lower levels of nutrients (N, P and K); and
7. Ryegrass vegetated in diesel contaminated and inoculated soil with higher levels of nutrients.

Each treatment was triplicated. Surface-sterilized ryegrass seeds were planted, i.e. 100 seeds pot^{-1}. After one week of seed germination, poor growing seedlings were removed and a population of 80 plants per pot was maintained. Two different levels of nutrient, i.e., low ($N_{80}P_{50}K_{30}$) and high ($N_{120}P_{80}K_{60}$) were applied for two times with an interval of 5 days after one week of germination. Nitrogen was provided in the form of urea while phosphorus and potassium were added as KH_2PO_4. Finally, the amount of phosphorus was balanced with Na_2HPO_4. For higher dose of nutrients, 3.85 g urea, 3.138 g KH_2PO_4 and 2.29 g Na_2HPO_4 were added in each pot. Whereas for lower dose of nutrients, 2.57 g urea, 1.56 g KH_2PO_4 and 1.81 g Na_2HPO_4 were added in each pot.

3. Plant growth

Plants were cut 1 cm above-ground after 100 days of sowing and shoot length were determined. Rhizosphere soil samples were obtained by gently agitating the roots. Bulk soil samples were collected and stored at −80°C for further analysis. Roots were washed several times with tap water and then with sterile distilled water. Fresh and dry biomass of shoot and root were estimated as explained previously [21]. Furthermore, different agronomic characters such as average shoot length, average diameter, and network length were measured using ImageJ software [22]. The plants were selected using ranked set sampling procedure [23,24].

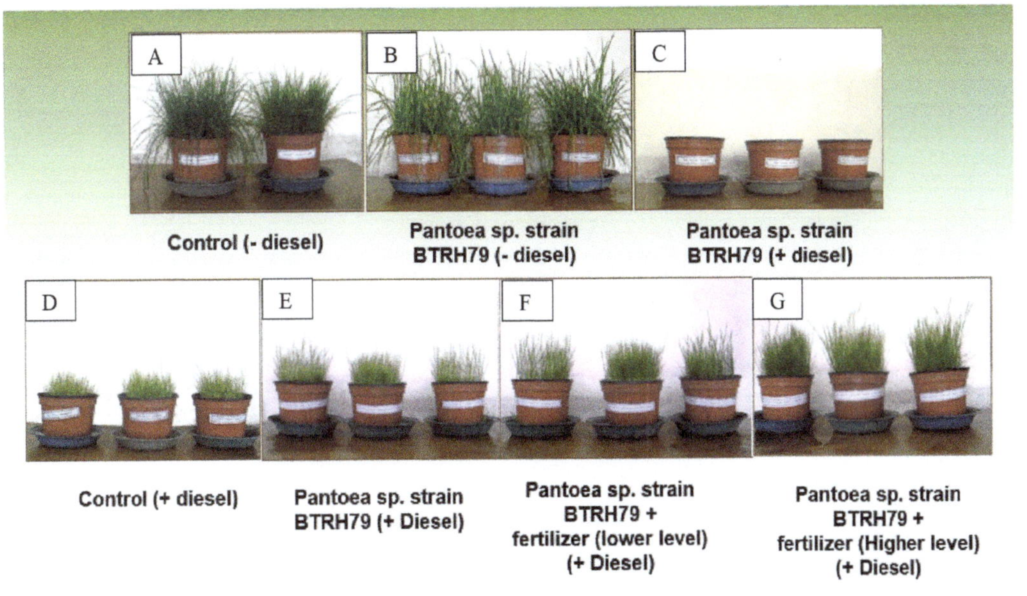

Figure 1. Experimental setup illustrating different treatments. Ryegrass vegetated in uncontaminated soil (A), ryegrass vegetated in uncontaminated soil and inoculated with *Pantoea* sp. strain BTRH79 (B), inoculation of *Pantoea* sp. strain BTRH79 in diesel contaminated soil (C), ryegrass vegetated in diesel contaminated soil (D), ryegrass vegetated in diesel contaminated soil and inoculated with *Pantoea* sp. strain BTRH79 (E), ryegrass vegetated in diesel contaminated soil treated with lower level of fertilize and inoculated with *Pantoea* sp. strain BTRH79 (F), and ryegrass vegetated in diesel contaminated soil treated with higher level of fertilizer and inoculated with *Pantoea* sp. strain BTRH79 (G).

Table 1. Effect of bacterial inoculation and nutrients on shoot length (SL), shoot and root fresh weight (FW) and dry weight (DW) of ryegrass vegetated in diesel contaminated soil after 100 days of sowing.

Treatment	SL (cm)	Shoot weight		Root weight	
		FW (g)	DW (g)	FW (g)	DW (g)
Control (− diesel)	27.6^{a} (1.6)	38.5^{a} (0.8)	8.5^{a} (0.4)	9.3^{a} (0.4)	3.94^{a} (0.03)
Pantoea sp. strain BTRH79 (− diesel)	29.1^{a} (1.3)	39.1^{a} (2.1)	8.8^{a} (0.2)	9.8^{a} (1.0)	4.15^{a} (0.04)
Control (+ Diesel)	7.9^{d} (0.6)	4.6^{e} (0.6)	1.7^{d} (0.1)	5.3^{d} (0.5)	1.45^{d} (0.03)
Pantoea sp. strain BTRH79 (+ Diesel)	15.1^{c} (1.7)	8.6^{d} (0.7)	2.2^{d} (0.1)	6.4^{c} (0.1)	2.90^{c} (0.08)
Pantoea sp. strain BTRH79 + fertilizer (lower level) (+ Diesel)	17.6^{c} (1.3)	10.9^{c} (0.4)	3.1^{c} (0.1)	7.6^{b} (0.4)	3.13^{c} (0.08)
Pantoea sp. strain BTRH79 + fertilizer (higher level) (+ Diesel)	21.7^{b} (1.6)	15.5^{b} (1.1)	4.2^{b} (0.5)	8.4^{b} (0.5)	3.61^{b} (0.04)

Each value is the mean of three replicates, means in the same column followed by the same letter are not significantly different at a 5% level of significance, and the standard error of three replicates is presented in parentheses.

4. Enumeration of inoculated bacterium

The population of *Pantoea* sp. strain BTRH79 in the rhizosphere soil was enumerated by viable cell count procedure. Soil slurry was prepared by mixing 3 g of rhizospheric soil in 9 mL of 0.9% (w/v) NaCl solution. After settlement of soil particles, supernatant was taken in sterilized falcon tubes and their ten-fold serial dilutions (up to 10^{-3}) were plated on M9 medium containing diesel as a sole source of carbon. Finally, the colony forming units (CFU) were enumerated by using Fotodyne's TotalLab Quant Analysis software [25]. From each treatment, ten colonies were randomly picked and the identity of the isolates with the inoculant strain was confirmed by restriction fragment length polymorphism analysis [21,26].

5. Quantification of the abundance and expression of CYP153 gene

For the quantification of gene abundance and gene expression, DNA and RNA were isolated from the rhizospheric soil using FastDNA Spin kit and FastRNA Pro Soil-Direct kit (Biomedical, USA), respectively. For expression analysis, cDNA was transcribed from the RNA using RevertAid First Strand cDNA Synthesis Kit.

The primers, P450fw1 (5′-GTSGGCGGCAACGACACSAC-3′) and P450rv3 (5′-GCASCGGTGGATGCCGAAGCCRAA-3′) were used for the amplification of the CYP153 gene in qPCR [27]. Gene abundance was determined by using real-time PCR, iCycler (IQ5) (BioRad). Reaction mixtures (25 μL) contained 12.5 μL of Q-Mix (Biorad), 2.5 μL of 10 mg/ml BSA, 1.0 μL DMSO, 0.5 μL of each forward and reverse primer, 5.5 μL of PCR water, 2.5 μL of template DNA and RNase-free water. Temperature profile began with 4 minutes at 94°C followed by 40 cycles of 94°C for 30 seconds, 58°C for 30 seconds, and 72°C for 45 seconds followed by a melt curve from 50 to 100°C. Besides melting curve analysis, PCR products were examined on 2% agarose gels. No primer-dimers were detected. Primer-dimer is a potential by-product and in quantitative PCR it may interfere with accurate quantification. Serial dilutions of DNA and cDNA were spiked with 10^6 copies of amplified CYP153 genes to check for real-time PCR inhibition. Highly linear standard curves (r^2 values >0.95, PCR efficiency > 98%) over the dilution range and a detection limit of 10^1 copies were obtained indicating no PCR inhibition. The CYP gene copy numbers were quantified relative to a standard curve of positive control and were normalized to the copy number of control plants [13,26,28].

6. Hydrocarbon analysis

Total petroleum hydrocarbons in soil samples were estimated by using FTIR as explained previously [20].

7. Statistical analysis

Plant growth, biomass production, residual hydrocarbon contents, gene abundance and expression data were analyzed by using Minitab software package. One-way analysis of variance (ANOVA) was used for the comparisons between treatments and Tukey's test was performed for ANOVA after testing homogeneity of variance.

Table 2. Effect of bacterial inoculation and nutrients on average shoot diameter and shoot network length of ryegrass vegetated in diesel contaminated soil after 100 days of sowing.

Treatments	Average shoot diameter (mm)	Shoot network length (cm)
Control (− diesel)	3.2^{a}	112.8^{a}
Pantoea sp. strain BTRH79 (− diesel)	3.3^{a}	123.6^{a}
Control (+ Diesel)	1.3^{d}	32.4^{d}
Pantoea sp. strain BTRH79 (+ Diesel)	1.8^{cd}	65.6^{c}
Pantoea sp. strain BTRH79 + fertilizer (lower level) (+ Diesel)	2.1^{bc}	83.1^{b}
Pantoea sp. strain BTRH79 + fertilizer (higher level) (+ Diesel)	2.4^{b}	96.4^{b}

Each value is the mean of three replicates, means in the same column followed by the same letter are not significantly different at a 5% level of significance, and the standard error of three replicates is presented in parentheses.

Table 3. Effect of nutrients on CFU, abundance and expression of *Pantoea* sp. strain BTRH79 in the rhizosphere of ryegrass.

Treatment	CFU g^{-1} soil $\times 10^4$	Gene abundance (copies g^{-1} dry soil) $\times 10^4$	Gene expression (copies g^{-1} dry soil) $\times 10^4$
Pantoea sp. strain BTRH79 (+ diesel) (unvegetated)	0.62 (0.23)^d	0.42^d (0.18)	0.16^d (0.11)
Pantoea sp. strain BTRH79 (− diesel) (vegetated)	0.73 (0.25)^d	0.24^d (0.14)	0.07^d (0.02)
Pantoea sp. strain BTRH79 (+ diesel) (vegetated)	1801 (240)^c	1230^c (371)	590^c (634)
Pantoea sp. strain BTRH79 (+ diesel) + fertilizers (lower level) (vegetated)	2780 (380)^b	2040^b (259)	1230^b (213)
Pantoea sp. strain BTRH79 (+diesel) + fertilizers (higher level) (vegetated)	4015 (830)^a	2750^a (455)	1860^a (462)

Each value is the mean of three replicates, means in the same column followed by the same letter are not significantly different at a 5% level of significance, the standard error of three replicates is presented in parentheses.

Results

1. Plant biomass

Growth parameters (root fresh and dry biomass, shoot fresh and dry biomass, average shoot length, average shoot diameter and shoot network length) were estimated to assess the influence of nutrients and bacterial application on plant growth and biomass production (Tables 1 and 2, Fig. 1). In polluted soil, plants showed significantly less growth and biomass production than the plants vegetated in uncontaminated soil. However, bacterial inoculation and nutrients application enhanced plant growth and biomass production. Maximum plant growth and biomass production were obtained when inoculum was applied in the soil containing high concentration of nutrients. Bacterial inoculum in uncontaminated soil improved plant growth but not significantly.

2. Abundance of inoculant strain and expression of alkane degrading gene

The survival and metabolic activity of the inoculated bacterium, *Pantoea* sp. strain BTRH79, in the rhizospheric soil were determined by both culture-dependent and independent approaches (Table 3). In uncontaminated soil with vegetation and in contaminated soil without vegetation, very poor survival and activity of the inoculated strain was observed. However, in contaminated soil, BTRH79 exhibited better colonization, and maximum colonization was observed when nutrients were applied in high concentration. Similarly, higher levels of the abundance and expression of CYP153 gene were found in the rhizospheric soil applied with higher concentration of nutrients.

3. Hydrocarbon degradation

In planted soil, more hydrocarbon degradation was found than unplanted soil. Hydrocarbon degradation was further increased by bacterial and nutrients application. Maximum hydrocarbon degradation was found in the vegetated inoculated soil containing higher levels of nutrients (Table 4).

Discussion

In this study, diesel contamination in soil reduced plant growth and development. It is well established that the presence of hydrocarbons adversely affects plant health and development due to their toxicity and hydrophobicity [4,29]. However, when bacterial strain, *Pantoea* sp. BTRH79, was inoculated in diesel contaminated soil, it enhanced plant growth and development. Enhanced plant growth and development might be attributed to plant growth promoting activities (including ACC deaminase) of the inoculated bacterium [20,21,27]. In principle, bacterial ACC deaminase activity reduces the contaminant induced stress symptoms in developing plant and improves plant health and growth in the presence of contaminants as shown in Fig. 2 [30–32]. Similarly, hydrocarbon-degrading activity of the inoculated microorganisms reduces the phytotoxicity of pollutants due to their potential to produce hydrocarbon-degrading enzymes [33]. Plant growth was further enhanced by the application of nutrients. Earlier studies also reported that the application of nutrients improved plant growth and development during remediation of polluted soil [18,34].

Table 4. Effect of inoculum and nutrients on hydrocarbon degradation.

Treatment	Hydrocarbon	
	Initial (g kg^{-1} soil)	Residual (g kg^{-1} soil)
Control (unvegetated soil)	15	12.54^a (1.36)
Control (vegetated)	15	9.39^c (0.99)
Pantoea sp. strain BTRH79 (unvegetated soil)	15	12.01^b (1.81)
Pantoea sp. strain BTRH79 (vegetated)	15	5.25^d (0.71)
Pantoea sp. strain BTRH79 + fertilizers (lower level) (vegetated)	15	3.83^e (0.65)
Pantoea sp. strain BTRH79 + fertilizers (higher level) (vegetated)	15	2.04^f (0.52)

Each value is the mean of three replicates, means in the same column followed by the same letter are not significantly different at a 5% level of significance; the standard error of three replicates is presented in parentheses.

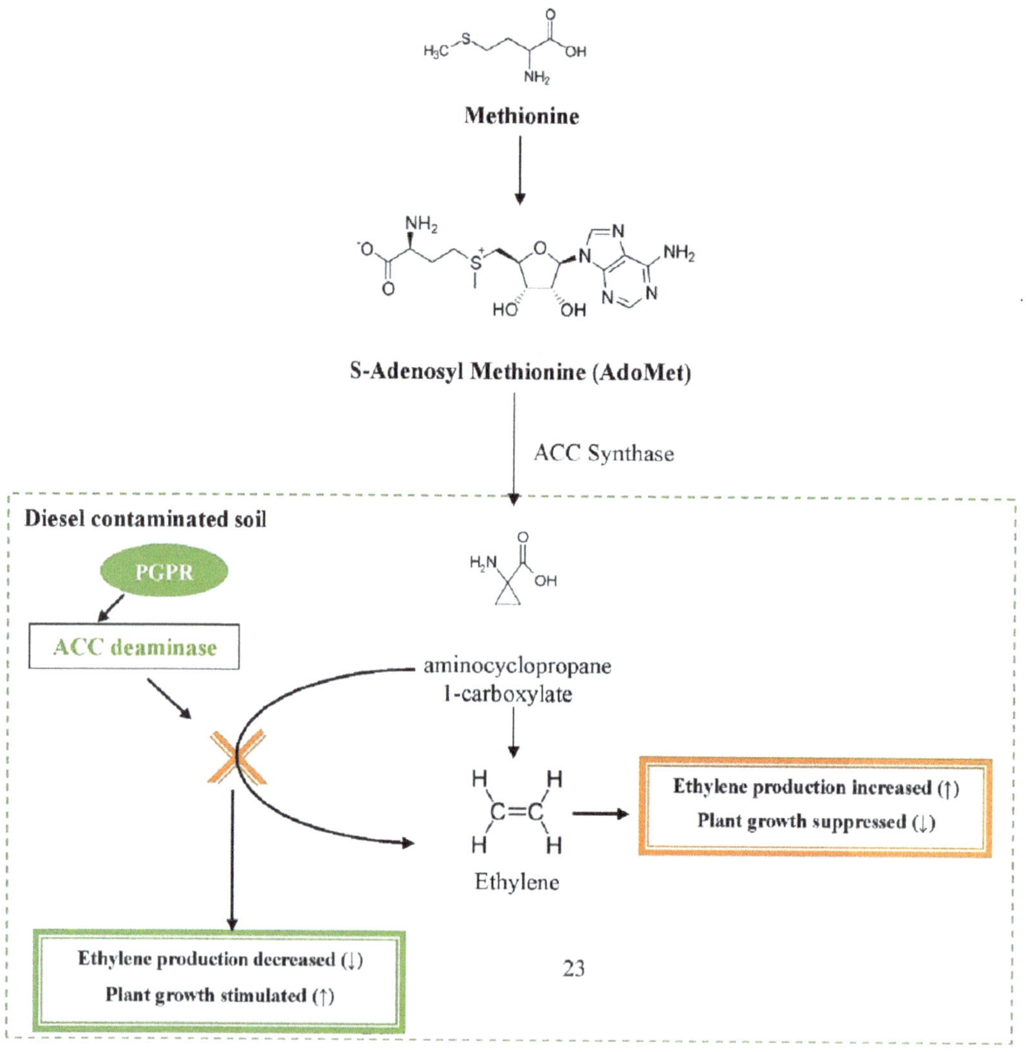

Figure 2. Schematic representation of how bacteria containing ACC deaminase activity lower the ethylene concentration and thereby prevent ethylene-caused inhibition of root elongation.

The higher levels of bacterial colonization, the abundance and expression of CYP153 gene were observed in the rhizospheric soil of ryegrass vegetated in diesel-contaminated soil. A strong positive correlation ($r = 0.991$) between gene abundance and expression further indicated that the inoculated bacterium not only colonized but was also metabolically active in hydrocarbon degradation. In contaminated soil without vegetation, the survival of *Pantoea* sp. strain BTRH79 and expression of CYP153 were significantly lower than in the vegetated soil. This might be due to the fact that the inoculated rhizobacterium could not proliferate and was not metabolically active in the soil without vegetation. Nutrients application enhanced bacterial survival, the abundance and expression of CYP153 gene in the rhizosphere. Similarly, in a previous study, the addition of N, P and K in a petroleum-contaminated soil enhanced microbial population and hydrocarbon mineralization [35]. Between two levels of applied nutrients, significantly more gene abundance and expression were seen in the rhizosphere of ryegrass grown in the soil treated with higher level of nutrients. Higher amount of applied nutrients was required for bacterial proliferation and their catabolic activity [36]. The ratio between gene expression and gene abundance suggests that there were more metabolically active cells in the presence of high concentration of nutrients in the soil (Fig. 3).

Similarly, more hydrocarbon mineralization was found in the vegetated soil than the unvegetated soil. It is well known that plants enhance population and pollutant-degrading activity of microorganisms in soil [37,38]. Hydrocarbon degradation was increased with bacterial inoculation and it might be attributed to hydrocarbon-degrading and plant growth-promoting activity of the inoculated strain, *Pantoea* sp. BTRH79. Many earlier studies have also reported that the application of pollutant-degrading and plant growth-promoting bacteria increased hydrocarbon degradation [6,21]. In this study, hydrocarbon mineralization was further increased by the application of nutrients (N, P and K) in addition to inoculum. Between two applied nutrients levels, more hydrocarbon degradation was observed with higher nutrients level. This might be due to the fact that higher amount of nutrients enhanced more plant growth and microbial population in the rhizosphere that consequently lead to more hydrocarbon degradation.

Statistically speaking, a strong positive correlation ($r = 0.934$) was found between plant biomass and hydrocarbon degradation.

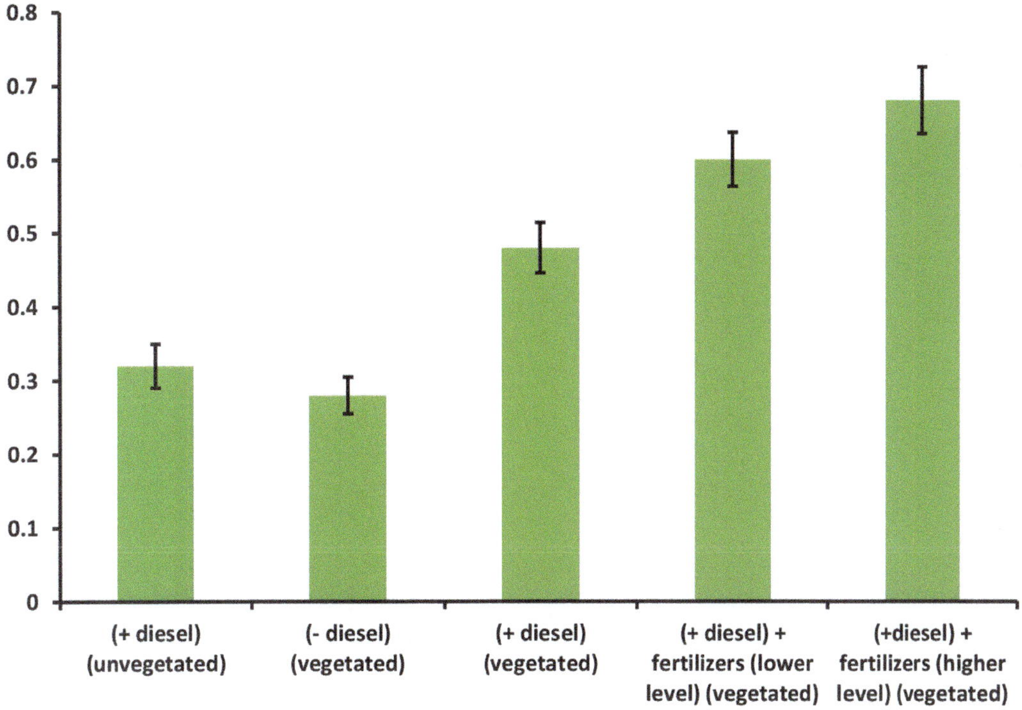

Figure 3. Ratio of gene expression and gene abundance in the vegetated and unvegetated soil.

This supports the idea of high degradation if we manage to increase the plant biomass. Furthermore, regression estimation illustrates that the increment of one unit of hydrocarbons statistically reduce the plant biomass up to 3.007 grams. Plant response (*Y*) in the presence of hydrocarbons can be written as follows:

$$Y = 45.281 - 3.007X$$

Hereby, 45.281 is the statistically calculated value of plant biomass when there was no diesel in the soil. Therefore, adjustment of the nutrients can optimize the remediation setup and improve the plant biomass that ultimately results into the degradation of petroleum hydrocarbons.

Conclusions

This study demonstrates that the nutrients level in soil affect plant growth, hydrocarbon degradation and bacterial colonization and activity. Nutrients can enhance the survival of inoculated bacteria, the abundance and expression hydrocarbon-degrading gene in the rhizosphere. Therefore, the combined use of pollutant-degrading bacteria and nutrients is a more promising approach to enhance phytoremediation efficiency. Further studies should be conducted to determine the individual role of these nutrients as well as the optimum dose of nutrients with a specific plant to get maximum bacterial colonization and activity of pollutant degrading genes.

Author Contributions

Conceived and designed the experiments: SI QMK M Afzal. Performed the experiments: M Arslan IA. Analyzed the data: M Arslan M Afzal IA. Contributed reagents/materials/analysis tools: M Afzal IA. Wrote the paper: M Arslan SI M Afzal.

References

1. Ololade IA (2010) Prediction of polycyclic aromatic hydrocarbons toxicity using equilibrium partitioning approach and narcosis model. Bull Environ Contam Toxicol 85: 238–242.
2. Khan S, Afzal M, Iqbal S, Khan QM (2013) Plant–bacteria partnerships for the remediation of hydrocarbon contaminated soils. Chemosphere 90: 1317–1332.
3. Batty LC, Dolan C (2011) The potential use of phytoremediation for sites with mixed organic and inorganic contamination. Crit Rev Env Sci Tec 43: 217–259.
4. Chen L, Liu X, Zhang X, Liu S, Wei J, et al. (2013) Response characteristics of seed germination and seedling growth of *Acorus tatarinowii* under diesel stress. Plant Soil 368: 355–363.
5. Afzal M, Yousaf S, Reichenauer TG, Sessitsch A (2013) Ecology of alkane-degrading bacteria and their interaction with the plant. FJd . Bruijn (Ed.), Molecular Microbial Ecology of the Rhizosphere. Hoboken, NJ, USA: John Wiley & Sons. pp. 975–989.
6. Khan S, Afzal M, Iqbal S, Mirza MS, Khan QM (2013) Inoculum pretreatment affects bacterial survival, activity and catabolic gene expression during phytoremediation of diesel contaminated soil. Chemosphere 91: 663–668.
7. Afzal M, Khan QM, Sessitsch A (2014) Endophytic bacteria: Prospects and applications for the phytoremediation of organic pollutants. Chemosphere 117: 232–242.
8. Yang C-H, Crowley DE (2000) Rhizosphere microbial community structure in relation to root location and plant iron nutritional status. Appl Environ Microb 66: 345–351.
9. Zhang N, Wang D, Liu Y, Li S, Shen Q, et al. (2014) Effects of different plant root exudates and their organic acid components on chemotaxis, biofilm formation and colonization by beneficial rhizosphere-associated bacterial strains. Plant Soil 374: 689–700.
10. Cébron A, Beguiristain T, Faure P, Norini M-P, Masfaraud J-F, et al. (2009) Influence of vegetation on the in situ bacterial community and polycyclic aromatic hydrocarbon (PAH) degraders in aged PAH-contaminated or thermal-desorption-treated soil. Appl Environ Microb 75: 6322–6330.
11. Weyens N, van der Lelie D, Taghavi S, Newman L, Vangronsveld J (2009) Exploiting plant–microbe partnerships to improve biomass production and remediation. Trends Biotechnol 27: 591–598.

12. Cunliffe M, Kawasaki A, Fellows E, Kertesz MA (2006) Effect of inoculum pretreatment on survival, activity and catabolic gene expression of *Sphingobium yanoikuyae* B1 in an aged polycyclic aromatic hydrocarbon-contaminated soil. FEMS Microb Ecol 58: 364–372.
13. Yousaf S, Afzal M, Reichenauer TG, Brady CL, Sessitsch A (2011) Hydrocarbon degradation, plant colonization and gene expression of alkane degradation genes by endophytic *Enterobacter ludwigii* strains. Environ Pollut 159: 2675–2683.
14. Juhanson J, Truu J, Heinaru E, Heinaru A (2009) Survival and catabolic performance of introduced *Pseudomonas* strains during phytoremediation and bioaugmentation field experiment. FEMS Microb Ecol 70: 446–455.
15. Tara N, Afzal M, Ansari TM, Tahseen R, Iqbal S, et al. (2013) Combined use of alkane-degrading and plant growth-promoting bacteria enhanced phytoremediation of diesel contaminated soil. Int J Phytorem 16: 1268–1277.
16. Nie M, Wang Y, Yu J, Xiao M, Jiang L, et al. (2011) Understanding plant-microbe interactions for phytoremediation of petroleum-polluted soil. PLOS ONE 6: e17961.
17. Hutchinson SL, Schwab AP, Banks MK (2004) Biodegradation of petroleum hydrocarbons in the rhizosphere. Phytoremediation. John Wiley & Sons, Inc. pp. 355–386.
18. Adesemoye A, Kloepper J (2009) Plant–microbes interactions in enhanced fertilizer-use efficiency. Appl Microbiol Biotechnol 85: 1–12.
19. Adesemoye AO, Torbert HA, Kloepper JW (2008) Enhanced plant nutrient use efficiency with PGPR and AMF in an integrated nutrient management system. Can J Microb 54: 876–886.
20. Yousaf S, Ripka K, Reichenauer TG, Andria V, Afzal M, et al. (2010) Hydrocarbon degradation and plant colonization by selected bacterial strains isolated from Italian ryegrass and birdsfoot trefoil. J Appl Microb 109: 1389–1401.
21. Afzal M, Yousaf S, Reichenauer TG, Sessitsch A (2011) The inoculation method affects colonization and performance of bacterial inoculant strains in the phytoremediation of soil contaminated with diesel oil. Int J Phytorem 14: 35–47.
22. Abràmoff MD, Magalhães PJ, Ram SJ (2004) Image processing with ImageJ. Biophotonics Int 11: 36–43.
23. Mehmood R, Riaz M, Does RJMM (2012) Control charts for location based on different sampling schemes. J Appl Stat 40: 483–494.
24. Patil G, Sinha A, Taillie C (1994) Ranked set sampling. Handbook of Statistics 12: 167–200.
25. Nerandzic MM, Cadnum JL, Eckart KE, Donskey CJ (2012) Evaluation of a hand-held far-ultraviolet radiation device for decontamination of *Clostridium difficile* and other healthcare-associated pathogens. BMC Infect Dis 12: 120.
26. Andria V, Reichenauer TG, Sessitsch A (2009) Expression of alkane monooxygenase (*alkB*) genes by plant-associated bacteria in the rhizosphere and endosphere of Italian ryegrass (*Lolium multiflorum* L.) grown in diesel contaminated soil. Environ Pollut 157: 3347–3350.
27. van Beilen JB, Funhoff EG, van Loon A, Just A, Kaysser L, et al. (2006) Cytochrome P450 alkane hydroxylases of the CYP153 family are common in alkane-degrading eubacteria lacking integral membrane alkane hydroxylases. Appl Environ Microb 72: 59–65.
28. Afzal M, Yousaf S, Reichenauer TG, Kuffner M, Sessitsch A (2011) Soil type affects plant colonization, activity and catabolic gene expression of inoculated bacterial strains during phytoremediation of diesel. J Hazard Mater 186: 1568–1575.
29. Afzal M, Khan S, Iqbal S, Mirza MS, Khan QM (2013) Inoculation method affects colonization and activity of *Burkholderia phytofirmans* PsJN during phytoremediation of diesel-contaminated soil. Int Biodeter Biodeg 85: 331–336.
30. Arshad M, Saleem M, Hussain S (2007) Perspectives of bacterial ACC deaminase in phytoremediation. Trends Biotechnol 25: 356–362.
31. Glick BR (2010) Using soil bacteria to facilitate phytoremediation. Biotechnol Adv 28: 367–374.
32. Glick BR, Stearns JC (2011) Making phytoremediation work better: Maximizing a plant's growth potential in the midst of adversity. Int J Phytorem 13: 4–16.
33. Beilen J, Funhoff E (2007) Alkane hydroxylases involved in microbial alkane degradation. Appl Microbiol Biotechnol 74: 13–21.
34. Carvalhais LC, Dennis PG, Fan B, Fedoseyenko D, Kierul K, et al. (2013) Linking plant nutritional status to plant-microbe interactions. PLOS ONE 8: e68555.
35. Merkl N, Schultze-Kraft R, Infante C (2005) Assessment of tropical grasses and legumes for phytoremediation of petroleum-contaminated soils. Water Air Soil Pollut 165: 195–209.
36. Merkl N, Schultze-Kraft R, Infante C (2004) Phytoremediation in the tropics—The effect of crude oil on the growth of tropical plants. Bioremediation J 8: 177–184.
37. Bais HP, Weir TL, Perry LG, Gilroy S, Vivanco JM (2006) The role of root exudates in rhizosphere interactions with plants and other organisms. Ann Rev Plant Biol 57: 233–266.
38. Ho CH, Applegate B, Banks MK (2007) Impact of microbial/plant interactions on the transformation of polycyclic aromatic hydrocarbons in rhizosphere of *Festuca arundinacea*. Int J Phytorem 9: 107–114.

10

De Novo Assembly of the Common Bean Transcriptome Using Short Reads for the Discovery of Drought-Responsive Genes

Jing Wu, Lanfen Wang, Long Li, Shumin Wang*

Key Laboratory of Crop Germplasm Resources and Utilization, Ministry of Agriculture; The National Key Facility for Crop Gene Resources and Genetic Improvement, Institute of Crop Science, the Chinese Academy of Agricultural Sciences, Beijing, China

Abstract

The common bean (*Phaseolus vulgaris* L.) is one of the most important food legumes, far ahead of other legumes. The average grain yield of the common bean worldwide is much lower than its potential yields, primarily due to drought in the field. However, the gene network that mediates plant responses to drought stress remains largely unknown in this species. The major goals of our study are to identify a large scale of genes involved in drought stress using RNA-seq. First, we assembled 270 million high-quality trimmed reads into a non-redundant set of 62,828 unigenes, representing approximately 49 Mb of unique transcriptome sequences. Of these unigenes, 26,501 (42.2%) common bean unigenes had significant similarity with unigenes/predicted proteins from other legumes or sequenced plants. All unigenes were functionally annotated within the GO, COG and KEGG pathways. The strategy for *de novo* assembly of transcriptome data generated here will be useful in other legume plant transcriptome studies. Second, we identified 10,482 SSRs and 4,099 SNPs in transcripts. The large number of genetic markers provides a resource for gene discovery and development of functional molecular markers. Finally, we found differential expression genes (DEGs) between terminal drought and optimal irrigation treatments and between the two different genotypes Long 22-0579 (drought tolerant) and Naihua (drought sensitive). DEGs were confirmed by quantitative real-time PCR assays, which indicated that these genes are functionally associated with the drought-stress response. These resources will be helpful for basic and applied research for genome analysis and crop drought resistance improvement in the common bean.

Editor: Zhong-Hua Chen, University of Western Sydney, Australia

Funding: This work was supported by grants from the Ministry of Agriculture of China (the earmarked fund for China Agriculture Research System [CARS-09]), National Key Technology Research and Development Program of the Ministry of Science and Technology of China (2013BAD01B03-18a) and the Agricultural Science and Technology Innovation Program (ASTIP) of CAAS. The funders had no role in study design, data collection and analysis, decision to publish, or preparation of the manuscript.

Competing Interests: The authors have declared that no competing interests exist.

* Email: wangshumin@caas.cn

Introduction

The common bean (*Phaseolus vulgaris* L.), a legume native to America, is now one of the most important crops worldwide and plays an important role in solving food shortages in poor areas and adjusting the diet structure in developed countries. The global bean harvest is approximately 23 million tons, with Asia being the most important producing region with 14 million hectares, representing almost half of the global output in 2012 (FAO). However, the growth and production of the common bean are usually limited by many abiotic stresses, among which drought is the most complex and devastating on a global scale. Approximately 60% of common bean production occurs in agricultural land prone to water deficit, without irrigation systems, where unexpected drought periods result in losses that may reach up to 80% of yield reduction [1,2]. Thus, improvement of drought resistance is a major goal for common bean breeders worldwide.

Drought is an increasingly important constraint of crop productivity and stability worldwide due to climate change. It is a physiologically complex trait and involves complex cross-talk between different regulatory levels, including adjustment of metabolism and gene expression for physiological and morphological adaptation. Previous studies are mainly focused on the traits related to drought resistance in the common bean, such as flower, seed filling, number of pods and seeds, seed weight and days to maturity [3,4]. Traits associated with drought tolerance were identified and drought-tolerant germplasm that differed in rooting depth was identified [5]. In addition, common bean germplasm that exhibits improved levels of drought tolerance has been identified [6,7]. Physiological analysis of common bean cultivars uncovers characteristics related to terminal drought resistance [8].

Molecular markers are powerful tools to analyze the genetic control of complex traits such as diseases resistance, seed iron and drought tolerance [9,10,11]. A reliable genetic map was developed to analyze the inheritance of yield traits under drought and fully irrigated conditions over three years of experiments [9]. Until now, some drought tolerance quantitative trait loci (QTL) in common beans were identified, and they were mainly associated with yield components, phenology, canopy biomass dry weight, biomass partitioning indices, stem and seed total nonstructural carbohydrate content, leaf area index, and leaf chlorophyll content and rooting pattern traits instead of photosynthate remobilization traits [9,12–15]. Drought-responsive genes were increasingly described in a number of plant reviews, such as P5CS, ADC, SPDS, ZMDREB2 and OSTPS1 [16–21]. However, there are few studies involving gene cloning and functional verification in the common bean. *P5CS2* was isolated from the common bean and analyzed for genetic diversity [22], and a root-specific bZIP transcription factor is responsive to water deficit stress in the common bean [23]. However, the mechanisms underlying gene regulation in drought response remains elusive. In other words, drought tolerance is a cumulative process with stepwise changes in gene regulation. Therefore, the drought-induced response of the transitional landraces between drought-resistant and drought-susceptible plants may provide a better understanding of drought tolerance.

In recent years, high-throughput next generation sequencing (NGS) technologies such as Roche 454, Illumina, Solexa and ABI SOLiD have made it possible to generate gene resources at the whole genome level and to advance crop genetics and breeding with relatively low cost [24,25]. Whole genome transcriptome analysis is an economical and effective way to exploit key factors for plant responses to biotic and abiotic stress that are involved in transcriptional and metabolic activities. These technologies have been effectively used to generate large-scale differentially expressed gene data in several plant species such as *Millettia pinnata*, cassava, hyacinth Bean, and Asian seabass [26–29]. Despite the common bean's growing edible and economic importance, investigation at a comprehensive transcriptome level has been lacking. Here, we report the expression profiling of the two genotypes, Long 22-0579 (drought-tolerant) and Naihua (drought-sensitive), in response to drought stress, using RNA-seq to explore the potential candidate drought-responsive genes. The data obtained will serve as an invaluable genomic reference to further our knowledge about the common bean at the molecular level, and can be applied to molecular breeding for plants with enhanced drought tolerance.

Materials and Methods

Plant growth and drought treatments

In this study, we included improved cultivars of two gene pools: Andean and Mesoamerican (Table S1). Common bean cultivars, Long 22-0579 (drought-tolerant genotype) and Naihua (drought-sensitive genotype), were obtained from the National Gene Bank (China, Beijing). The seedlings were grown in plastic pots (23 cm×18 cm×18 cm) under a 14/10 h photoperiod at 25°C (day) and 20°C (night) in a greenhouse at the Institute of Crop Science, the Chinese Academy of Agricultural Sciences (China, Beijing, 116°46′E, 39°92′N). The water content of each pot was measured three times a week, and the water lost was supplemented in the pots to keep equivalent levels according to treatment requirements. A completely random block design with two treatments, terminal drought and optimal irrigation was used. Twenty-five plants were used in each treatment. All pots were irrigated to field capacity until 4 weeks after seeding. For terminal drought treatment, watering was restricted to 25% of field capacity in the pot media from 5 weeks after seeding. For optimal irrigation, pots were kept to field capacity throughout the experiment.

Drought resistance index

Field experiments were carried out following the randomized complete block design with three replications for both control and drought stress conditions at the Institute of Crop Science, the Chinese Academy of Agricultural Sciences (China, Beijing, 116°46′E, 39°92′N). Twenty-five seeds for each line were planted, with the distance of 20 cm between the plants within a plot, and 50 cm between adjacent plots. Grain production was determined harvesting plants from the central part of the rows of each cultivar, excepting two plants as a border at each end of the row. The drought resistance index (DRI) was calculated on the basis of a multiple regression of the grain yield of stressed and unstressed plants for each cultivar [30]. DRI for individual cultivars was computed as: $DRI = Y_{DS} (Y_{DS}/Y_{WW})/\overline{Y_{DS}}$ where Y_{DS} is the grain yield obtained under unstressed per cultivar and Y_{WW} is the yield under stressed conditions. $\overline{Y_{DS}}$ is average grain yield obtained under unstressed all cultivars.

Sample collection and RNA preparation

The leaves were sampled when the leaves begin to wilt after the application of optimal irrigation and terminal drought treatment. Five leaves at the shoot apex were simultaneously collected from each individual plant and were frozen in liquid nitrogen and stored at −80°C prior to RNA extraction. The total RNA was isolated using TRIzol reagent (Tiangen, Beijing) following the manufacturer's instructions. Equal amount of total RNAs from fifteen plants were pooled for a single combined sample. In total, four combined samples were collected and denoted as LOI, LTD, NOI, and NTD according to cultivar (Long 22-0579 or Naihua) and the treatments (optimal irrigation or terminal drought) of their sampling sources. For each sample, at least 20 μg of total RNA was used for Illumina Hiseq 2500 sequencing conducted at the Beijing Berry Genomics company. All sequence data have been deposited in the Short Read Archive (SRA) at the NCBI database under the project accession number SRR1523069.

Sequences assembly

After sequencing, the raw sequence data were first purified by trimming adapter sequences and removing low-quality sequences. The resulting clean reads were assembled using the trinity software [31]. Trinity assembler was used with the inchworm k-mer method, and all of the server resources (stack size, CPU time, file size, data size, core dump size, memory usage, and virtual memory usage) were set to unlimited.

Data analysis

SSR identification and GC content analysis were performed using in-house perl scripts. The perl script program MISA (MIcroSAtellite; http://pgrc.ipk-gatersleben.de/misa/) and SAM-tools [32] was used for identification of SSRs and SNPs, respectively.

For assignments of gene descriptions, the all-unigenes were searched against the Nr database using BLASTx with an E-value cut-off of 10^{-5}. Based on their annotations, the all-unigenes were assigned GO annotations using blast2GO, followed by functional classification using the WEGO software. Moreover, the putative metabolic pathways for the all-unigenes were assigned by performing BLASTx against the KEGG and COG pathway database with an E-value cut-off of 10^{-5}. The RPKMs (reads per kilobase per million reads) were applied to measure the gene expression levels. The differentially expressed genes (DEGs) between the optimal irrigation and terminal drought samples were identified using 'DEGseq'. Four separate differential expression tests implemented by DEGseq and the corresponding significance thresholds used were likelihood ratio test (LRT), Fisher's exact test (FET) and the MA-plot-based method with random sampling model (MARS) (p-value≤0.001) and fold-change threshold on the MA-plot (FC) $\log_2$ normalized fold change ≥2. The GO enrichment analysis and KEGG pathway enrichment analysis for the DEGs were both performed by conducting hypergeometric tests with the whole common bean transcriptome set as the background.

Quantitative real-time PCR analysis

qRT-PCR was conducted using the common bean actin gene (EU369188.1) as the control. The first-strand cDNAs were synthesized from 1 μg of total RNAs using the SuperScript II reverse transcriptase kit (Invitrogen). The gene-specific primers were designed using Primer Premier software (version 5.0). Real-time PCR was performed on an ABI PRISM 7300 Sequence Detection System (Applied Biosystems) using 1 μl of first-strand cDNA and SYBR Premix Ex Taq (TAKARA). All reactions were performed in triplicate. The relative expression levels for each gene were calculated using the $2^{-\Delta\Delta CT}$ method with normalization to the internal control.

Results

Comparison of drought tolerance between Long 22-0579 and Naihua

First, we evaluated drought resistance of one hundreds common bean cultivars by DRI value. According to the DRI values, cultivars Long 22-0579 and Naihua were the most contrasting (Table S2). Long 22-0579 in terms of yield under terminal drought was tolerant to optimal irrigation showed the highest DRI value (1.25) (Table S2). In contrast, Naihua which had the lowest seed yield under stress had lower DRI value (0.19) (Table S2). Based on screening experiments, we selected one drought-tolerant genotype (Long 22-0579) and one drought-sensitive genotype (Naihua). Further screening experiments revealed that the young seedling of the Naihua plants treated with terminal drought began to wilt after treatment. Meanwhile, the Long 22-0579 plants showed no symptoms during the course of the terminal drought treatments. These results indicate that the Long 22-0579 plants exhibit stronger drought tolerance than Naihua plants (Figure 1). Based on these results, these two genotypes (Long 22-0579 and Naihua) were chosen for RNA-seq.

Generation and assembly of transcript reads

To obtain a general overview of the common bean transcriptome and an initial comparison between drought-resistant and drought-susceptible bean transcripts, four libraries (LOI, LTD, NOI, and NTD) were constructed for paired end (PE) sequencing. The Illumina sequencing was then performed separately for four cDNA libraries and four sub-transcriptomes with 100-bp raw reads were generated. After filtration of low-quality and adapter sequences, a total of 54 055 718, 72 287 786, 71 518 270 and 72 100 736 raw reads were generated from the Illumina GAIIx sequencer, accounting for approximately 5.4 Gb, 7.2 Gb, 7.2 Gb and 7.2 Gb of sequence data for the library cultivars LOI, LTD, NOI, and NTD, respectively. These processed paired-end reads with high quality were used for further analysis (Table S3). In total, these clean reads constitute ~27.0 GB of sequence data. The GC contents of raw reads were from approximately 45% to approximately 47%.

Nearly 270 million clean reads from LOI, LTD, NOI, and NTD sub-transcriptomes were assembled and 139,487 transcripts were obtained. The longest transcript length was 25,393 bp and the mean transcript length was 1,487 bp. The average GC content of the common bean transcript was 39.3%. Clustering resulted in 62,828 unigenes. The longest unigene length was 14,799 bp and the mean unigenes length was 777 bp (Table 1). Among these unigenes, 34,858 (55.5%) were from 200 bp to 500 bp, 13,062 (20.8%) were longer than 500 bp, 9,648 (15.4%) were longer than 1,000 bp, and 5,260 (8.4%) were longer than 2,000 bp. Figure S1 shows that all-unigene length distributions range from 200 bases to more than 2,000 bases. The average GC content of the common bean unigenes (36%), and soybean unigenes (legume reference, 40.9%) was slightly lower than that of *Arabidopsis* (dicot reference, 42.5%) and rice (monocot reference, 55%) as reported previously [33,34].

Frequency and distribution of EST-SSRs and SNPs in the common bean transcriptome

Transcriptome sequencing generated a high quantity of data in which different types of polymorphisms (e.g., EST-SSRs and SNPs) can be observed, thus providing valuable resources for the development of molecular markers. The molecular markers are important resources for marker-assisted breeding, determining functional genetic variation and map-based cloning genes. Here, we investigated two types of putative markers from common bean leaves: EST-SSRs and SNPs. Both need future validation for practical use in common bean breeding and research.

EST-SSRs are highly polymorphic, easier to develop and serve as rich resources for diversity. We identified a total of 10,482 SSR loci in transcripts of the common bean with a frequency of one SSR per 4.70 kb of sequence (Table S4). The mononucleotide SSRs represented the largest fraction (58.5%) of SSRs identified, followed by dinucleotide (20.5%) and tri-nucleotide (19.8%) SSRs. Although only a small fraction of tetra- (106), penta- (13) and hexa-nucleotide (10) SSRs were identified in common bean transcripts, the number is quite significant.

The frequencies of EST-SSRs with different numbers of tandem repeats were calculated as shown in Table S6. The most common SSRs were those with six tandem repeats (33.6%), followed by five tandem repeats (29.6%), seven tandem repeats (17.4%), eight tandem repeats (8.7%), nine tandem repeats (5.7%), ten tandem repeats (3.6%), and more than 10 tandem repeats (1.5%). The dominant repeat motif in EST-SSRs was AG/CT (32.9%), followed by AAG/CTT (16.8%), AT/TA (10.5%), AC/GT (7.5%), and AGT/ACT (7.3%) (Table S5). However, very few CG/CG (0.1%) repeats were identified in this study.

SNP markers offer the promise of higher map resolution, higher throughput, lower cost and a lower error rate. Between Long 22-0579 and Naihua, 4,099 SNPs were predicted. Approximately 82.7% of the SNPs were single-base changes of which 44.0% were transitions and 38.8% were transversions. The remaining 17.3% SNPs were indels. Of a total of 1,802 transitions, a similar proportion of bi-allelic types were detected, i.e., 896 for A/G and 906 for C/T. In 1,589 transversion mutations, four types of base substitution (T/G, G/C, A/T, A/C) were detected (Table S6).

Gene annotation and functional classification

To identify the putative function of common bean unigenes, they were BLASTed against the NCBI non-redundant (Nr) protein sequences database. A total of 26,501 (42.2%) common bean unigenes that showed significant similarity to the proteins in Nr database were assigned Nr annotations. The majority of these annotated unigenes had the highest homology to genes from plants, and only 263 (0.99%) annotated unigenes were annotated with sequences from the non-plant sources, such as *Candida*

Figure 1. Phenotype of Long 22-0579 and Naihua genotypes after drought-stress treatment.

albicans, *Actinomyces turicensis*, etc. Three-quarters of annotated unigenes retrieved annotations from *Glycine max*, reflecting the evolutionary relationship between the common bean and soybean. Moreover, there were also 1,098 annotated unigenes with matched accessions from *Phaseolus vulgaris*. Small percentages of annotated unigenes showed similarity with proteins from other legumes, such as *Cicer arietinum* (5.43%), *Medicago truncatula* (3.75%), *Lotus japonicas* (1.03%) and *Vigna* (0.46%), etc. In addition, many common bean unigenes showed homology to uncharacterized proteins annotated as unknown hypothetical and expressed proteins (Table S7).

GO enrichment analysis was carried out to classify the gene functions of the unigenes identified. A total of 11,580 (18.4%) unigenes were assigned and classified into 50 terms from the three main categories: biological process category, molecular function category and cellular component category. The remaining unigenes failed to be assigned GO terms, which may be attributed to the lack of information on these genes (e.g., unknown, or hypothetical and expressed proteins). Among those, 20,309, 14,166 and 14,208 unigenes were assigned at least one GO term in the biological process category, molecular function category and cellular component category, respectively. In the cellular component categories, the cell (39.65%) and cell parts (39.65%) were most abundantly represented. In contrast, rare unigenes were sorted into extracellular region parts (0.03%) or extracellular region parts (0.03%). Among the various biological processes, the two most highly represented lineages were metabolic process (58.30%) and cellular process (51.20%). The genes involved in other important biological processes, such as establishment of localization, localization, pigmentation and biological regulation, were also identified through GO annotations. Similarly, catalytic activity (50.85%) and binding (54.9%) were most represented among the various molecular functions (Figure 2a).

Using COG functional classification, 8,156 (30.8%) unigenes aligned to the COG database and were classified into 24 functional categories, among which the general function prediction was the largest group (3,035 genes, 37.2%). Followings are the groups of 1) posttranslational modification, protein turnover, chaperones (1,421 genes, 17.4%); 2) replication, recombination and repair (1,371 genes, 16.8%); 3) transcription (1,295 genes, 15.9%); 4) amino acid transport and metabolism (1,271 genes, 15.6%), 5) inorganic ion transport and metabolism (1,228 genes,

Table 1. Statistics for the unigenes of the common bean.

Length of unigenes (bp)	Number of unigenes
200–500	34,858 (55.5%)
500–1000	13,062 (20.8%)
1000–2000	9,618(15.4%)
≥2000	5,260 (8.4%)
Total	62,828
Longest length	14,799 bp
Mean length	777 bp
N50	1,321 bp
Total length	48,789,691 bp

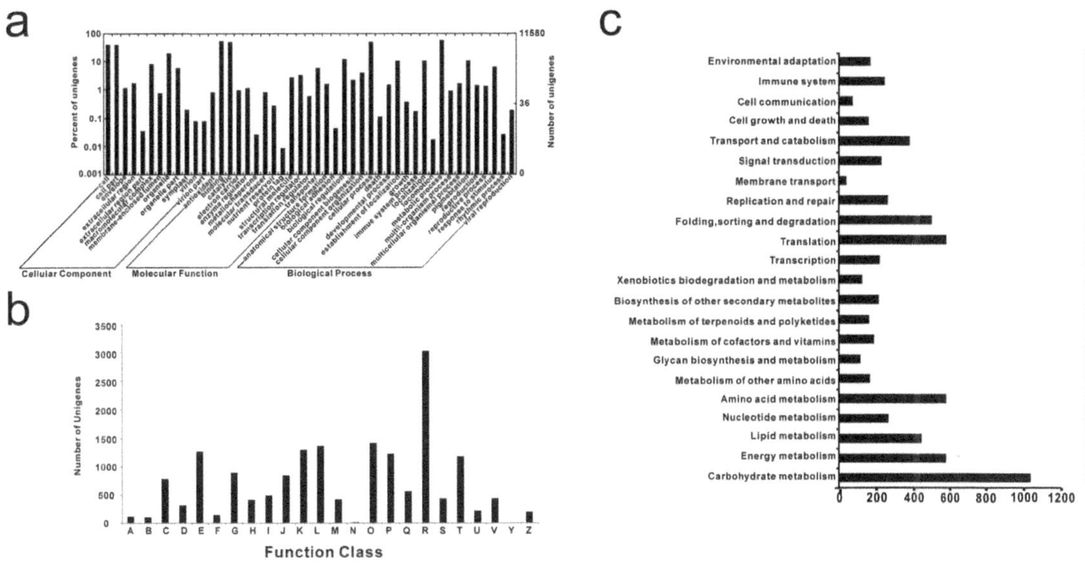

Figure 2. Annotation of common bean unigenes. a), GO annotation of common bean unigenes. The results are summarized in three main categories: biological process, cellular component, and molecular function. In total, 11,580 genes have been assigned 50 GO terms. In some cases, one gene has multiple terms. b), COG functional annotations of common bean unigenes. A, RNA processing and modification; B, Chromatin structure and dynamics; C, Energy production and conversion; D, Cell cycle control, cell division, chromosome partitioning; E, Amino acid transport and metabolism; F, Nucleotide transport and metabolism; G, Carbohydrate transport and metabolism; H, Coenzyme transport and metabolism; I, Lipid transport and metabolism; J, Translation, ribosomal structure and biogenesis; K, Transcription; L, Replication, recombination and repair; M, Cell wall/ membrane/envelope biogenesis; N, Cell motility; O, Posttranslational modification, protein turnover, chaperones; P, Inorganic ion transport and metabolism; Q, Secondary metabolites biosynthesis, transport and catabolism; R, General function prediction only; S, Function unknown; T, Signal transduction mechanisms; U, Intracellular trafficking, secretion, and vesicular transport; V, Defense mechanisms; Y, Nuclear structure; Z, Cytoskeleton. c), Histogram presentation of KEGG classification of unigenes. The all unigenes were assigned X pathways within X clades under five major categories: I, Metabolism; II, Genetic information processing; III, Environmental information processing; IV, Cellular processes; V, Organismal systems.

15.1%); and 6) signal transduction mechanisms (1,178 genes, 14.4%). Genes annotated as "cell motility" (21 genes, 0.3%) and "nuclear structure" (7, 0.1%) represent the smallest groups predicted by COG. (Figure 2b).

To characterize the active biological pathways in the common bean, the Kyoto Encyclopedia of Genes and Genomes (KEGG) was used to analyze the pathway annotations of unigene sequences. In total, 3,444 (13.00%) sequences were aligned with the KEGG database and were assigned to 173 KEGG pathways. These pathways belonged to 22 clades under five major KEGG categories, including 'metabolism', 'genetic information processing', 'environmental information processing', 'cellular processes', and 'organismal systems' (Figure 2c). Among them, 'ribosome'(246), 'plant hormone signal transduction'(233), 'protein processing in endoplasmic reticulum'(182), 'starch and sucrose metabolism'(154) and 'purine metabolism'(149) were the top five pathways most represented by all-unigenes. These results provide an advantageous resource for investigating specific processes, functions and pathways in plant drought research.

Different expression genes (DEGs) analysis under drought conditions

First, we performed whole transcriptome sequencing to explore whether steady-state levels of certain protein-encoding RNAs differ under drought conditions using DEGseq. We collated a list of 4,139 genes according to the DEGseq statistical tests, FET, LRT, MARS (p-value$\leq$0.001) and FC (log2 normalized fold change $\geq$2), representing differences in RNA expression between samples LOI and LTD, and 6,989 genes between NOI and NTD. In this study, DEGs with higher expression levels in terminal drought-treated samples when compared with optimal irrigation-treated samples were denoted as 'up-regulated' versus 'down-regulated'. Overall, there are much more DEGs in the Naihua than in the Long 22-0579. Between samples LOI and LTD, the number of down-regulated DEGs (2,013) was higher than the up-regulated DEGs (2,126). By contrast, more up-regulated DEGs (1,185) than down-regulated DEGs (5,804) were identified when comparing NOI and NTD. Only a small portion of DEGs (659 up-regulated and 1,528 down-regulated) shared common tendency of expression changes between optimal irrigation and terminal drought in Long 22-0579 or Naihua. Five DEGs were up-regulated between LOI and LTD but down-regulated between NOI and NTD, and 4 DEGs were down-regulated between LOI and LTD but up-regulated between NOI and NTD (Figure 3).

In this study, the KEGG pathway analysis helped us understand the biological function of DEGs under different treatments. Eight hundred thirty DEGs were assigned to 235 pathways, mostly associated with plant hormone signal transduction (69 DEGs), carbon metabolism (63 DEGs) and starch sucrose metabolism (60 DEGs) (Table S8). We also conducted GO enrichment analysis for DEGs with the whole transcriptome set as the background. In the biological process category, molecular function category and cellular component category, 2,899, 1,817 and 2,022 DEGs were assigned at least one GO term, respectively. The significantly overrepresented GO categories of biological processes were from single-organism metabolic pathways, oxidation-reduction pathways and carbohydrate metabolic pathways (Table S9).

After removing DEGs between treatments, the identified DEGs were analyzed using edgeR with a p-value$\leq$0.001 and fold change $\geq$2 representing differences in RNA expression between different cultivars (Long 22-0579 or Naihua). DEGs with higher expression levels in Long 22-0579 when compared with Naihua were denoted

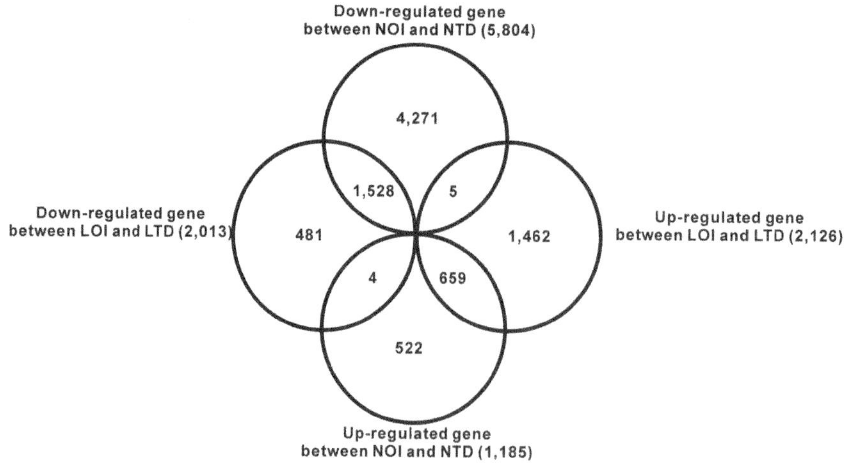

Figure 3. Number of DEGs in the different genotypes. The numbers of DEGs that were exclusively up- or down-regulated in one genotype are shown in each circle. The numbers of DEGs with common or opposite tendency of expression changes between different genotypes are shown in the overlapping regions. The total numbers of up- or down-regulated gene in each genotype are shown outside of the circles.

as 'up-regulated' versus 'down-regulated'. Overall, we collated a list of 473 genes representing differences in RNA expression between Long 22-0579 and Naihua, with more up-regulated DEGs (281) than down-regulated DEGs (192) (Table S10). With GO enrichment analysis for DEGs with the whole transcriptome set as the background, 80,151 and 3 were assigned at least one GO term in the biological processes category, molecular functions category and cellular components category, respectively. The significantly overrepresented GO terms were catalytic activity (72), oxidoreductase activity (15) and oxidation-reduction process (15) (Table S11). Twelve unigenes are related to stress response. Of the 473 DEGs, 31 were aligned with the KEGG database and were assigned to 68 KEGG pathways. The majority of the KEGG pathways included one gene, and most contained only four genes (Table S12).

Experimental verification of DEGs

The 16 candidate DEGs obtained by RNA-seq analysis were further validated using RT-PCR (Figure 4). These candidates mainly included genes associated with, or involved in drought stress response in other plant species, such as genes involved in plant hormone pathways (auxin, gibberellins, and ethylene) or zinc finger proteins and genes encoding transcription factors (WRKY, NAC and DREB). In addition, there were three candidates with unknown functions and two candidates with no annotations, which represented a new resource for genes related to drought response. All these candidates fall into three classes, first, four of these genes (*comp13117_c0_seq1*, *comp19490_c0_seq1*, *comp18282_c0_seq1* and *comp40404_c0_seq1*) were shown to have differential expression between the control and treatment (Long 22-0579 or Naihua). In the case of drought, the expression level of *comp13117_c0_seq1* all up-regulated between the control and treatment (Long 22-0579 or Naihua, respectively. However, *comp19490_c0_seq1* and *comp40404_c0_seq1* genes were up-regulated after drought stress. These genes exhibited up or down-regulated under drought stress, moreover, these four genes expression pattern were similar between two different drought tolerant genotypes after drought stress. These four genes may not be responsible for the different levels of drought tolerance in Long 22-0579 and Naihua. More studies will be required before we know what the functions of these genes are and whether they relate to drought. Second, eight DEGs (*comp18253_c0_seq1*, *comp18888_c3_seq1*, *comp18293_c0_seq1*, *comp19585_c0_seq1*, *comp16965_c0_seq1*, *comp40354_c0_seq1*, *comp19389_c0_seq1*, and *comp38170_c0_seq1*) were shown to have differential expression between LOI and LTD. These genes whose expression levels changed about two to ten-fold in response to drought treatment. Thirdly, four DEGs (*comp11891_c0_seq1*, *comp18632_c0_seq1*, *comp19229_c0_seq1* and *comp18242_c0_-seq1*) were shown to have differential expression between NOI and NTD. Significantly, the expression patterns of the most of the genes did not display clear correlations with the differences in drought tolerance observed in the Long 22-0579 and Naihua. Nevertheless, it is clear that the genes differ considerably between the two cultivars, suggesting that the DEGs could not totally explain the drought higher tolerance observed in Long 22-0579. Finally, we compared relative gene expression levels between drought-treated and control leaves. The expression profiles of the 16 candidates were generally in agreement with the predictions from the RNA-seq results (Table S13). These results suggested that the data obtained from the DEGs analysis were credible.

Discussion

Common bean transcriptome characterization

The common bean is one of the most important legume crop plants, with a high protein content and high amounts of fiber, complex carbohydrates and other dietary element proteins, making it a very important target for genomic studies, thus being used as a diploid model for soybean research [2]. For legumes such as soybean, *Medicago*, lotus, chickpea and pigeonpea, genome sequencing has been nearly completed and a vast collection of ESTs are available for functional genomic studies [35–40]. However, very few genomic resources, including transcript sequences, EST sequences and molecular markers are available for the common bean. We have generated more than 270 million sequence reads, approximately 27 Gb from the common bean whole plant transcriptome, representing approximately 42× sequencing depth of the common bean genome. This sequencing depth provides a greater number of repeat sequence reads for individual gene and greater accuracy and reliability of sequencing data, especially during the assembly process. Based on the final assembly results, we obtained a non-redundant set of 62,828

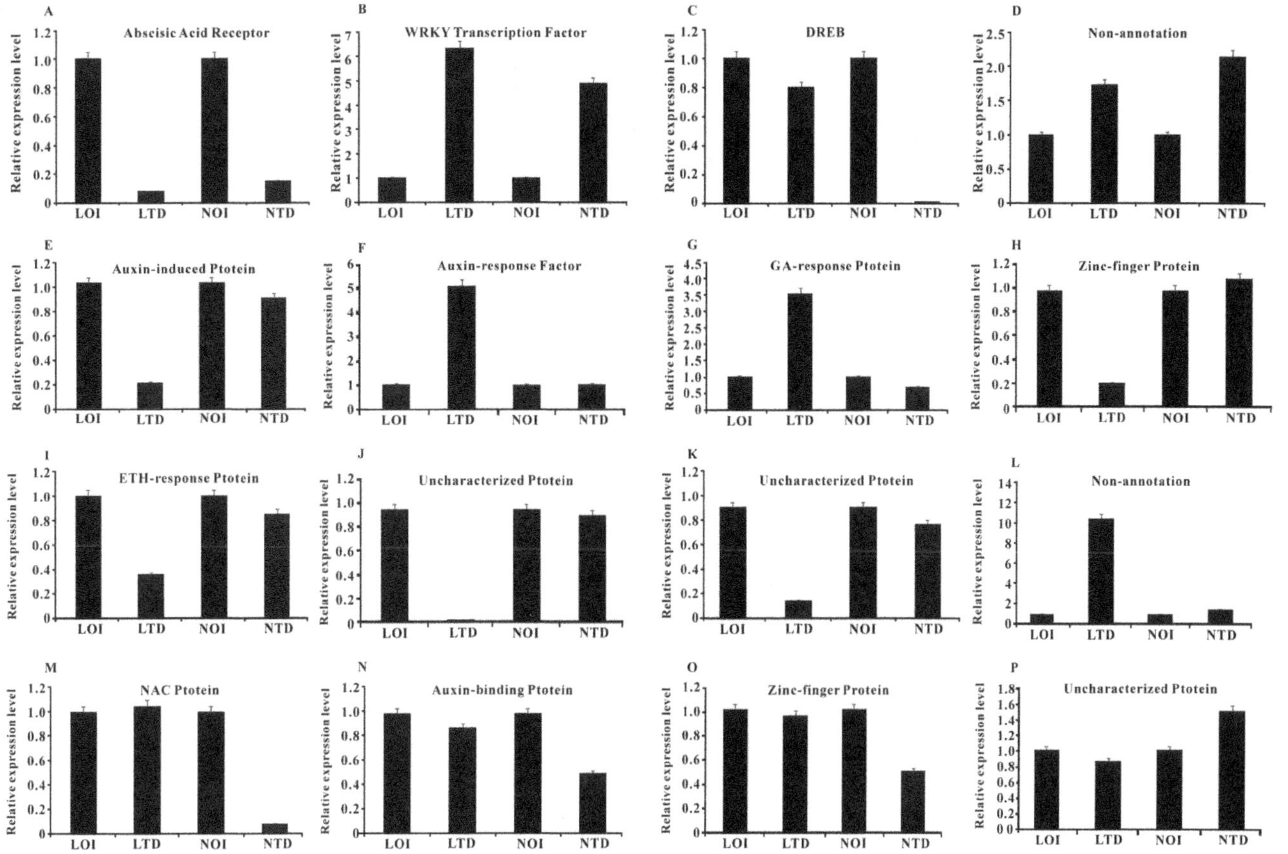

Figure 4. Relative expression levels of 16 DEGs. The relative gene expression levels as expressed by $2^{-\Delta\Delta CT}$ were determined separately for each treatment as the mean±S. A, *comp13117_c0_seq1*; B, *comp19490_c0_seq1*; C, *comp18282_c0_seq1*; D, *comp40404_c0_seq1*; E, *comp18253_c0_-seq1*; F, *comp18888_c3_seq1*; G, *comp18293_c0_seq1*; H, *comp19585_c0_seq1*; I, *comp16965_c0_seq1*; J, *comp40354_c0_seq1*; K, *comp19389_c0_seq1*; L, *comp38170_c0_seq1*; M, *comp11891_c1_seq1*; N, *comp18632_c0_seq1*; O, *comp19229_c0_seq1*; P, *comp18242_c0_seq1*.

unigenes representing approximately 49 Mb sequences. More than half of the assembled gene sizes are less than 500 bp (55.5%), which might not represent the full-length of all genes in the database. The current minimum gene size is satisfactory for future EST probe applications. Finally, we successfully constructed the first common bean whole plant unigene database deploying normalization and Illumina Hiseq 2500 sequencing technologies.

The *de novo* transcriptome assemblies are useful to develop molecular markers for further efforts in common bean breeding or molecular studies. For this purpose, a large number of EST-SSR and SNP loci were generated in this study. In contrast to the several previous publications describing the development EST-SSRs and SNPs primers [41–43], this study provides the first large set of potential EST-SSRs (10,482) and SNPs (4,099) using transcriptome sequences in the common bean. The number of dinucleotide SSRs was much higher than tri-nucleotide SSRs in this study, which is in contrast to a previous study reporting the larger number of tri-nucleotide (53.8%) SSRs than dinucleotide (30.9%) SSRs in common bean ESTs [41]. In support of our study, the larger number of dinucleotide (31.3%) SSRs than tri-nucleotide (22.3%) SSRs has also been reported in common bean ESTs [44]. Among the tri-nucleotide and dinucleotide motifs, it appears that AG/CT, AAG/CTT was more prevalent than other motifs, as reported by Blair et al [41]. In this study, the proportions of transitions and transversions were 44.0 and 38.8%, respectively. The proportion of SNP classes detected here was similar to that previously reported in the common bean (43.57% transitions, 38.41% transversions, and 18.02% indels), but differed significantly from what was previously reported in the common bean: 43.57% transitions, 38.41% transversions, and 18.02% indels or 55.70% transitions, 44.30% transversions, and 15.00% indels [45,46]. These genome-wide markers will enrich the existing common bean resources, and could especially be of great value for constructing high-density linkage maps, marker-trait association, diversity analysis, etc.

Identification of drought-responsive genes

In this study, RNA-Seq DEG technology was used to identify DEGs of the common bean under optimal irrigation and terminal drought conditions. This is the first report to identify common bean drought-responsive regulatory proteins using drought-tolerant and drought-sensitive genotypes, despite the reports that the identification of DEGs related to drought stress in *Arabidopsis*, rice, and other legume plants [29,47–50]. The identification of candidate genes would aid the research of common bean drought-tolerant molecular mechanisms and, more broadly, drought stress responses in legume plants. In this paper, we detected DEGs by different method. Nevertheless, it could help us to found more DEGs about drought. This strategy may be detected two type DEGs, which were identified to be differentially expressed following drought treatment in drought tolerant or sensitive genotype, and between different drought tolerant genotypes.

Drought tolerance is a complex trait that involves multiple complex molecular mechanisms to avoid or tolerate periods of water deficit. Based on the comparison of gene expression levels, we identified 9,298 drought-responsive candidate genes, approximately 14.8% of the total all-unigenes, which can be divided into two types: DEGs between treatment and control and DEGs between different genotypes. Among them, there were many homologs of drought-inducible genes that were identified previously in other plant species, such as *Arabidopsis* and soybean. According to the GO and KEGG annotations, many DEGs are involved in important drought-related metabolic processes, such as cell metabolic processes and cell wall and carbohydrate biosynthetic processes (comp34280_c0_seq1, comp3320_c0_seq1 and comp34280_c0_seq1, etc.). Cell wall genes encode proteins that function to enhancing mechanical resistance of drought-exposed cells [51]. Carbohydrate metabolism-related genes responded greatly to drought stress and were reported to enhance the capacity of the leaf lamina of *Arabidopsis thaliana* to endure a transient water deficit [52]. In this study, we also detected several proline-related genes: comp25511_c0_seq1, comp37160_c1_seq1, comp18574_c1_seq1, etc. Proline is an osmo-protectant critical for adaptation to drought [53]. Many proline-related genes have been cloned and verified to improve drought resistance of plants [54,55]. Some transcription factor (TF) were enriched in MYB, WRKY, C2H2, NAC, and DREB family of DEGs, such as comp32316_c1_seq1, comp40771_c0_seq1, comp7618_c0_seq1, comp27508_c1_seq1 and comp34578_c1_seq1. Previous studies have shown that TF families, such as MYB, DREB, NAC, bZIP and WRKY, are directly or indirectly involved in the regulation of plant response to drought stress [56–58]. Recently, it has been shown that GmDREB2, GmERF3, GmWRKY13, GmWRKY21, GmWRKY54, and GsZFP1 confer the tolerance to drought in transgenic plants [59–62] differentially. In addition, several plant hormone-related genes, such as gibberellin biosynthesis genes (comp34755_c0_seq1), auxin related protein (comp17800_c0_-seq1), ABA related genes (comp24144_c1_seq1), and ethylene related genes (comp40878_c0_seq1) were identified to be significantly and differentially between drought treatment and control. Overexpression of dehydration-responsive element-binding proteins (SlDREB) in tomato suppresses GA biosynthesis and promotes drought resistance [63]. Similarly, tomato plants overexpressing *Arabidopsis* gibberellin methyl transferase 1 (*AtGAMT1*) also exhibited increased tolerance to drought [64]. Other plant hormone-related genes were reported to play an important role under terminal drought condition [20,65–66]. In addition to these genes, many other important genes were up- or down-regulated under drought, such as signaling and cell communication-related genes, including Ca^{2+}-binding (comp18500_c0_seq1) and GTP binding protein (comp16864_c1_seq1), kinase and protein phosphatase (comp5659_c0_seq1), etc. These kinase and phosphatase transcripts have been suggested to impact the drought process in plants [67–69]. Some of the differentially expressed genes during the terminal drought process could not be annotated; therefore, their functions remain to be investigated in future study.

In conclusion, we have obtained a comprehensive transcriptome of the common bean using the Illumina sequencing technology. This study will contribute a significant, non-redundant set of 62,828 unigenes in the common bean. The detailed analyses have identified a large number of EST-SSRs and SNPs. Our work will be a great help to the development of genomic resources for the common bean and will accelerate functional genomics studies and breeding programs. In addition, the candidate drought-responsive genes identified in the common bean will be a new resource for molecular breeding in legumes or other crops. Moreover, our work has demonstrated the high reliability of the sequencing-based approach for identifying stress-responsive genes in legumes.

Supporting Information

Figure S1 Length distributions of unigenes.

Table S1 Classification and agronomic traits of two common bean cultivars.

Table S2 Drought resistance index of each cultivar.

Table S3 Summary of sequencing outputs.

Table S4 Statistics of SSRs identified in common bean transcripts.

Table S5 Frequency of di- and tri-nucleotide EST-SSR repeat motifs in the common bean.

Table S6 Statistics for SNPs between Long 22–0579 and Naihua.

Table S7 Annotation sources for the unigenes of the common bean.

Table S8 KEGG annotation of DEGs.

Table S9 GO annotation of DEGs.

Table S10 DEGs between Long 22–0579 and Naihua.

Table S11 GO annotation of DEGS.

Table S12 KEGG classification of DEGs.

Table S13 qRT-PCR verification of 16 DEGs in the drought-treated leaves compared to the control.

Author Contributions

Conceived and designed the experiments: JW SW. Performed the experiments: JW LW LL. Analyzed the data: JW. Contributed reagents/materials/analysis tools: LW LL. Contributed to the writing of the manuscript: JW SW.

References

1. Broughton WJ, Hernández G, Blair M, Beebe S, Gepts P, et al. (2003) Beans (*Phaseolus spp.*)-model food legumes. Plant Soil 252, 55–128.
2. Miklas PN, Kelly JD, Beebe SE, Blair MW (2006) Common bean breeding for resistance against biotic and abiotic stresses, from classical to MAS breeding, Euphytica, 147, 105–131.

3. Nielsen DC, Nelson N (1998) Black bean sensitivity to water stress at various growth stages. Crop Sci 38, 422–427.
4. Ramirez-Vallejo P, Kelly JD, (1998) Traits related to drought resistance in common bean. Euphytica 99, 127–138.
5. Henry A, Rosas JC, Beaver JS, Lynch JP (2010) Multiple stress response and belowground competition in multilines of common bean (*Phaseolus vulgaris* L.). Field Crops Res 117, 209–218.
6. Singh SP, Terán H, Gutierrez JA (2001) Registration of SEA5 and SEA13 drought tolerant dry bean germplasm. Crop Sci 41, 276–277.
7. Terán H, Singh SP (2002) Comparison of sources and lines selected for drought tolerance in common bean. Crop Sci., 42, 64–70.
8. Rosales MA, Ocampo E, Rodríguez-Valentín R, Olvera-Carrillo Y, Acosta-Gallegos J, et al. (2012) Physiological analysis of common bean (*Phaseolus vulgaris* L.) cultivars uncovers characteristics related to terminal drought resistance, Plant Physiol Biochem 56, 24–34.
9. Blair MW, Galeano CH, Tovar E, Muñoz Torres MC, Castrillón AV, et al. (2012) Development of a Mesoamerican intra-gene pool genetic map for quantitative trait loci detection in a drought tolerant × susceptible common bean (*Phaseolus vulgaris* L.) cross. Mol Breed 29, 71–88.
10. Cichy K, Caldas G, Snapp S, Blair M (2008) QTL Analysis of seed iron, zinc, and phosphorus levels in an andean bean population. Crop Sci 49, 1742–1750.
11. Prophete E, Demosthenes G, Godoy-Lutz G, Porch TG Beaver JS (2014) Registration of PR0633-10 and R0737-1 red mottled dry bean germplasm lines with resistance to BGYMV, BCMV, BCMNV, and common bacterial blight. J Plant Reg 8, 49–52.
12. Asfaw A, Blair M (2012) Quantitative trait loci for rooting pattern traits of common beans grown under drought stress versus non-stress conditions. Mol Breed 30, 681–695.
13. Asfaw A, Blair M, Struik P (2012) Multi environment quantitative trait loci analysis for photosynthate acquisition, accumulation, and remobilization traits in common bean under drought stress. G3 (Bethesda), 5, 579–595.
14. Schneider KA, Brothers ME, Kelly JF (1997) Marker-assisted selection to improve drought resistance in common bean. Crop Sci 37, 51–60.
15. Schneider KA, Rosales-Serna R, Ibarra-Perez F, Cazares-Enriquez B, Acosta-Gallego JA, et al. (1997) Improving common bean performance under drought stress. Crop Sci 37, 43–50.
16. Alcazar R, Planas J, Saxena T, Zarza X, Bortolotti C, et al. (2010) Putrescine accumulation confers drought tolerance in transgenic Arabidopsis plants overexpressing the homologous arginine decarboxylase 2 gene. Plant Physiol Biochem 48, 547–552.
17. Alet AI, Sanchez DH, Cuevas JC, Del Valle S, Altabella T, et al. (2011) Putrescine accumulation in *Arabidopsis thaliana* transgenic lines enhances tolerance to dehydration and freezing stress. Plant Signal Behav 6, 278–286.
18. Kasukabe Y, He L, Nada K, Misawa S, Ihara I, et al. (2004) Overexpression of spermidine synthase enhances tolerance to multiple environmental stresses and up-regulates the expression of various stress-regulated genes in transgenic *Arabidopsis thaliana*. Plant Cell Physiol 45, 712–722.
19. Li HW, Zang BS, Deng XW, Wang XP (2011) Overexpression of the trehalose-6-phosphate synthase gene *OsTPS1* enhances abiotic stress tolerance in rice. Planta 234, 1007–1018.
20. Liu S, Wang X, Wang H, Xin H, Yang X, et al. (2013) Genome-wide analysis of ZmDREB genes and their association with natural variation in drought tolerance at seedling stage of *Zea mays* L. PLoS Genetics 9, e1003790.
21. Yamada M, Morishita H, Urano K, Shiozaki N, Yamaguchi-Shinozaki K, et al. (2005) Effects of free proline accumulation in petunias under drought stress. J Exp Bot 56, 1975–1981.
22. Chen J, Zhang X, Jing R, Blair MW, Mao X, et al. (2010) Cloning and genetic diversity analysis of a new P5CS gene from common bean (*Phaseolus vulgaris* L.). Theor Appl Genet 120, 1393–1404.
23. Rodriguez-Uribe L, O'Connell MA (2006) A root-specific bZIP transcription factor is responsive to water deficit stress in tepary bean (*Phaseolus acutifolius*) and common bean (*P. vulgaris*). J Exp Bot 57, 1391–1398.
24. Mardis ER (2008) Next-generation DNA sequencing methods. Annu Rev Genomics Hum Genet 9, 387–402.
25. Varshney RK, Nayak SN, May GD, Jackson SA (2009) Next generation sequencing technologies and their implications for crop genetics and breeding. Trends Biotechnol 27, 522–530.
26. Huang J, Lu X, Yan H, Chen S, Zhang W, et al. (2012) Transcriptome characterization and sequencing-based identification of salt-responsive genes in *Millettia pinnata*, a semi-mangrove plant. DNA Res 19, 195–207.
27. Utsumi Y, Tanaka M, Morosawa T, Kurotani A, Yoshida T, et al. (2012) Transcriptome analysis using a high-density oligomicroarray under drought stress in various genotypes of cassava: an important tropical crop. DNA Res 19, 335–345.
28. Xia JH, Liu P, Liu F, Lin G, Sun F, et al. (2013) Analysis of stress-responsive transcriptome in the intestine of asian seabass (*lates calcarifer*) using RNA-seq. DNA Res 20, 449–460.
29. Yao LM, Wang B, Cheng LJ, Wu TL (2013) Identification of key drought stress-related genes in the hyacinth bean. Plos One 8, e58108.
30. Wu XS, Chang XP, Jing RL (2008) Screening indexes for drought resistance of wheat at grain-filling stage. J Triticeae Crops 28, 626–632.
31. Haas BJ, Papanicolaou A, Yassour M, Grabherr M, Blood PD, et al. (2013) De novo transcript sequence reconstruction from RNA-seq using the Trinity platform for reference generation and analysis. Nat Protoc 8, 1494–1512.
32. Li H, Handsaker B, Wysoker A, Fennell T, Ruan J, et al. (2009) The Sequence alignment/map (SAM) format and SAM tools. Bioinformatics 25, 2078–2079.
33. Carels N, Bernardi G (2000) Two classes of genes in plants. Genetics 154, 1819–1825.
34. Garg R, Patel RK, Tyagi AK, Jain M (2011) De novo assembly of chickpea transcriptome using short reads for gene discovery and marker identification. DNA Res 18, 53–63.
35. Cannon SB, May GD, Jackson SA (2009) Three sequenced legume genomes and many crop species, rich opportunities for translational genomics. Plant Physiol 151, 970–977.
36. Cheung F, Haas BJ, Goldberg SM, May GD, Xiao Y, et al. (2006) Sequencing *Medicago truncatula* expressed sequenced tags using 454 life sciences technology. BMC Genomics 7, 272.
37. Sato S, Nakamura Y, Kaneko T, Asamizu E, Kato T, et al. (2008) Genome structure of the legume, *Lotus japonicas*. DNA Res 15, 227–239.
38. Schmutz J, Cannon SB, Schlueter J, Ma J, Mitros T, et al. (2010) Genome sequence of the palaeopolyploid soybean. Nature 463, 178–183.
39. Varshney RK, Chen W, Li Y, Bharti AK, Saxena RK, et al. (2012) Draft genome sequence of pigeonpea (*Cajanus cajan*), an orphan legume crop of resource-poor farmers. Nat Biotechnol 31, 83–89.
40. Varshney RK, Song C, Saxena RK, Azam S, Yu S, et al. (2013) Draft genome sequence of chickpea (*Cicer arietinum*) provides a resource for trait improvement, Nat Biotechnol 31, 240–246.
41. Blair MW, Hurtado N, Chavarro CM, Muñoz-Torres MC, Giraldo MC, et al. (2011) Gene-based SSR markers for common bean (*Phaseolus vulgaris* L.) derived from root and leaf tissue ESTs, an integration of the BMc series. BMC Plant Biol 11, 50.
42. Galeano CH, Fernández AC, Gómez M, Blair MW (2009) Single strand conformation polymorphism based SNP and indel markers for genetic mapping and synteny analysis of common bean (*Phaseolus vulgaris* L.), BMC genomics 10, 629.
43. Goretti D, Bitocchi E, Bellucci E, Rodriguez M, Rau D, et al. (2014) Development of single nucleotide polymorphisms in *Phaseolus vulgaris* and related *Phaseolus* spp. Mol Breed 33, 531–544.
44. Garcia RA, Rangel PN, Brondani C, Martins WS, Melo LC, et al. (2011) The characterization of a new set of EST-derived simple sequence repeat (SSR) markers as a resource for the genetic analysis of *Phaseolus vulgaris*. BMC Genet 12, 41.
45. Gaitán-Solís E, Choi IY, Quigley C, Cregan P, Tohme J (2008) Single nucleotide polymorphisms in common bean, their discovery and genotyping using a multiplex detection system. Plant Genome 1, 125–134.
46. Souza TLPO, de Barros EG, Bellato CM, Hwang EY, Cregan PB, Pastor-Corrales MA (2012) Single nucleotide polymorphism discovery in common bean. Mol Breed 30, 419–428.
47. Degenkolbe T, Do PT, Zuther E., Repsilber D, Walther D, et al. (2009) Expression profiling of rice cultivars differing in their tolerance to long term drought stress. Plant Mol Biol 69, 133–153.
48. Gong PJ, Zhang JH, Li HX (2010) Transcription profiles of drought-responsive genes in Transcription profiles of drought-responsive genes in modulating transcription signal transduction, and biochemical pathways in tomato. J Exp Bot 61, 3563–3575.
49. Oono Y, Seki M, Nanjo T, Narusaka M, Fujita M, et al. (2003) Monitoring expression profiles of Arabidopsis gene expression during rehydration process after dehydration using ca. 7000 full-length cDNA microarray. Plant J 34, 868–887.
50. Zhou YJ, Gao F, Liu B, Feng JC, Li HJ (2012) De novo sequencing and analysis of root transcriptome using 454 pyrosequencing to discover putative genes associated with drought tolerance in *Ammopiptanthus mongolicus*. BMC Genomics 13, 266.
51. Lopez CG, Banowetz GM, Peterson CJ, Kronstad WE (2003) Dehydrin expression and drought tolerance in seven wheat cultivars. Crop Sci 43, 577–582.
52. Zeeman SC, Thorneycroft D, Schupp N, Chapple A, Weck M, et al. (2004) Plastidial a-glucan phosphorylase is not required for starch degradation in Arabidopsis leaves but has a role in the tolerance of abiotic stress. Plant Physiol 135, 849–858.
53. Verslues PE, Agarwal M, Katiyar-Agarwal S, Zhu JH, Zhu JK (2006) Methods and concepts in quantifying resistance to drought, salt and freezing, abiotic stresses that affect plant water status. Plant J 45, 523–539.
54. Choudhary NL, Sairam RK, Tyagi A (2005) Expression of delta1-pyrroline-5-carboxylate synthetase gene during drought in rice (*Oryza sativa* L.). Indian J Bioch Bio 42, 366–370.
55. De Ronde JA, Cress WA, Krüger GH, Strasser RJ, Van Staden J (2004) Photosynthetic response of transgenic soybean plants, containing an Arabidopsis P5CR gene, during heat and drought stress. J Plant Physiol 161, 1211–1224.
56. Huang G, Ma S, Bai L, Zhang L, Ma H, et al. (2010) Signal transduction during cold, salt, and drought stresses in plants. Mol Biol Rep 39, 969–987.
57. Ren X, Chen Z, Liu Y, Zhang H, Zhang M, Liu Q, et al. (2010) ABO3, a WRKY transcription factor, mediates plant responses to abscisic acid and drought tolerance in Arabidopsis. Plant J 63, 417–429.
58. Singh K, Foley RC, Onate-Sanchez L (2002) Transcription factors in plant defense and stress responses, Curr Opin Plant Biol 5, 430–436.

59. Chen M, Wang QY, Cheng XG, Xu ZS, Li LC, et al. (2007) GmDREB2, a soybean DRE-binding transcription factor, conferred drought and high salt tolerance in transgenic plants. Biochem Bioph Res Co 353, 299–305.
60. Luo X, Bai X, Zhu D, Li Y, Ji W, et al. (2011) GsZFP1, a new Cys2/His2-type zinc-finger protein, is a positive regulator of plant tolerance to cold and drought stress. Planta 235, 1–15.
61. Zhang G, Chen M, Li L, Xu Z, Chen X, et al. (2009) Overexpression of the soybean GmERF3 gene, an AP2/ERF type transcription factor for increased tolerances to salt, drought, and diseases in transgenic tobacco. J Exp Bot 60, 3781–3796.
62. Zhou QY, Tian AG, Zou HF, Xie ZM, Lei G, et al. (2008) Soybean WRKY-type transcription factor genes, GmWRKY13, GmWRKY21, and GmWRKY54, confer differential tolerance to abiotic stresses in transgenic Arabidopsis plants Plant Biotechnol J 6, 486–503.
63. Li J, Sima W, Ouyang B, Wang T, Ziaf K, et al. (2012) Tomato SlDREB gene restricts leaf expansion and internode elongation by downregulating key genes for gibberellin biosynthesis. J Exp Bot 63, 6407–6420.
64. Nir I, Moshelion M, Weiss D (2014) The Arabidopsis GIBBERELLIN METHYL TRANSFERASE 1 suppresses gibberellin activity, reduces whole-plant transpiration and promotes drought tolerance in transgenic tomato. Plant Cell Environ 37, 113–123.
65. Mittal A, Gampala SS, Ritchie GL, Payton P, Burke JJ, et al. (2014) Related to ABA-Insensitive3 (ABI3)/Viviparous1 and AtABI5 transcription factor coexpression in cotton enhances drought stress adaptation. Plant Biotechnol J 12, 578–589.
66. Xu ZS, Chen M, Li LC, Ma YZ (2008) Functions of the ERF transcription factor family in plants. Botany 86, 969–977.
67. Liu J, Ishitani M, Halfter U, Kim CS, Zhu J (2000) The Arabidopsis thaliana SOS2 gene encodes a protein kinase that is required for salt tolerance. Proc Natl Acad Sci USA 97, 73730–73734.
68. Marshall A, Aalen RB, Audenaert D, Beeckman T, Broadley MR, et al. (2012) Tackling drought stress, receptor-like kinases present new approaches. Plant Cell 24, 2262–2278.
69. Merlot S, Gosti F, Guerrier D, Vavasseur A, Giraudat J (2001) The ABI1 and ABI2 protein phosphatases 2C act in a negative feedback regulatory loop of the abscisic acid signalling pathway. Plant J 25, 295–303.

11

Tomato *ABSCISIC ACID STRESS RIPENING* (*ASR*) Gene Family Revisited

Ido Golan[1], Pia Guadalupe Dominguez[2], Zvia Konrad[1], Doron Shkolnik-Inbar[1], Fernando Carrari[2], Dudy Bar-Zvi[1]*

1 Department of Life Sciences and Doris and Bertie Black Center for Bioenergetics in Life Sciences, Ben-Gurion University of the Negev, Beer-Sheva, Israel, **2** Instituto de Biotecnología, Instituto Nacional de Tecnología Agropecuaria, Buenos Aires, Argentina

Abstract

Tomato *ABSCISIC ACID RIPENING 1* (*ASR1*) was the first cloned plant *ASR* gene. *ASR* orthologs were then cloned from a large number of monocot, dicot and gymnosperm plants, where they are mostly involved in response to abiotic (drought and salinity) stress and fruit ripening. The tomato genome encodes five *ASR* genes: *ASR1, 2, 3* and *5* encode low-molecular-weight proteins (ca. 110 amino acid residues each), whereas *ASR4* encodes a 297-residue polypeptide. Information on the expression of the tomato *ASR* gene family is scarce. We used quantitative RT-PCR to assay the expression of this gene family in plant development and in response to salt and osmotic stresses. *ASR1* and *ASR4* were the main expressed genes in all tested organs and conditions, whereas *ASR2* and *ASR3/5* expression was two to three orders of magnitude lower (with the exception of cotyledons). *ASR1* is expressed in all plant tissues tested whereas ASR4 expression is limited to photosynthetic organs and stamens. Essentially, *ASR1* accounted for most of *ASR* gene expression in roots, stems and fruits at all developmental stages, whereas *ASR4* was the major gene expressed in cotyledons and young and fully developed leaves. Both *ASR1* and *ASR4* were expressed in flower organs, with *ASR1* expression dominating in stamens and pistils, *ASR4* in sepals and petals. Steady-state levels of *ASR1* and *ASR4* were upregulated in plant vegetative organs following exposure to salt stress, osmotic stress or the plant abiotic stress hormone abscisic acid (ABA). Tomato plants overexpressing *ASR1* displayed enhanced survival rates under conditions of water stress, whereas *ASR1*-antisense plants displayed marginal hypersensitivity to water withholding.

Editor: Sara Amancio, ISA, Portugal

Funding: This study was supported in part by the I-CORE Program of the Planning and Budgeting Committee and the Israel Science Foundation (Center No. 757). DBZ is the incumbent of The Israel and Bernard Nichunsky Chair in Desert Agriculture, Ben-Gurion University. The funders had no role in study design, data collection and analysis, decision to publish, or preparation of the manuscript.

Competing Interests: The authors have declared that no competing interests exist.

* Email: barzvi@bgu.ac.il

Introduction

The first member of the tomato *ABSCISIC ACID STRESS RIPENING* (*ASR*) gene family, *ASR1*, was identified by screening a tomato fruit cDNA library with cDNA from stressed leaves [1], hence its name. Since then, a large number of *ASR* orthologs have been cloned from many plant species, including gymnosperms and angiosperms (reviewed by [2]). *ASR* gene families are found in the genomes of both monocots and dicots, but they are missing in the model plant *Arabidopsis* [2]. Interestingly, no *ASR* orthologs have been found in organisms outside the plant kingdom. *ASR* genes have been shown to be induced by abscisic acid (ABA) and abiotic stress, mainly salinity and drought [1,3–18]. They are also highly expressed in ripening fruit [1,4,14,19–23], and during potato-tuber development [24,25].

The tomato ASR gene family consists of five genes localized in one cluster on chromosome 4. Four members of the *ASR* gene family have been cloned from tomato [4,26–30]. These genes encode highly homologous proteins and possess a single intron of different size, but conserved location [4,27,28]. Upon completion of the tomato genome sequence [31], a fifth *ASR* gene was annotated, whose exon nucleotide sequence is highly similar to that of *ASR3* (88% identity in coding sequences, see also [32]). The loci of *ASR1–ASR5* genes in the tomato genome are Solyc04g071610, Solyc04g071580, Solyc04g071590, Solyc04g071620 and Solyc04g071600, respectively. Four of the genes (*ASR1–ASR3, ASR5*) encode low-molecular-weight proteins, whereas the polypeptide encoded by *ASR4* is approximately double the size of the other proteins [30]. Wild tomato species also encode this five member ASR gene family. suggesting that this family was not lost during tomato domestication and breeding [30,33]. Furtheremore, *ASR1, 2* and *4* genes from wild tomato species were also induced by drought and cold. The majority of plant ASR genes encode low molecular weight proteins. In addition, genomes of a number of plant species contain in addition a gene encoding higher molecular ASR polypeptides [25,34,35].

ASR proteins have been proposed to belong to the hydrophylin group of proteins [36]. Tomato ASR1 was shown to be a natively unordered protein [37] that possesses chaperone-like activity [38], and was localized to both the cytosol and nucleus [39]. Dual subcellular localization was also shown for the lily pollen ASR protein [40]. Rice ASR1 was also shown to possess chaperone-like activity [41,42]. ASR1 has Zn^{2+}-dependent DNA-binding activity [39,43]. Upon binding of Zn^{2+}, ASR1 becomes structured and dimerizes [26,37], and is translocated to the nucleus [37,44,45]. Zn^{2+} and Fe^{3+} ions affected the structure of a soybean ASR [46]. Nuclear ASR proteins modulate gene expression via binding to specific promoter sequences [14,39,47,48].

ASR genes have been shown to play a central role in drought and salinity stress. Overexpression of *ASR* genes resulted in increased tolerance of the transgenic plants to water/osmotic [42,48–51], salinity [48,50,52,53] and cold [45,54] stresses. However, until now these responses have been seen only in heterologous systems. Transgenic *Arabidopsis* plants expressing ASR proteins from other plant species [42,54] demonstrated increased tolerance to abiotic stresses. *Arabidopsis* plants do not encode ASR proteins. Ectopic expression of tomato *ASR1* in *Arabidopsis* was shown to affect the plant's response to ABA, glucose and tolerance to abiotic stress via competition for DNA binding with the transcription factor ABI4 [55]. Thus, expression studies in heterologous organisms that do not naturally have the studied gene(s) should be analyzed with caution, especially for regulatory proteins, as results may not directly reflect the biological role of the analyzed gene. In addition, *ASR* gene involvement in carbohydrate signaling, sugar trafficking and metabolism has been shown [14,24,56,57], as has its influence on the biogenesis of branched-chain amino acids [58].

In tomato, the best-studied member of the *ASR* family is *ASR1*, followed by *ASR2*. Information on *ASR3*, *ASR4* and *ASR5* is scarce. Although a few studies have compared the expression of some members of the tomato *ASR* gene family [59,60], results are from northern blot, semi-quantitative PCR, or histological staining studies analyzing two or three of the family's genes, under rather restricted conditions. In this work, we revisited the tomato *ASR* gene family using the highly accurate quantitative RT-PCR technology to determine the expression patterns of its members in vegetative and reproductive organs. Because of their high sequence homology, we could not design gene-specific primers for *ASR3* and *ASR5*, we determined the summed expression of these genes (*ASR3/5*). We found that *ASR1* and *ASR4* are highly expressed in vegetative tissues of nonstressed plants, whereas the steady-state levels of *ASR2* and *ASR3/5* transcripts are two to three orders of magnitude lower. *ASR1* was the major member expressed in roots, stems, stamens, pistils and fruits. *ASR4* accounted for more than two-thirds of total *ASR* transcripts in leaves, shoot vegetative meristem, sepals and petals. Both of these genes were induced by ABA, osmotic stress and salt stress. Tomato plants overexpressing *ASR1* (*ASR1*-OE) survived better under water stress than wild-type (WT) plants, where *ASR1*-antisense (*ASR1*-AS) plants had slightly lower survival rates than the WT.

Materials and Methods

Plant material and growth conditions

Generation and selection of transgenic lines. The 348-bp coding region of the tomato *ASR1* gene (GenBank U86130.1) was cloned in sense or antisense orientation into the multiple cloning site of the vector [61] between the *Cauliflower Mosaic Virus* (*CaMV*) 35S promoter and the *octopine synthase* (*ocs*) terminator. Tomato plants (*Solanum esculentum* cv. Moneymaker) were transformed by *Agrobacterium* as previously described [62]. Emerging shoots were excised and selected on Murashige and Skoog (MS) medium containing kanamycin (100 mg/l), and then transferred to the greenhouse for selection by qPCR in T1 plants. Transgenic plants were numbered XX-YY, where XX represents an independent transformation event, and YY represents a sub-line of the founder of T2 generation seeds. Plants were self-pollinated and seeds were collected. T3 generation plants were used in this study. Western blot analyses showed that the *ASR1*-OE lines have higher levels of *ASR1* than WT plants (Figure S1 in File S1).

Plant growth. Plants were grown in pots in the greenhouse or hydroponically in the growth room in aerated half-strength Hoagland mineral solution at 28°C and 70% relative humidity, under a diurnal cycle of 18 h light, or in the greenhouse at an average temperature of 28°C, and >50% relative humidity. Seeds were germinated in water-soaked vermiculite. Ten-day-old seedlings were transferred to aerated containers with half-strength Hoagland mineral solution [63], or to pots containing equal volumes of planting mix and vermiculite. Hydroponic growth medium was replaced 1 week after transfer and every 3–4 days thereafter.

Water-stress tolerance and survival assays. Seedlings were grown in pots under optimal conditions for 3 weeks. Water was then withheld for 22 days, followed by rewatering. Plant survival was scored 17 days later. At least 20 plants from each line were used. Plants were grown in random order and pot location was changed every few days.

NaCl, polyethylene glycol (PEG) and ABA treatment

WT seedlings were germinated on vermiculite and transferred to aerated 0.5X Hoagland's solution as described above. After 1 week acclimation, plants were transferred to fresh mineral solution containing, where indicated, 40 μM ABA, 0.2 M NaCl, or 8% (w/v) PEG 8000. Plants were harvested 24 h after treatments.

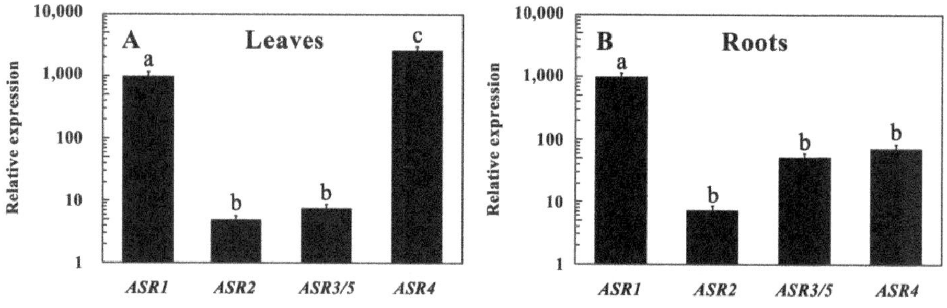

Figure 1. Relative expression of the tomato *ASR* genes. Steady-state levels of the indicated genes were determined in leaves and roots of 10-day-old hydroponically grown tomato seedlings. Expression of *ASR1* in each tissue was defined as 1000. Data shown are average ± SE. Bars with different letters represent statistically different values by Tukey's HSD post-hoc test (P≤0.05).

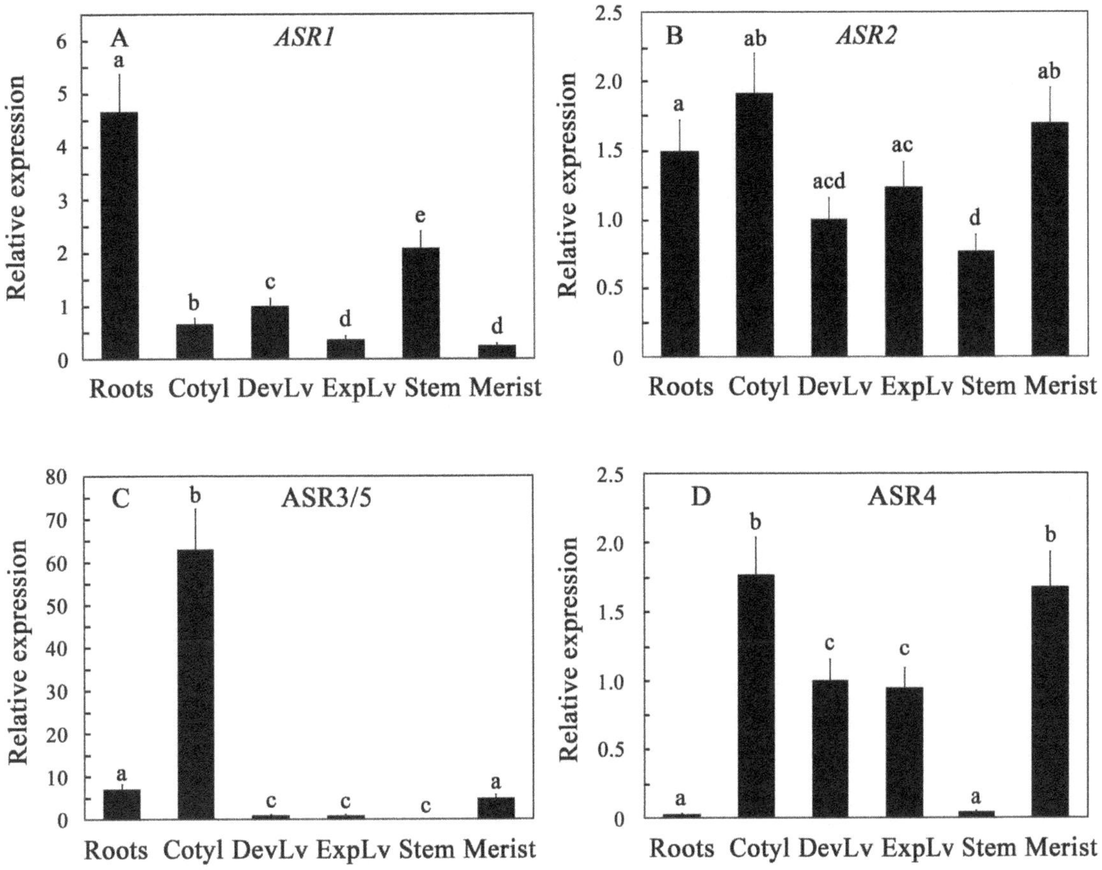

Figure 2. Expression of the tomato *ASR* genes in vegetative tissues. Steady-state levels of the indicated genes were determined in hydroponically grown tomato plants at the 8-true-leaf stage. The expression levels were normalized for each of the genes to their expression in young developing leaves, defined as 1. Cotyl, cotyledons; DevLv, developing leaves; ExpLv, fully expanded leaves; Merist, shoot meristem. Data shown are average ± SE. Bars with different letters represent statistically different values by Tukey's HSD post-hoc test (P≤0.05).

mRNA level assay

Total RNA was extracted from the indicated plant tissues using EZ-RNA (Biological Industries, Israel) according to the manufacturer's protocol. This protocol uses improved RNA-extraction methods as described by Chomczynski and Sacci [64]. RNA quality integrity was checked spectrally (at 230 nm, 260 nm and 280 nm) and by running samples on denaturing agarose gels electrophoresis. Relative steady-state transcript levels were assayed by RT-qPCR as described previously [50,55,65,66]. cDNA was synthesized from DNase-treated RNA using high-capacity cDNA reverse transcription kit (Applied Biosystems). Primers were designed by Primer-Express software Vers. 2.0 (Applied Biosystems). When possible, one of the primers in each set was anchored at an exon–exon border to reduce possible amplification from contaminating genomic DNA. All amplicon lengths were between 75 and 90 bp. Primer sequences are presented in Table S1 in File S1. Transcript levels were assayed using the 7300 Real-Time PCR System (Applied Biosystems), with 18S rRNA as the internal standard. PCR efficiency was close to 100%. RNA relative quantification analyses were performed using 7300 System SDS software (Applied Biosystems). The list of primers used is shown in Table S1 in File S1. The data represent the mean ± SE of $n = 3$ independent experiments. Each data point was determined in triplicates in each of the three biological replicates and presented as mean ± SE. Data presented in a single graph were carried in a single run. Differences between groups were analyzed by Tukey's HSD post-hoc test (P≥0.05).

Results and Discussion

Relative expression levels of the tomato *ASR* genes

Relative steady-state levels of the members of the *ASR* gene family were determined in leaves and roots of hydroponically grown seedlings. In general, *ASR1* and *ASR4* were the most highly expressed genes in this family (Figure 1). In young leaves, *ASR4* levels were 2.6 times higher than those of *ASR1*, whereas transcript levels of *ASR2* and *ASR3/5* were more than two orders of magnitude lower. In tomato roots, *ASR1* was the most abundantly expressed gene, whereas transcript levels of *ASR3/5* and *ASR4* were approximately one order of magnitude lower than that of *ASR1*, and that of *ASR2* two orders of magnitude lower than the steady-state levels of *ASR1*. These results are in agreement with Frankel et al. [30] who reported that *ASR2* and *ASR3/5* transcripts could not be detected in tomato leaves by northern blot analysis. On the other hand, previous studies from the same laboratory reported that *ASR2* transcripts are highly abundant in roots of stressed tomato plants [60] and that the *ASR2* promoter can drive transcription in both tomato and other Solanaceae plant cells [60,67]. RNA Seq also suggest that *ASR1* and *ASR4* transcripts in tomato leaves and fruits are relatively

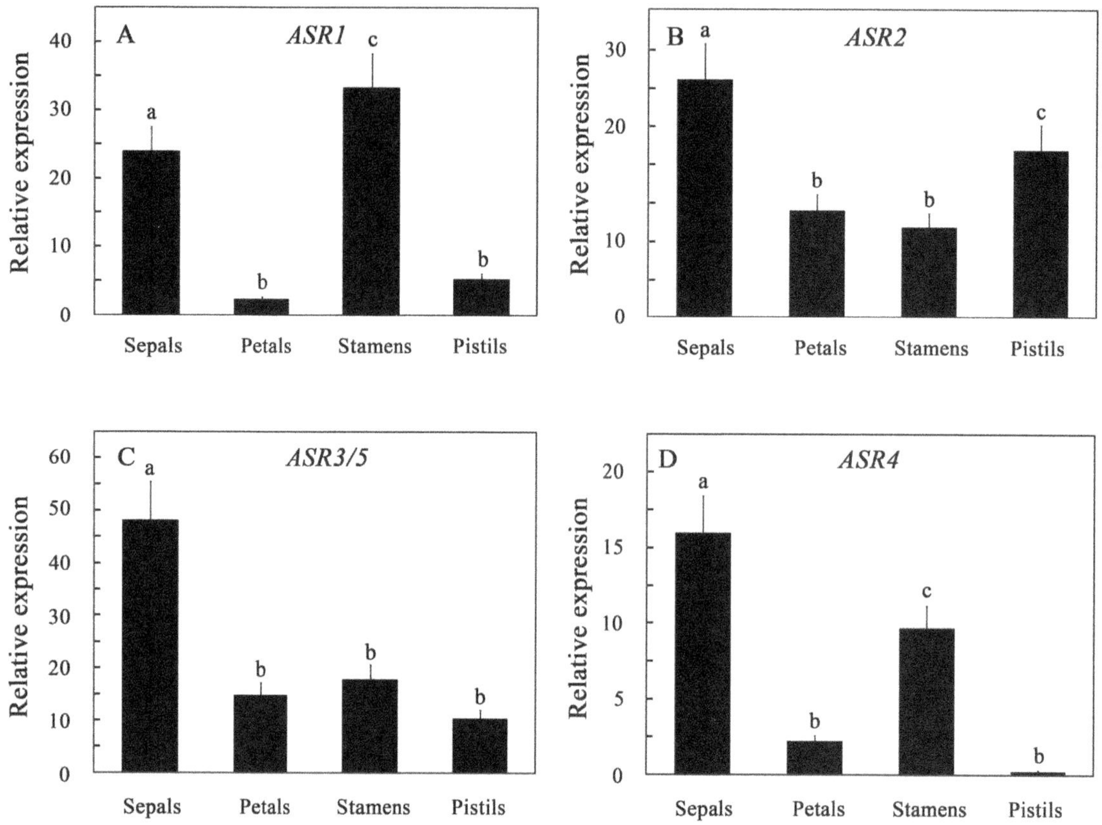

Figure 3. Expression of the tomato *ASR* genes in flower organs. Flowers and developing leaves were harvested from greenhouse-grown tomato plants. Flowers were dissected, and steady-state levels of the indicated genes were determined. The expression levels were normalized for each of the genes to their expression in young developing leaves, defined as 1. Data shown are average ± SE. Bars with different letters represent statistically different values by Tukey's HSD post-hoc test (P≤0.05).

abundant, whereas transcripts from *ASR2*, *ASR3* and *ASR5* are hardly found [31].

(data presented at the tomato Sol Genomic Network database (http://solgenomics.net)).

Expression of tomato *ASR* genes in vegetative tissues

Expression of tomato *ASR* genes was determined in vegetative tissues of hydroponically grown tomato (Figure 2). Transcript levels of each of the genes were normalized to the expression of the same gene in developing leaves. This reference tissue was selected since it is present in all plant ages used in the study. Highest levels of *ASR*1 transcript were found in roots and stems (ca. 4.5 and 2 times that in the leaves, respectively). *ASR1* transcript levels decreased with leaf development (Figure 2A), in agreement with Amitai-Zeigerson et al. [68]. *ASR1* expression in stems and roots suggests its role in the plant's vascular system [59]. *ASR2* transcript levels were only slightly different in vegetative organs (Figure 2B), and its steady-state levels in all vegetative organs were marginal (see Figure 1). *ASR3/5* levels were highest in the cotyledons (Figure 2C), reaching transcript levels in the same order of magnitude of that of *ASR1* in cotyledons, suggesting that *ASR3* and/or *ASR5* my play a role in advanced stages of seed development. ASR proteins were detected immunologically in tomato seeds using anti-ASR1 antiserum [37]. Since there is high amino acid conservation between the different members of the tomato ASR proteins, is it likely that this antibody crossreacts with other ASR proteins such as ASR3 and ASR5. *ASR4* was expressed mainly in leaves, cotyledons and meristem, with relatively low expression rates in roots and stems (Figure 2D). After normalization of the relative expression of each gene in young leaf tissues (Figure 2A), *ASR1* seemed to be the main expressed gene in the roots and stems. On the other hand, *ASR4* was the highest expressed *ASR* gene in cotyledons, and young and fully expanded leaves. *ASR1* accounted for up to one quarter of of *ASR* gene family expression in these tissues-. Expression of *ASR2* and *ASR3/5* in leaves was negligible. Interestingly, and Our results indicate that *ASR1* is the primarily expressed gene in tomato roots, stems and fruits, whereas both *ASR1 and ASR4* are both expressed in cotyledons, leaves and meristems, where *ASR4* levels exceed that of *ASR1*. Transcript levels of *ASR2* and *ASR3/5* were marginal in vegetative tissues, with the exception of *ASR3/5* in cotelydons. Our analyzes determine averaged transcript levels in the entire organ, rather than cell specific expression. Thus, it will be interesting to find out if in these organs, *ASR1* and *ASR4* coexpress in the same cells, or in different cell types.

Expression of the tomato ASR genes in reproductive tissues

The *ASR* gene family showed differential expression in flower organs. Although *ASR2* and *ASR3/5* expression varied in the different flower organs (Figure 3B, C), their expression levels can be estimated to be two order of magnitude order lower than that of *ASR1* and *ASR4* of *ASR1* and *ASR4* were highly expressed in the sepals and stamens, and to a lower extent in petals and pistils (Figure 3A, D), where *ASR4* estimated transcript levels were higher than *ASR1* in the sepals and petals. One the other hand,

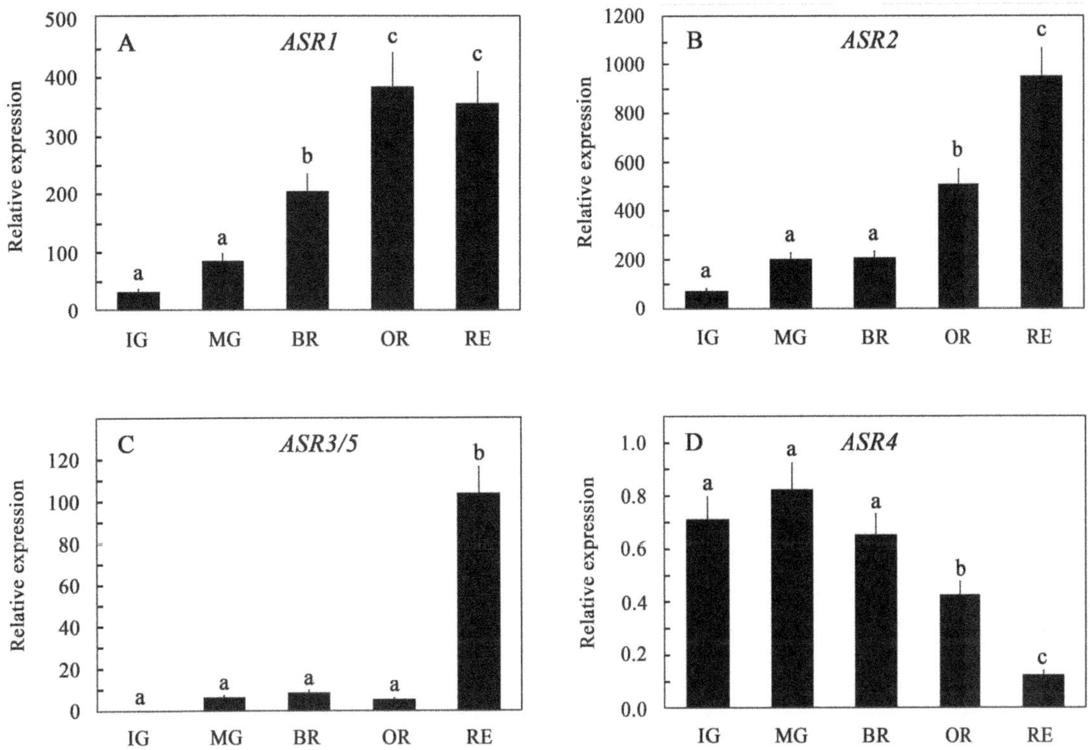

Figure 4. Expression of the *ASR* genes in fruit development. Fruits and developing leaves were harvested from greenhouse-grown tomato plants, and steady-state levels of the indicated genes were determined. The expression levels were normalized for each of the genes to their expression in young developing leaves, defined as 1. Fruit stages were defined according to [75]: IG, immature green; MG, mature green; BR, breaker; OR, orange; RE, red. Data shown are average ± SE. Bars with different letters represent statistically different values by Tukey's HSD post-hoc test ($P \leq 0.05$).

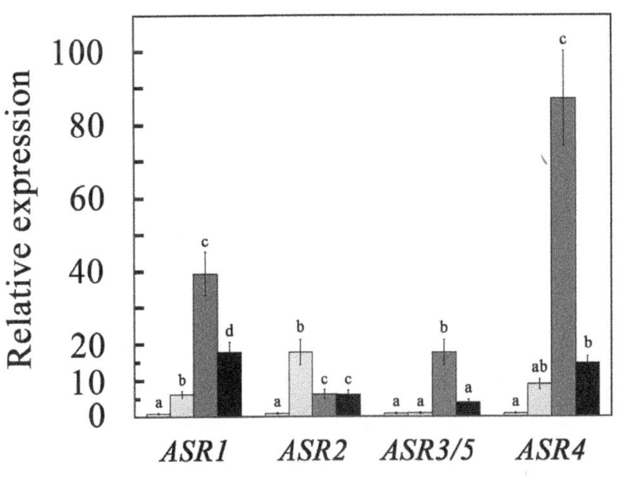

Figure 5. Effects of ABA, NaCl and PEG on the steady-state transcript levels of *ASR* genes. One-week-old hydroponically grown seedlings were transferred to fresh Hoagland solution containing: no addition (white bars), 40 μM ABA (light gray bars), 0.2 M NaCl (dark gray bars), or 8% PEG 8000 (black bars). Leaves were harvested 24 h later and expression levels were determined as described in Materials and Methods. Data were normalized for each of the genes to their expression in young developing leaves, defined as 1. Data shown are average ± SE. Expression of each gene in response to plant treatment was analyzed separately using Tukey's HSD post-hoc test. Bars with different letters represent statistically different values by ($P \leq 0.05$).

ASR1 is the most highly expressed *ASR* gene in the stamen and pistils.

Although *ASR1* expression increased during fruit development (Figure 4A), it was already higher than that in all other tissues in young immature green fruits. *ASR1* was essentially the main fruit-expressed *ASR* gene at all fruit developmental (Figure 4). Steady-state levels of *ASR2 and ASR3/5* transcripts in fruit tissues were also the highest measured in any plant tissue. Nevertheless, their levels in fruits are approximately two order of magnitude lower of that of *ASR1*. An increase in *ASR1* transcript levels in tomato fruit development and ripening is in agreement with Gilad et al. [4] and Iusem et al. [1], but not with Maskin et al. [60]. Increase in *ASR1* during tomato fruit ripeing also correlates with increase in its protein levels [20]. Increased steady state transcript levels during fruit ripening of *SlASR1* orthologous genes were reported in a number of plant species [21–23,69,70].

Tomato ASR genes are differentially responsive to ABA and abiotic stress

Steady-state levels of *ASR1* and *ASR4* transcript increased following plant exposure to salt stress, osmotic stress (PEG) or to the hormone ABA (Figure 5). The relative induction levels by these treatments were rather similar for these two highly expressed *ASR* genes. On the other hand, the less expressed genes *ASR2* and *ASR3/5* showed different responses: *ASR2* responded most strongly to ABA treatment and to a lesser extent to salt or osmotic stress, whereas *ASR3/5* was not affected by ABA and was relatively highly expressed after NaCl treatment, suggesting that they are induced by an ABA-independent pathway. A large

Figure 6. ASR1-overexpressing tomato plants have enhanced tolerance to water stress. Plants were grown in pots in the greenhouse using optimal irrigation for 17 days. Water was withheld for 22 days, following by rewatering for 17 days. Panels A and B show representative plants (left to right): wild type, *ASR1*-OE-31, *ASR1*-OE-12, *ASR1*-OE-16 after 22 days of dehydration (A) and 17 days of rewatering (B). Panel C, quantitative data of survival of three lines of *ASR1*-overexpressing (*ASR1*-OE) plants and four lines of *ASR1*-antisense (*ASR1*-AS) plants, measured at the end of the rewatering stage. Data shown are average ± SE. Bars with different letters represent statistically different values by Tukey's HSD post-hoc test ($P \leq 0.01$).

number of *ASR* genes have been identified in different plant species based on their response to abiotic stresses such as water, salinity and osmotic stress [9,10,14,71,72]. Expression results for the tomato *ASR1* gene are in agreement with previous studies [4,59,60,68]. In contrast, activity of the tomato *ASR2* promoter was enhanced by ABA when expressed in papaya and tobacco, but not in tomato or potato [67].

Genetic manipulation of *ASR1* in tomato affects water-stress tolerance

Tomato plants overexpressing *CaMV 35S:ASR1* (*ASR1*-OE) or *CaMV 35S:Reverse ASR1* (*ASR1*-AS) were tested for tolerance to water and salt stresses. The modified plants did not perform significantly differently from WT plants when treated with NaCl (not shown). However, *ASR1*-OE plants survived water stress better than WT plants (Figure 6A and 6B). The response of *ASR1*-AS plants was highly variable: many lines showed hypersensitivity to water stress, whereas others showed no difference, or even better tolerance (Figure S2 in File S1). Averaging the recovery rates of all lines showed that *ASR1*-OE plants were significantly more tolerant to water stress than WT controls and *ASR1*-AS plants, whereas the latter were slightly more sensitive to lack of water than WT plants (Figure 6C). Transgenic plants overexpressing *ASR* genes have been reported to be more tolerant to abiotic stresses such as water/osmotic stress [12,18 51], salinity stress [48,50,52,53] and cold stress [44,54]. However, most of those studies expressed the studied gene in a heterologous system [42,48–50,52–54]. In some of those studies, the biological species of the gene of origin and the transgenic plant belonged to the same botanical genus or family [24,53,56], but only a few studies have been performed within a single species [45,51]. The biological relevance of studying the role of a regulatory protein on a genetic background that naturally lacks it has been questioned [55]: the phenotype of *Arabidopsis* plants expressing tomato *ASR1* was shown to result from competition for DNA binding between the ectopically expressed tomato gene and the Arabidopsis transcription factor ABI4, essentially resulting in an *abi4*-mutant-like phenotype [55]. Thus, expressing the studied gene on the same genetic background ([45,51] and this study), or in closely related species [24,53,56], is more likely to shed light on the actual role of the studied gene and its protein product. The increased survival of transgenic tomato plants overexpressing tomato *ASR1* following water stress (Figure 6) is in agreement with reports on the expression of plant *ASR* genes on similar or different genetic backgrounds in response to stress [42,48–51]. Results obtained using the *ASR1*-AS lines were more variable: lines ASR1-AS5-4 and ASR1-AS18-4 were significantly hypersensitive to water stress as compared to WT plants (Figure S2 in File S1), whereas other lines were not significantly different. On the other hand, no significant differences were found in the sensitivity of *ASR1*-OE and *ASR1*-AS plants to NaCl stress (not shown), most probably due to the relative tolerance of the tomato WT line. ASR proteins are localized in the cytosol and nucleus. In the latter organelle, they are associated with DNA [14,39,40,48,73]. Tomato and grape ASR proteins have been shown to bind specific DNA sequences [14,53], suggesting that they regulate the expression of genes involved in abiotic stress responses or sugar metabolism [4,14,24,54,57,74]. In addition, ASR proteins have been shown to possess protein-chaperone-like activity [38,41], increasing protein stability. Thus, the increased water-stress tolerance of *ASR1*-OE plants most likely results from an increase in the transcription of ASR1-regulated genes and from ASR1 chaperon activity.

Conclusions

Two of the five genes in the *ASR* gene family (*ASR1* and *ASR4*) are significantly expressed in tomato plants, whereas the expression levels of the other three member of the gene family are less pronounced. *ASR1* and *ASR4* encode 115- and 297-amino acid polypeptides, respectively, thus most likely encoding proteins whose activity may not be fully redundant. ASR1 is expressed in all plant tissues tested: it is most highly expressed in the stem, roots and reproductive organs–stamen, pistils and fruit at all developmental stages. *ASR4* is mainly expressed in photosynthetic organs, in in sepals and petals. The steady-state levels of *ASR1* and *ASR4* increased following salt stress, osmotic stress and treatment with ABA. *ASR2* expression is negligible in all tested tissues. *ASR3* and *ASR5*, being highly similar genes, were assayed together, with significant expression detected only in the cotyledons. These results suggest that *ASR2/3/5* may be very low expressing genes, or that their expression is limited to specific low abundant cells, thus resulting in low transcript activity when assayed in the whole tissue. Tomato plants overexpressing *ASR1* show increased tolerance to water stress, whereas *ASR1*-AS plants show a certain degree of hypersensitivity to water withholding. Our data suggest that *ASR1* and *ASR4* may be expressed in different cell types. The differential expression patterns of the tomato ASR gene family under non-stress and stress conditions may be used for genetic manipulation of tomato (as well as other crop plants) to affect vegetative and fruit parameters under non-stress and stress conditions.

Author Contributions

Conceived and designed the experiments: IG DSI DBZ. Performed the experiments: IG ZK DSI. Analyzed the data: IG ZK DSI DBZ. Contributed to the writing of the manuscript: IG PGD DSI FC DBZ. Constructed the transgenic tomato plants: PGD FC.

References

1. Iusem ND, Bartholomew DM, Hitz WD, Scolnik PA (1993) Tomato (*Lycopersicon esculentum*) transcript induced by water deficit and ripening. Plant Physiol 102: 1353–1354.
2. Carrari F, Fernie AR, Iusem ND (2004) Heard it through the grapevine? ABA and sugar cross-talk: The ASR story. Trends Plant Sci 9: 57–59.
3. Chang S, Puryear JD, Dias MADL, Funkhouser EA, Newton RJ, et al. (1996) Gene expression under water deficit in loblolly pine (*Pinus taeda*): Isolation and characterization of cDNA clones. Physiol Plantar 97: 139–148.
4. Gilad A, Amitai-Zeigerson H, Bar-Zvi D, Scolnik PA (1997) *ASR1*, a tomato water-stress regulated gene: Genomic organization, developmental regulation and DNA-binding activity. Acta Hortic 447: 447–453.
5. Padmanabhan V, Dias DMAL, Newton RJ (1997) Expression analysis of a gene family in loblolly pine (*Pinus taeda* L.) induced by water deficit stress. Plant Mol Biol 35: 801–807.
6. Schneider A, Salamini F, Gebhardt C (1997) Expression patterns and promoter activity of the cold-regulated gene *ci21A* of potato. Plant Physiol 113: 335–345.
7. Wang CS, Liau YE, Huang JC, Wu TD, Su CC, et al. (1998) Characterization of a desiccation-related protein in lily pollen during development and stress. Plant Cell Physiol 39: 1307–1314.
8. De Vienne D, Leonardi A, Damerval C, Zivy M (1999) Genetics of proteome variation for QTL characterization: Application to drought-stress responses in maize. J Exp Bot 50: 303–309.
9. Vaidyanathan R, Kuruvilla S, Thomas G (1999) Characterization and expression pattern of an abscisic acid and osmotic stress responsive gene from rice. Plant Sci 140: 21–30.
10. Dóczi R, Csanaki C, Bánfalvi Z (2002) Expression and promoter activity of the desiccation-specific *Solanum tuberosum* gene, StDS2. Plant Cell Environ 25: 1197–1203.
11. Salekdeh GH, Siopongco J, Wade LJ, Ghareyazie B, Bennett J (2002) A proteomic approach to analyzing drought- and salt-responsiveness in rice. Field Crops Res 76: 199–219.
12. Sugiharto B, Ermawati N, Mori H, Aoki K, Yonekura-Sakakibara K, et al. (2002) Identification and characterization of a gene encoding drought-inducible protein localizing in the bundle sheath cell of sugarcane. Plant Cell Physiol 43: 350–354.
13. Wang JT, Gould JH, Padmanabhan V, Newton RJ (2002) Analysis and localization of the water-deficit stress-induced gene (*lp3*). J Plant Growth Regul 21: 469–478.
14. Çakir B, Agasse A, Gaillard C, Saumonneau A, Delrot S, et al. (2003) A grape ASR protein involved in sugar and abscisic acid signaling. Plant Cell 15: 2165–2180.
15. Frankel N, Hasson E, Iusem ND, Rossi MS (2003) Adaptive evolution of the water stress-induced gene *Asr2* in *Lycopersicon* species dwelling in arid habitats. Mol Biol Evol 20: 1955–1962.
16. Liu HY, Dai JR, Feng DR, Liu B, Wang HB, et al. (2010) Characterization of a novel plantain *Asr* gene, *MpAsr*, that is regulated in response to infection of *Fusarium oxysporum* f. sp *cubense* and abiotic stresses. J Integ Plant Biol 52: 315–323.
17. Philippe R, Courtois B, McNally KL, Mournet P, El-Malki R, et al. (2010) Structure, allelic diversity and selection of *Asr* genes, candidate for drought tolerance, in *Oryza sativa* L. and wild relatives. Theor Applied Genet 121: 769–787.
18. Joo J, Lee YH, Kim YK, Nahm BH, Song SI (2013) Abiotic stress responsive rice *ASR1* and *ASR3* exhibit different tissue-dependent sugar and hormone-sensitivities. Mol Cells 35: 421–435.
19. Srivastava A, Handa AK (2005) Hormonal regulation of tomato fruit development: A molecular perspective. J Plant Growth Regul 24: 67–82.
20. Rocco M, D'Ambrosio C, Arena S, Faurobert M, Scaloni A, et al. (2006) Proteomic analysis of tomato fruits from two ecotypes during ripening. Proteomics 6: 3781–3791.
21. Xu BY, Su W, Liu JH, Wang JB, Jin ZQ (2007) Differentially expressed cDNAs at the early stage of banana ripening identified by suppression subtractive hybridization and cDNA microarray. Planta 226: 529–539.
22. Chen JY, Liu DJ, Jiang YM, Zhao ML, Shan W, et al. (2011) Molecular characterization of a strawberry *FaASR* gene in relation to fruit ripening. PLoS One 6: e24649.
23. Luo C, Dong L, He XH, Yu HX, Ou SJ, et al. (2014) Molecular cloning and characterisation of a cDNA encoding an abscisic acid-, stress-, and ripening-induced (ASR) protein in mango (*Mangifera indica* L.). J Hort Sci Biotechnol 89: 352–358.
24. Frankel N, Nunes-Nesi A, Balbo I, Mazuch J, Centeno D, et al. (2007) *ci21A/Asr1* expression influences glucose accumulation in potato tubers. Plant Mol Biol 63: 719–730.
25. Silhavy D, Hutvagner G, Barta E, Banfalvi Z (1995) Isolation and characterization of a water-stress-inducible cDNA clone from *Solanum chacoense*. Plant Mol Biol 27: 587–595.
26. Iusem ND, Maskin L, Frankel N, Gudesblat G, Demergasso MJ, et al. (2007) Dimerization and DNA-binding of *ASR1*, a small hydrophilic protein abundant in plant tissues suffering from water loss. Biochem Biophys Res Commun 352: 831–835.
27. Rossi M, Lijavetzky D, Bernacchi D, Hopp HE, Iusem N (1996) *Asr* genes belong to a gene family comprising at least three closely linked loci on chromosome 4 in tomato. Mol Gen Genet 252: 489–492.
28. Amitai-Zeigerson H, Scolnik PA, Bar-Zvi D (1994) Genomic nucleotide sequence of tomato *Asr2*, a second member of the stress/ripening-induced *Asr1* Gene Family. Plant Physiol 106: 1699–1700.
29. Rossi M, Iusem ND (1995) Sequence of *Asr2*, a member of a gene family from *Lycopersicon esculentum* encoding chromosomal proteins: homology to an intron of the polygalacturonase gene. DNA Seq 5: 225–227.
30. Frankel N, Carrari F, Hasson E, Iusem ND (2006) Evolutionary history of the *Asr* gene family. Gene 378: 74–83.
31. The Tomato Genome Consortium (2012) The tomato genome sequence provides insights into fleshy fruit evolution. Nature 485: 635–641.
32. Fischer I, Camus-Kulandaivelu L, Allal F, Stephan W (2011) Adaptation to drought in two wild tomato species: The evolution of the *Asr* gene family. New Phytol 190: 1032–1044.
33. Fischer I, Steige KA, Stephan W, Mboup M (2013) Sequence evolution and expression regulation of stress-responsive genes in natural populations of wild tomato. PLoS One 8: e78182.
34. Hara M, Kumagai K, Kuboi T (2002) Characterization and expression of a water stress responsive gene from a seashore plant *Calystegia soldanella*. Plant Biotechnol 19: 277–281.
35. Yamada T, Ichimura K, Kanekatsu M, Doorn W (2007) Gene expression in opening and senescing petals of morning glory (*Ipomoea nil*) flowers. Plant Cell Reports 26: 823–835.
36. Battaglia M, Olvera-Carrillo Y, Garciarrubio A, Campos F, Covarrubias AA (2008) The enigmatic LEA proteins and other hydrophilins. Plant Physiol 148: 6–24.
37. Goldgur Y, Rom S, Ghirlando R, Shkolnik D, Shadrin N, et al. (2007) Desiccation and zinc binding induce transition of tomato Abscisic Acid Stress Ripening 1, a water stress- and salt stress-regulated plant-specific protein, from unfolded to folded state. Plant Physiol 143: 617–628.
38. Konrad Z, Bar-Zvi D (2008) Synergism between the chaperone-like activity of the stress regulated ASR1 protein and the osmolyte glycine-betaine. Planta 227: 1213–1219.
39. Kalifa Y, Gilad A, Konrad Z, Zaccai M, Scolnik PA, et al. (2004) The water- and salt-stress-regulated *Asr1* (abscisic acid stress ripening) gene encodes a zinc-dependent DNA-binding protein. Biochemical Journal 381: 373–378.
40. Wang HJ, Hsu CM, Guang YJ, Wang CS (2005) A lily pollen ASR protein localizes to both cytoplasm and nuclei requiring a nuclear localization signal. Physiol Plantar 123: 314–320.
41. Kim IS, Kim YS, Yoon HS (2012) Rice ASR1 protein with reactive oxygen species scavenging and chaperone-like activities enhances acquired tolerance to abiotic stresses in *Saccharomyces cerevisiae*. Mol Cells 33: 285–293.
42. Dai JR, Liu B, Feng DR, Liu H, He Y, et al. (2011) *MpAsr* encodes an intrinsically unstructured protein and enhances osmotic tolerance in transgenic Arabidopsis. Plant Cell Report 30: 1219–1230.
43. Rom S, Gilad A, Kalifa Y, Konrad Z, Karpasas MM, et al. (2006) Mapping the DNA- and zinc-binding domains of ASR1 (abscisic acid stress ripening), an abiotic-stress regulated plant specific protein. Biochimie 88: 621–628.
44. Ricardi MM, Guaimas FF, Gonzalez RM, Burrieza HP, Lopez-Fernandez MP, et al. (2012) Nuclear Import and dimerization of tomato ASR1, a water stress-inducible protein exclusive to plants. PLoS One 7: e41008.
45. Kim SJ, Lee SC, Hong SK, An K, An G, et al. (2009) Ectopic expression of a cold-responsive *OsAsr1* cDNA gives enhanced cold tolerance in transgenic rice plants. Mol Cells 27: 449–458.
46. Li RH, Liu GB, Wang H, Zheng YZ (2013) Effects of Fe^{3+} and Zn^{2+} on the structural and thermodynamic properties of a soybean ASR protein. Biosci Biotechnol Biochem 77: 475–481.
47. Saumonneau A, Agasse A, Bidoyen MT, Lallemand M, Cantereau A, et al. (2008) Interaction of grape ASR proteins with a DREB transcription factor in the nucleus. FEBS Lett 582: 3281–3287.
48. Yang CY, Chen YC, Jauh GY, Wang CS (2005) A lily ASR protein involves abscisic acid signaling and confers drought and salt resistance in Arabidopsis. Plant Physiol 139: 836–846.

49. Hu W, Huang C, Deng XM, Zhou SY, Chen LH, et al. (2013) *TaASR1*, a transcription factor gene in wheat, confers drought stress tolerance in transgenic tobacco. Plant Cell Environ 36: 1449–1464.
50. Shkolnik-Inbar D, Bar-Zvi D (2010) ABI4 mediates abscisic acid and cytokinin inhibition of lateral root formation by reducing polar auxin transport in Arabidopsis. Plant Cell 22: 3560–3573.
51. Jeanneau M, Gerentes D, Foueillassar X, Zivy M, Vidal J, et al. (2002) Improvement of drought tolerance in maize: towards the functional validation of the *Zm-Asr1* gene and increase of water use efficiency by over-expressing C4-PEPC. Biochimie 84: 1127–1135.
52. Jha B, Lal S, Tiwari V, Yadav SK, Agarwal PK (2012) The *SbASR-1* gene cloned from an extreme halophyte *Salicornia brachiata* enhances salt tolerance in transgenic tobacco. Marine Biotechnol 14: 782–792.
53. Kalifa Y, Perlson E, Gilad A, Konrad Z, Scolnik PA, et al. (2004) Over-expression of the water and salt stress-regulated *Asr1* gene confers an increased salt tolerance. Plant Cell Environ 27: 1459–1468.
54. Hsu YF, Yu SC, Yang CY, Wang CS (2011) Lily ASR protein-conferred cold and freezing resistance in Arabidopsis. Plant Physiol Biochem 49: 937–945.
55. Shkolnik D, Bar-Zvi D (2008) Tomato ASR1 abrogates the response to abscisic acid and glucose in Arabidopsis by competing with ABI4 for DNA binding. Plant Biotechnol J 6: 368–378.
56. Dominguez PG, Frankel N, Mazuch J, Balbo I, Iusem N, et al. (2013) ASR1 mediates glucose-hormone cross talk by affecting sugar trafficking in tobacco plants. Plant Physiol 161: 1486–1500.
57. Saumonneau A, Laloi M, Lallemand M, Rabot A, Atanassova R (2012) Dissection of the transcriptional regulation of grape *ASR* and response to glucose and abscisic acid. J Exper Bot 63: 1495–1510.
58. Virlouvet L, Jacquemot MP, Gerentes D, Corti H, Bouton S, et al. (2011) The *ZmASR1* protein nfluences branched-chain amino acid biosynthesis and maintains kernel yield in maize under water-limited conditions. Plant Physiol 157: 917–936.
59. Maskin L, Maldonado S, Iusem N (2008) Tomato leaf spatial expression of stress-induced *Asr* genes. Mol Biol Report 35: 501–505.
60. Maskin L, Gudesblat GE, Moreno JE, Carrari FO, Frankel N, et al. (2001) Differential expression of the members of the *Asr* gene family in tomato (*Lycopersicon esculentum*). Plant Sci 161: 739–746.
61. Liu X, Prat S, Willmitzer L, Frommer W (1990) cis regulatory elements directing tuber-specific and sucrose-inducible expression. Mol Gen Genet 223: 401–406.
62. Obiadalla-Ali H, Fernie AR, Lytovchenko A, Kossmann J, Lloyd JR (2004) Inhibition of chloroplastic fructose 1,6-bisphosphatase in tomato fruits leads to decreased fruit size, but only small changes in carbohydrate metabolism. Planta 219: 533–540.
63. Hoagland DR, Arnon DI (1950) The water-culture method for growing plants without soil. California Agricultural Experiment Station Circular 347: 1–32.
64. Chomczynski P, Sacchi N (1987) Single-step method of RNA isolation by acid guanidinium thiocyanate phenol chloroform extraction. Anal Biochem 162: 156–159.
65. Adler G, Blumwald E, Bar-Zvi D (2010) The sugar beet gene encoding the sodium/proton exchanger 1 (BvNHX1) is regulated by a MYB transcription factor. Planta 232: 187–195.
66. Shkolnik-Inbar D, Adler G, Bar-Zvi D (2013) ABI4 downregulates expression of the sodium transporter *HKT1;1* in Arabidopsis roots and affects salt tolerance. Plant J 73: 993–1005.
67. Rossi M, Carrari F, Cabrera-Ponce JL, Vazquez-Rovere C, Herrera-Estrella L, et al. (1998) Analysis of an abscisic acid (ABA)-responsive gene promoter belonging to the Asr gene family from tomato in homologous and heterologous systems. Mol Gen Genet 258: 1–8.
68. Amitai-Zeigerson H, Scolnik PA, Bar-Zvi D (1995) Tomato *Asr1* mRNA and protein are transiently expressed following salt stress, osmotic stress and treatment with abscisic acid. Plant Sci 110: 205–213.
69. Hong SH, Kim IJ, Yang DC, Chung WI (2002) Characterization of an abscisic acid responsive gene homologue from *Cucumis melo*. J Exper Bot 53: 2271–2272.
70. Liu J, Jia C, Dong F, Wang J, Zhang J, et al. (2013) Isolation of an abscisic acid senescence and ripening inducible gene from litchi and functional characterization under water stress. Planta 237: 1025–1036.
71. Doczi R, Kondrak M, Kovacs G, Beczner F, Banfalvi Z (2005) Conservation of the drought-inducible, DS2 genes and divergences from their ASR paralogues in solanaceous species. Plant Physiol and Biochem 43: 269–276.
72. Riccardi F, Gazeau P, de Vienne D, Zivy M (1998) Protein changes in response to progressive water deficit in maize - Quantitative variation and polypeptide identification. Plant Physiol 117: 1253–1263.
73. Maskin L, Frankel N, Gudesblat G, Demergasso MJ, Pietrasanta LI, et al. (2007) Dimerization and DNA-binding of ASR1, a small hydrophilic protein abundant in plant tissues suffering from water loss. Biochem Biophys Res Commun 352: 831–835.
74. Ricardi MM, Gonzalez RM, Zhong S, Dominguez PG, Duffy T, et al. (2014) Genome-wide data (ChIP-seq) enabled identification of cell wall-related and aquaporin genes as targets of tomato ASR1, a drought stress-responsive transcription factor. BMC Plant Biol 14: 29.
75. Gillaspy G, Bendavid H, Gruissem W (1993) Fruits - A Developmental perspective. Plant Cell 5: 1439–1451.

12

Illumina Amplicon Sequencing of 16S rRNA Tag Reveals Bacterial Community Development in the Rhizosphere of Apple Nurseries at a Replant Disease Site and a New Planting Site

Jian Sun, Qiang Zhang, Jia Zhou, Qinping Wei*

Institute of Forestry and Pomology, Beijing Academy of Agriculture and Forestry Sciences, Beijing, China

Abstract

We used a next-generation, Illumina-based sequencing approach to characterize the bacterial community development of apple rhizosphere soil in a replant site (*RePlant*) and a new planting site (*NewPlant*) in Beijing. Dwarfing apple nurseries of 'Fuji'/SH6/Pingyitiancha trees were planted in the spring of 2013. Before planting, soil from the apple rhizosphere of the replant site (*ReSoil*) and from the new planting site (*NewSoil*) was sampled for analysis on the Illumina MiSeq platform. In late September, the rhizosphere soil from both sites was resampled (*RePlant* and *NewPlant*). More than 16,000 valid reads were obtained for each replicate, and the community was composed of five dominant groups (Proteobacteria, Acidobacteria, Bacteroidetes, Gemmatimonadetes and Actinobacteria). The bacterial diversity decreased after apple planting. Principal component analyses revealed that the rhizosphere samples were significantly different among treatments. Apple nursery planting showed a large impact on the soil bacterial community, and the community development was significantly different between the replanted and newly planted soils. Verrucomicrobia were less abundant in *RePlant* soil, while *Pseudomonas* and *Lysobacter* were increased in *RePlant* compared with *ReSoil* and *NewPlant*. Both *RePlant* and *ReSoil* showed relatively higher invertase and cellulase activities than *NewPlant* and *NewSoil*, but only *NewPlant* soil showed higher urease activity, and this soil also had the higher plant growth. Our experimental results suggest that planting apple nurseries has a significant impact on soil bacterial community development at both replant and new planting sites, and planting on new site resulted in significantly higher soil urease activity and a different bacterial community composition.

Editor: Gabriele Berg, Graz University of Technology (TU Graz), Austria

Funding: This work was supported by China Agriculture Research System, CARS-28, Minister of Agriculture (PRC), (http://english.agri.gov.cn/) (QPW), and Beijing Technology Foundation for Selected Overseas Chinese Scholar, Human Resource and Social Security Bureau of Beijing (http://www.bjld.gov.cn/) (JS). The funders had no role in study design, data collection and analysis, decision to publish, or preparation of the manuscript.

Competing Interests: The authors have declared that no competing interests exist.

* Email: qinpingwei@gmail.com

Introduction

Apple trees are among the most important fruit trees in the world. China has the world's highest apple tree acreage (2.060 million hectares, 42.54% of the world's supply; FAOSTAT, 2012) and production (37.00 million tons, 48.44% of the world's supply; FAOSTAT, 2012). However, apple trees in most of China's dominant production areas experience a full fruit period and senescence phase; currently, approximately 70% apple orchards in China are over 20 years old. The need for apple orchard renewal is more than 140 thousand hectares per year. Complicating this renewal is apple replant disease (ARD), which, because of a lack of land resources, is becoming a serious problem in fruit tree nurseries and old orchards.

The term ARD refers to the poor growth of young apple trees, which occurs after replanting on a site that was previously planted with apple. The phenomenon is common to all major apple growing regions of the world, including Asia, Europe, North America, Africa etc. Compared with new planting sites, directly replanting nurseries on old sites can result in decreased tree growth and a significantly lower seedling survival rate [1]. Trees may take several years to recover from the initial growth depression and eventually reach the size and annual yields of unaffected trees, and the cumulative yields and profitability in ARD-affected orchards are usually much lower than in unaffected orchards [2]. In China, the traditional method to avoid ARD is allowing a fallow period of over three years, while using plants from the grass family such as wheat to 'clean' the soil [3]. The efficacy of growing wheat for reducing apple root infection by species of *Rhizoctonia* and *Pythium* was confirmed by greenhouse and field trials [4,5]. However, the lost productivity is unaffordable for famers because land is limited and expensive.

Typically, in ARD-affected orchards, the root systems of apple seedlings are small, with discolored feeder roots and few functional root hairs [6]. In a previous field trial, leaf analysis for macro- and micronutrients showed most elements in ARD-affected and ARD-unaffected orchards exhibited no significant difference; thus, the observed growth responses were not associated with any nutritional effect [7]. Replant disease of fruit trees has been studied for

Table 1. The invertase, urease and cellulase activity of soil of different treatments and plant growth mass of *RePlant* and *NewPlant*.

	Invertase	Urease	Cellulase	New shoots (DW, g)	New roots (DW, g)
RePlant	8.93±0.93^c	0.73±0.06^a	0.164±0.032^b	150.5±22.9	22.3±6.2
NewPlant	6.59±0.14^b	1.02±0.22^b	0.053±0.015^a	261.4±34.6	44.1±9.7
ReSoil	8.87±0.22^c	0.55±0.18^a	0.182±0.064^b	NA	NA
NewSoil	5.16±0.65^a	0.60±0.13^a	0.073±0.040^a	NA	NA

"DW" means dry weight. Averages of replicates ± standard error; means followed by different letters are significantly different at $P<0.05$.
NA: not applicable.

many decades and although reported to be attributed to certain abiotic elements including phytotoxins [8], nutrient imbalance, low or high pH, and lack or excess of moisture [9], the preponderance of evidence indicates that the disease in large part is due to biotic factors [10–12]. Soil sterilization and fumigation with methyl bromide were effective in treating ARD [13,14], thus demonstrating the role of biotic components of soil as prominent determinants of ARD in orchards.

Soil microorganisms are recognized as the key factor inducing ARD, but it is a challenging task to fully characterize soil microbial communities. The information from culture-dependent methods is limited because only a small fraction of soil microorganisms is culturable [15]. Recently, culture-independent methods have been developed for investigating microbial communities, including molecular analyses of nucleic acids extracted from soil, including denaturing gradient gel electrophoresis (DGGE) and terminal restriction fragment length polymorphism (T-RFLP) after PCR [16,17], as well as community profiling based on fatty acid methyl esters (FAME) and phospholipid fatty acid (PLFA) [18]. Next-Generation Sequencing (NGS) is a new DNA sequencing method, which relies on the detection of pyrophosphate release upon nucleotide incorporation, rather than chain termination with dideoxynucleotides. Compared with DGGE and T-RFLP, NGS may provide more detailed information about the community because each DNA molecule is sequenced as an individual read and because the identification of individual species group is more accurate [19]. Of NGS, both Pyrosequencing and Illumina Miseq are used for characterize soil microbial community structure including rhizosphere microbiome [16,20], however in recent studies, Illumina Miseq platform has been more frequently used since the around 10-fold increase in read depth, similar sequencing quality together with much lower cost [21].

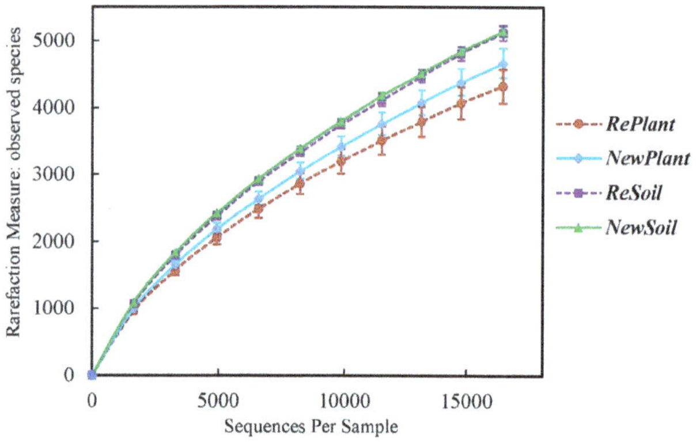

Figure 1. Rarefaction on species-abundance data. Average value of 3 replicates and error bar were showed.

Growth of fruit plants is affected by soil enzyme activity [22,23], and soil enzyme activity is highly related to soil microbes [24], which are involved in nutrient cycling and plant nutrient availability, and are in turn influenced by plant species. Sun et al. [25] used pyrosequencing to characterize the bacterial community structure of apple rhizosphere soil with different manure ratios, finding that certain levels of manure treatment resulted in significantly higher soil enzyme activity and a more diverse bacterial community composition. Past surveys of microbial communities of apple replant sites in specialized growing areas in Europe [13], Australia [26] and the USA [2] have confirmed a complex of biotic pathogens as causal agents of this etiology. It is also suggested that genotype-specific interactions with soil microbial consortia are linked with apple rootstock tolerance or susceptibility to ARD [27], while some CG series dwarfing rootstock showed high tolerance to replant disease compared to M26 [28]. However, apple orchards in China have significant differences compared with other areas in the world, including soil with low organic matter content and the specific rootstock, such as Pingyitiancha, which is commonly used in the Bohai Bay region, one of the two leading apple-producing areas in China. Little is known about the status of soil microbial communities in ARD orchards in China. For this reason, a specific survey of bacterial community development in the rhizosphere of apple trees at a replant disease site and a new planting site was conducted with the following objectives: (i) to compare tree growth rate and soil enzyme activity between a replant disease site and a new planting site, (ii) to reveal bacterial community development under ARD and non-ARD conditions and (iii) to define the relative importance of biotic components in replant disease etiology.

Methods

Ethics statement

The experiment was carried out in our scientific research field for pomology studies which is owned by our institute, therefore, no specific permissions were required for these locations/activities, and the field studies did not involve endangered or protected species.

Soil sampling

The test soil was collected from an orchard operated by the Institute of Pomology and Forestry, Beijing Academy of Agricultural Sciences. The soil type of local area was sandy loam soil, while the pH was 6.0–6.5. The orchard site was originally planted with apple trees approximately 1985 with trees grafted on *Malus Robusta* rootstock. In September 2012, the old trees were removed, and in April 4, 2013, the soil from a depth of 0–20

Table 2. Comparison of the estimated operational taxonomic unit (OTU) richness, diversity indexes and Pielous evenness of the 16S rRNA gene libraries for clustering at 97% identity as obtained from the pyrosequencing analysis.

Treatments	Observed OTUs	Shanon	Chao1	Pielous evenness (%)
RePlant	4328.1[a]	10.53[a]	9010.8[a]	87.17[a]
NewPlant	4509.9[a]	10.67[a]	9942.7[ab]	88.09[a]
ReSoil	5125.8[b]	11.02[b]	11371.8[c]	89.45[a]
NewSoil	5176.9[b]	11.12[b]	10614.9[bc]	88.53[a]

Averages of replicates ± standard error; means followed by different letters are significantly different at $P<0.05$.

cm was collected and mixed as replant soil sample (*ReSoil*). Another soil sample from alongside the orchard intensively cultivated with vegetables was collected as a new plant soil sample (*NewSoil*). Soil was put into 64 liter cubic Plexiglas boxes with rainproof shelter, and the nursery was planted right after the soil sampling. In late September, soil from a depth of 0–20 cm in three different locations at 20-cm distances from the center of trunk was collected from both sites and the youngest part of roots and the adjacent soil were resampled as *RePlant* and *NewPlant*.

Rootstock variety and tree growth

The two-year-old apple saplings were planted on April 4, 2013. Trees were of the scion variety 'Fuji' and were first grafted onto SH6 inter stock and then onto *Malus hupehensis* Var. Pingyiensis Jiang rootstock. The rootstock Pingyiensis Jiang has been widely used in Chinese orchards since the 1970 s. Trees were planted into either the replant soil or the new plant soil, with 10 replicates/treatment, and trunk heights were handed to 1.0–1.2 m. On September 20, 2013, new shoots and new roots of these trees were collected separately, dried at 105°C for 30 min and then dried at 70°C until a constant weight was reached in a forced-air oven.

Soil enzyme activity characterization

Soil samples were collected from depths of 0–20 cm in three different locations at 20-cm distances from the center of nurseries using a 5-cm diameter soil auger and transferred on ice to the laboratory both before nursery planting, and at the beginning of autumn 2013, just after the growth of autumn-shoot ceased. The soil samples were sieved through a 2-mm screen and homogenized prior to the analysis. One portion of the composite soil was stored

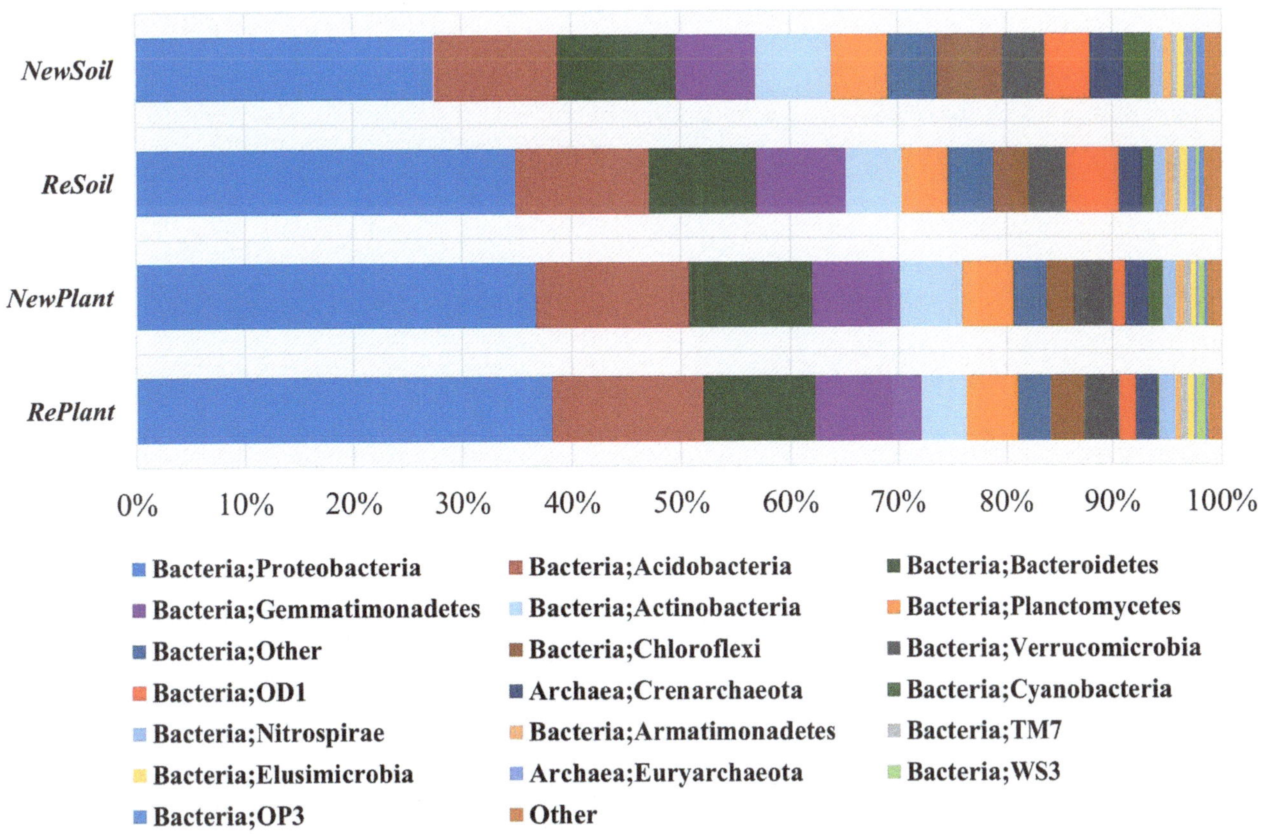

Figure 2. Comparison of the bacterial communities at the phylum level. Relative read abundance of different bacterial phyla within the different communities. Sequences that could not be classified into any known group were labeled "Other".

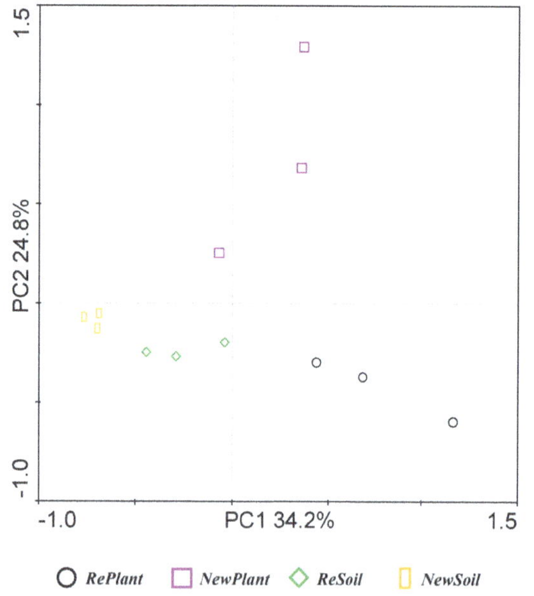

Figure 3. Principal component analysis (PCA) on the relative abundance of bacterial genera using Canoco 4.5. Principal components (PCs) 1 and 2 explained 34.2% and 24.8% of the variance, respectively.

in DNA-free polythene bags and kept on dry ice for the molecular analysis, while another portion was used for enzyme activity measurements.

Soil urease, invertase and cellulase activities were estimated according to a previous report [25]. Soil urease activity was detected using improved sodium phenate and sodium hypochlorite colorimetry. Invertase and cellulase activities were estimated colorimetrically by determining the reduction of 3,5-dinitrosalicylic acid from reducing sugars after the soil was incubated with a buffered sucrose and sodium carboxymethylcellulose solution and toluene at 37°C for 24 h and 72 h, respectively.

Soil DNA extraction

Three replicate samples were randomly picked for one treatment and used for DNA extraction. Soil DNA was extracted from the 1 g of soil after sieving using a Fast DNA SPIN Kit for soil (MP Biomedicals, Santa Ana, CA), according to the manufacturer's instructions. The extracted soil DNA was dissolved in 100 µl TE buffer, quantified by NanoDrop and stored at −70°C before use.

Bacterial 16S rRNA gene amplification and Illumina Sequencing

To determine the diversity and composition of the bacterial communities in each of these samples, we used the protocol described in Caporaso et al. [29]. PCR amplifications were conducted in with the 515f/806r primer set that amplifies the V4 region of the 16S rDNA gene. The primer set was selected as it exhibits few biases should yield accurate phylogenetic and taxonomic information. The reverse primer contains a 6-bp error-correcting barcode unique to each sample. DNA was amplified following the protocol described previously [30]. Amplicon pyrosequencing was performed on the Illumina MiSeq platforms at Novogene Bioinformatics Technology Co., Ltd, Beijing, China. Complete data sets are submitted to the NCBI Short Read Archive under accession no. SRX337490.

Pairs of reads from the original DNA fragments were merged by using FLASH [29] -a very fast and accurate software tool which was designed to merge pairs of reads when the original DNA fragments were shorter than twice the length of reads. Sequencing reads was assigned to each sample according to the unique barcode of each sample. Sequences were analyzed with the QIIME [31] software package (Quantitative Insights Into Microbial Ecology) and UPARSE pipeline [32], in addition to custom Perl scripts to analyze alpha (within sample) and beta (between sample) diversity.

First, the reads were filtered by QIIME quality filters. Default settings for Illumina processing in QIIME was used ($r = 3$ $p = 0.75$ total read length; $q = 3$; $n = 0$).

(p) min_per_read_length: minimum number of consecutive high-qualitybase calls to retain read(as percentage of totalread length).

(r) max_bad_run_length: maximum number of consecutive low-quality base calls allowed before truncating a read.

(n) sequence_max_n: maximum number of ambiguous (N) characters allowed in a sequence.

(q) phred_quality_score: last quality score considered low quality.

Then we use UPARSE pipeline to picking operational taxonomic units (OTUs) through making OTU table. Sequences were assigned to OTUs at 97% similarity. We picked a representative sequences for each OTU and used the RDP classifier [33] to assign taxonomic data to each representative sequence. In order to compute Alpha Divesity, we rarified the OTU table and calculated three metrics: Chao1 metric estimates the richness, the Observed OTUs metric was simply the count of unique OTUs found in the sample, and shannon index. Rarefaction curves were generated based on these three metrics.

Statistical and bioinformatics analysis

Heatmap figures were generated using custom R scripts. Canoco 4.5 was used to run principal component analysis (PCA). Analysis of variance and Spearman's rank correlations were performed using SPSS Statistics 18 (IBM, Armonk, New York, USA). The community richness index, community diversity index, data preprocessing, operational taxonomic unit-based analysis and hypothesis tests were performed using mothur (http://www.mothur.org/). The histogram was created using Microsoft Excel 2010 (Microsoft, Redmond, Washington, USA). Significance was accepted at $p<0.05$, unless otherwise noted.

Results

Seedling biomass accumulation

In the spring of 2013, dwarfing apple nurseries of 'Fuji'/SH6/Pingyitiancha trees were planted a replant site (*RePlant*) and a new planting site (*NewPlant*). Before planting, soil from the replant site (*ReSoil*) and from the new planting site (*NewSoil*) was sampled. After 1 year of growth, significant differences ($P<0.05$) were found between *RePlant* and *NewPlant* sites in dry mass accumulation (**Table 1**). Seedling growth was significantly inhibited in replant soil comparing with non-replant soil; the inhibition levels on root dry weight and shoot dry weight were 49.4% and 42.3%, respectively.

Soil enzyme activity

Soil urease activity was highest in *NewPlant* soil, whereas there was no significant difference in urease activity in soil from

Table 3. The genera showing significant differences among the samples.

Taxon	*RePlant* (%)	*NewPlant* (%)	*ReSoil* (%)	*NewSoil* (%)
Pseudomonas	0.207 ± 0.048^{b}	0.120 ± 0.041^{a}	0.063 ± 0.031^{a}	0.057 ± 0.009^{a}
Lysobacter	0.941 ± 0.260^{b}	0.471 ± 0.065^{a}	0.276 ± 0.061^{a}	0.277 ± 0.037^{a}
Phenylobacterium	0.337 ± 0.007^{c}	0.258 ± 0.045^{b}	0.317 ± 0.022^{c}	0.135 ± 0.013^{a}
Ramlibacter	0.276 ± 0.128^{a}	0.728 ± 0.109^{b}	0.345 ± 0.061^{a}	0.336 ± 0.046^{a}
Chthonomonas	0.051 ± 0.013^{a}	0.171 ± 0.048^{b}	0.12 ± 0.055^{ab}	0.202 ± 0.038^{b}
Flavisolibacter	0.217 ± 0.060^{a}	0.847 ± 0.350^{b}	0.364 ± 0.095^{a}	0.354 ± 0.048^{a}
Opitutus	0.378 ± 0.022^{a}	0.400 ± 0.057^{a}	0.644 ± 0.120^{b}	0.524 ± 0.060^{ab}
Bdellovibrio	0.096 ± 0.014^{a}	0.161 ± 0.058^{ab}	0.256 ± 0.025^{b}	0.239 ± 0.031^{b}
Novosphingobium	0.185 ± 0.136^{ab}	0.266 ± 0.049^{b}	0.063 ± 0.018^{a}	0.058 ± 0.009^{a}
Planctomyces	0.167 ± 0.025^{a}	0.152 ± 0.032^{a}	0.195 ± 0.032^{ab}	0.239 ± 0.002^{b}
Kaistobacter	2.463 ± 0.555^{b}	2.302 ± 0.461^{b}	2.17 ± 0.311^{b}	0.978 ± 0.083^{a}
Cellvibrio	0.124 ± 0.173^{a}	0.100 ± 0.064^{a}	0.144 ± 0.051^{a}	0.378 ± 0.06^{b}
Kribbella	0.012 ± 0.012^{a}	0.031 ± 0.011^{a}	0.026 ± 0.004^{a}	0.081 ± 0.025^{b}
Pseudonocardia	0.035 ± 0.014^{a}	0.045 ± 0.004^{ab}	0.045 ± 0.028^{ab}	0.095 ± 0.034^{b}
Rubrobacter	0.118 ± 0.066^{a}	0.226 ± 0.082^{ab}	0.144 ± 0.023^{a}	0.311 ± 0.092^{b}
Luteimonas	0.388 ± 0.071^{b}	0.268 ± 0.158^{ab}	0.325 ± 0.052^{ab}	0.120 ± 0.029^{a}
Steroidobacter	0.902 ± 0.136^{ab}	0.701 ± 0.085^{a}	0.996 ± 0.104^{b}	0.761 ± 0.065^{ab}
Candidatus Nitrososphaera	1.282 ± 0.400^{a}	2.054 ± 0.372^{ab}	2.04 ± 0.357^{ab}	3.044 ± 0.683^{b}
Cenarchaeaceae;g__	0.476 ± 0.180^{b}	0.043 ± 0.043^{a}	0.061 ± 0.064^{a}	0.017 ± 0.010^{a}
Hyphomonadaceae;g__	0.721 ± 0.078^{b}	0.360 ± 0.069^{a}	0.646 ± 0.055^{b}	0.382 ± 0.117^{a}
Rhodospirillaceae;g__	0.721 ± 0.104^{c}	0.348 ± 0.012^{a}	0.577 ± 0.099^{bc}	0.455 ± 0.129^{ab}
Haliangiaceae;g__	1.520 ± 0.475^{b}	0.793 ± 0.028^{a}	0.811 ± 0.143^{a}	0.879 ± 0.108^{a}
Syntrophobacteraceae;g__	0.914 ± 0.046^{b}	0.614 ± 0.212^{a}	0.563 ± 0.095^{a}	0.587 ± 0.051^{a}
Rhodothermaceae;g__	0.033 ± 0.018^{a}	0.102 ± 0.037^{b}	0.033 ± 0.013^{a}	0.031 ± 0.016^{a}
Saprospiraceae;g__	0.366 ± 0.114^{a}	0.671 ± 0.142^{b}	0.378 ± 0.054^{a}	0.807 ± 0.093^{b}
Erythrobacteraceae;g__	0.553 ± 0.150^{ab}	0.469 ± 0.080^{a}	0.815 ± 0.229^{b}	0.309 ± 0.042^{a}
Polyangiaceae;g__	0.033 ± 0.015^{a}	0.061 ± 0.044^{a}	0.073 ± 0.016^{a}	0.145 ± 0.012^{b}
NB1-j;g__	0.341 ± 0.056^{b}	0.157 ± 0.113^{a}	0.083 ± 0.044^{a}	0.108 ± 0.013^{a}
Chromatiales;g__	0.106 ± 0.025^{b}	0.020 ± 0.020^{a}	0.047 ± 0.029^{a}	0.027 ± 0.003^{a}
Sphingomonadales;g__	0.159 ± 0.048^{ab}	0.240 ± 0.015^{c}	0.220 ± 0.043^{a}	0.099 ± 0.009^{a}
Sphingomonadaceae;g__	0.679 ± 0.076^{a}	1.026 ± 0.234^{b}	0.689 ± 0.064^{a}	0.423 ± 0.051^{a}
Micrococcaceae;g__	0.183 ± 0.062^{a}	0.563 ± 0.167^{b}	0.228 ± 0.013^{a}	0.673 ± 0.105^{b}

"g__" represents genus not grouped into any known genera within these families/groups. Averages of replicates ± standard error; means followed by different letters are significantly different at $P<0.05$.

RePlant, *ReSoil* and *NewSoil* samples, which were 28.4%~46.1% lower than *NewPlant* soil (**Table 1**). Cellulase activities in *NewPlant* and *NewSoil* were lower than in *RePlant* and *ReSoil*, but planting apple nurseries did not significantly change soil cellulase activity. Regardless of whether a new apple nursery had yet been planted, the soil from the old orchard had about three-fold higher cellulase activity than soil from a site that had never been planted. Invertase activity was lower in *NewSoil* and *NewPlant* samples than in *RePlant* and *ReSoil* samples.

Richness

More than 16,000 valid reads were obtained for each replicate through a sequence optimization process, and the bacterial community richness index was calculated as shown in **Table 2**. After quality filtering, median sequence length of each read was 252 bp. In *ReSoil* and *NewSoil* samples, more than 600 additional OTUs were observed compared with *RePlant* and *NewPlant* soil (**Fig. 1**). Higher Shannon and Chao 1 indices before planting indicated that planting of apple nurseries reduced diversity within the bacterial community. However, Pielou's evenness values were indicating approximately equally distributed OTU abundances among community members (**Table 2**).

Taxonomic coverage

All of the sequences were classified into 26 phyla or groups by the mothur program. The overall bacterial composition of the different samples was similar, while the distribution of each phylum or group varied (**Fig. 2**). In all samples, Proteobacteria, Acidobacteria, Bacteroidetes, Gemmatimonadetes and Actinobacteria were the five most dominant phyla, accounting for >60% of the reads. Significantly more unclassified species were detected in *ReSoil* and *NewSoil* samples, which was in accordance with their higher diversity indices. Compared with other samples, *NewSoil* had a significantly higher percentage of GN02 (2.7–5.3-fold), OP3

(1.8–3.1-fold), Chloroflexi (1.8–2.4-fold), and Verrucomicrobia (1.1–1.3-fold), and a lower percentage of Proteobacteria (*NewSoil*: 27.2%, *ReSoil*: 34.9%, *RePlant*: 38.2% and *NewPlant*: 36.8%). The percentages of CyanoBac and Verrucomicrobia were lowest in *RePlant*. More Firmicutes were detected in *NewPlant* than *RePlant* or *ReSoil* samples, and the lowest level was found in *NewSoil*. The OD1 group was approximately two- to three-fold higher in *RePlant* and *NewPlant* compared with *ReSoil* and *NewSoil*.

On a genus level, all 260 detected genera were shared by the four samples, except for *Chryseobacterium,* which was not detected in *ReSoil*, and *Marinobacter*, which was not detected in *RePlant*. Several cold-tolerant species belong to *Chryseobacterium* [34], while *Marinobacter* has been reported to be halophilic and is found in seawater [35].

To further compare the microbiota among the different samples, we performed PCA on the relative abundance of bacterial genera using Canoco 4.5 (**Fig. 3**). Data are presented as a 2D plot to better illustrate the relationship. *ReSoil* and *NewSoil* were relatively similar, but planting of an apple nursery had a significant impact on the soil microbial community. *RePlant* had a significantly higher PC1 value, and *NewPlant* had a higher PC2 value. 32 genera showing significant differences among the samples are listed in **Table 3**.

Pseudomonas, Lysobacter and *Phenylobacterium* were significantly higher in *RePlant* compared with *NewPlant,* while *Ramlibacter, Chthonomonas* and *Flavisolibacter* were higher in *NewPlant*. *Phenylobacterium* and *Kaistobacter* were higher in *ReSoil* compared with *NewSoil*, while *Cellvibrio, Kribbella* and *Rubrobacter* were higher in *NewSoil*. On both the replant site and the new plant site, apple nurseries showed significant impacts on the bacterial community structure. On the new plant site, *Phenylobacterium, Ramlibacter, Flavisolibacter, Novosphingobium* and *Kaistobacter* had increased abundance, and *Planctomyces, Cellvibrio* and *Kribbella* decreased in abundance. On the replant site, only *Pseudomonas* and *Lysobacter* increased, while *Opitutus* and *Bdellovibrio* decreased. Compared with the *NewPlant* sample, planting in the replant site also resulted in a greater abundance of certain genera of Cenarchaeaceae, Hyphomonadaceae, Rhodospirillaceae, Haliangiaceae and Syntrophobacteraceae families, however these genera weren't grouped into known genera within these families.

Discussion

Replanted soil significantly inhibited root and shoot development and exhibited different soil enzyme activity and a different bacterial community pattern. The levels of inhibition on root dry weight and shoot dry weight were 49.4% and 42.3%. Rosette disease and decreasing photosynthetic efficiency were also observed in the replant site, as were fewer main branches (data not shown). This observation is in accordance with the negative impact of replanting on apple growth that is widely reported [9,13,36].

Soil enzymes are involved with biological cycling and the development of fertility, so they are crucial indicators of the soil biochemistry. Urease catalyzes the hydrolysis of urea to produce ammonia and carbamate, and it is thus recognized as an important indicator of soil health. In this study, rhizosphere soil of *NewPlant* showed significantly increased urease activity, but rhizosphere soil of RePlant did not, indicating that in new planting sites, the root exudates might support a new and different functional microbial community which was responsible for this apparent increase in mineralization and result in a better supply of available nutrients. This result was supported by data on plant dry mass (**Table 1**), where *NewPlant* exhibited more than 40% greater plant growth compared with *RePlant*. Xun et al. [22] reported that soil urease activity increased during apple orchard maturation, and in our previous study on manure refinement of apple orchards, urease was the key indicator of soil health and highly correlative to tree growth, no matter whether the soil type was sandy [24] or loam [25]. Therefore, soil urease is an ideal indicator of apple orchard maturation and the lack of a significant increase in urease activity at the replant site may explain the decreased growth and late fruiting. However, soil invertase, which is an important factor affecting hydrolysis of sucrose into glucose and fructose, and cellulase, which is involved in breaking down cellulase, were higher in replant soil rather than new soil, and planting of apple nurseries had no further impact on these two enzyme activities. This could be related to the residual small root tissues of previous trees. Invertase also increased with an overdose of manure refinement in apple orchards, but was not closely associated with tree growth [37].

The soil microbial community composition of the replant site and the new plant site were distinct, and planting of an apple nursery significantly increased the difference. The diversity within the bacterial community was reduced after planting an apple nursery in both replanted and newly planted soil, which was in accordance with the fewer unclassified species observed after nurseries had been growing for one year. It has been reported that many plant species reduce the microbial diversity of rhizosphere soil compared with surrounding sites, including maize [38] and switchgrass [39]. Such a reduced bacterial richness in the plant rhizosphere is known as the 'rhizosphere effect'. This is typically characterized by a selective enrichment of root specialized guilds and reduction of rhizosphere bacterial richness in comparison to unplanted soil. However, although the diversity at both sites decreased after planting, there was no significant difference in bacterial diversity between the replant and new plant sites, either before or after the nursery had been planted, which means the rhizosphere effect of apple trees is the critical factor determining bacterial community diversity.

Although Proteobacteria, Acidobacteria, Bacteroidetes, Gemmatimonadetes and Actinobacteria predominated in all of the samples, *NewSoil* had a unique phyla composition compared with the other samples. Higher percentages of GN02, OP3, Chloroflexi and Verrucomicrobia and lower percentages of Proteobacteria were observed. This could also be explained by the rhizosphere effect' because there must be some root tissue left in the *ReSoil* site, even without a new nursery being planted. It is worth noticing that the OD1 group was approximately two- to three-fold higher in the *RePlant* and *NewPlant* samples as compared with *ReSoil* and *NewSoil*, and the *RePlant* soil had a higher percentage of WS3 compared with the other treatments. Verrucomicrobia were less common in the *RePlant* sample compared with *NewPlant*. Cultivation-independent approaches detect representatives of the Verrucomicrobia phylum in a wide range of environments, including soils, water and human feces [40], suggesting the phylum is widespread, but it is still poorly characterized. A few species that are extremely acidophilic [41] or ectosymbionts of protists [42] belong to this phylum. In our previous study [25], on loam soil of apple orchards with manure refinement, an optimal manure ratio resulted in an increase of Verrucomicrobia compared with soil with no manure applied, with the abundance increasing from 1.10% to 1.51%. However, on sandy soil [24], Verrucomicrobia decreased monotonically from 2.62% to 1.71%, 1.36% and 0.97% following application of 5%, 10% and 15% (which was optimal for tree growth) manure, respectively, and increased back to 1.28% when 20% manure applied. More research is required to determine whether this phylum is related to the rhizosphere effect of apple trees. Because most species of GN02, OP3, OD1 and WS3 groups are known only from metagenomics study and remain uncultivated, little information is available for functional discussion.

Principal component analysis on the genus level showed that apple nurseries had significant impacts on the soil microbial community, and the changes from *ReSoil* to *RePlant* and from *NewSoil* to *NewPlant* differed significantly. The bacterial community difference between *RePlant* and *NewPlant* was much greater than between *ReSoil* and *NewSoil*.

Phenylobacterium and *Kaistobacter* were higher in *ReSoil* compared with *NewSoil*, while *Cellvibrio, Kribbella* and *Rubrobacter* were higher in *NewSoil*. The genus *Phenylobacterium* comprises a single species called *P. immobile*, which has previously been described as growing optimally only on artificial compounds such as chloridazon [43]. More *P. immobile* in replant soil could be related to herbicide applied in the orchard. Unfortunately, little information is available in the literature concerning the genus *Kaistobacter*. *Cellvibrio* is a genus of gamma proteobacteria, which can oxidize cellulose to oxycellulose, but *NewSoil* showed lower cellulase than *ReSoil*, probably because in an aerobic environment, soil cellulase mainly derives from fungi, rather than bacteria. Some species of *Kribbella* have been isolated from the rhizosphere of *Typhonium giganteum* [44] and from tissues such as apricot leaves [45] and roots of *Lupinus angustifolius* [46]. *Kribbella antibiotic* [47] of the genus was reported to have a strong inhibitory activity toward *Botrytis* sp., *Rhizoctonia solani* and *Pyricularia oryzae*. *Rubrobacter* is a genus of Actinobacteria, which are radiotolerant [48], and a novel DNA repair enzyme was isolated from *Rubrobacter radiotolerans* [49], but little information is available regarding its presence in orchard soil.

On the replant site, *Lysobacter* and *Pseudomonas* increased in abundance, while *Opitutus* and *Bdellovibrio* decreased. *Pseudomonas* and *Lysobacter* were significantly higher in *RePlant* compared with *NewPlant*, while *Ramlibacter, Chthonomonas* and *Flavisolibacter* were more abundant in the *NewPlant* sample. *Pseudomonas* has been proposed to play a role in replant disease etiology of peach [50] and apple [51] trees through the production of hydrogen cyanide (HCN). However, *Pseudomonas* also includes several soil bacterial species with plant growth promoting activity, including *P. fluorescens* [10,52], and *Pseudomonas* and *Bacillus* are two of the most common biocontrol agent sources [53]. *Lysobacter* [54] has been a rich source for novel antibiotics, and some species have potential as biological control agents for plant diseases. Considering the previous study of manure refinement of apple orchard, in which an optimal manure ratio for nursery growth resulted in a decrease of *Pseudomonas* and *Lysobacter* [24,37], we speculate that in the *RePlant* sample, the higher percentage of *Pseudomonas* was related to replant disease and the increase of *Lysobacter* and was probably induced by the high percentages of Pathogenic fungi. However, further culture-based experiments would be needed to confirm this. There was only one species of *Chthonomonas*, *G. calidirosea*, which is an aerobic, pigmented, thermophilic micro-organism [55], while two species of *Flavisolibacter* were isolated from ginseng cultivating soil [56]. However, little was known about their function in soil or their relationship with planting until now.

Compared with *NewPlant*, planting in the replant site also resulted in more of certain genera of Cenarchaeaceae, Hyphomonadaceae, Rhodospirillaceae, Haliangiaceae and Syntrophobacteraceae families; however, these genera weren't grouped into any known genera within these families. This is in accordance with the fact that pyrosequencing can detect many uncultured microbes.

Conclusions

Our study has documented that replanting has a large negative impact on growth of apple nurseries in Beijing, China. Planting apple nurseries raised soil urease activity at the new planting site but not the replant site, while no significant impact on invertase and cellulase was observed. Apple nurseries had a significant impact on the soil bacterial community. *Lysobacter* and *Pseudomonas* were increased at the replant site, and the bacterial communities of the new and replant sites responded differently, resulting in more distinct community patterns.

Author Contributions

Conceived and designed the experiments: JS QPW. Performed the experiments: JS QZ JZ. Analyzed the data: JS QPW. Contributed reagents/materials/analysis tools: QZ JZ QPW. Contributed to the writing of the manuscript: JS QPW.

References

1. Mazzola M, Granatstein DM, Elfving DC, Mullinix K, Gu Y (2002) Cultural management of microbial community structure to enhance growth of apple in replant soils. Phytopathology 92: 1363–1366.
2. Mazzola M (1998) Elucidation of the microbial complex having a causal role in the development of apple replant disease in Washington. Phytopathology 88: 930–938.
3. Fan H, Zhao Z, Liu H, Zhao G, Zhang X, et al. (2008) Changes of Soil Nutrition in Root Zone and Their Effects on Growth of the Replanted Apple. Acta Horticulturae Sinica 12: 3.
4. Mazzola M, Gu Y (2000) Impact of wheat cultivation on microbial communities from replant soils and apple growth in greenhouse trials. Phytopathology 90: 114–119.
5. Mazzola M, Mullinix K (2005) Comparative field efficacy of management strategies containing *Brassica napus* seed meal or green manure for the control of apple replant disease. Plant Dis 89: 1207–1213.
6. Wilson S, Andrews P, Nair TS (2004) Non-fumigant management of apple replant disease. Sci Hortic-Amsterdam 102: 221–231.
7. Hofmann A, Wittenmayer L, Arnold G, Schieber A, Merbach W (2012) Root exudation of phloridzin by apple seedlings (*Malus* x *domestica* Borkh.) with symptoms of apple replant disease. Journal of Applied Botany and Food Quality 82: 193–198.
8. Utkhede RS, Smith EM (1993) Biotic and abiotic causes of replant problems of fruit trees. III International Symposium on Replant Problems 363. 25–32.
9. Rumberger A, Yao S, Merwin IA, Nelson EB, Thies JE (2004) Rootstock genotype and orchard replant position rather than soil fumigation or compost amendment determine tree growth and rhizosphere bacterial community composition in an apple replant soil. Plant Soil 264: 247–260.
10. Mazzola M, Manici LM (2012) Apple replant disease: role of microbial ecology in cause and control. Annu Rev Phytopathol 50: 45–65.
11. Yim B, Smalla K, Winkelmann T (2013) Evaluation of apple replant problems based on different soil disinfection treatments-links to soil microbial community structure? Plant Soil 1–15.
12. Covey Jr RP, Benson NR, Haglund WA (1979) Effect of soil fumigation on the apple replant disease in Washington [USA]. Phytopathology 69.
13. Manici LM, Kelderer M, Franke-Whittle IH, Rühmer T, Baab G, et al. (2013) Relationship between root-endophytic microbial communities and replant disease in specialized apple growing areas in Europe. Appl Soil Ecol 72: 207–214.
14. Cardenas E, Tiedje JM (2008) New tools for discovering and characterizing microbial diversity. Curr Opin Biotech 19: 544–549.
15. Smalla K, Oros-Sichler M, Milling A, Heuer H, Baumgarte S, et al. (2007) Bacterial diversity of soils assessed by DGGE, T-RFLP and SSCP fingerprints of PCR-amplified 16S rRNA gene fragments: do the different methods provide similar results? J Microbiol Meth 69: 470–479.
16. Hirsch PR, Mauchline TH, Clark IM (2010) Culture-independent molecular techniques for soil microbial ecology. Soil Biology and Biochemistry 42: 878–887.
17. Drenovsky RE, Elliott GN, Graham KJ, Scow KM (2004) Comparison of phospholipid fatty acid (PLFA) and total soil fatty acid methyl esters (TSFAME)

for characterizing soil microbial communities. Soil Biology and Biochemistry 36: 1793–1800.
18. Ronaghi M, Uhlén M, Nyrén P (1998) A sequencing method based on real-time pyrophosphate. Science 281: 363–365.
19. Lee OO, Wang Y, Yang J, Lafi FF, Al-Suwailem A, et al. (2010) Pyrosequencing reveals highly diverse and species-specific microbial communities in sponges from the Red Sea. The ISME journal 5: 650–664.
20. Mazzola M, Strauss SL (2013) Resilience of orchard replant soils to pathogen re-infestation in response to Brassicaceae seed meal amendment. Aspects of Applied Biology 119: 69–77.
21. Gołębiewski M, Deja-Sikora E, Cichosz M, Tretyn A, Wróbel B (2014) 16S rDNA Pyrosequencing Analysis of Bacterial Community in Heavy Metals Polluted Soils. Microbial Ecol 1–13.
22. Qian X, Gu J, Sun W, Li Y, Fu Q, et al. (2014) Changes in the soil nutrient levels, enzyme activities, microbial community function, and structure during apple orchard maturation. Appl Soil Ecol 77: 18–25.
23. Jia T, Chao S, Mingyan Y, Xiaoqi Z (2012) Studies on the activities of three kinds of soil enzyme, organic matters, microbes and the yields and quality of apple in different tree-aged apple orchards in Loess plateau. Journal of Agricultural Science and Technology (Beijing) 14: 115–122.
24. Zhang Q, Sun J, Liu S, Wei Q (2013) Manure Refinement Affects Apple Rhizosphere Bacterial Community Structure: A Study in Sandy Soil. PloS one 8: e76937.
25. Sun J, Zhang Q, Zhou J, Wei Q (2014) Pyrosequencing technology reveals the impact of different manure doses on the bacterial community in apple rhizosphere soil. Appl Soil Ecol 78: 28–36.
26. Dullahide SR, Stirling GR, Nikulin A, Stirling AM (1994) The role of nematodes, fungi, bacteria, and abiotic factors in the etiology of apple replant problems in the Granite Belt of Queensland. Animal Production Science 34: 1177–1182.
27. St. Laurent A, Merwin I, Fazio G, Thies J, Brown M (2010) Rootstock genotype succession influences apple replant disease and root-zone microbial community composition in an orchard soil. Plant Soil 337: 259–272.
28. Atucha A, Emmett B, Bauerle T (2014) Growth rate of fine root systems influences rootstock tolerance to replant disease. Plant Soil 376: 337–346.
29. Caporaso JG, Kuczynski J, Stombaugh J, Bittinger K, Bushman FD, et al. (2010) QIIME allows analysis of high-throughput community sequencing data. Nature methods 7: 335–336.
30. Magoč T, Salzberg SL (2011) FLASH: fast length adjustment of short reads to improve genome assemblies. Bioinformatics 27: 2957–2963.
31. Wang Q, Garrity MG, Tiedje MJ, James RC (2007) Naïve Bayesian Classifier for Rapid Assignment of rRNA Sequences into the New Bacterial Taxonomy. Applied and Environmental Microbiology 73: 5261–5267.
32. Edgar RC (2013) UPARSE: highly accurate OTU sequences from microbial amplicon reads. 10: 996–998.
33. Caporaso JG, Lauber CL, Walters WA, Berg-Lyons D, Lozupone CA, et al. (2011) Global patterns of 16S rRNA diversity at a depth of millions of sequences per sample. Proc Natl Acad Sci U S A 108 Suppl 1: 4516–4522.
34. Kämpfer P, Dreyer U, Neef A, Dott W, Busse H (2003) Chryseobacterium defluvii sp. nov., isolated from wastewater. Int J Syst Evol Micr 53: 93–97.
35. Gauthier MJ, Lafay B, Christen R, Fernandez L, Acquaviva M, et al. (1992) *Marinobacter hydrocarbonoclasticus* gen. nov., sp. nov., a new, extremely halotolerant, hydrocarbon-degrading marine bacterium. International Journal of Systematic Bacteriology 42: 568–576.
36. Sewell G (1981) Effects of *Pythium* species on the growth of apple and their possible causal role in apple replant disease. Ann Appl Biol 97: 31–42.
37. Sun J, Zhang Q, Liu S, Wang X, Liu J, et al. (2013) Comparison of soil microbe community structure between organic and conventional apple orchards. Journal of Fruit Science 30: 230–234.
38. García Salamanca A, Molina Henares MA, Dillewijn P, Solano J, Pizarro Tobías P, et al. (2013) Bacterial diversity in the rhizosphere of maize and the surrounding carbonate-rich bulk soil. Microbial biotechnology 6: 36–44.
39. Mao Y, Li X, Smyth EM, Yannarell AC, Mackie RI (2014) Enrichment of specific bacterial and eukaryotic microbes in the rhizosphere of switchgrass (*Panicum virgatum* L.) through root exudates. Environmental Microbiology Reports.
40. Hou S, Makarova KS, Saw JH, Senin P, Ly BV, et al. (2008) Complete genome sequence of the extremely acidophilic methanotroph isolate V4, *Methylacidiphilum infernorum*, a representative of the bacterial phylum Verrucomicrobia. Biol Direct 3: 11.
41. Dunfield PF, Yuryev A, Senin P, Smirnova AV, Stott MB, et al. (2007) Methane oxidation by an extremely acidophilic bacterium of the phylum Verrucomicrobia. Nature 450: 879–882.
42. Sato T, Kuwahara H, Fujita K, Noda S, Kihara K, et al. (2013) Intranuclear verrucomicrobial symbionts and evidence of lateral gene transfer to the host protist in the termite gut. The ISME journal.
43. Kanso S, Patel BK (2004) *Phenylobacterium lituiforme* sp. nov., a moderately thermophilic bacterium from a subsurface aquifer, and emended description of the genus *Phenylobacterium*. Int J Syst Evol Micr 54: 2141–2146.
44. Xu Z, Xu Q, Zheng Z, Huang Y (2012) *Kribbella amoyensis* sp. nov., isolated from rhizosphere soil of a pharmaceutical plant, *Typhonium giganteum* Engl. Int J Syst Evol Micr 62: 1081–1085.
45. Kaewkla O, Franco CM (2013) *Kribbella endophytica* sp. nov., an endophytic actinobacterium isolated from the surface-sterilized leaf of a native apricot tree. Int J Syst Evol Micr 63: 1249–1253.
46. Trujillo ME, Kroppenstedt RM, Schumann P, Martínez-Molina E (2006) *Kribbella lupini* sp. nov., isolated from the roots of *Lupinus angustifolius*. Int J Syst Evol Micr 56: 407–411.
47. Wang D, Zhang Y, Jiang Y, Wu W, Jiang C (2003) A study on polyphasic taxonomy of one antifungal actinomycete strain YIM31530~(T). Journal of Yunnan University (Natural Sciences) 26: 265–269.
48. Kausar J, Ohyama Y, Terato H, Ide H, Yamamoto O (1997) 16S rRNA gene sequence of *Rubrobacter radiotolerans* and its phylogenetic alignment with members of the genus *Arthrobacter*, gram-positive bacteria, and members of the family Deinococcaceae. International journal of systematic bacteriology 47: 684–686.
49. Asgarani E, Terato H, Asagoshi K, Shahmohammadi HR, Ohyama Y, et al. (2000) Purification and characterization of a novel DNA repair enzyme from the extremely radioresistant bacterium *Rubrobacter radiotolerans*. J Radiat Res 41: 19–34.
50. Yang J, Ruegger PM, McKenry MV, Becker JO, Borneman J (2012) Correlations between root-associated microorganisms and peach replant disease symptoms in a California soil. PloS one 7: e46420.
51. Rumberger A, Merwin IA, Thies JE (2007) Microbial community development in the rhizosphere of apple trees at a replant disease site. Soil Biology and Biochemistry 39: 1645–1654.
52. Himani S, Mohinder K, Prakash VP, Sheetal R (2011) Phenotypic and genotypic characterization of plant beneficial fluorescent *Pseudomonas* species associated with apple to overcome replant problem. Plant Disease Research 26.
53. Santoyo G, Orozco-Mosqueda MDC, Govindappa M (2012) Mechanisms of biocontrol and plant growth-promoting activity in soil bacterial species of *Bacillus* and *Pseudomonas*: a review. Biocontrol Sci Techn 22: 855–872.
54. Xie Y, Wright S, Shen Y, Du L (2012) Bioactive natural products from *Lysobacter*. Nat Prod Rep 29: 1277–1287.
55. Lee KC, Dunfield PF, Morgan XC, Crowe MA, Houghton KM, et al. (2011) *Chthonomonas calidirosea* gen. nov., sp. nov., an aerobic, pigmented, thermophilic micro-organism of a novel bacterial class, *Chthonomonadetes classis* nov., of the newly described phylum Armatimonadetes originally designated candidate division OP10. Int J Syst Evol Micr 61: 2482–2490.
56. Yoon M, Im W (2007) *Flavisolibacter ginsengiterrae* gen. nov., sp. nov. and *Flavisolibacter ginsengisoli* sp. nov., isolated from ginseng cultivating soil. Int J Syst Evol Micr 57: 1834–1839.

13

Salicylic Acid Alleviates the Adverse Effects of Salt Stress in *Torreya grandis* cv. Merrillii Seedlings by Activating Photosynthesis and Enhancing Antioxidant Systems

Tingting Li[1,2]⁹, Yuanyuan Hu[1,2]⁹, Xuhua Du[3], Hui Tang[1,2], Chaohua Shen[1,2], Jiasheng Wu[1,2]*

1 Nurturing Station for the State Key Laboratory of Subtropical Silviculture, Zhejiang A & F University, Lin'an, Hangzhou, P. R. China, **2** School of Forestry & Biotechnology, Zhejiang A & F University, Lin'an, Hangzhou, P. R. China, **3** China National Bamboo Research Center, Hangzhou, P. R. China

Abstract

Background: Salt stress is a major factor limiting plant growth and productivity. Salicylic acid (SA) has been shown to ameliorate the adverse effects of environmental stress on plants. To investigate the protective role of SA in ameliorating salt stress on *Torreya grandis* (*T. grandis*) trees, a pot experiment was conducted to analyze the biomass, relative water content (RWC), chlorophyll content, net photosynthesis (Pn), gas exchange parameters, relative leakage conductivity (REC), malondialdehyde (MDA) content, and activities of superoxide dismutase (SOD) and peroxidase (POD) of *T. grandis* under 0.2% and 0.4% NaCl conditions with and without SA.

Methodology/Principal Findings: The exposure of *T. grandis* seedlings to salt conditions resulted in reduced growth rates, which were associated with decreases in RWC and Pn and increases in REC and MDA content. The foliar application of SA effectively increased the chlorophyll (chl (a+b)) content, RWC, net CO_2 assimilation rates (Pn), and proline content, enhanced the activities of SOD, CAT and POD, and minimized the increases in the REC and MDA content. These changes increased the capacity of *T. grandis* in acclimating to salt stress and thus increased the shoot and root dry matter. However, when the plants were under 0% and 0.2% NaCl stress, the dry mass of the shoots and roots did not differ significantly between SA-treated plants and control plants.

Conclusions: SA induced the salt tolerance and increased the biomass of *T. grandis* cv. by enhancing the chlorophyll content and activity of antioxidative enzymes, activating the photosynthetic process, and alleviating membrane injury. A better understanding about the effect of salt stress in *T. grandis* is vital, in order gain knowledge over expanding the plantations to various regions and also for the recovery of *T. grandis* species in the future.

Editor: Jin-Song Zhang, Institute of Genetics and Developmental Biology, Chinese Academy of Sciences, China

Funding: This project was supported in part by the Fruit Innovation Team Project of Zhejiang Province (2009R50033-7) and the Major Project of National Spark Plan of China (2012GA700001). The funders had no role in study design, data collection and analysis, decision to publish, or preparation of the manuscript.

Competing Interests: The authors have declared that no competing interests exist.

* Email: wujs@zafu.edu.cn

⁹ These authors contributed equally to this work.

Introduction

Salt stress is one of the major abiotic stress factors that limit plant growth and productivity [1]. Salt stress alters various biochemical and physiological responses in plants, thereby affecting almost all plant processes [2]. The major effects of salinity are significant reductions in growth, such as decreased leaf area, leaf length and root and shoot dry weight [3]. High salt stress can also restrict photosynthesis by decreasing the abundance of green pigments and causing stomatal closure and oxidative stress, thereby resulting in the formation of reactive oxygen species (ROS) [4,5,6]. Excessive amounts of ROS can enhance membrane lipid peroxidation, cause electrolyte leakage, and damage the chloroplast, thus inhibiting photochemical reactions and decreasing photosynthesis [7,8]. To alleviate the harmful effects of these ROS, plants have evolved an effective scavenging system composed of non-enzymatic antioxidants (carotenoids, ascorbate, and tocopherol) and enzymatic antioxidants, such as catalase (CAT) and peroxidase (POD). Gossett et al. [9] reported that treatment with 150 mM NaCl resulted in a 72% increase in the POD activity in salt-tolerant cotton. Therefore, it is necessary to improve the salinity tolerance of plants and exploit various compounds to alleviate plant stress.

As an important signaling molecule, salicylic acid (SA) influences various physiological and biochemical functions in plants and has diverse effects on the tolerance to biotic and abiotic stress [10,11,12]. The role of SA in plant tolerance to abiotic stresses such as ozone, UV-B, heat, heavy metal, and osmotic stress has been reported [12,13]. Khan et al. [14] reported that SA application alleviated the adverse effects of salt stress in mungbean through the improvement of plant photosynthesis and growth and also enhancing antioxidant system. However, the improvement of salt tolerance by exogenous SA depends on genotype and the concentration of SA used. For example, Arfan et al. [10] reported that exogenous SA counteracted the salt stress-induced growth inhibition in a salt-tolerant wheat genotype, although no improvement occurred in a salt-sensitive cultivar. Therefore, the questions of whether the effectiveness of SA in the alleviation of salt stress is dependent on the plant species must be further elucidated.

The genus Torreya is an endangered and primitive member of the yew family (Taxaceae) and consists only of six species with a restricted worldwide distribution [15]. The conservation and recovery of these species have received increasing attention in recent years. *Torreya grandis* (*T. grandis*) cv. Merrillii, belonging to the genus Torreya family Taxaceae, is an important economic tree species and has drupe-like fruits with nut seeds that were used as food for thousands years in China. *T. grandis* has also important medicinal value due to its anthelmintic, antitussive, carminative, laxative, antifungal, antibacterial, and antitumor activity [16]. Currently, *T. grandis* is only planted on hills and mountain slopes in south China. The expansion of *T. grandis* plantations on new sites is important for the recovery of *T. grandis* species. However, throughout the entire country, 99 million hm^2 arable land is adversely affected by high salt concentrations, and the soil salinization is forecasted to increase drastically in the coming decades due to the worldwide ongoing climate change [17]. Therefore, it is necessary to improve the salinity tolerance of *T. grandis* species to expand its plantation under stress conditions. For the past years, numerous studies have been focused over *T. grandis* app:addword:concentratespecies for their ecological characterization; chemical constituents, fatty acid composition and its distribution in the seeds and the therapeutic use of the seed extracts [18,19,20,21]. Unfortunately, few published studies are available on the effect of salt stress on the growth and physiology of *T. grandis* trees. Additionally, it remains unclear whether SA plays a significant protective role in ameliorating the influence of salt stress on *T. grandis* trees to increase its salt tolerance.

The objective of this study was to investigate the effects of exogenous SA application on *T. grandis* growth and physiology under 0.2% and 0.4% NaCl conditions by analyzing the biomass, photosynthesis, membrane injury, and antioxidative enzyme activity. The information obtained is essential for propagation and cultivation of T. grandis and the other Torreya species under salt stress condition.

Materials and Methods

Plant materials and growth conditions

The pot experiments under different salt stress treatments were conducted in a controlled environment room of the College of Forest and Biotechnology, Zhejiang Agriculture and Forestry University, Zhejiang Province, China (N30°23′, E119°72′) in 2012. In March, two-year-old healthy and homogenous *T. grandis* seedlings were transplanted from Chunan, Zhejiang Province, China, and transferred to plastic pots (13.5 cm inner-diameter and 16 cm height with holes in the bottom, one seedling per pot) that were filled with 4 kg mixture of silt and perlite soil (3:1, v/v, pH 6.40). The soil was a loam with an organic matter content of 2.90%, total N content of 0.255%, total P content of 0.125%, and total K content of 1.574%. Four weeks after transplanting, a completely randomized design with 3 replications per treatment and 5 plants per replication was adopted. Six treatments with different salt concentrations (0, 0.2 and 0.4% NaCl) and different SA concentrations (0 and 0.5 mmol) were implemented as follows: T1 (distilled water without SA); T2 (distilled water with 0.5 mmol SA); T3 (0.2% NaCl without 0.5 mmol SA); T4 (0.2% NaCl with 0.5 mmol SA); T5 (0.4% NaCl without 0.5 mmol SA); and T6 (0.4% NaCl with 0.5 mmol SA). To avoid osmotic shock, the NaCl solution was gradually added to the soil in eight steps to achieve the final concentrations of 0.2% and 0.4% in 4 days. SA was dissolved in ethanol and then added to "Tween-20" (0.1% dilute solution) to facilitate spreading of the solution on the plant-leaf surface. The desired SA concentration was attained using double-distilled water (DDW). SA was sprayed over the leaves 3 days before NaCl treatment and also continued during the NaCl application. SA was sprayed over the adaxial and abaxial surfaces of a leaf until dripping a common insecticidal atomizer and it was done for twice a day at the time interval of 07:00 and18:00. When SA solution was sprayed, the surface of the potting soil was covered with plastic film in case SA solution dripped to the root. The plants were then collected for biomass analysis after 60 days, and measurements of other indexes were collected after 30 days of treatments. Twice in week irrigation were done for different treatmental plants with their corresponding solution in order to maintain the field capacity at 70–75%. The environmental conditions were maintained as same throughout the experiment.

Analysis of growth and biomass

Sixty days after the treatments, one intact plant from each replicate of treatments was uprooted for biomass analysis. The green plant tissues and roots were oven-dried at 65°C to a constant weight and weighed using an electronic scale to determine the biomass.

Determination of the leaf relative water content (RWC) and proline content

A total of 0.5 g of fresh leaf tissue was immediately weighed (fresh weight, WF) after rehydrating for 24 h in the dark (saturated weight, WS) and after oven-drying at 85°C for 24 h to a constant weight (WD). The RWC was calculated using the following formula:

$$RWC(\%) = [(WF - WD)/(WS - WD)] \times 100.$$

Proline accumulation was determined as described by Bates et al. [22] with slight modifications. Approximately 0.2 g of fresh leaf material was homogenized in 5 mL 3% aqueous sulfosalicylic acid. The samples were mixed, heated in boiling water for 30 min, and filtered through Whatman's No. 2 filter paper. A total of 2 mL of filtrate was mixed with 2 mL of acid-ninhydrin and 2 mL of glacial acetic acid in a test tube. The mixture was placed in a water bath for 1 h at 100°C and cooled to room temperature. The reaction mixture was extracted with 5 mL of toluene, and the chromophore containing toluene was aspirated. The absorbance was measured at 520 nm using toluene as a blank. The proline concentration was calculated using proline standards (0–50 mg/mL).

Analysis of gas exchange

The youngest healthy, fully developed leaves from a randomly selected branch were used for the gas exchange measurements on recently expanded leaves using a portable photosynthesis measuring instrument (LI-6400, LiCor, Inc. Lincoln, NE, USA) equipped with an 6400-05 conifer chamber at concentration of 400 μmol mol^{-1} CO_2, 21% O_2, and 50% relative humidity with a natural light source. The temperature of the chamber was maintained at

29–32°C. Photosynthetic gas exchange measurements were made during 9:00–10:30 when the light intensity was between 1200–1600 μmol $m^{-2}s^{-1}$. This irradiance was higher than the saturation irradiance for current year leaf of *T. grandis* (452–514 μmol m^{-2} s^{-1}, data not shown). Therefore, light was not the limitation for photosynthesis in these gas exchange measurements.

Determinations of the chlorophyll concentration and soluble protein content

A total of 0.1 g of leaf discs was weighed, placed into a test tube, and then repeatedly extracted with 8 mL of 95% (V/V) ethanol. The photosynthetic pigments were extracted at 4°C for 48 h in the darkness. The absorbance of the supernatant was measured at 646 and 663 nm. The chlorophyll a (Chla) and Chlb concentration was determined according to the methods of Lichtenthaler [23].

The soluble protein content was measured according to the procedure of Bradford [24] with a slight change. A 0.5-g frozen sample was extracted in sodium phosphate buffer (50 mmol, pH 7.8). The extracts were centrifuged at 4000 ×*g* for 10 min at 4°C. The soluble proteins were quantified using the supernatants of samples with bovine serum albumin as a standard.

Determinations of the relative electrolyte leakage rate and lipid peroxidation

To measure the electrolyte leakage, 10 leaf discs (10 mm diameter) from young, fully expanded leaves were placed in 50-mL glass vials and rinsed with distilled water to remove electrolytes that were released during the leaf disc excision. The vials were then filled with 30 mL of distilled water and allowed to stand in the dark for 24 h at room temperature. The electrical conductivity (EC_1) of the bathing solution was determined at the end of incubation period. The vials were heated in a temperature-controlled water bath at 95°C for 20 min and then cooled to room temperature, after which the electrical conductivity (EC_2) was measured. The relative electrolyte leakage (REC) was calculated using the following formula [25]:

$$REC(\%)=\left(\frac{EC_1}{EC_2}\right)\times 100.$$

The degree of lipid peroxidation was assessed based on the malondialdehyde (MDA) contents according to Zheng et al. [26]. A volume of 2 mL of the supernatant sample was combined with 2 mL of thiobarbituric acid (TBA) incubated in boiling water for 30 min and then quickly cooled in an ice bath. The mixture was centrifuged at 5000 × g for 10 min, and the absorbance of supernatant was monitored at 532, 600 and 450 nm. The MDA concentration was expressed as μmol g^{-1} using the following formula:

$$MDA(\mu molg^{-1})=6.45\times(A_{532}-A_{600})-0.56\times A_{450}$$

.

Determination of antioxidative enzyme activities

The enzymes were extracted at 4°C from approximately 0.3 g of tissue from the second leaves using a mortar and pestle with 5 mL of extraction buffer, containing 0.1 mol phosphate buffer (pH 7.8), 0.1 mol ethylenediaminetetraacetic acid (EDTA), and 1% polyvinylpyrrolidone (PVP). The extracts were centrifuged at 6000 × g for 30 min, and the supernatants were used for the enzyme assays.

The superoxide dismutase (SOD) activity was assayed by the inhibition of the photochemical reduction of β-nitroblue tetrazolium chloride (NBT). The reaction mixture (3 mL) consisted of 50 mol potassium phosphate buffer (pH 7.8) containing 0.3 μmol EDTA, 39.15 μmol methionine, 0.225 μmol nitroblue tetrazolium, 0.006 μmol riboflavin, and 0.05 mL of enzyme extract. The reaction was conducted for 15 min at 25°C under 4000 l×. One unit of SOD activity was defined as the amount of enzyme that inhibited the NBT reduction by 50% at 560 nm [27].

The CAT activity was assayed by monitoring the disappearance of hydrogen peroxide (H_2O_2) [28]. This disappearance can be detected by measuring the decrease in the absorbance at 240 nm of a reaction mixture consisting of 1.5 mL of 50 mmol sodium phosphate buffer (pH 7.8), 0.3 mL of 100 mmol H_2O_2, and 0.2 mL of enzyme extract. One CAT unit was defined as the amount of enzyme that was necessary to decompose 1 mmol H_2O_2 per minute under the aforementioned assay conditions, and the specific activity was expressed as U g^{-1} FW min^{-1}.

The POD activity was determined based on guaiacol oxidation measured at 470 nm [29]. The reaction mixture (3 mL) contained 0.05 mL of enzyme extract, 2.75 mL of 50 mmol phosphate buffer (pH 7.0), 0.1 mL of 1% H_2O_2, and 0.1 mL of a 4% guaiacol solution. The increase in absorbance at 470 nm due to guaiacol oxidation was recorded for 2 min. One unit of enzyme activity was defined as the amount of the enzyme that caused a change of 0.01 absorbance unit per minute. The specific POD activity was expressed as U g^{-1} FW min^{-1}.

Statistical analyses

The experiment was conducted according to a simple randomized block design. Each treatment was replicated three times. The data were statistically analyzed by an analysis of variance (ANOVA) using SAS 9.2 Institute Inc. (2008) software. The least significant difference (LSD) was calculated to separate the means. A simple correlation analysis was performed using the SAS program to determine the relationship between each physiological variable.

Results

Dry mass of the shoots and roots of *T. grandis* under different salt stress and SA treatments

Salt stress significantly inhibited the plant growth, leading to a pronounced reduction in the dry mass of the shoots and roots. Compared with the non-salt-treated *T. grandis* seedlings (T1), the 0.2% and 0.4% NaCl treatments significantly reduced the dry mass by 12.1% and 29% in the shoots and by 9.9% and 25.3% in the roots after 60 days, respectively (Table 1). The SA treatment reduced the decrease in the growth of the salt-stressed plants. Compared with the salt-stressed plants, SA treatment increased the dry mass of the shoots, roots, and shoot+roots by 6.2% ($P<0.05$), 5.6% ($P<0.05$) and 5.9% ($P<0.05$) under 0.2% NaCl conditons, and 16.8% ($P<0.05$), 18.2% ($P<0.05$), and 17.7% ($P<0.05$) under 0.4% NaCl conditions, respectively (Table 1). However, there were no significant differences in the dry mass of shoots and roots between different treatments with or without SA treatment under 0% NaCl stress condition. In addition, the salt-stressed plants grown under SA were larger than those without SA (Fig. 1).

Table 1. Effects of SA on the dry matter of the shoots, roots, and roots+shoots in *T. grandis* grown under salt stress (means±SD).

Treatments	Shoots (g)	Roots (g)	Shoots+Roots (g)
T1	18.4±1.15^{a}	11.9±0.68^{a}	30.3±0.49^{a}
T2	18.7±0.93^{a}	12.0±0.39^{a}	30.7±0.98^{a}
T3	16.2±1.17^{b}	10.7±0.37^{b}	26.9±1.54^{bc}
T4	17.2±1.31^{ab}	11.3±0.74^{ab}	28.5±1.62^{ab}
T5	13.1±0.92^{c}	8.88±0.56^{c}	22.0±1.47^{d}
T6	15.3±1.34^{b}	10.5±0.45^{b}	25.9±1.10^{c}

Treatments: T1, distilled water without SA; T2, distilled water with 0.5 mmol SA; T3, 0.2% NaCl without SA; T4, 0.2% NaCl with 0.5 mmol SA; T5, 0.4% NaCl without SA; and T6, 0.4% NaCl with 0.5 mmol SA. Numbers followed by different letters indicate significant differences ($P<0.05$) according to an LSD test; the same letter indicates no significant differences between the treatments, n = 5.

Relative water content and proline content of *T. grandis* under different salt stress and SA treatments

The RWC is a useful parameter when evaluating the physiological water status of plants [24]. Compared to the distilled water-treated seedlings (T1), the RWC in the *T. grandis* seedling leaves decreased by 4.50% ($P<0.05$) under 0.2% salt stress conditions and 16.70% ($P<0.05$) under 0.4% salt stress conditions. The 30-day treatment with 0.5 mmol SA reduced the effects of salt stress on the relative water content decrease in the presence of 0% and 0.2% NaCl; however, these changes were not significant ($P>0.05$). Under 0.4% NaCl stress conditions, the plants that were treated with SA had a higher RWC than those without SA treatment (Fig. 2A, $P<0.05$). The salt stress dramatically induced the accumulation of proline in leaves. SA treatment further increased the proline content. In the SA treatments, the proline content in the 0.2% and 0.4% salt-stressed plants increased by 27.6% ($P<0.05$) and 21.8% ($P<0.05$), respectively (Fig. 2B).

Gas exchange parameters in *T. grandis* under different salt and SA treatments

As shown in Table 2, salt stress reduced the gas exchange parameters, with decreases in the net assimilation rate (Pn), stomatal conductance (Gs), and transpiration rate (Tr) of the leaves of 16.2% ($P<0.05$), 50% ($P<0.05$), and 30.2% ($P<0.05$), respectively, under 0.2% NaCl conditions and 65.0% ($P<0.05$), 50% ($P<0.05$), and 59.3% ($P<0.05$) under 0.4% NaCl condition, respectively. Interestingly, the intercellular CO_2 (Ci) did not change significantly with 0.2% NaCl ($P>0.05$) but was markedly increased with 0.4% NaCl ($P<0.05$). The SA treatment increased the Pn, Tr, and Gs in leaves under salt stress. The Pn, Tr, and Gs in SA-treated plants increased by approximately 65%, 25%, and 100%, respectively, under 0.2% NaCl conditions and 146%, 95%, and 100%, respectively, under 0.4% NaCl conditions compared with the non SA-treated plants. The Ci in the SA-treated plants under 0.2% NaCl conditions was similar to that in the non SA-treated plants ($P>0.05$). However, the Ci in the SA-treated plants under 0.4% NaCl stress was markedly lower (22%) than that of the plants without SA treatment ($P<0.05$). Additionally, there were no significant differences in the Pn, Tr, Gs, and Ci values between the SA- and distilled water-treated plants.

Contents of chlorophyll and soluble protein of *T. grandis* under different salt stress and SA treatments

Chlorophyll is a biologically important pigment that is used for the photosynthetic conversion of inorganic molecules or ions into organic bio-molecules. Salt stress caused a decrease in the chl(a+b) content compared with that of the non-salt-treated seedlings. The contents of chl(a+b) decreased by 9.9% under 0.2% saline conditions and 24.1% under 0.4% saline conditions app:addword:respectively (Fig. 3A). The foliar spray of SA under salt stress conditions significantly reduced the decreases in the contents of chl(a+b) by approximately 10% ($P<0.05$) under 0.2% saline conditions and 17% ($P<0.05$) under 0.4% saline conditions compared that of the non SA-treated plants (Fig. 3A). In contrast to the change in the chlorophyll contents under saline condition, the salt stress increased the total soluble protein content. The 0.2% NaCl treatment significantly increased the soluble protein content ($P<0.05$); however, no obvious increase was found between the 0.4% NaCl- and non NaCl-stressed plants ($P>0.05$) (Fig. 3B). The addition of the SA solution significantly increased the soluble protein contents under NaCl conditions, and the soluble protein content in the SA-treated plants increased by 13.3% ($P<0.05$) under 0.2% saline conditions and 15.9% ($P<0.05$) under 0.4% saline conditions (Fig. 3B). However, the SA treatment did not affect the contents of chl(a+b) and soluble protein compared with those of the SA- and distilled water-treated seedlings ($P>0.05$).

Activities of antioxidative enzymes and cellular membrane damage of *T. grandis* under different salt stress and SA treatments

The activities of SOD, CAT, and POD in *T. grandis* seedlings were significantly affected by the salt stress and SA treatments. With the increasing extent of salt stress, the activities of SOD, CAT, and POD increased markedly and then significantly

Figure 1. The appearance of whole plants in *T. grandis* seedlings. Treatments: T1, distilled water without SA; T2, distilled water with 0.5 mmol SA; T3, 0.2% NaCl without SA; T4, 0.2% NaCl with 0.5 mmol SA; T5, 0.4% NaCl without SA; and T6, 0.4% NaCl with 0.5 mmol SA.

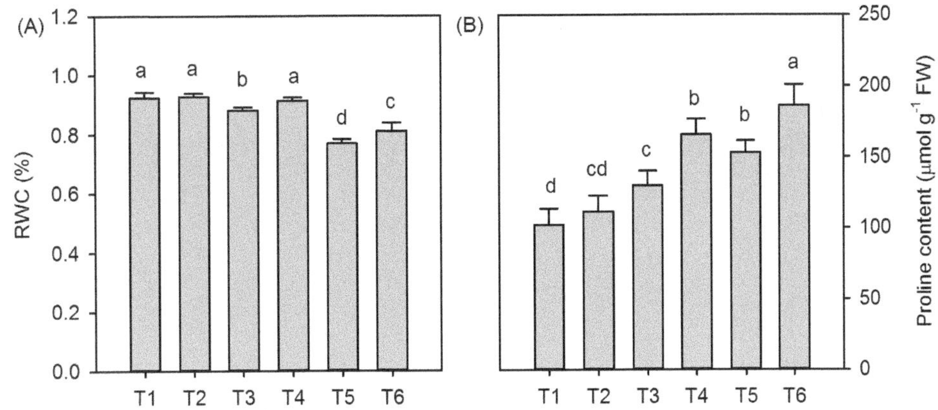

Figure 2. Effects of SA on the RWC (A) and proline content (B) in *T. grandis* grown under salt stress conditions (means ± SD). Treatments: T1, distilled water without SA; T2, distilled water with 0.5 mmol SA; T3, 0.2% NaCl without SA; T4, 0.2% NaCl with 0.5 mmol SA; T5, 0.4% NaCl without SA; and T6, 0.4% NaCl with 0.5 mmol SA. Different letters indicate significant differences ($P<0.05$) according to an LSD test; the same letter indicates no significant differences between the treatments, n = 5.

decreased (Fig. 4). The SA treatment significantly enhanced the activities of SOD, CAT, and POD under salt stress, and the increases in the activities of SOD, CAT, and POD were approximately 9.8%, 20.6%, and 14.4% under 0.2% NaCl conditions and 13.4%, 38.2%, and 18.9% at 0.4% under NaCl conditions, respectively (Fig. 4).

Salt stress caused membrane injury in the leaves of *T. grandis* seedlings (Fig. 5). The relative electrolyte leakage rate (REC) increased by 6.9% and 17.4% under 0.2% and 0.4% NaCl conditions, respectively, and the MDA content increased by 13.5% and 43.3% under 0.2% and 0.4% NaCl conditions, respectively. The SA treatments prevented lipid peroxidation and then alleviated the membrane injury of *T. grandis* seedlings. Compared with the non-SA-treated plants, the reductions in the REC and MDA content in the SA-treated seedlings were approximately 9.0% ($P<0.05$) and 12.6% ($P<0.05$) under 0.2% NaCl conditions and 12.0% ($P<0.05$) and 26.4% ($P<0.05$) under 0.4% NaCl conditions, respectively. Interestingly, no significant difference was found between the REC and MDA content between the SA- and distilled water-treated seedlings ($P>0.05$).

Relationships between antioxidant enzymatic activities and cellular membrane damage

MDA content shows linear and negative correlation with SOD and POD activities (Fig. 6A, C, $P<0.05$) likewise, REC content also shows negative correlation with SOD and POD activities (Fig. 6D, F, $P<0.05$). However, there were no significant negative relationships between MDA content and CAT activity, and also between membrane permeability and CAT activity, respectively (Fig. 6B, E, $P>0.05$).

Discussion

Salt stress constrains plant growth by adversely affecting various physiological and biochemical processes, such as photosynthesis, antioxidant phenomena, proline metabolism, and osmolyte accumulation [30,31,32,33]. Given that SA plays key roles in the regulation of plant growth, development, the interaction with other organisms, and the responses to environmental stresses [34,35]. However, there is some controversy in the literature regarding whether SA can counteract the adverse effects of salt stress [10,30]. In this study, a marked reduction in the growth of both the shoots and roots of *T. grandis* seedlings was observed under increasing salt stress (Fig. 1). These studies show that SA treatment recovered the growth rate of *T. grandis* seedlings from high salt conditions (0.4% NaCl), when compared with their corresponding non-SA applied plants (Table 1). This is in consistent with the reports of Idrees et al. [31] where it was shown that SA treatment ameliorated the adverse effects of salt stress in terms of growth parameters in *Catharanthus roseus*. Coronado et al. [36] also showed that applying aqueous solutions

Table 2. Effects of SA treatments on the net photosynthetic rate (Pn), internal carbon dioxide concentration (Ci), transpiration rate (Tr), and stomatal conductance (Gs) in *T. grandis* grown under salt stress conditions.

Treatments	Pn (μmol $m^{-2}s^{-1}$)	Ci (μmol mol^{-1})	Tr (μmol $m^{-2}s^{-1}$)	Gs (μmol $m^{-2}s^{-1}$)
T1	1.44 ± 0.14^{bc}	251.8 ± 15.8^{ab}	1.37 ± 0.098^{a}	0.02 ± 0.002^{a}
T2	1.50 ± 0.11^{b}	257.5 ± 10.7^{ab}	1.43 ± 0.086^{a}	0.02 ± 0.016^{a}
T3	1.20 ± 0.12^{d}	230.5 ± 25.4^{bc}	0.96 ± 0.068^{c}	0.01 ± 0.001^{c}
T4	1.98 ± 0.12^{a}	243.0 ± 16.0^{bc}	1.20 ± 0.129^{b}	0.02 ± 0.001^{a}
T5	0.50 ± 0.05^{e}	276.2 ± 22.5^{a}	0.56 ± 0.035^{d}	0.01 ± 0.001^{c}
T6	1.23 ± 0.15^{cd}	214.7 ± 14.7^{c}	1.09 ± 0.108^{bc}	0.02 ± 0.002^{b}

Treatments: T1, distilled water without SA; T2, distilled water with 0.5 mmol SA; T3, 0.2% NaCl without SA; T4, 0.2% NaCl with 0.5 mmol SA; T5, 0.4% NaCl without SA; and T6, 0.4% NaCl with 0.5 mmol SA. Numbers followed by different letters indicate significant differences ($P<0.05$) according to an LSD test, n = 5.

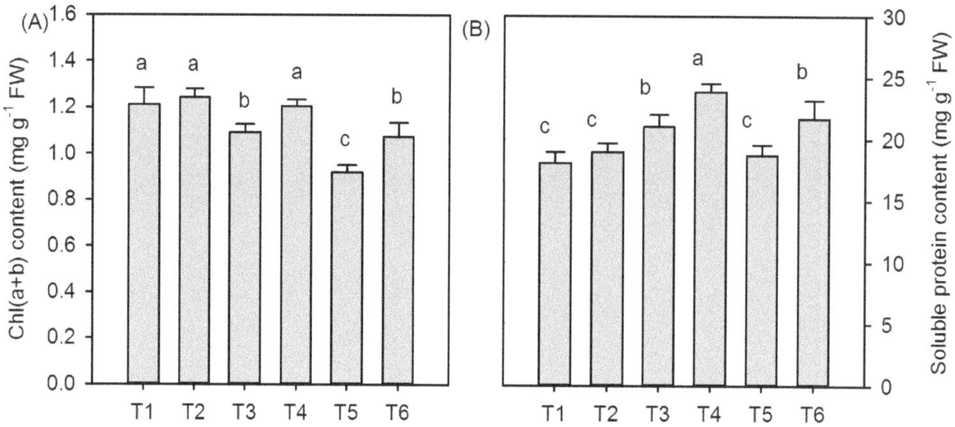

Figure 3. Effects of SA on the chl (a+b) (A) and soluble protein contents (B) in *T. grandis* grown under salt stress conditions (means ± SD). Treatments: T1, distilled water without SA; T2, distilled water with 0.5 mmol SA; T3, 0.2% NaCl without SA; T4, 0.2% NaCl with 0.5 mmol SA; T5, 0.4% NaCl without SA; and T6, 0.4% NaCl with 0.5 mmol SA. Different letters indicate significant differences ($P<0.05$) according to an LSD test; the same letter indicates no significant differences between the treatments, n = 5.

of SA as a spray solution to the shoots of soybean significantly increase the growth of the shoots and roots under greenhouse or field conditions. Interestingly, no significant differences were found in the dry mass of the shoots and roots between the plants with and without SA treatment under 0% NaCl stress conditions (Table 1), suggesting that exogenous SA did not positively affect the growth of *T. grandis* seedlings that were grown under non-saline and low-saline conditions. This result was not consistent with the report of Kaydan et al. [37], who found that a pre-sowing soaking treatment of the seeds that were treated with SA positively affected the shoot and root dry mass in wheat seedlings under both saline and non-saline conditions.

Growing leaf cells can be maintained through the accumulation of solutes. Proline accumulation is one of the most frequently reported adjustments that is induced by salt stress in plants and is often considered to be involved in maintaining the water content. Increases in the leaf RWC and accumulation of free proline in *T. grandis* seedlings after 30 days of salt stress under SA treatment might be an adaptive feature in improving its succulence and maintaining the water balance in response to salinity-induced osmotic stress (Fig. 2). A similar result was reported by Misra and Saxena [32], who found that SA alleviated the salt stress of lentil by improving the proline metabolizing system. Moreover, proline enhances the salt-tolerance of *Pancratium maritimum* L. by up-regulating stress-protective proteins [38]. The application of SA induced an increase in the protein content under both saline stress conditions (Fig. 3B). Therefore, it has been concluded that SA ameliorates the stress generated by NaCl through proline accumulation, which was involved in the synthesis of protective proteins that are necessary for the stress response. However, to test this hypothesis, further studies are necessary to determine which stress-protective protein would increase under a stress environment.

Stomatal closure is one of the earliest responses to salt stress. Compared to the non-salt-treated *T. grandis* seedlings, both the Pn and Gs in *T. grandis* seedlings under 0.2% NaCl conditions decreased by approximately 16% and 39%, respectively, suggesting that the decrease in the photosynthetic rate was associated with the closure of stomata under 0.2% NaCl stress. SA alleviated the negative effects of 0.2% NaCl stress on leaf photosynthesis by increasing Pn and Gs. Fariduddin et al. [33] also found that the exogenous application of SA could enhance the net photosynthetic rate, stomatal conductance, and transpiration rate in *B. juncea*. However, non-stomatal factors, such as enzyme-related activities at the chloroplast level, the efficiency of PSII, the concentration and activity of the Rubisco enzyme, and the supply of ATP and NADPH to the photosynthetic carbon reduction cycle might also affect the Pn in the leaves under stressful conditions [39]. Furthermore, Brugnoli and Bjökman [40] noted that both stomatal and non-stomatal limitations accounted for a reduction in photosynthesis. In this study, compared with the 0.2% NaCl treatment, the Pn further decreased, although the Gs did not obviously change in *T. grandis* seedlings under 0.4% NaCl conditions, suggesting that other non-stomatal limiting factors might affect photosynthesis under 0.4% NaCl conditions. The enhancement of the effects of SA on the Pn was attributable to its stimulatory effects on the pigment contents and Rubisco enzyme activity [41]. In this study, the SA-treated seedling leaves exhibited higher Pn and lower Ci contents, consistent with higher levels of Gs, chlorophyll, and soluble proteins than those of the 0.4% NaCl-treated plants that were treated with SA (Table 2; Fig. 3A,B). These results indicate that the application of SA induced the severely salt-stressed *T. grandis* seedlings to effectively use the available CO_2 inside the leaves and increased the Pn by up-regulating the related photosynthetic enzyme activities at the chloroplast level.

The photosynthetic electron transport system is a major source of ROS in plant tissues and has the potential to generate singlet oxygen $^1O_2$ and superoxide ($O_2.^-$). Consequently, the accumulation of reduced intermediates in the photosynthetic electron transport chain results in the formation of ROS, thereby damaging the membrane system and photosynthetic complexes [42]. Plants can activate antioxidative defense systems to protect themselves from the harmful effects of salinity-induced oxidative stress. In this study, salinity stress induced drastic changes in the activities of antioxidative enzymes, including SOD, CAT, and POD. There were dramatic increases in the activities of SOD, CAT, and POD under 0.2% salt stress conditions, although there was only a slight increase under 0.4% salt stress conditions compared to the distilled water-treated seedlings (Fig. 3), suggesting that with increasing salt stress, the ability of *T. grandis* seedlings to resist salt stress decreased. SOD and CAT, which act synergistically to resist oxidative damage induced by salt stress in *T. grandis* seedlings, were also observed in other plant tissues under NaCl stress [43].

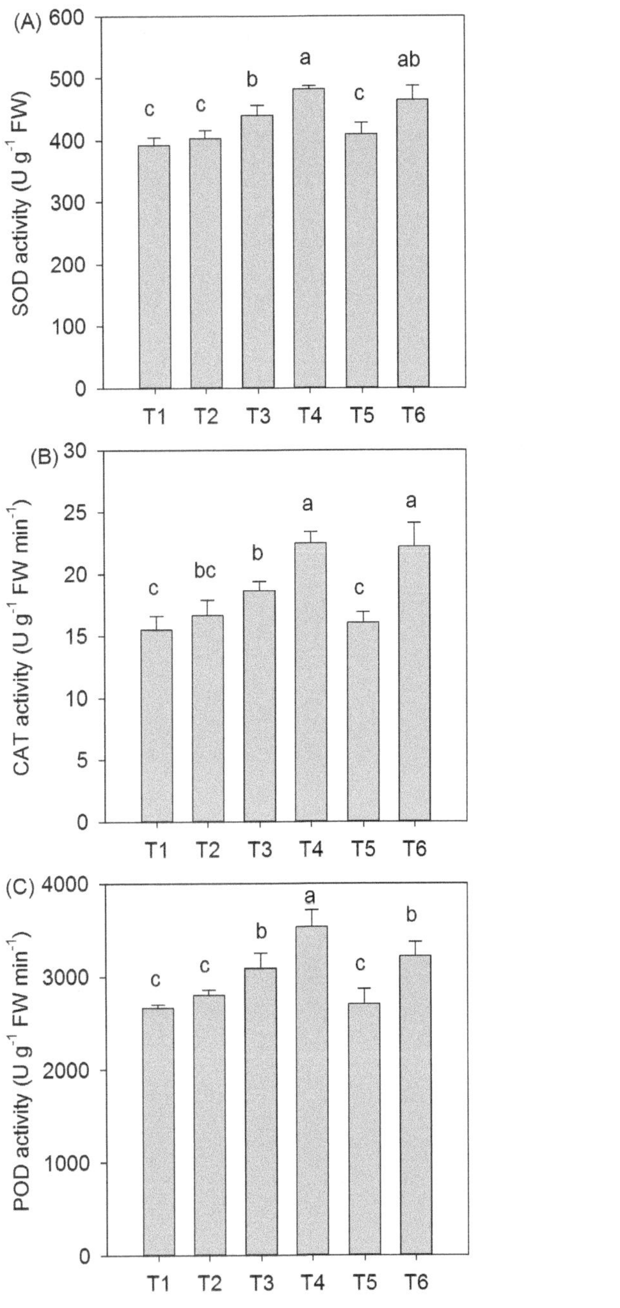

Figure 4. Influence of the SA treatments on the SOD (A), CAT (B), and POD (C) activities in *T. grandis* grown under salt stress conditions. Treatments: T1, distilled water without SA; T2, distilled water with 0.5 mmol SA; T3, 0.2% NaCl without SA; T4, 0.2% NaCl with 0.5 mmol SA; T5, 0.4% NaCl without SA; and T6, 0.4% NaCl with 0.5 mmol SA. Different letters indicate significant differences (*P<0.05*) according to an LSD test, n = 5.

After the application of exogenous SA, the plants showed higher activities of SOD, CAT, and POD than those of the non-SA-treated plants grown under 0.2% and 0.4% salt stress conditions (Fig. 4). Similarly, SA enhanced the activities of CAT, POD, and SOD when sprayed exogenously on salinity-stressed *B. juncea* plants [44].

One of the major injuries due to salinity stress is the damage of plant cell membranes, which can impair their ion selectivity. This deficiency results in the exosmosis of intracellular ions and organic matter and the infiltration of extracellular toxic ions, ultimately disrupting plant physiological and biochemical processes. The maintenance of cellular membrane integrity under salt stress is considered to be an integral part of the salinity tolerance mechanism [45]. The membrane damage caused by different abiotic stresses, including salinity, is largely mediated through the end product of lipid peroxidation (MDA) [46]. In this study, a significant positive correlation ($r = 0.98$, $p = 0.0029$) was observed between the MDA concentration and membrane permeability. An increase in the electrical conductivity with increasing salt stress indicated an elevated leakiness of ions due to a loss of membrane integrity (Fig. 4), which is in agreement with results described by Parida and Das [47]. The application of exogenous SA could ameliorate the membrane deterioration in plants under salt stress and facilitate the maintenance of membrane functions [7,45]. We also found that the SA treatment significantly reduced the increase in the electrolyte leakage rate and MDA content in *T. grandis* seedlings under salt stress (Fig. 5).

The photosynthetic pigment chlorophyll plays an important role in light absorption and energy transduction and is essential for photosynthesis. The reduction of leaf chlorophyll under high salinity has been attributed to the destruction of pigments and the instability of the pigment-protein complex [48]. In this study, NaCl treatment reduced the chlorophyll content of *T. grandis* seedlings, and there was a significantly positive and linear relationship between Pn and the chlorophyll content ($r = 0.96$, $p = 0.0099$) (Fig. 3A). The addition of SA to NaCl-stressed plants markedly increased the chlorophyll content, suggesting that the increase in chloroplast pigments under the SA treatment might be due to the ability of SA to increase the activity of certain enzymes, thereby stimulating chlorophyll biosynthesis or reducing chlorophyll degradation, leading to increased net photosynthesis under salt stress tolerance. The reduction in chlorophyll due to osmotic stress has been ascribed to the strong damage to chloroplast membranes, which increases the membrane permeability or loss of membrane integrity [49]. Indeed, a strong negative correlation between the chlorophyll and MDA contents ($r = 0.99$, $p = 0.001$) was observed in this study. Membrane oxidative damage can be prevented by antioxidative enzymes. Correlation studies were demonstrated for MDA content and REC content against the enzymatic activities of SOD, CAT and POD in *T. grandis* seedlings grown under 0.2% and 0.4% salt stress conditions. We found that both MDA content and membrane permeability shows significant negative correlation against the activities of SOD and POD (Fig. 6A,C,D,F, $P<0.05$), while there is no significant negative correlation between MDA content and CAT activity, and also between membrane permeability and CAT activity (Fig. 6B,E, $P>0.05$). Therefore, the observed increase in the growth and CO_2 assimilation under salt stress and SA treatment might be due to the enhanced activities of SOD and POD, which protect membrane integrity, leading to an increase in the amount of photosynthesis pigments and net photosynthesis rate.

In conclusion, our study showed that salt stress decrease the vegetative growth and CO_2 assimilation in *T.grandis* together with increased membrane damage and lipid peroxidation. SA efficiently ameliorated the negative effects of salt stress over the growth and photosynthesis of *T. grandis* by increasing the chlorophyll content and also enhancing the plants antioxidant enzyme mechanisms, thus alleviate the membrane oxidative damage from salinity. To our knowledge, this is the first report of salt tolerance mechanisms in *T. grandis* seedlings. The results of this study can also provide a useful guide for the conservation and large-scale plantation of *T. grandis*.

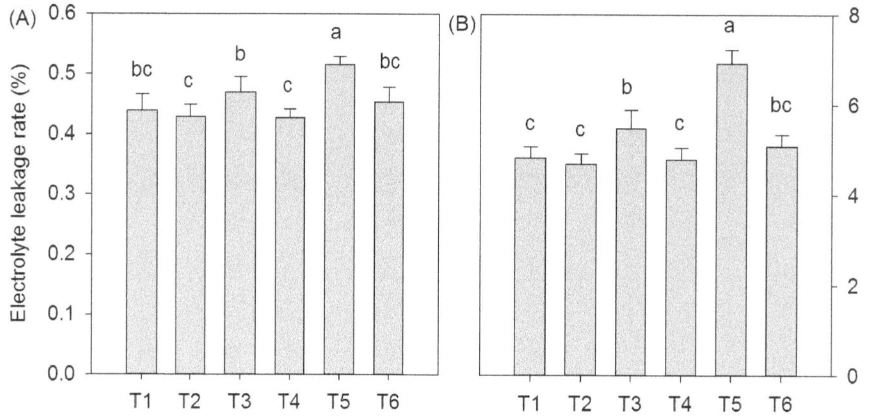

Figure 5. Effects of the SA treatments on the electrolyte leakage rate (A) and MDA contents (B) in *T. grandis* seedlings grown under salt stress conditions. Treatments: T1, distilled water without SA; T2, distilled water with 0.5 mmol SA; T3, 0.2% NaCl without SA; T4, 0.2% NaCl with 0.5 mmol SA; T5, 0.4% NaCl without SA; and T6, 0.4% NaCl with 0.5 mmol SA. Different letters indicate significant differences ($P<0.05$) according to an LSD test, n = 5.

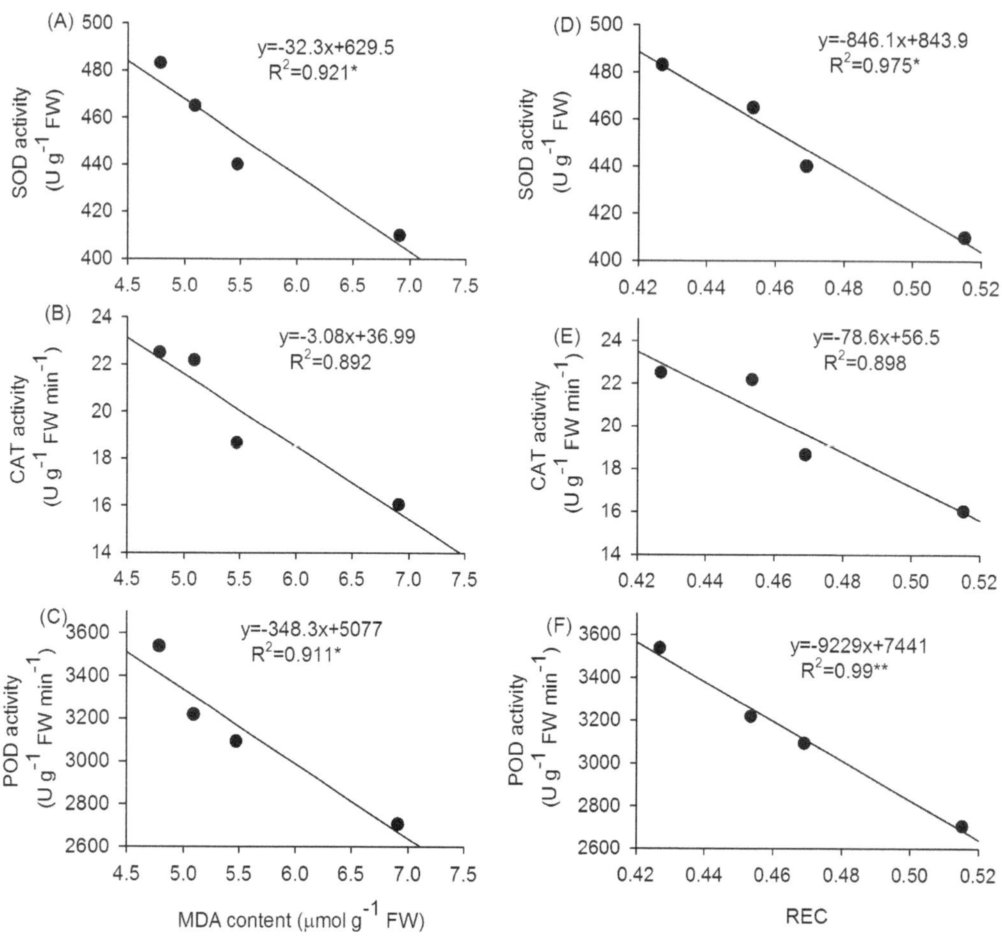

Figure 6. Relationship between MDA content and activities of SOD (A), CAT (B) and POD (C), and between REC and activities of SOD (D), CAT (E) and POD (F), respectively in *T. grandis* seedlings grown under 0.2% and 0.4% salt stress conditions. Each point represents the mean of 5 seedlings. *and **Significant at $P<0.05$ and $P<0.01$, respectively.

Acknowledgments

This work was funded by the Fruit Innovation Team Project of Zhejiang Province (2009R50033-7) and the Major Project of National Spark Plan of China (2012GA700001).

Author Contributions

Conceived and designed the experiments: JSW. Performed the experiments: TTL YYH XHD HT CHS. Analyzed the data: TTL YYH. Contributed reagents/materials/analysis tools: TTL YYH JSW. Wrote the paper: TTL YYH JSW.

References

1. Allakhverdiev SI, Sakamoto A, Nishiyama Y, Inaba M, Murata N (2000) Ionic and osmotic effects of NaCl-induced inactivation of photosystems I and II in *Synechococcus* sp. Plant Physiol 123: 1047–1056.
2. Iqbal M, Ashraf M, Jamil A, Rehman S (2006) Does seed priming induce changes in the levels of some endogenous plant hormones in hexaploid wheat plants under salt stress? Journal of Integrative Plant Biol 48: 181–189.
3. Saleh B (2012) Salt stress alters physiological indicators in cotton (Gossypium hirsutum L.). Soil Environ 31(2): 113–118.
4. He Y, Yu CL, Zhou L, Chen Y, Liu A, et al. (2014) Rubisco decrease is involved in chloroplast protrusion and Rubisco-containing body formation in soybean (*Glycine max.*) under salt stress. Plant Physiol Biochem 74: 118–124.
5. Nazar R, Iqbal N, Syeed S, Khan NA (2011) Salicylic acid alleviates decreases in photosynthesis under salt stress by enhancing nitrogen and sulfur assimilation and antioxidant metabolism differentially in two mungbean cultivars. J plant physiol 168(8): 807–815.
6. Sudhir P, Murthy SDS (2004) Effects of salt stress on basic processes of photosynthesis. Photosynthetica 42 (2): 481–486.
7. Gunes A, Inal A, Alpaslan M, Eraslan F, Bagci EG, Cicek N (2007) Salicylic acid induced changes on some physiological parameters symptomatic for oxidative stress and mineral nutrition in maize (*Zea mays* L.) grown under salinity. J Plant Physiol 164: 728–736.
8. Steduto P, Albrizio R, Giorio P, Sorrentino G (2000) Gas exchange response and stomatal and non-stomatal limitations to carbon assimilation of sunflower under salinity. Environ Exp Bot 44: 243–55.
9. Gossett DR, Millhollon EP, Lucas MC (1994) Antioxidant Response to NaCl Stress in Salt-Tolerant and Salt-Sensitive Cultivars of Cotton. Crop Sci 34: 706–714.
10. Arfan M, Athar HR, Ashraf M (2007) Does exogenous application of salicylic acid through the rooting medium modulate growth and photosynthetic capacity in two differently adapted spring wheat cultivars under salt stress? J Plant Physiol 6: 685–694.
11. Syeed S, Anjum NA, Nazar R, Iqbal N, Masood A et al. (2011) Salicylic acid-mediated changes in photosynthesis, nutrients content and antioxidant metabolism in two mustard (*Brassica juncea* L.) cultivars differing in salt tolerance. Acta Physiol Plant 33 (3): 877–886.
12. Wang LJ, Fan L, Loescher W, Duan W, Liu GJ et al. (2010) Salicylic acid alleviates decreases in photosynthesis under heat stress and accelerates recovery in grapevine leaves. BMC Plant Biol 10: 34.
13. El-Tayeb MA (2005) Response of barley grains to the interactive effect of salinity and salicylic acid. Plant Growth Regul 45: 215–24.
14. Khan MIR, Asgher M, Khan NA (2014) Alleviation of salt-induced photosynthesis and growth inhibition by salicylic acid involves glycinebetaine and ethylene in mungbean (*Vigna radiata* L.). Plant Physiol Biochem 80: 67–74.
15. Zhou XR, Zhang W, He XJ, Zhou SD, Yu Y (2012) Resources and Conservation of Torreya fargesii. J Northeast For Univ 40 (2): 42–46.
16. Huang Y, Wang J, Li G, Zheng Z, Su W (2001) Antitumor and antifungal activities in endophytic fungi isolated from pharmaceutical plants Taxus mairei, Cephalataxus fortunei and Torreya grandis. FEMS Immunol Med Mic 31: 163–167.
17. Liu AR, Zhang YB, Wang JF, Jiao M, Liu XM (2013) Appropriate nitrogen fertilizer strengthens growth and antioxidative ability of Festuca arundinacea under salt stress. Trans Chin Soc Agric Eng 29 (15), 126–135. (in Chinese).
18. Chen BQ, Cui XY, Zhao X, Zhang YH, Piao HS et al. (2006) Antioxidative and acute antiinflammatory effects of *Torreya grandis*. Fitoterapia 77:262–267.
19. Cheng XJ, Li ZJ, Yu WW, Dai WS, Fu QG (2007) Distribution and ecological characteristics of *Torreya grandis* in China. Journal of Zhejiang Forestry College 24: 383–388. (In Chinese).
20. Saeed MK, Deng Y, Parveen Z, Dai R, Ahmad W et al. (2007) Studies on the chemical constituents of *Torreya grandis* Fort. Ex Lindl. J App Sci 7: 269–273.
21. Wolff RL, Pédrono F, Marpeau AM, Christie WW, Gunstone FD (1998) The seed fatty acid composition and the distribution of Δ5-olefinic acids in the triacylglycerols of some taxaceae (*Taxus and Torreya*). J Am Oil Chem Soc 75: 1637–1641.
22. Bates LS, Waldren RP, Teare ID (1973) Rapid determination of proline for water stress studies. Plant Soil 39: 205–207.
23. Lichtenthaler HK (1987) Chlorophylls and carotenoids: Pigments of photosynthetic biomembranes. Method enzymol 148: 350–382.
24. Bradford MM (1976) A rapid and sensitive method for the quantitation of microgram quantities of protein utilizing the principle of protein-dye binding. Anal biochem 72 (1): 248–254.
25. Shi QH, Bao ZY, Zhu ZJ, Ying QS, Qian QQ (2006) Effects of different treatments of salicylic acid on heat tolerance, chlorophyll fluorescence, and antioxidant enzyme activity in seedlings of Cucumis sativa L. Plant Growth Regul 48: 127–135.
26. Zheng B, Jing S, Yan Y (2006) Research methodology of modern plant physiology and biochemistry. China Meteorological Press, Beijing, pp 189–198.
27. Costa H, Gallego SM, Tomaro ML (2002) Effect of UV-B radiation on antioxidant defense system in sunflower cotyledons. Plant Sci 162(6): 939–945.
28. Fu J, Huang B (2001) Involvement of antioxidants and lipid peroxidation in the adaptation of two cool-season grasses to localized drought stress. Environ Exp Bot. 45: 105–114.
29. González L, González-Vilar M (2003) Determination of relative water content. In: Handbook of plant ecophysiology techniques. Springer Netherlands, pp 207–212.
30. Borsani O, Valpuesta V, Botella MA (2001) Evidence for a role of salicylic acid in the oxidative damage generated by NaCl and osmotic stress in *Arabidopsis* seedlings. Plant Physiol 126: 1024–1030.
31. Idrees M, Naeem M, Aftab T, Khan MMA, Moinuddin (2011) Salicylic acid mitigates salinity stress by improving antioxidant defense system and enhances vincristine and vinblastine alkaloids production in periwinkle (*Catharanthus roseus* L. G. Don). Acta Physiol Plant 33:987–999.
32. Misra N, Saxena P (2009) Effect of salicylic acid on proline metabolism in lentil grown under salinity stress. Plant Sci 177: 181–189.
33. Fariduddin Q, Hayat S, Ahmad A (2003) Salicylic acid influences net photosynthetic rate, carboxylation efficiency, nitrate reductase activity and seed yield in Brassica juncea. Photosynthetica 41: 281–284.
34. Hayat Q, Hayat S, Irfan M, Ahmad A (2010) Effect of exogenous salicylic acid under changing environment: A review. Environmental and Experimental Botany 68: 14–25.
35. Senaratna T, Touchell D, Bunn E, Dixon K (2000) Acetyl salicylic acid (aspirin) and salicylic acid induce multiple stress tolerance in bean and tomato plants. Plant Growth Regul 30: 157–161.
36. Coronado MAG, Lopez CT, Saavedra AL (1998) Effects of salicylic acid on the growth of roots and shoots in soybean, Plant Physiol Biochem 8: 563–565.
37. Kaydan D, Yagmur M, Okut N (2007) Effects of Salicylic acid on the growth and some physiological characters in salt stressed wheat (*Triticum aestivum* L.). Tarim Bilim Derg 13 (2): 114–119.
38. Khedr AHA, Abbas MA, Wahid AAA, Quick WP, Abogadallah GM (2003) Proline induces the expression of salt-stress-responsive proteins and may improve the adaptation of *Pancratium maritimum* L. to salt-stress. J Exp Bot 54 (392): 2553–2562.
39. Lawlor DW, Cornic G (2002) Photosynthetic carbon assimilation and associated metabolism in relation to water deficits in higher plants. Plant Cell Environ 25: 275–94.
40. Brugnoli E, Björkman O (1992) Growth of cotton under continuous salinity: influence on allocation pattern, stomatal and light energy. Planta 187: 335–347.
41. Khodary SEA (2004) Effect of salicylic acid on the growth, photosynthesis and carbohydrate metabolism in salt-stressed maize plants. Inter J Agric Biol 6(1): 5–8.
42. Asada K (2006) Production and scavenging of reactive oxygen species in chloroplasts and their functions. Plant Physiol 141: 391–396.
43. Tang D, Shi S, Li D, Hu C, Liu Y (2007). Physiological and biochemical responses of Scytonema javanicum (*cyanobacterium*) to salt stress. J Arid Environ 71(3): 312–320.
44. Yusuf M, Hasan SA, Ali B, Hayat S, Fariduddin Q et al. (2008) Effect of salicylic acid on salinity-induced changes in Brassica juncea. J Integr Plant Biol 50: 1096–1102.
45. Stevens J, Senaratna T, Sivasithamparam K (2006) Salicylic acid induces salinity tolerance in tomato (*Lycopersicon esculentum cv. 'Roma'*): associated changes in gas exchange, water relations and membrane stabilisation. Plant Growth Regul 49: 77–83.
46. Mishra A, Choudhuri MA (1999) Effects of salicylic acid on heavy metal-induced membrane deterioration mediated by lipoxygenase in rice. Biol Plantarum 42: 409–15.
47. Parida AK, Das AB (2005) Salt tolerance and salinity effects on plants: a review. Ecotox Environ Safe 60: 324–349.
48. Jaleel CA, Sankar B, Sridharan R, Panneerselvam R (2009) Soil salinity alters growth, chlorophyll content, and secondary metabolite accumulation in *Catharanthus roseus*. Turk J Biol 32(2): 79–83.
49. Kaiser WM, Heber U (1981) Photosynthesis under osmotic stress: effect of high solute concentrations on the permeability properties of the chloroplast envelope and on activity of stroma enzymes. Planta 153: 423–429.

14

Identification and Characterization of Wilt and Salt Stress-Responsive MicroRNAs in Chickpea through High-Throughput Sequencing

Deshika Kohli[1], Gopal Joshi[2], Amit Atmaram Deokar[1], Ankur R. Bhardwaj[2], Manu Agarwal[2], Surekha Katiyar-Agarwal[3], Ramamurthy Srinivasan[1], Pradeep Kumar Jain[1]*

1 NRC on Plant Biotechnology, IARI Campus (PUSA), New Delhi, India, **2** Department of Botany, North campus, University of Delhi, Delhi, India, **3** Department of Plant Molecular Biology, South Campus, University of Delhi, Delhi, India

Abstract

Chickpea (*Cicer arietinum*) is the second most widely grown legume worldwide and is the most important pulse crop in the Indian subcontinent. Chickpea productivity is adversely affected by a large number of biotic and abiotic stresses. MicroRNAs (miRNAs) have been implicated in the regulation of plant responses to several biotic and abiotic stresses. This study is the first attempt to identify chickpea miRNAs that are associated with biotic and abiotic stresses. The wilt infection that is caused by the fungus *Fusarium oxysporum* f.sp. *ciceris* is one of the major diseases severely affecting chickpea yields. Of late, increasing soil salinization has become a major problem in realizing these potential yields. Three chickpea libraries using fungal-infected, salt-treated and untreated seedlings were constructed and sequenced using next-generation sequencing technology. A total of 12,135,571 unique reads were obtained. In addition to 122 conserved miRNAs belonging to 25 different families, 59 novel miRNAs along with their star sequences were identified. Four legume-specific miRNAs, including miR5213, miR5232, miR2111 and miR2118, were found in all of the libraries. Poly(A)-based qRT-PCR (Quantitative real-time PCR) was used to validate eleven conserved and five novel miRNAs. miR530 was highly up regulated in response to fungal infection, which targets genes encoding zinc knuckle- and microtubule-associated proteins. Many miRNAs responded in a similar fashion under both biotic and abiotic stresses, indicating the existence of cross talk between the pathways that are involved in regulating these stresses. The potential target genes for the conserved and novel miRNAs were predicted based on sequence homologies. miR166 targets a HD-ZIPIII transcription factor and was validated by 5′ RLM-RACE. This study has identified several conserved and novel miRNAs in the chickpea that are associated with gene regulation following exposure to wilt and salt stress.

Editor: Xianlong Zhang, National Key Laboratory of Crop Genetic Improvement, China

Funding: This study was supported by a grant from ICAR-Network Project on Trasgenic Crops (Functional Genomics Component) no. 2049/3003. The funder had no role in study design, data collection and analysis, decision to publish, or preparation of the manuscript.

Competing Interests: The authors have declared that no competing interests exist.

* Email: pkjain@nrcpb.org

These authors contributed equally to this work.

Introduction

MicroRNAs (miRNAs) are small, endogenous, non-coding RNAs that are present in animals, plants and some viruses. These RNAs participate in the regulation of target genes by binding to complementary mRNAs, resulting in either their cleavage or translational repression. miRNAs are involved in diverse processes in different organisms, including developmental timing in worms, cell death and fat metabolism in flies, hematopoiesis in mammals and leaf development, floral patterning and environmental stress responses in plants [1].

MIRNA genes are transcribed as independent transcriptional units by RNA polymerase II enzymes to generate primary miRNAs (pri-miRNAs). pri-miRNAs form imperfect folded structures that are processed by Dicer-like1 nuclease (a member of the RNase III endonuclease family) to precursor miRNAs (pre-miRNAs). The secondary structures of these precursors are well conserved in plants. The pre-miRNA contains a miRNA-star miRNA (miRNA*) intermediate duplex from which the miRNA* eventually is degraded. However, recent studies have revealed the higher accumulation of miRNA* under certain conditions in plants, indicating the probable role of miRNAs in modulating plant growth and development [2]. Mature miRNAs are 19 to 24 nucleotides (nt) in length and interact with an RNA-induced silencing complex (RISC) to cleave specific target mRNAs or inhibit their translation. This complementarity plays an important role in determining the fate of the mRNA. When the complementarity between the miRNA and mRNA is perfect or near perfect, the mRNA is cleaved; however, if there are many mismatches between them, translational repression occurs. There are also instances in which miRNAs and mRNAs with perfect

Table 1. Distribution of the sequenced reads in the control, wilt- and salt-stressed chickpea libraries.

Library	Control		Wilt Stress		Salt Stress	
	Total	Unique	Total	Unique	Total	Unique
Total number of sequences	15,744,289	6,103,870	7,007,282	2,677,947	6,418,892	3,353,754
Sequences remaining after 3' adaptor removal (TCGTAT)	14,782,514	5,033,168	6,412,190	2,130,060	5,761,024	2,683,044
Sequences remaining after size-range filtering (16 to 30 nt)	14,689,499	5,005,568	6,192,469	2,054,560	5,584,159	2,627,840
SSR/TR	1414	757	222	216	359	307
t/rRNA	720,912	35,136	645,529	37,632	538,049	32,987
Putative small RNA population	**13,940,841**	**4,946,095**	**5,546,371**	**2,016,408**	**5,034,576**	**2,584,961**

complementarities lead to the repression of translation and not to the usual cleavage.

The first identified miRNAs were lin4 and let7 in *Caenorhabditis elegans*, which is a model nematode [3,4]. The first plant miRNAs were identified in Arabidopsis [5] and later in other plants. Currently, 7,321 mature miRNAs have been reported in 72 plant species (miRBase version 20) [6]. Among dicots, the maximum number of miRNAs occurs in the legume family (1,460), followed by Brassicaceae (863). Although the legume family has the best representation in terms of the number of miRNAs, chickpea is a notable omission from the list.

Chickpea (*Cicer arietinum*) is the world's second most widely grown legume and is cultivated in more than 40 countries. The Indian subcontinent is the principal chickpea-producing and consuming region, contributing almost 70% of the world's total production [7]. Chickpea seeds are a rich source of protein and starch for the human population and the records of chickpea cultivation date back to 6,000 BC. Globally, chickpea is grown on 11.5 million hectares (ha) to produce 10.4 million tons with an average yield of approximately 0.9 t/ha, which is far below its yield potential of 6 t/ha under optimal growth conditions [7]. The disparity between the actual and potential yields can be explained by large numbers of biotic and abiotic stresses that adversely affect its productivity. Among the biotic stresses, wilt infection that is caused by the fungus *Fusarium oxysporum* f.sp. *ciceris* is a major concern. Abiotic stress conditions, such as terminal drought and salt stress, also lead to major losses. ICC4958 is a drought tolerant chickpea cultivar and gets affected at terminal drought, which occurs at the pod filling and seed-developing stage of the crop [8,9]. However, recent studies on salinity tolerance and ion accumulation in chickpea have revealed it as a highly sensitive crop to salinity when compared to other species in cropping systems [10,11]. Thus, salinity is another major constraint in chickpea yield. A better understanding of genes and their interactions with the environment can play a very important and determinant role in tackling these stress conditions. The recently available transcriptome and genome sequences that have been reported by independent groups are important resources that will facilitate the attainment of these goals in the chickpea [12,13,14]. Hu et al. (2013) identified 28 potential miRNA candidates belonging to 20 families from 16 ESTs and 12 GSSs in the chickpea using a comparative genome-based computational analysis [15]. A total of 664 miRNA targets were predicted, including genes encoding transcription factors (TFs) in addition to those that function in the stress response, signal transduction, methylation and a variety of other metabolic processes. These findings lay the foundation for the elucidation of miRNA function in the development and stress responses of the chickpea.

miRNAs have been discovered primarily using direct cloning and bioinformatic approaches. All of the miRNAs in plants have been identified via the cloning of small RNAs or a computational approach, in which the homologs of known miRNAs are searched. We have generated small RNA libraries corresponding with the control conditions, Fusarium wilt infection and salt stress, which were sequenced using the Illumina sequencing platform to identify miRNAs in the chickpea. This study is the first report in which small RNA libraries have been constructed and sequenced to identify miRNAs in the chickpea.

Results

Sequence analyses

Three separate small RNA libraries that were constructed from the total RNA of the control, Fusarium wilt-infected and salt-stressed plants were subjected to Illumina Solexa sequencing. This sequencing generated 29,170,463 raw reads, which after processing by UEA sRNA workbench 2.4- Plant version sequence file pre-processing tool (http://srna-tools.cmp.uea.ac.uk/), produced approximately 12,135,571 total unique reads (Table 1). After removing the adaptor sequences, filtering the low-quality tags and eliminating the t/rRNA sequences, the putative small RNA population accounted for approximately 88.5%, 79.1% and 78.4% in the control, wilt-infected and salt-stressed libraries, respectively (Figure S1). The majority of small RNAs (approximately 50%) from the control and salt-stressed libraries were 24 nt in length (Figure 1), which is similar to other plant species, such as *Arabidopsis thaliana*, *Solanum lycopersicum* and *Medicago truncatula* [16,17,18]. Notably, in the wilt-infected library, small RNAs that were 20 nt in length accounted for 20% of the population, but when unique reads were analyzed, the small RNA distribution revealed that 24 nt was the major size class. Similar patterns have been reported in cucumber and soybean [19,20]. In soybean, the unique and redundant sequence classes possessed 24 nt and 23 nt long small RNAs, respectively, as the most abundant sequences. For the differential expression analysis, the total numbers of miRNA reads in each given sample were normalized as transcripts per million, and the fold changes between the treated and control samples were calculated. Out of 122 conserved miRNAs, 44 were upregulated in response to wilt stress, but in the case of salt stress, the number of down regulated miRNAs was greater than that which was observed in response to wilt stress. However, the differential expression of novel miRNAs under both of these stresses showed relatively similar patterns, with

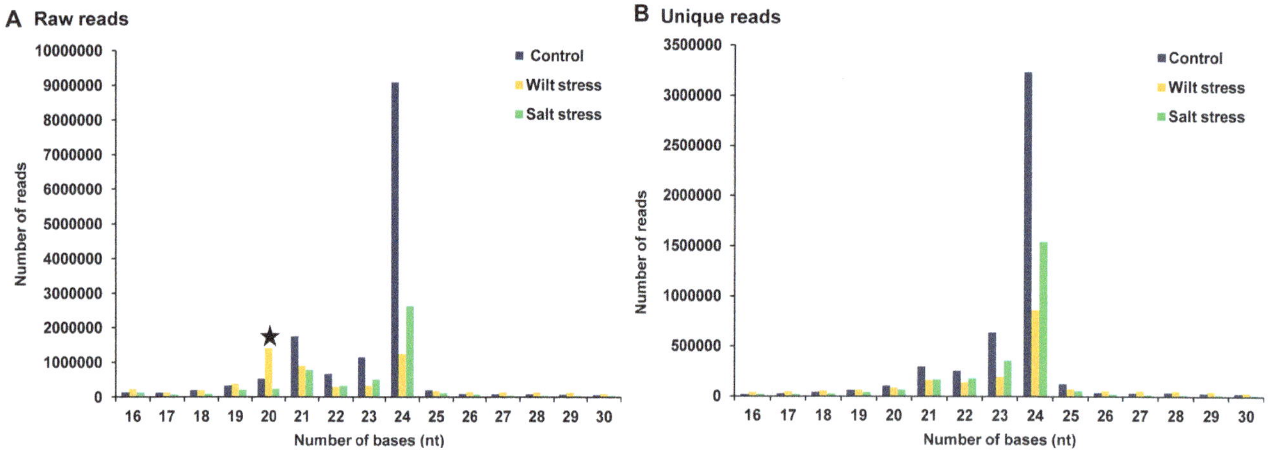

Figure 1. Length distribution of small RNA population. Size distributions of the miRNAs in the three chickpea libraries. In the wilt stress library, 20 nt miRNAs are more frequent than 24 nt miRNAs. However, in the other two libraries 24 nt miRNAs are more frequent.

approximately 60% of the miRNAs being down regulated under either wilt and/or salt stress (Figure 2a, b).

Identification of conserved miRNAs in chickpea

The unique reads that were obtained from the miRCat analysis tool (UEA small RNA workbench) were mapped to the miRNAs that were available in miRBase version 18 (http://www.mirbase.org/) [21,22,23]. The small RNA sequences that matched the known miRNAs from the miRBase database were identified as conserved miRNAs in the chickpea. The sequence analyses revealed the presence of 122 miRNAs belonging to 25 conserved families. The most abundant family was miR156 with 14 members. Among the others, miR171 (12 members), miR169 and miR172 (9 members each), miR166 and miR167 (8 members each), miR319 and miR399 (6 members each) and miR396 (5 members) were present. The remaining miRNA families had less than five members, with some families, such as miR530, miR162, miR5232 and miR408, being represented by only one member (Table 2; Table S1). In a recent report of the chickpea genome, the sequences of 20 unique miRNA families were reported, of which MIR169_2 and MIR159 were the most abundant [14]. In our study, out of 25 conserved miRNAs families, 16 possessed miRNA* sequences, thus providing additional evidence in support of the authenticity of the miRNAs. However, no miRNA* sequences were obtained for miR2111, miR162, miR164, miR390, miR394, miR397, miR530, miR408 or miR5213. Detailed information regarding the precursor structures of the conserved miRNAs is provided in Table S2.

Identification of legume-specific miRNAs in chickpea library

We obtained four legume-specific miRNAs, including miR2111, miR2118, miR5213 and miR5232, in our libraries that were previously reported in another legume, Medicago [24,25]. To date, miR5232 has only been reported in Medicago in a study involving miRNA regulation during arbuscular mycorrhizal symbiosis [25]. Accordingly, miR5232 may be a legume-specific miRNA that is involved in the biotic stress response. The multiple sequence alignment of the mature miRNAs in addition to the precursor sequences of these four legume-specific miRNAs revealed that they were most closely similar to Medicago and consequently has been conserved throughout evolution (Figure 3a, 3b). However, in recent studies, sequences that are similar to miR2118 have been reported in other non-leguminous plant systems, such as tomato and rice [26]. Apart from Fabaceae, miR2118 family members are most abundant in the Rutaceae and Solanaceae plant families [27]. Even the nomenclature of the miR2118 family is inconsistent in miRBase: the miR2118-like sequences have been disparately named miR482 (sly-miR482), miR5300 (as in tomato) and miR2809. The variation in the miR2118 sequence is species specific. Thus, miR2118 sequences in the chickpea are more similar with mtr-miR2118a [24] in comparison with other plant systems.

Identification of novel miRNAs in chickpea

We identified 59 novel miRNAs using the miRCat module of the UEA sRNA workbench, which aligned the pooled reads from all three of the libraries to the chickpea genome (NCBI Genome: PRJNA175619) [13], the ESTs database from NCBI and transcriptome data from the chickpea transcriptome database [28], and applied prediction criteria for plant miRNAs [29] (Table 3; Table S3). The low abundance of novel miRNAs in our data supports the earlier notion of the lower expression levels of novel miRNAs compared with those of conserved miRNAs [30]. The precursor miRNA candidates were then tested using RandFold with a cutoff of 0.1. The minimum free energy that was required to form the predicted hairpin structure for the precursor was in the range of −97.2 to −26.03 Kcal/mol, which is similar to the values that were reported for the precursors of other plant species (Table S4). The secondary structures of the precursors of five validated novel miRNAs were evaluated using the Mfold software (Figure 4) [31]. The data analysis revealed the presence of miRNA* sequences for all of the 59 novel miRNAs of the chickpea. The miRNA* supports the release of the miRNA duplex from the predicted hairpin structure [32]; therefore, the presence of miRNA* sequences further supports the identity of these small RNA sequences in our libraries as novel miRNAs.

Expression patterns of known and novel miRNAs in chickpea

Total RNA from the tissues of control, wilt-infected and salt-stressed plants were used to validate the miRNAs. The poly (A)RNA of these three samples was reverse transcribed into cDNA for the validation of the expression of eleven conserved and five

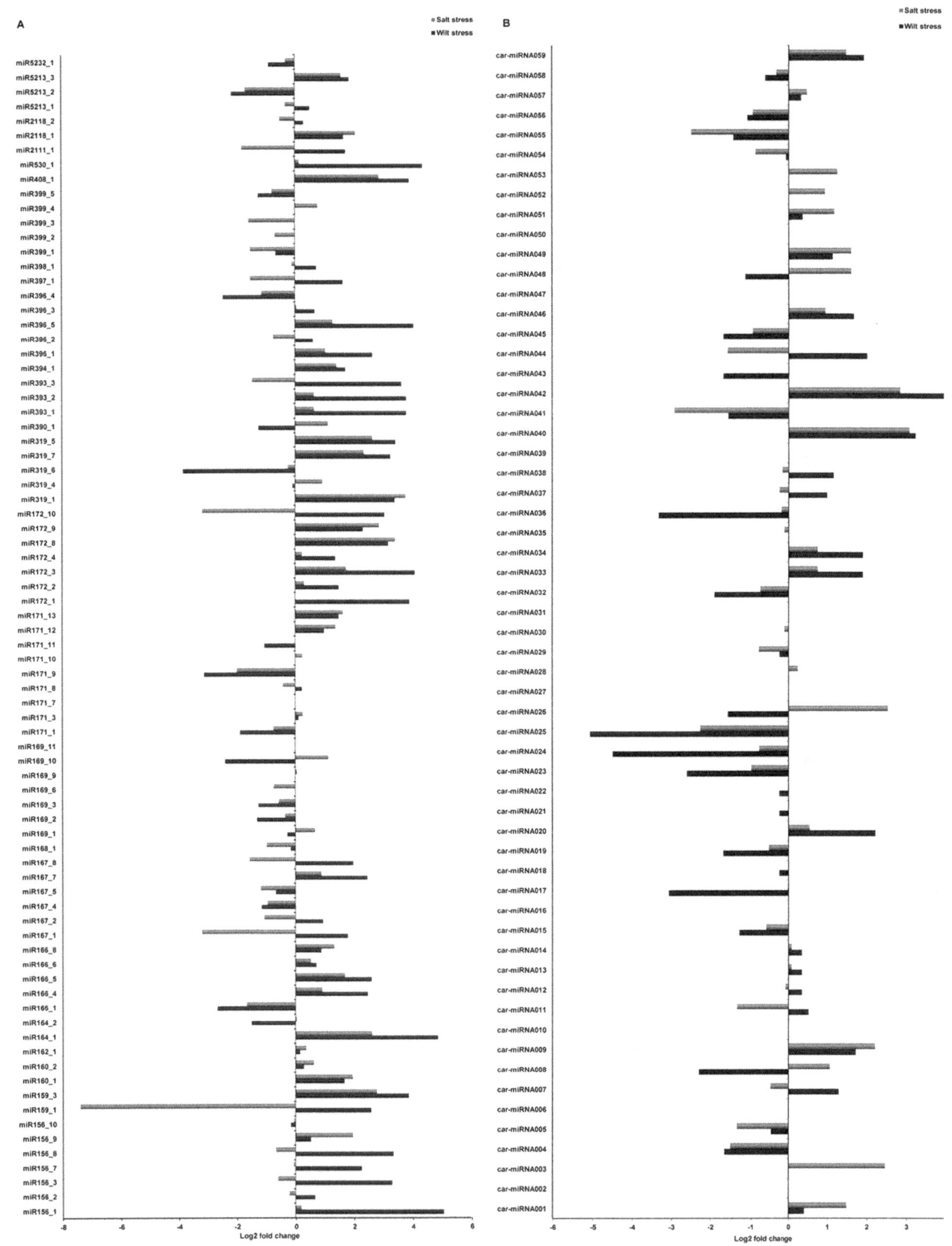

Figure 2. Differential expression patterns of chickpea miRNAs under wilt and salt stresses. (A) Conserved miRNAs, (B) novel miRNAs.

Table 2. Conserved miRNAs in chickpea.

miR_ID	miR Family	Sequence	L	Conserved in					miRNA*	Start	End	PL	MFE	Adjusted	Hairpin
				gma	mtr	vun	ath	os	Y/N					MFE	G/C%
miR156_1	mir156	UUGACAGAAGAGAGAGAGCAC	21	+	-	-	-	-	N	52925	52945	85	-41.8	-49.176468	44.705883
miR156_2		UUGACAGAAGAUAGAGGGCAC	21	+	-	-	-	-	N	17629788	17629808	109	−42.6	−39.08257	38.532112
miR156_3		UGACAGAAGAGAGUGAGCAC	20	+	+	+	+	+	Y	44874629	44874648	92	−53.1	−57.71739	44.56522
miR156_4		UGACAGAAGAGAGUGAGCAC	20						Y	35818615	35818634	91	−52.8	−58.021973	43.956043
miR156_5		UGACAGAAGAGAGUGAGCAC	20						N	35226603	35226622	85	−39.6	−46.588234	42.352943
miR156_6		UGACAGAAGAGAGUGAGCAC	20						N	1284221	1284240	145	−65.1	−44.896553	37.931034
miR156_7		UGACAGAAGAGGGUGAGCAC	20	+	+	+	-	-	N	3228450	3228469	77	−30.7	−39.870132	32.467533
miR156_8		UGACAGACGAGAGUGAGCAC	20	+	-	+	-	-	N	7424	7443	91	−52.7	−57.912086	42.857143
miR156_9		UUGACAGAAGAUAGAAAGCAC	21	+	+	+	+	-	N	2202	2222	94	−42.7	−45.42553	39.361702
miR156_10		UUGACAGAAGAUAGAGAGCAC	21	+	+	+	-	-	Y	134	154	155	−60.3	−38.903225	32.258064
miR156_11		UUGACAGAAGAUAGAGAGCAC	21						Y	2528278	2528298	81	−47.1	−58.148148	39.506172
miR156_12		UUGACAGAAGAUAGAGAGCAC	21						N	25200808	25200828	82	−45.9	−55.97561	32.92683
miR156_13		UUGACAGAAGAUAGAGAGCAC	21						N	14260671	14260691	100	−46.4	−46.4	34
miR156_14		UUGACAGAAGAUAGAGAGCAC	21						N	14160024	14160044	119	−51.1	−42.941174	35.294117
miR159_1	mir159	UUUGGAUUGAAGGGAGCUCUA	21	+	+	-	+	-	Y	14884652	14884672	195	−97.2	−49.846153	38.46154
miR159_2		UUUGGAUUGAAGGGAGCUCUA	21						Y	11393460	11393480	195	−94.2	−48.30769	38.97436
miR159_3		AUUGGAGUGAAGGGAGCUCCA	21	+	+	-	-	-	N	13935752	13935772	188	−83.4	−44.361702	41.489365
miR160_1	mir160	UGCCUGGCUCCCUGAAUGCCA	21	-	+	-	-	+	N	6658138	6658158	85	−42.8	−50.352943	40
miR160_2		UGCCUGGCUCCCUGUAUGCCA	21	+	+	+	+	+	Y	32984804	32984824	86	−45.6	−53.023254	48.837208
miR160_3		UGCCUGGCUCCCUGUAUGCCA	21	+	+	+	+	+	Y	10315830	10315850	86	−47.1	−54.76744	47.674416
miR160_4		UGCCUGGCUCCCUGUAUGCCA	21	+	+	+	+	+	N	19971530	19971550	86	−48.8	−56.744186	48.837208
miR162_1	mir162	UCGAUAAACCUCUGCAUCCAG	21	+	+	+	+	+	N	7679640	7679660	124	−47.5	−38.30645	42.741936
miR164_1	mir164	UGGAGAAGCAGGGCACAUGCU	21	-	+	-	-	-	N	33442678	33442698	75	−34.34	−45.786667	45.333336
miR164_2		UGGAGAAGCAGGGCACGUGCA	21	+	+	+	+	+	N	20759462	20759482	175	−68.1	−38.914284	39.42857
miR164_3		UGGAGAAGCAGGGCACGUGCA	21						N	38007519	38007539	86	−37.3	−43.372093	41.860466
miR164_4		UGGAGAAGCAGGGCACGUGCA	21						N	7727761	7727781	118	−45.93	−38.92373	47.457626
miR166_1	mir166	UCGGACCAGGCUUCAUUCCCC	21	+	+	-	+	+	Y	47802209	47802229	101	−42.15	−41.732674	47.524754
miR166_2		UCGGACCAGGCUUCAUUCCCC	21						Y	34613199	34613219	80	−41.9	−52.375	47.5
miR166_3		UCGGACCAGGCUUCAUUCCCC	21						N	2982935	2982955	98	−46.2	−47.142857	41.836735
miR166_4		UCGGACCAGGCUUCAUUCCCG	21	+	-	-	-	-	Y	609780	609800	163	−59.7	−36.625767	34.969322
miR166_5		UCGGACCAGGCUUCAUUCCCU	21	-	-	-	-	+	N	10985854	10985874	168	−60.8	−36.190475	39.88095
miR166_6		UCGGACCAGGCUUCAUUCCUC	21	-	+	-	-	+	N	45773545	45773565	94	−39.78	−42.31915	46.80851
miR166_7		UCGGACCAGGCUUCAUUCCUC	21						N	788963	788983	99	−48.07	−48.555557	40.40404

Table 2. Cont.

miR_ID	miR Family	Sequence	L	Conserved in					miRNA*	Start	End	PL	MFE	Adjusted	Hairpin
				gma	mtr	vun	ath	os	Y/N					MFE	G/C%
miR166_8		UCUCGGACCAGGCUUCAUUCC	21	+	-	-	-	-	Y	609782	609802	163	−59.7	−36.625767	34.969322
miR167_1	mir167	UGAAGCUGCCAGCAUGAUCU	20	+	+	-	-	-	N	32102772	32102791	71	−37.4	−52.676056	39.43662
miR167_2		UGAAGCUGCCAGCAUGAUCUA	21	+	+	-	+	+	N	30873622	30873642	161	−67.2	−41.73913	37.8882
miR167_3		UGAAGCUGCCAGCAUGAUCUA	21						N	1691349	1691369	101	−48.9	−48.415844	39.603962
miR167_4		UGAAGCUGCCAGCAUGAUCUG	21	+	+	-	-	+	N	34009600	34009620	210	−79.6	−37.90476	32.380955
miR167_5		UGAAGCUGCCAGCAUGAUCUGA	22	+	-	-	-	-	Y	5038949	5038970	102	−49.1	−48.137253	39.215687
miR167_6		UGAAGCUGCCAGCAUGAUCUGA	22						Y	38728619	38728640	108	−44.93	−41.601852	48.148148
miR167_7		UGAAGCUGCCAGCAUGAUCUGG	22	-	-	-	+	-	Y	2509	2530	158	−57.8	−36.58228	42.405064
miR167_8		UGAAGCUGCCAGCAUGAUCUUA	22						N	57870929	57870950	75	−36.3	−48.4	45.333336
miR168_1	mir168	UCGCUUGGUGCAGGUCGGGAA	21	+	+	+	+	-	Y	7328597	7328617	136	−65.9	−48.455883	55.88235
miR169_1	mir169	AGCCAAGGAUGACUUGCCGG	20	+	+	+	+	+	N	19659099	19659118	86	−38.3	−44.53488	46.51163
miR169_2		CAGCCAAGGAUGACUUGCCGA	21	+	+	-	+	+	N	10422873	10422893	200	−83.7	−41.85	34
miR169_3		CAGCCAAGGAUGACUUGCCGG	21	+	+	+	+	+	N	2353405	2353425	95	−40.7	−42.842106	46.31579
miR169_4		CAGCCAAGGAUGACUUGCCGG	21						N	3920369	3920389	135	−63.5	−47.037037	44.444447
miR169_5		CAGCCAAGGAUGACUUGCCGG	21						Y	2526196	2526216	119	−50.17	−42.15966	30.252102
miR169_6		CAGCCAAGGGUGAUUUGCCGG	21	+	+	-	-	-	N	19606166	19606186	134	−57.6	−42.985073	40.298508
miR169_9		GAGCCAAGGAUGACUUGCCGG	21	-	+	-	-	-	N	19659098	19659118	86	−38.3	−44.53488	46.51163
miR169_10		UGAGCCAGGAUGACUUGCCGG	21	-	+	-	-	-	Y	19609060	19609080	76	−37.4	−49.21053	42.105263
miR169_11		CAGCCAAGGAUAACUUGCCGG	21	+	+	+	+	+	N	4770	4790	94	−38.5	−40.957447	43.617023
miR171_1	mir171	UGAUUGAGCCGCGUCAAUAUC	21	-	+	-	-	-	N	14153792	14153812	102	−47.7	−46.76471	40.19608
miR171_3		UGAUUGAGCCGUGCCAAUAUC	21	+	+	-	+	+	Y	1004434	1004454	97	−49.6	−51.13402	36.082474
miR171_4		UGAUUGAGCCGUGCCAAUAUC	21						N	5540425	5540445	78	−34.4	−44.102566	34.615387
miR171_5		UGAUUGAGCCGUGCCAAUAUC	21						N	15207158	15207178	94	−40.2	−42.765957	39.361702
miR171_6		UGAUUGAGCCGUGCCAAUAUC	21						N	6250414	6250434	117	−45.7	−39.05983	30.769232
miR171_7		UGAUUGAGUCGUGCCAAUAUC	21	-	+	-	-	-	N	73	93	77	−32.6	−42.337658	35.064934
miR171_8		AGAUAUUGGUGCGGUUCAAUC	21	+	-	-	-	-	Y	36014922	36014942	102	−52.4	−51.37255	38.235294
miR171_9		CGAUGUUGGUGAGGUUCAAUC	21	+	-	-	-	-	Y	27995981	27996001	95	−39.8	−41.894737	40
miR171_10		UUGAGCCGCGCCAAUAUCAC	20	-	-	-	-	-	N	1567139	1567158	92	−41.5	−45.108696	47.826088
miR171_11		UUGAGCCGCGCCAAUAUCACU	21	+	-	-	-	-	Y	6636919	6636939	94	−42.6	−45.319145	39.361702
miR171_12		UUGAGCCGUGCCAAUAUCAC	20	-	-	-	-	-	N	14104770	14104789	85	−35.7	−42	36.47059
miR171_13		UUGAGCCGUGCCAAUAUCACA	21	+	-	-	-	-	N	5540422	5540442	78	−34.4	−44.102566	34.615387
miR172_1	mir172	AGAAUCCUGAUGAUGCUGCAG	21	-	+	-	-	-	N	33800439	33800459	134	−69.8	−52.089554	38.059704
miR172_2		AGAAUCUUGAUGAUGCUGCA	20	-	-	-	+	-	N	1018482	1018501	106	−48.7	−45.943398	40.56604
miR172_3		AGAAUCUUGAUGAUGCUGCAG	21	-	-	-	+	-	Y	1018481	1018501	106	−48.7	−45.943398	40.56604

Table 2. Cont.

miR_ID	miR Family	Sequence	L	Conserved in					miRNA*	Start	End	PL	MFE	Adjusted	Hairpin
				gma	mtr	vun	ath	os	Y/N					MFE	G/C%
miR172_4		AGAAUCUUGAUGAUGCUGCAU	21	+	+	+	+	+	Y	11960794	11960814	108	−47.4	−43.88889	29.62963
miR172_5		AGAAUCUUGAUGAUGCUGCAU	21						Y	28871865	28871885	112	−47.51	−42.419643	39.285713
miR172_6		AGAAUCUUGAUGAUGCUGCAU	21						N	11265321	11265341	82	−39.1	−47.682926	34.146343
miR172_8		GCAGCAGCAUCAAGAUUCACA	21	+	-	-	-	-	Y	2893096	2893116	184	−71.7	−38.96739	39.673912
miR172_9		GGAGCAUCAUCAAGAUUCACA	21	-	-	-	-	-	Y	2969008	2969028	126	−58	−46.031746	43.650795
miR172_10		GAAUCUUGAUGAUGCUGCAG	20	+	+	+	+	-	Y	2969100	2969119	124	−57.1	−46.048386	44.35484
miR319_1	mir319	UUGGACUGAAGGGAGCUCCCU	21	+	+	+	+	+	Y	4935256	4935276	216	−74.03	−34.273148	39.814816
miR319_2		UUGGACUGAAGGGAGCUCCCU	21						N	27801302	27801322	216	−93.6	−43.333332	35.185184
miR319_4		UUGGACUGAAGGGGCCUCUU	20	+	-	-	-	-	N	15212722	15212741	207	−90.71	−43.821255	44.927536
miR319_5		UGGACUGAAGGGGAGCUCCUUC	22	+	-	-	-	-	N	46353126	46353147	213	−90.2	−42.347416	35.21127
miR319_6		GAGCUUCCUUCAGUCCACUC	20	+	+	+	+	-	Y	31787184	31787203	199	−86.6	−43.517586	38.693466
miR319_7		UGGACUGAAGGGAGCUCCUUC	21	+	+	+	+	+	N	9436191	9436211	93	−49.3	−53.01075	47.31183
miR390_1	mir390	AAGCUCAGGAGGGAUAGCGCC	21	+	+	-	+	-	N	26531589	26531609	82	−47	−57.317074	43.90244
miR390_2		AAGCUCAGGAGGGAUAGCGCC	21						N	26325697	26325717	71	−39.4	−55.49296	40.84507
miR393_1	mir393	UCCAAAGGGAUCGCAUUGAUCC	22	+	-	-	+	-	N	30299002	30299023	121	−53.2	−43.96694	34.710743
miR393_2		UCCAAAGGGAUCGCAUUGAUCC	22						N	1754148	1754169	77	−32.4	−42.077923	28.57143
miR393_3		AUCAUGCUAUCCCUUUGGAUU	21	+	+	-	+	+	Y	34480633	34480653	141	−57.7	−40.921986	39.00709
miR394_1	mir394	UUGGCAUUCUGUCCACCUCC	20	+	-	-	+	+	N	49015076	49015095	129	−62.3	−48.294575	41.08527
miR394_2		UUGGCAUUCUGUCCACCUCC	20						N	31533893	31533912	67	−26.03	−38.85075	47.761192
miR394_3		UUGGCAUUCUGUCCACCUCC	20						N	8950170	8950189	125	−36.35	−29.079998	40.8
miR396_1	mir396	UUCCACAGCUUUCUUGAACUG	21	+	+	-	+	+	N	35366632	35366652	114	−45.6	v39.999996	39.473686
miR396_2		UUCCACAGCUUUCUUGAACUU	21	+	+	-	+	+	N	556766	556786	85	-35.9	-42.2353	35.294117
miR396_3		CUCAAGAAAGCUGUGGGAGA	20	+	+	-	-	-	Y	6590654	6590673	108	-47.9	-44.351852	37.962963
miR396_4		GCUCAAGAAAGCUGUGGGAGA	21	+	+	-	-	-	Y	6590654	6590674	108	-47.9	-44.351852	37.962963
miR396_5		UUCCACAGUUUUCUUGAACUG	21	+	+	-	+	+	N	31162290	31162310	121	-47.8	-39.50413	41.322315
miR397_1	mir397	UCAUUGAGUGCAGCGUUGAUG	21	+	-	-	+	+	N	1821127	1821147	153	-60.1	-39.281044	30.065361
miR398_1	mir398	UGUGUUCUCAGGUCGCCCCUG	21	+	+	-	-	+	N	33289822	33289842	80	−29.04	−36.3	50
miR398_2		UGUGUUCUCAGGUCGCCCCUG	21						N	33373257	33373277	108	−47.8	−44.25926	50
miR399_1	mir399	UGCCAAAGAAGAUUUGCCCCG	21	-	+	-	-	-	N	103	123	79	−33.4	−42.278484	41.772152
miR399_2		UGCCAAAGGAGAGCUGCCCUA	21	-	+	-	-	-	N	254628	254648	124	−48.8	−39.354836	35.48387
miR399_3		UGCCAAAGGAGAGCUGCCCUG	21	-	+	-	-	+	N	216204	216224	110	−47.8	−43.454544	36.363636
miR399_4		UGCCAAAGGAGAGCUGCUCUU	21	+	+	+	+	+	N	29007702	29007722	172	−60.23	−35.01744	29.651161
miR399_5		UGCCAAAGGAGAGUUGCCCUG	21	+	+	+	+	+	Y	441789	441809	103	−44.9	−43.592236	43.68932
miR399_6		UGCCAAAGGAGAGUUGCCCUG	21						N	12886367	12886387	89	−43.3	−48.651684	41.573032

Table 2. Cont.

miR_ID	miR Family	Sequence	L	Conserved in					miRNA*	Start	End	PL	MFE	Adjusted	Hairpin
				gma	mtr	vun	ath	os	Y/N					MFE	G/C%
miR408_1	mir408	AUGCACUGCCUCUUCCCUGGC	21	+	+	+	+	+	N	21952158	21952178	82	−36.6	−44.634144	51.219513
miR530_1	mir530	UGCAUUUGCACCUGCACUUUA	21	+	+	-	-	+	N	40319328	40319348	181	−79.3	−43.812157	38.674034
miR2111_1	mir2111	UAAUCUGCAUCCUGAGGUUUA	21	+	+	-	+	-	N	37594	37614	81	−30.5	−37.654324	33.333336
miR2111_2		UAAUCUGCAUCCUGAGGUUUA	21						N	93808	93828	66	−27.5	−41.666664	36.363636
miR2111_3		UAAUCUGCAUCCUGAGGUUUA	21						N	19125	19145	67	−33.1	−49.40298	31.343285
miR2111_4		UAAUCUGCAUCCUGAGGUUUA	21						N	37881	37901	81	−35.7	−44.074078	33.333336
miR2118_1	mir2118	UUACCGAUUCCACCCAUUCCUA	22	-	+	-	-	-	N	112	133	155	−41.2	−26.580647	40
miR2118_2		GGAUAUGGGAGGGUCGGUAAAG	22	+	-	-	-	-	Y	15537249	15537270	150	−60.54	−40.36	38
miR5213_1	mir5213	UACGUGUGUCUUCACCUCUGAA	22	+	-	-	-	-	N	16737686	16737707	119	−42.1	−35.37815	36.974792
miR5213_2		UACGUGUGUCUUCACCUCUGA	21	-	+	-	-	-	N	16737686	16737706	119	−42.1	−35.37815	36.974792
miR5213_3		UACGGGUGUCUUCACCUCUGA	21	-	+	-	-	-	Y	31172531	31172551	114	−47.7	−41.842106	38.59649
miR5232_1	mir5232	UACAUGUCGCUCUCACCUGGA	21	-	+	-	-	-	Y	29987519	29987539	167	−69.8	−41.79641	44.31138

gma- *Glycin max;* mtr- *Medicago tranculata;* vun- *Vigna unguiculata;* ath- *Arabidopsis thaliana;* os- *Oryza sativa.*
MFE- Minimum Folding Energy; L: Length; PL: Precursor Length.

novel miRNAs using qRT-PCR. The expression levels of the chickpea miRNAs under wilt stress were significantly altered compared with those of the control conditions. In contrast, those that were observed under salt stress did not greatly change. Among the validated conserved miRNAs, miR530 was upregulated seventeen-fold during wilt stress, suggesting that it is an important candidate miRNA that is involved in the plant wilt stress response. miR156_1 and miR156_10 were slightly upregulated under both the wilt and salt stresses. miR2118, which is one of the legume-specific miRNAs, was also upregulated by approximately 0.5-fold during wilt stress compared with the control seedlings (Figure 5a, 5b). Conversely, no significant expression patterns were detected with respect to the known miRNAs in response to salt stress in the chickpea. The expression analysis of the novel miRNAs revealed that three out of five (car-miR008, car-miR011 and car-miR015) were approximately three fold upregulated on average during salt stress (novel chickpea miRNAs have been designated as "car-miRNA" throughout the manuscript, in which "car" is an abbreviation for *Cicer arietinum*). However, the expression patterns that were observed during wilt stress revealed limited information because little significant changes occurred.

Prediction and validation of miRNA targets in chickpea

The putative miRNA targets in chickpea were predicted using the psRNATarget program [33]. The predicted target genes (approximately 358 different transcripts) were extensively involved in different biological processes involving a large number of gene families. Some of these genes encoded TFs, DNA replication proteins and those that are involved in cellular metabolism in addition to a variety of stress response-associated proteins. The target prediction analysis revealed the involvement of some of the miRNAs in regulating metabolic processes through the target genes. In chickpea, miR159 is involved in the metabolism of amino acids, fatty acids and lipids. One of the target genes of miR159 encodes acyltransferase, which is essential for ester biosynthesis. The chickpea miR156 and miR166 target genes encode squamosa promoter-binding protein and homeobox-leucine zipper protein, respectively, as previously reported [34,35]. Table 4 describes details of the target genes of validated miRNAs; a complete list is provided as supporting information (Table S5; Table S6). The most widely targeted class of genes is the protein kinases, which are associated with plant defense mechanisms via cell signaling-related processes. The novel car-miRNA008 targets the chalcone synthase (CHS) gene. Chalcone, which is an intermediate in flavonoid biosynthesis, is involved in natural defense mechanisms. CHS expression is also involved in salicylic acid defense pathways. car-miR2118 and car-miR5213 target two defense-response chickpea genes encoding Toll/Interleukin-1 receptor-nucleotide binding site-leucine-rich repeats (TIR-NBS-LRR). Members of the TIR-NBS-LRR gene family are genuine targets for miR2118 [24]. Additionally, miR5213 suppresses defense-response genes in Medicago. The cleavage of such transcripts as mediated by miR5213 is notably conserved in AM symbiosis-capable plants, such as *Medicago truncatula, Glycine max, Lotus japonicus, Populus trichocarpa* and *Cicer arietinum*, but not in plants for which this symbiosis is not observed, such as *Arabidopsis thaliana* [25].

Members of the miR166/165 family target HD-ZIP III TF genes by cleaving the mRNA at complementary base pairs in leguminous plants [34,36,37]. These results are similar to those of earlier predicted reports involving other plant systems. The target gene of miR166 was experimentally validated by modified 5′RLM-RACE [38,39]. All of the positive clones were sequenced,

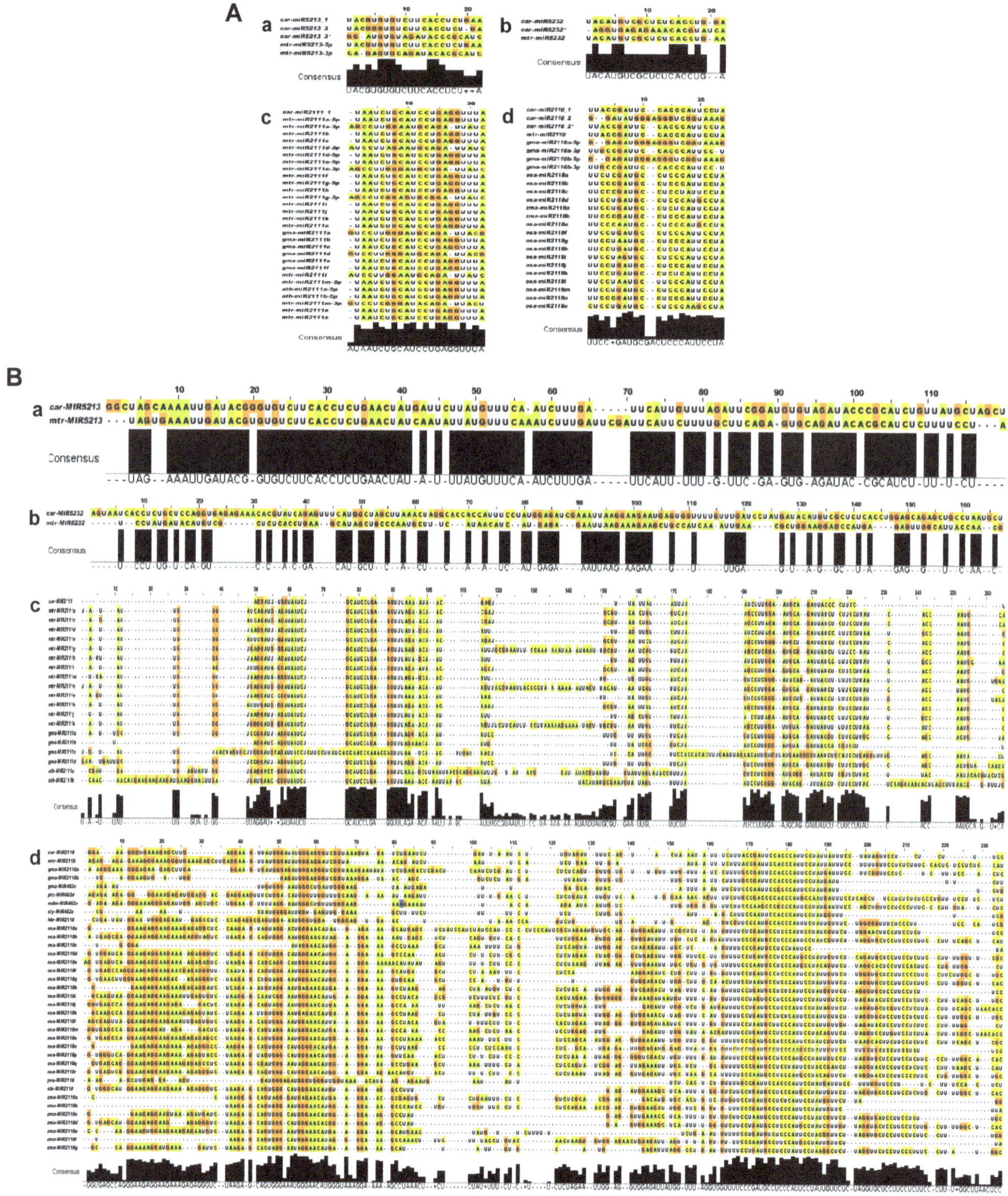

Figure 3. Multiple sequence alignments of legume-specific miRNAs. (A) Mature miRNAs, (B) precursor miRNAs. Four legume-specific miRNAs, including a) MIR5213, b) MIR5232, c) MIR2111and d) MIR2118, were used for the multiple sequence alignments by ClustalW2 in the different plants. car- *Cicer arietinum*, mtr- *Medicago truncatula*, gma- *Glycine max*, ath- *Arabidopsis thaliana*, osa- *Oryza sativa*, zma- *Zea mays*, sbi- *Sorghum bicolor*, sly- *Solanum lycopersicum*, hbr- *Hevea brasiliensis*, pvu- *Phaseolus vulgaris*, vun- *Vigna unguiculata*, ptc- *Populus trichocarpa* and mdm- *Malus domestica.*

and cleavage was observed at the 17th and 18th positions of the mRNA by the 5′ end of miR166 (Figure 6), unlike the previously reported miRNA-target recognition parameters [40]. Although our results are not in agreement with previous studies, such as those involving the soybean, in which miR166 target validation by 5′RACE and degradome sequencing confirmed cleavage at the

Table 3. List of novel chickpea miRNAs with their miRNA*.

New miRID	Sequence	L	miRNA*	Start	End	PL	MFE	Adjusted MFE	Hairpin G/C%
car-miRNA001	AACCAGGCUCUGAUACCAUGA	21	AUGGUAUCAGGUCCUGCUUCA	20306919	20306939	87	−37.34	−42.91954	42.528736
car-miRNA002	AAGAUUGAUCUUGACCUUCUGC	22	UUAUGGCAUAAACAAGGAUAAU	588	609	93	−29	−31.182796	37.634407
car-miRNA003	AAGCAGGCUCUGAUACCAUGA	21	UGGUAUCAGGUCCUGCUUCA	696	716	95	−42.3	−44.526314	47.368423
car-miRNA004	AAUAGAUUGUCCAAUCGAUUGU	22	CAAUCGAUUUCCCAAUCGAUUU	343	364	159	−47.8	−30.062893	33.962265
car-miRNA005	AAUCACGGUGAGCCACUGUGA	21	AAUCACGGUGGCUCACCGUGA	251	271	91	−54.2	−59.56044	39.56044
car-miRNA006	ACCGGAAGCUGGGUUACGGUC	21	CGCGACCUAUACCCGGCCGU	598	618	199	−75.01	−37.693466	56.281406
car-miRNA007	ACGACUGUUACAUCAUACAAC	21	UGUAUGGUGCAACAGUCGCAG	23710719	23710739	137	−65.8	−48.029198	43.065693
car-miRNA008	ACGAGACAGAUGGACACGACGG	22	CGUACGUUGUCGGAUAUGUCGC	338	359	97	−28.5	−29.381443	47.42268
car-miRNA009	AGCGAUCUCGUACUAAACGAA	21	CUUCGAUAGGCGAGAGGUGUA	23987481	23987501	78	−24.9	−31.923077	47.435898
car-miRNA010	AGGAGAAAGUCUUUGCAACCG	21	UGUGUUGCUGAGACAUGCGCC	273099	273119	65	−21.7	−33.384617	52.307693
car-miRNA011	AUGGUUGAGAGGGUGACUUGA	21	AAGUCACUUUCUCAAUCUUA	1161	1181	155	−72.3	−46.645164	39.35484
car-miRNA012	CAGGGAACAGGCUGAGCAUGG	21	AUGCACUGCCUCUUCCCUGGC	171	191	87	−47.7	−54.827587	51.724136
car-miRNA013	CAGGGAACAGGCUGAGCAUGG	21	AUGCACUGCCUCUUCCCUGGC	149	169	87	−47.7	−54.827587	51.724136
car-miRNA014	CAGGGAACAGGCUGAGCAUGG	21	AUGCACUGCCUCUUCCCUGGC	285	305	87	−47.7	−54.827587	51.724136
car-miRNA015	CGAGACAGAUGGACACGACGG	21	CGUACGUUGUCGGAUAUGUCG	336	356	97	−28.5	−29.381443	46.391754
car-miRNA016	CGAUUGCGGCGACGUGGGCG	20	CUGCCCGCGACGUUGUGAGA	50	69	66	−32.8	−49.696968	57.575756
car-miRNA017	CGGAAUACAAGCUCUGUACCGGAA	24	CGGAACACUCUUCUGUACCGGAAA	18012181	18012204	116	−42.41	−36.560345	39.655174
car-miRNA018	CUGACUUAGCUUGUAGUCGAC	21	UAGUCGACUACAGAUGGGUGU	533	553	67	−30.5	−45.52239	46.268658
car-miRNA019	CUGGGUUGGGUCGAUCGGUCC	21	CACCGGUUGGCUCGUCCCUU	842	862	75	−32.8	−43.73333	65.33333
car-miRNA020	CUGUAGCAUCACUAUAGCCGC	21	CGGCUAUAGUGGCGCUAUAGC	37423709	37423729	95	−45	−47.368423	42.105263
car-miRNA021	GAAACGGGUAGCUGAGGGUU	20	CACUCUAAACAGCAGCUCCGU	580	599	113	−34.3	−30.353981	40.707962
car-miRNA022	GAAACGGGUAGCUGAGGGUU	20	CACUCUAAACAGCAGCUCCGU	499	518	113	−32.3	−28.584068	39.823006
car-miRNA023	GAAAUGGACGGCAAUGAAUCUA	22	UUGAAGGUUUUGCUGACCUUU	157	178	88	−26.52	−30.136364	42.045452
car-miRNA024	GAACGAGACAGAUGGACACGA	21	UACGUUGUCGGAUAUGUCGCGA	337	357	98	−29.9	−30.510204	45.918365
car-miRNA025	GAGUUCACUGUUGGAGAUGUGCCA	24	GCACAACUCCAACGGUGAACCCAC	30794763	30794786	220	−138.6	−63.000004	45.909092
car-miRNA026	GCCGGCCUGUCAGACCUAAUAGGC	24	UCAAGCCAUAGGCCUCUGACGGAC	9008149	9008172	162	−62.4	−38.518517	41.975307
car-miRNA027	GCGAAGCUAUCGUGCGUUGGAU	22	UUCGCACAAUUGGUCAUCGCG	138313	138334	95	−32.3	−34	50.526314
car-miRNA028	GGGUUGGGUCGAUCGGUCCA	20	CACCGGUUGGCUCGUCCCUU	677	696	69	−33.8	−48.985504	63.76812
car-miRNA029	GGGUUGGGUCGAUCGGUCCGCC	22	GGUGUGCACCGGUUGGCUCGU	18682982	18683003	69	−35.5	−51.449276	66.66667
car-miRNA030	GGGUUGGGUCGAUCUGUCCGCC	22	GGUGUGCACCGGUUGGCUCGU	619	640	69	−29.3	−42.463768	65.21739
car-miRNA031	GGUUGGGUCGAUCGGUUCGCCU	22	GUGUGCACCGGUCUGCUCGUC	383	404	69	−30.9	−44.782608	65.21739
car-miRNA032	GUUCUAGAUCGACGGUGGCAU	21	GUCACCACCGUCGUCUCGCA	103	123	66	−28.6	−43.333332	56.060608
car-miRNA033	GUUCUAGAUCGACGGUGGCAU	21	GUCACCACCGUCGUCUCGCA	111	131	66	−28.6	−43.333332	56.060608

Table 3. Cont.

New miRID	Sequence	L	miRNA*	Start	End	PL	MFE	Adjusted MFE	Hairpin G/C%
car-miRNA034	GUUCUAGAUCGACGGUGGCAU	21	CACCACCGUCGUCUCGCAGCU	313	333	67	−27.6	−41.19403	56.71642
car-miRNA035	UAACUCUGAUGAAGUUGUGCA	21	GCUCAAUUUGUAUCUGGGACAU	147	167	61	−19.3	−31.639343	37.704918
car-miRNA036	UAAGUCGGUGACGUCUACGUAUAC	24	UCUAUACGGAAGAUGCAUGGACUA	21957421	21957444	82	−24.2	−29.512197	41.463413
car-miRNA037	UAGCGACACGGAACGUCCAAC	21	GGGGUUGGACGCUCCGUGCCA	215357	215377	112	−81.4	−72.67857	51.785713
car-miRNA038	UAUGUGAACGAGACAGAUGGA	21	UCGGAUAUGUCGCGACACAAA	342	362	98	−29.9	−30.510204	45.918365
car-miRNA039	UCAUAUUUGUUGGACAUUUGA	21	UCAAUGUAUUGAUGGGUAUGU	942	962	113	−29.7	−26.283186	38.938053
car-miRNA040	UCGCGUGAGUGAAGAAGGGCA	21	AGCUCUUUCGUCGAGUGCGCG	69427	69447	66	−23.3	−35.30303	50
car-miRNA041	UCGCUGUUGCGUUGGCGAUUA	21	AUCGCCACCGCAACAGCGAAG	21404687	21404707	114	−59.83	−52.482456	53.50877
car-miRNA042	UCGGACCAGGCUUCAUUCUUC	21	AAUGAGGUUUGAUCCAAGAUC	1317	1337	197	−57.8	−29.3401	38.071068
car-miRNA043	UGAACUAUUCGAUCUUCGUUC	21	GAUGAAGAUCAAACGGUUCAU	101	121	147	−67.8	−46.122448	32.65306
car-miRNA044	UGAAGCUGCCAGCAUGAUCUUA	22	AGAUCAUGUGGCAGUUUCACC	1047	1068	73	−36.4	−49.863018	47.945206
car-miRNA045	UGAUUGUAUAAUCGAUUAGGCA	22	AUUGGCAAAUCGAUUGUGCACA	26	47	121	−47.8	−39.50413	38.842976
car-miRNA046	UGAUUGUAUAAUCGAUUAGGCA	22	AUUGGCAAAUCGAUUGUGCACA	26	47	121	−47.8	−39.50413	38.842976
car-miRNA047	UGCCAAGCGCUGUAGUAGGUCA	22	AUAAGGCUUUUACAGCGCUUG	606	627	94	−50.1	−53.29787	44.68085
car-miRNA048	UGGACUAAAAUUCUGUUUGGAGAC	24	GUCCCCCAAGCAGAAUUUUGGUCC	30591970	30591993	218	−91.99	−42.197247	31.192661
car-miRNA049	UGGACUAAAAUUCUGUUUGGAGAC	24	GUCCCCCAAGCAGAAUUUUGGUCC	47072	47095	220	−95.82	−43.554543	31.818182
car-miRNA050	UGGAUUGAGAUCGAAUGGUGC	21	GACCGUCCGGUCUUGAUCCAAG	347	367	137	−60.2	−43.941605	37.956203
car-miRNA051	UGGGACAAUCGAUUUGGACAUC	22	UUGGAAAUCGAUUGAUUCAGUG	260	281	72	−22.4	−31.111109	34.72222
car-miRNA052	UGGUCUGUGAGAGACUGCACGGUA	24	UCGUGCUGGUCUGUGAGAGACUGC	52521	52544	96	−40.7	−42.395832	53.125
car-miRNA053	UGGUUGGGUCGAUCGGUCCGCC	22	GGUGUGCACCGGUUGGCUCGU	677	698	85	−37.1	−43.647057	63.529415
car-miRNA054	UGUGGAUGAUGCAGGAGCUGA	21	AGCUGCUGACUCAUUCAUUCA	27572322	27572342	152	−59.5	−39.144737	35.526314
car-miRNA055	UGUUGCAAUCGACCAGGACUAC	22	AGUCUUGAUCGAUGUAACUGA	385	406	79	−39.5	−50	32.911392
car-miRNA056	UUAACCAGGCUCUGAUACCAU	21	UGGUAUCAGGUCCUGCUUCA	21161303	21161323	88	−37.4	−42.5	50
car-miRNA057	UUCGAAUCCUGCCGUCCACGCC	22	AGUGGACGUGCCGGAGUGGUUA	29887519	29887540	94	−37.2	−39.574467	55.31915
car-miRNA058	UUGAGCCGCGUCAAUAUCUUG	21	CAAGGUAUUGGCGCGCCUCAA	24546067	24546087	92	−53.4	−58.04348	41.304348
car-miRNA059	UUGAUCUUUCGAUGUCGGCU	20	GUUUAGACCGUCGUGAGACA	988	1007	116	−45.4	−39.13793	50.86207

MFE: Minimum Folding Energy; L: Length; PL: Precursor Length.

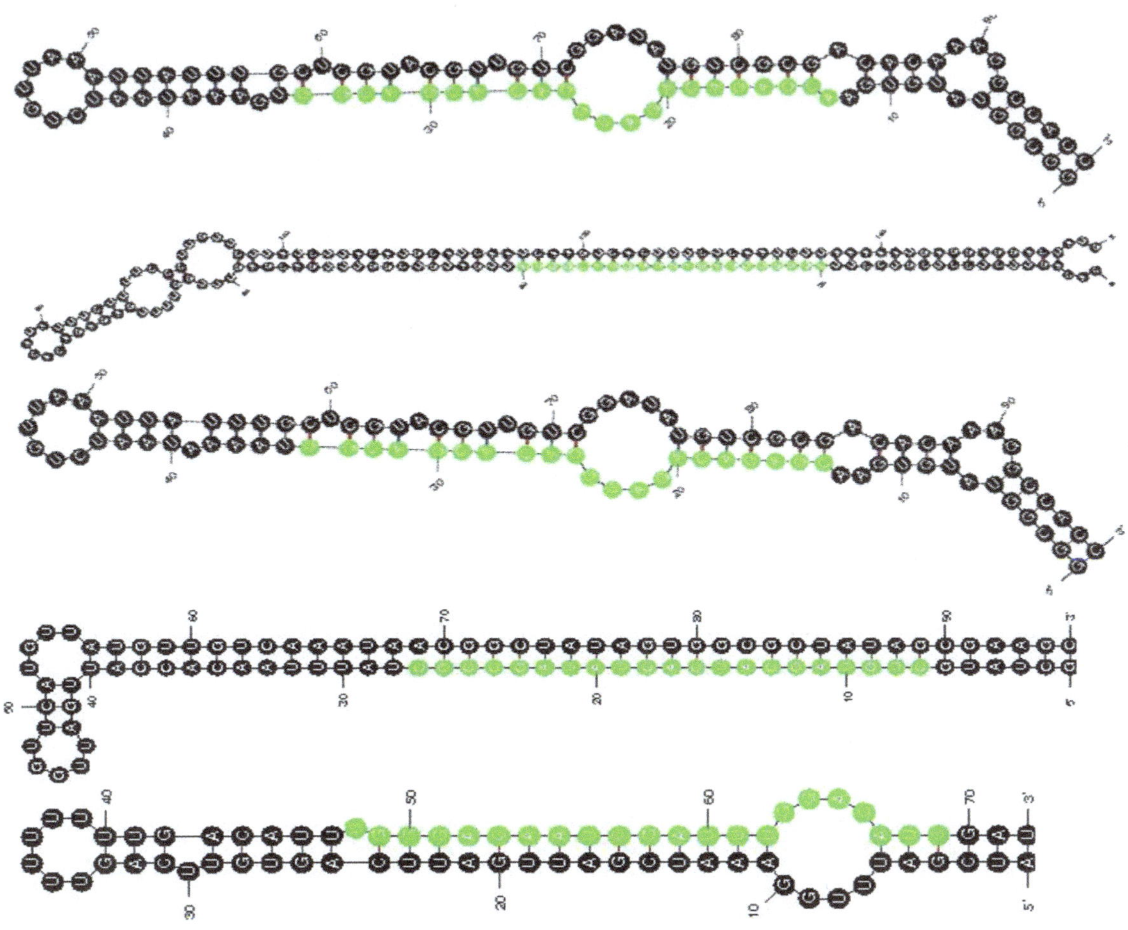

Figure 4. Predicted secondary structures of five validated novel miRNA precursors in chickpea using Mfold.

10th and 11th positions [41], there have been reports of the miRNA (belonging to different families)-mediated cleavage of target mRNA, thus defying the recognition rule. A total of 18 miRNA/target pairs of *Pinus taeda* possessed non-conventional cleavage sites, such as pta-miR951:AW065026, which is cleaved at the 16th and 17th positions [42]. Similar results have been reported in other plant species, such as mtr-miR397:AC135467 [24], ath-miR168:AGO1 [43], pvu-miR171:gi62704692 [44] and ath-miR398a:CSD1 [39]. Thus, it appears that the sequence of the target gene and the miRNA sequence determine the cleavage site apart from the conventional complimentary region-based target cleavage. Therefore, it is quite possible that chickpea has a different cleavage site for miR166 (pair miR166:TC04758) compared with other plant species.

Analyses of GO terms and KEGG pathways

The GO terms of the target genes were annotated according to their biological processes, molecular functions or involvement as cellular components. The enzyme mapping of the annotated sequences was performed using direct GO for the enzyme mapping and the Kyoto Encyclopedia of Genes and Genomes (KEGG) for the definitions of the KEGG orthologs. The miRNA-targeted genes belonged to various biological processes, cellular components and molecular functions as depicted in Figure 7. The maximum numbers of target genes were involved in biological processes, including both metabolic and cellular processes. However, the target genes that were involved in binding were the most abundant (80%) within the molecular functions category.

Discussion

In this study, high-throughput deep sequencing was used to gain in-depth knowledge of gene regulation by miRNAs in the chickpea under biotic and abiotic stresses. Salt stress is one of the major constraints to increasing chickpea productivity. Soil salinity levels affect germination in plants. Under salt stress conditions, chickpea plants show high levels of anthocyanin pigmentation in their foliage and reduced growth rates [45]. Among the biotic stresses, Fusarium wilt is one of the major soil/seed-borne diseases severely affecting chickpea growth. Its causative agent is *Fusarium oxysporum* f.sp. *ciceris*, which is a fungal pathogen.

Most of the miRNAs that were obtained in our library have a preference for the 5′-U as has been reported in other plants, which is in accordance with the defined structures of the mature miRNAs [1,46]. The lengths of the chickpea precursors ranged from 61 to 220 nt, which were similar to those of the soybean (55 to 239 nt) and peanut (75 to 343 nt) [30,35]. The calculation of the minimum free energy (MFE) values further added credence to these predicted hairpin structures as putative miRNA precursors. The chickpea precursors had minimum free energy values ranging from −97.2 to −26.03 Kcal/mol with an average of −50.1419 Kcal/mol, which was similar to the −50.01 Kcal/mol that was observed in *Arachis hypogaea* and the reported value

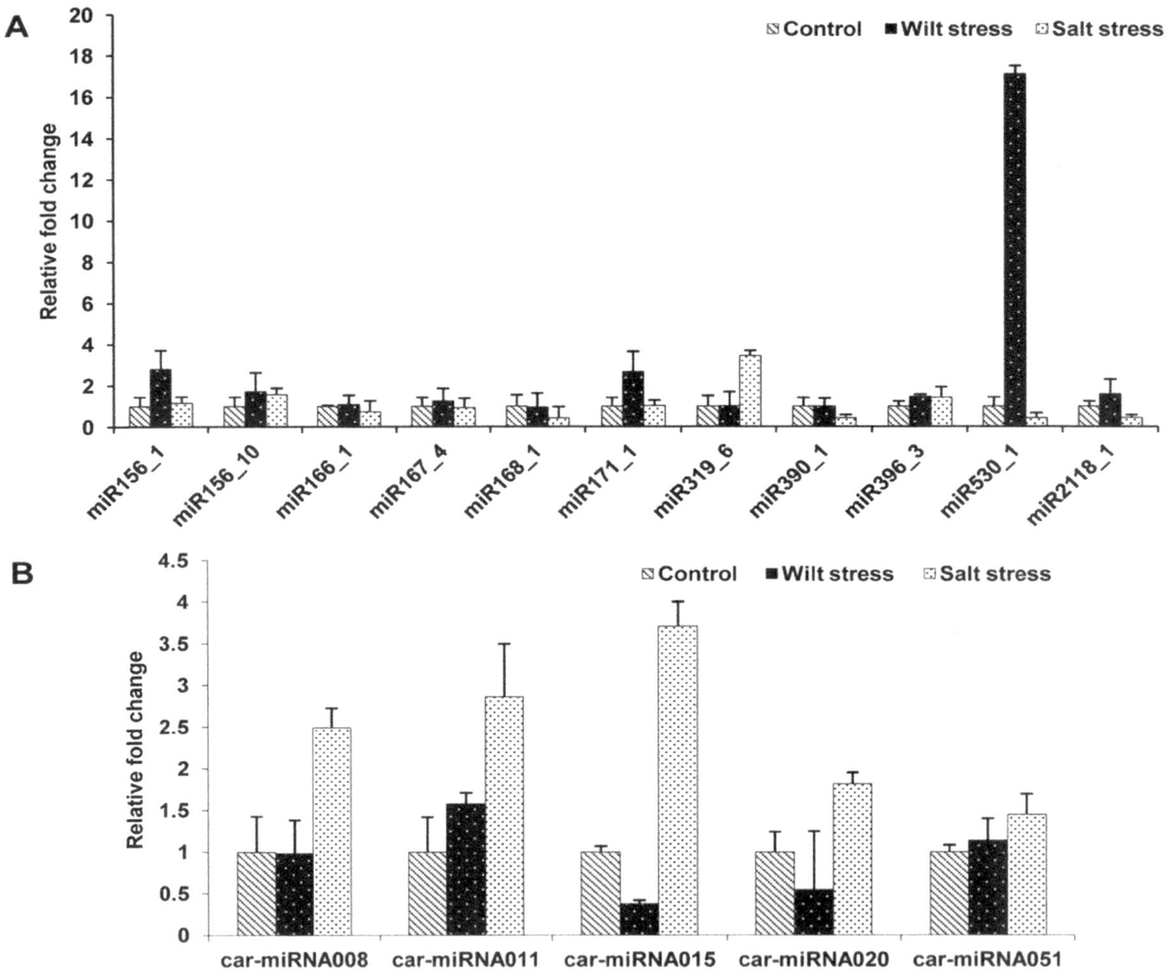

Figure 5. Expression analyses of selected miRNAs under wilt and salt stresses as evaluated by qRT-PCR. The relative expression levels are shown as fold changes with the standard errors (SE) of three biological replicates. (A) Expression profiling of conserved miRNAs under control, wilt and salt stress conditions. (B) Expression profiling of novel miRNAs under control, wilt and salt stress conditions.

of −59.5 Kcal/mol in *Arabidopsis thaliana* (Table S2) [30]. Greater increases in miRNA expression were observed following wilt stress compared with salt stress, suggesting the significant role of small RNAs in the response to pathogen attack. The total number of miRNAs was greater in the wilt stress library than in the salt stress library. Four legume-specific miRNAs were identified in the chickpea libraries, including miR2111, miR2118, miR5232 and miR5213, which were previously reported in Medicago. The sequence conservation among the different legumes and the precursor sequence similarity of these four chickpea miRNAs further substantiate their accurate identification in this study. car-miR5232 cleaves only two transcripts encoding an ATPase E1-E2 type and an expressed protein of unknown function, in concordance with a similar study in Medicago, in which miR5232 targets were experimentally confirmed by degradome sequencing [25].

The significance of miRNA* in authenticating the presence of miRNA has previously been established. A comparison of chickpea miRNA* and mature miRNA data revealed that they vary in abundance in response to the different stress treatments, which has also been previously reported [47,48]. Our target search analysis indicated that miRNA* act upon different transcripts than do their parental miRNAs (data not shown), which has been observed in plants, animals and humans [25,49,50]. For example, miR393 and its miRNA* counterpart regulated the expression of genes belonging to entirely different protein families; i.e., TIR1 and SNARE, respectively [51,52].

Expression patterns during biotic stress

This study is the first attempt to identify miRNAs that are associated with fungal attack in the chickpea. Alterations in the expression of genes that are involved in defense during pathogen attack have been previously reported. These genes are regulated by small RNAs. miR393 was the first miRNA whose role in pathogen attack was demonstrated [51]. Eleven conserved and five novel miRNAs were analyzed in the chickpea under wilt and salt stress. Interestingly, miR530 was significantly upregulated during wilt stress. This observation suggests that its target genes are expressed at lower levels, which included the zinc knuckle proteins and microtubule-associated proteins. Zinc knuckle proteins are involved in the regulation of morning-specific growth in Arabidopsis [53]. The target of miR530 varies in different plants under different conditions and tissues. In *Populus trichocarpa*, this miRNA targets zinc knuckle (CCHC type) family proteins along with a homeobox TF [54], whereas in soybean, it targets genes that encode the CONSTANS interacting protein and nuclear

Table 4. Predicted target genes of miRNAs in chickpea.

miRNA family	Target	Putative Functions of Predicted Targets
Conserved miRNAs		
mir156_1	TC12891, TC03863, TC05745,TC07318,	Squamosa promoter-binding TF family protein, SCP1-like
	TC03684, TC29077, TC07318, TC15422,	small phosphatase
	TC04572	
mir156_10	TC29077, TC15422, TC12891, TC03863,	SCP1-like small phosphatase, squamosa promoter-
	TC05745, TC03684, TC07318, TC19303,	binding protein, cationic amino acid transporter,
	TC10437, TC05493, TC18211	allantoinase 1-like protein
mir166_1	TC04758, TC15765, TC08004	ClassIII HD-ZIP, REVOLUTA
mir167_4	TC21867, TC03743, TC03697	Monosaccharide transport protein 1, MFS, tubulin-folding
		cofactor E
mir168_1	TC06138, TC16221, TC07642	GTP-binding protein, RNA binding (RRM/RBD/RNP motifs)
mir171_1	TC15816, TC01767, TC07982	HAIRY MERISTEM 3 (HAM3), cdk protein kinase, ClpX3
mir319_6	TC03909	Putative xylogalacturonanxylosyltransferase
mir390_1	TC12049, TC05305, TC19589	Protein kinase, CZF1
mir396_3	TC18749, TC21342, TC02165, TC16760,	RNA-directed DNA polymerase, NAD(P)-binding
	TC09727, TC02085	Rossmann-fold
mir530_1	TC01544, TC20787, TC01795, TC01794	Zinc knuckle protein, expressed protein
mir2118_1	TC01089, TC09480, TC00082, TC21040,	NB-ARC disease resistance protein, expressed protein,
	TC23505	TIR-NBS-LRR
Novel chickpea miRNAs		
car-miRNA008	TC06967, TC05545	RING/U-box superfamily protein, chalcone synthase (CHS)
car-miRNA011	TC02274, TC14659, TC17732, TC08052,	SERPIN family protein, amelogenin, RNA binding
	TC16830, TC06852, TC05883	(RRM/RBD/RNP motifs), LEA, anion channel protein family
car-miRNA015	TC17182, TC10107	Complex 1 protein (LYR family), ribosomal L23/L15e
		family protein
car-miRNA020	TC33381, TC29465, TC00653, TC28744,	TPR-like superfamily protein, ARM superfamily protein,
	TC05383	FAD/NAD(P)-binding oxidoreductase, Protein of unknown
		function (DUF1423)
car-miRNA051	TC11550, TC31151, TC21283	SMG7, HAD superfamily protein, unique electron
		transfer flavoprotein

transcription factor Y [55]. In *Eugenia uniflora*, miR530 targets wall-associated receptor kinase-like 14, S-acyltransferase tip-1 and a protein of unknown function in rice [56,57]. In a recent study in maize plants that were resistant to the fungus *Exserohilum turcicum*, miR530 was identified as a novel miRNA and was predicted to target genes that are involved in kinase activities in addition to DNA-binding TFs [58]. Based on the significant upregulation of miR530 in response to Fusarium infection and its unique target genes in the chickpea, it appears to be involved in the response to pathogen attack.

The three legume-specific miRNAs (miR2111, miR2118 and miR5213) play critical roles during pathogen attack. In the chickpea, miR2111 targets a Kelch repeat-containing F-box protein. F-box proteins are responsible for the controlled ubiquitin-dependent degradation of cellular regulatory proteins and are involved in defense responses, auxin responses and floral organ development [59,60,61]. Targets of F-box proteins are central regulators of key cellular events and include G1 cyclins and inhibitors of cyclin-dependent kinases [62]. It appears that miR2111 and F-box proteins act together to regulate the defense response in chickpea following biotic stress. Other than F-box proteins, miR2111 also targets TIR domain-containing NBS-LRR disease resistance proteins. miR2118 and miR5213 also target the same class of R genes. TIR, which is an F-box protein, is a receptor for the plant hormone auxin [63,64,65,51], and LRR consists of tandem Kelch repeats [66]. Interestingly, the chickpea miR2118 was upregulated in response to wilt infection and down regulated following salt stress. miR2118 has also been shown to be suppressed after Verticillium fungal attack in cotton [67]. Fusarium wilt leads to symptoms that are similar to those of Verticillium wilt, whose common host plant is cotton. miR2118 functions through three novel target transcripts encoding TIR-NBS-LRR disease resistance proteins, but its functional regulation remains unclear. In the soybean, miR2118 targets the protein family that is associated with disease resistance in addition to zinc finger proteins [55] and replication termination factor 2 in response to biotic (Asian soybean rust) and abiotic (water deficiency) stresses [35].

Other miRNAs also target disease resistance genes. For example, novel car-miRNA023 target proteins are involved in disease resistance. The highly conserved miRNA171 family targets more than 20 genes that are involved in different processes and

Figure 6. Mapping of target mRNA cleavage site of miR166 by modified 5′ RACE. The target of miR166 (TC04758) encodes a transcription factor belonging to class III of the HD-ZIP family protein. The arrow indicates the cleavage site, and the numbers above the arrow denote the frequencies of the sequenced clones.

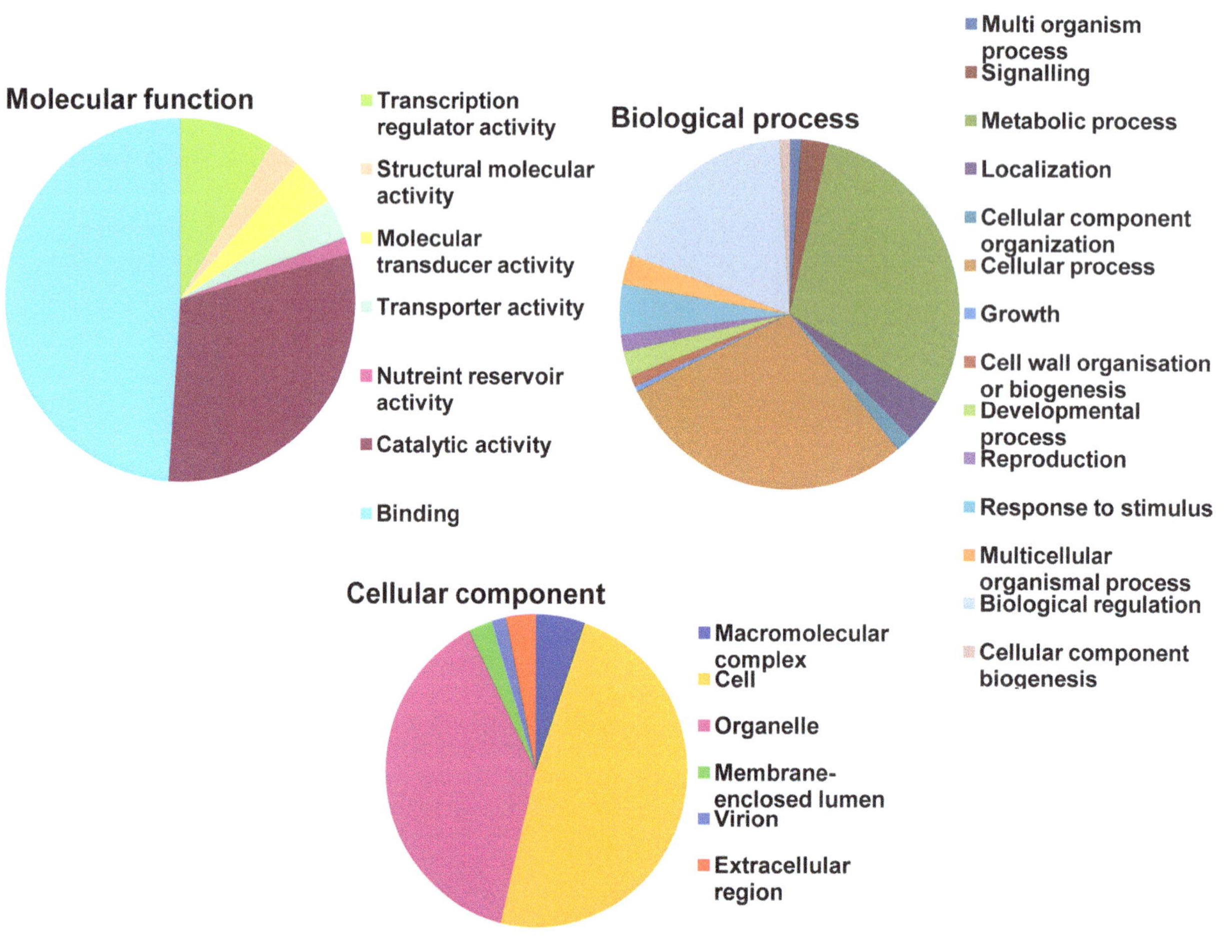

Figure 7. Gene ontology categories of predicted target transcripts for chickpea miRNAs. The miRNA-target genes were categorized according to the molecular function, biological process and cellular component sub-ontologies.

pathways in the chickpea. One particular member, miR171_7, targets a disease resistance-responsive dirigent-like protein (DIR). The conspicuous involvement of disease resistance genes in the response to pathogen attack has been previously established. ESTs encoding dirigent proteins were identified in the SSH library of a chickpea that was infected with Fusarium wilt [68]. Dirigent proteins impart disease resistance through their involvement in lignification during biotic stress. Similar studies have been reported involving *Gossypium barbadense* that was infected with Verticillium fungus, in which two DIR genes were isolated from the SSH library [69].

Many of the genes that are targeted by miRNAs are involved in disease resistance and growth-related processes. Therefore, it can be surmised that these miRNAs are involved in the regulation of plant development and pathogen growth by acting both as positive and negative regulators, depending on their target genes.

Expression patterns during abiotic stress

Our library allowed for the identification of a large number of conserved salt-responsive miRNAs, including miR390, miR172, miR171, miR169, miR408, miR159, miR396, miR2111, miR5213, miR397, miR393, miR162, miR168, miR166, miR167, miR156, miR530, miR399, miR160, miR319, miR164, miR398, miR2118 and miR394. Among these miRNAs, miR156, miR396 and miR319 were upregulated in response to salt stress, which was confirmed using qRT-PCR. Our results agreed with a previous study involving Arabidopsis, in which 10 salt-responsive miRNAs (miR156, miR165, miR319, miR393, miR396, miR167, miR168, miR171, miR152 and miR394) were reported to be involved in the high salinity stress response [70]. In the chickpea, the transcript levels of miR156 family members were elevated in response to salt stress compared with those of miR166 and others as has been reported in previous studies. Some of the miRNAs that are regulated under salt stress in other plant systems were not found in our library. This phenomenon may be due to different stages or stress conditions; i.e., particular treatment methods or species-specific responses.

Previous studies have demonstrated that miR169 family members are associated with high salt stress [71]. From our target prediction analysis, miR169-targeted genes belong to the nuclear TF family, which contains a CCAAT-binding complex. This CCAAT-binding complex is a eukaryotic promoter element that is evolutionary conserved [72]. Recent studies have demonstrated that these proteins play significant roles in abiotic stress-response pathways [39,73]. The genes that are targeted by miR169 function in transcriptional regulation, suggesting their significant involvement in the salt stress response.

In this study, the salt-responsive miRNA miR390 explicitly targeted protein kinases and the CZF1 TF. The CZF TF is associated with intracellular signal transduction, is involved in the negative regulation of programmed cell death and responds to fungal attack via plant defense mechanisms. CZF1 contains a zinc finger with a CCCH-type domain and has been reported in *Arabidopsis thaliana* to be salt-inducible. A parallel study in upland cotton reported that the LZF TF acted in response to salt stress [74], and its network of protein-protein interactions was deduced. The chickpea miR396 exhibited higher expression levels under salt stress and was also reported to be salt-responsive in rice. Additionally, transgenic lines over expressing osa-mir396c showed reduced tolerances to salt and alkali stresses compared with wild type plants [75].

In our analysis of the miRNA expression data under both biotic and abiotic stresses, few were upregulated under both types of stresses. miR396 and a member of the miR156 family were upregulated in response to both the wilt and salt stresses at levels of approximately 1.5-fold, indicating the relative similarity between fungal infection- and salinity stress-responses in the chickpea, which was stated in a previous report, in which the chickpea responded to fungal infection (*Ascochyta blight*) more similarly to high salinity stress than to drought or cold stresses [76]. Additionally, cross talk exists between the stress-signaling pathways that involve several kinases and TFs that are important targeting candidates for several miRNAs under wilt and salt stresses [77,78]. Our data indicate that miR172, miR319, miR171, miR390 and miR396 have serine/threonine protein kinases and MAPK protein kinases as their target genes, which involve signaling pathways. It can be presumed that together, these miRNAs might mediate defense mechanisms under stress conditions via transcriptional regulators. miRNAs also target genes that are directly or indirectly involved in the defense against various stresses. For example, car-miR08 targets a chalcone synthase gene, which is an intermediate in flavonoid biosynthesis. Flavonoids are secondary metabolites that serve variable functions, including those involving pigmentation, UV protection and antifungal defense. Therefore, it can be conferred that these miRNAs come into play during stress management in plants by targeting the genes that are involved either directly or indirectly.

The explicit role of miRNAs in regulating defense mechanisms by the complementary binding of target genes is evident through exhaustive literature reviews. This study will aid in the elucidation of the stress response mechanisms that are utilized by the chickpea. Further, there is limited available knowledge describing comprehensive studies of miRNA expression in the chickpea in response to particular stresses.

Materials and Methods

Plant materials and stress treatments

The chickpea cultivar ICC4958 was used throughout the study. ICC4958 is a Fusarium wilt-resistant and salt-sensitive chickpea cultivar [79,45]. The plants of the ICC4958 cultivar were grown on a 16-h day/8-h night photoperiod cycle at $25\pm2°C$. Fourteen-day-old seedlings were subjected to the wilt and salt stresses separately. The stress treatments were performed as follows: for wilt stress, two-week-old plants that were grown under hydroponic conditions were exposed to a toxin that was isolated from the fungus *Fusarium oxysporum* f.sp. *ciceris* for one day. For salt stress, the roots of two-week-old seedlings were immersed in a 150 mM NaCl solution for 12 h. All of the tissues (control, wilt-stressed and salt-stressed) were harvested at their respective time points, snap-frozen in liquid nitrogen and maintained at $-80°C$ for further analyses.

Small RNA library preparation and sequencing

Total RNA was isolated using the TRIzol reagent (Invitrogen, Carlsbad, CA, USA) according to the manufacturer's protocol. For the construction of the small RNA library, low molecular weight (LMW) RNA was enriched by the LiCl method. Equal amounts of RNA were pooled from the root and shoot tissues for each group to generate a LMW RNA library. The RNA was run on a 15% polyacrylamide gel, and the 20 to 30 nt small RNA fraction was extracted and eluted. A preadenylated adaptor was ligated to the 5′ end of the small RNAs with T4 ligase. The ligation product was eluted, and subsequently, 3′ end adaptor ligation was performed [80] followed by RT-PCR. The PCR products were checked for quality and quantified using a Bioanalyzer (Agilent, Germany). The samples were then sequenced using the Illumina Genome Analyzer IIx (Illumina Inc., USA).

Computational sequence analysis for identification of miRNAs

The total reads were trimmed and filtered using the UEA small RNA workbench 2.4- Plant version sequence file pre-processing tool (http://srna-tools.cmp.uea.ac.uk/) [81]. The unique tags were generated following a series of processing steps, which included adaptor trimming (using the adaptor removal tool), the elimination of low-quality sequences and the removal of contaminated and other non-coding RNAs, including tRNAs, rRNAs, etc. The UEA sRNA toolkit-Plant version filter pipeline (http://srna-tools.cmp.uea.ac.uk/) was used to exclude the low-complexity and low-quality sequences and eliminate the t/r RNA population by mapping them to plant t/r RNAs from the "Rfam" database, Arabidopsis tRNAs from "The Genomic tRNA Database" and plant t/rRNA sequences from the "EMBL" release 95. Then, the miRCat pipeline (miRNA categorization) was used to predict novel miRNAs and their precursors using default parameters [82]. The secondary structures of the small RNA sequences were folded using RNAfold (http://rna.tbi.univie.ac.at/cgi-bin/RNAfold.cgi) to predict potential miRNA precursors. The small RNA sequences that had characteristic hairpin structures, together with additional minimal folding free-energy indices (MFEI) [83,84], were considered to be candidate miRNAs by miRCat. The small RNA sequences that matched the following criteria were considered to be valid miRNA precursors: i) no more than 3 consecutive mismatches between the miRNA and miRNA*; ii) at least 17 of the 25 nt surrounding the miRNA must be involved in base pairing; iii) the hairpin must be at least 75 nt in length; and iv) at least 50% of the bases in the hairpin should be paired. The folding structures of the precursors of the new miRNA with the miRNA* were carried out using the UEA sRNA toolkit-RNA hairpin folding and annotation tool, which uses the Vienna Package to obtain the secondary structure of a precursor sequence, highlighting the miRNA/miRNA* sequences on the hairpin structure [85]. The data discussed in this publication has been deposited in Gene Expression Omnibus [86] repository under the accession number GSE57857 (http://www.ncbi.nlm.nih.gov/geo/query/acc.cgi?acc=GSE57857).

miRNA validation by poly(A) tail assay-based quantitative real-time PCR (qRT-PCR)

The predicted chickpea miRNAs were validated by performing poly(A)-tailed RT-PCR on sixteen miRNAs, including eleven conserved and five novel miRNAs. The total RNAs from the treated and control samples were extracted using the TRIzol reagent (Invitrogen, Carlsbad, CA, USA) according to the manufacturer's instructions. A 1-μg aliquot of this RNA was used for the poly(A) tailing using the Poly(A) Tailing Kit (Ambion, USA) according to the manufacturer's instructions and then purified using the RNeasyMinElute Cleanup Kit (QIAGENGmBH, Germany). The poly(A) RNA (2 μg) was then reverse-transcribed into cDNA that was primed by a standard poly(T) anchor adaptor using an RTQ primer. For the RT-PCR reaction, the conditions were as follows: 65°C for 10 min, 4°C for 2 min, 50°C for 60 min and 70°C for 15 min. Three biological replicates per sample were used for the analyses.

The poly(T) cDNA was diluted 10-fold and used to perform qRT-PCR using KAPA FAST SYBR Green chemistry (Kapa Biosystems, USA). For the qRT-PCR, the sequences of the specific miRNAs that were validated served as the forward primer and RTQ uni-primer, having an adaptor sequence as the reverse primer (Table S7). The 5S rRNA was used as the reference gene for all of the reactions. Three biological replicates were used per sample in addition to three technical replicates along with a no-template control and no-RT enzyme control. The data were analyzed using the 2[-Delta DeltaC(T)] method [87] and reported as the means ± standard errors (SE) of three biological replicates.

Prediction and validation of chickpea miRNA target genes

The chickpea transcript dataset, which was downloaded from the chickpea transcriptome database (CTDB), was used to determine the potential target mRNA candidates for the miRNAs using the psRNATarget program with default parameters (http://plantgrn.noble.org/psRNATarget/). To reduce the false-positive prediction rate, the cut-off threshold was set at 0 to 3.0 points. Thus, all of the sequences with ≤3.0 points were considered to be miRNA targets. The functional annotations of the predicted target transcripts were performed using the NCBI nucleic acid and protein databases. Based on the predicted data, miRNA166 was validated using modified 5′ RACE. For this validation, the FirstChoice RLM-RACE Kit (Ambion, USA) was used with minor modifications, and the cDNA amplification was carried out using 1 μg of total RNA. A single PCR fragment was cloned into the pGEM-T Easy Vector (Promega, USA) and sequenced to identify the 5′end of the target gene.

Analyses of GO terms and KEGG pathways

The GO terms of the target genes were annotated according to their biological processes, molecular functions or involvement as cellular components using Blast2GO [88]. The enzyme mapping of the annotated sequences was performed directly using the GO terms, and the Kyoto Encyclopedia of Genes and Genomes (KEGG) was used to define the KEGG orthologs.

Supporting Information

Figure S1 Elimination summary of the reads.

Table S1 Conserved miRNAs that were identified in the three libraries with their detailed information.

Table S2 Precursor sequences of conserved miRNAs with their predicted secondary structures.

Table S3 Detailed information regarding novel chickpea miRNAs.

Table S4 Precursor sequences of novel miRNAs with their predicted secondary structures.

Table S5 List of the target genes that were identified for all of the conserved miRNAs.

Table S6 List of the target genes that were identified for all of the novel miRNAs.

Table S7 List of the primer sequences that were used in this study.

Acknowledgments

The authors would like to thank Dr. Rajeev Varshney, ICRISAT, India, and Dr. C. Bharadwaj, Genetics, IARI, India, for providing the seeds of the chickpea cultivar ICC4958. We are also thankful to Dr. Pooja Choudhary for providing the toxin that was isolated from the fungus *Fusarium oxysporum* f.sp. *ciceris*. Small RNA sequencing was performed at DBT-funded High-throughput Sequencing Facility at University of Delhi South Campus, New Delhi.

Author Contributions

Conceived and designed the experiments: PKJ SKA RS MA. Performed the experiments: DK ARB. Analyzed the data: GJ DK AAD. Contributed reagents/materials/analysis tools: GJ SKA PKJ. Wrote the paper: DK PKJ.

References

1. Bartel DP (2004) MicroRNAs: genomics, biogenesis, mechanism, and function. Cell 116: 281–297.
2. Sunkar R, Li YF, Jagadeeswaran G (2012) Functions of microRNAs in plant stress responses. Trends Plant Sci 17: 196–203.
3. Lee RC, Feinbaum RL, Ambros V (1993) The *C. elegans* heterochronic gene lin-4 encodes small RNAs with antisense complementarity to lin-14. Cell 75: 843–854.
4. Reinhart BJ, Slack FJ, Basson M, Pasquinelli AE, Bettinger JC, et al. (2000) The 21-nucleotide let-7 RNA regulates developmental timing in *Caenorhabditis elegans*. Nature 403: 901–906.
5. Reinhart BJ, Weinstein EG, Rhoades MW, Bartel B, Bartel DP (2002) MicroRNAs in plants. Genes Dev 16: 1616–1626.
6. Kozomara A, Griffiths-Jones S (2011) miRBase: integrating microRNA annotation and deep-sequencing data. NAR 39(Database Issue): D152–D157.
7. FAOSTAT (2012) Faostat.fao.org/site/567/DesktopDefault.aspx?PageID = 567 anchor.
8. Boominathan P, Shukla R, Kumar A, Manna D, Negi D, et al. (2004) Long term transcript accumulation during the development of dehydration adaptation in Cicer arietinum. Plant Physiol 135: 1608–1620.
9. Deokar AA, Kondawar V, Jain PK, Karuppayil SM, Raju NL, et al. (2011) Comparative analysis of expressed sequence tags (ESTs) between drought-tolerant and -susceptible genotypes of chickpea under terminal drought stress. BMC Plant Biol 11: 70.
10. Flowers TJ, Gaur PM, Gowda CLL, Krishnamurthy L, Srinivasan S, et al. (2010) Salt sensitivity in chickpea. Plant Cell Environ 33: 490–509.
11. Turner NC, Colmer TD, Quealy J, Pushpavalli R, Krishnamurthy L, et al. (2013) Salinity tolerance and ion accumulation in chickpea (Cicer arietinum L.) subjected to salt stress. Plant Soil 365: 347–361.
12. Agarwal G, Jhanwar S, Priya P, Singh VK, Saxena MS, et al. (2012) Comparative analysis of kabuli chickpea transcriptome with desi and wild chickpea provides a rich resource for development of functional markers. PLoS One 7: e52443.
13. Varshney RK, Song C, Saxena RK, Azam S, Yu S, et al. (2013) Draft genome sequence of chickpea (*Cicer arietinum*) provides a resource for trait improvement. Nature Biotech 31: 240–246.
14. Jain M, Misra G, Patel RK, Priya P, Jhanwar S, et al. (2013) A draft genome sequence of the pulse crop chickpea (*Cicer arietinum L.*). Plant J. doi: 10.1111/tpj.12173.
15. Hu J, Sun L, Ding Y (2013) Identification of conserved microRNAs and their targets in chickpea (*Cicer arietinum L.*). Plant Signal Behav. 18: 8.
16. Rajagopalan R, Vaucheret H, Trejo J, Bartel DP (2006) A diverse and evolutionarily fluid set of microRNAs in Arabidopsis thaliana. Genes Dev 20: 3407–3425.
17. Moxon S, Jing R, Szittya G, Schwach F, Rusholme Pilcher RL, et al. (2008) Deep sequencing of tomato short RNAs identifies microRNAs targeting genes involved in fruit ripening. Genome Res18: 1602–1609.
18. Szittya G, Moxon S, Santos DM, Jing R, Fevereiro MP, et al. (2008) High-throughput sequencing of *Medicago truncatula* short RNAs identifies eight new miRNA families. BMC Genomics 9: 593.
19. Martinez G, Forment J, Llave C, Pallas V, Gomez G (2011) High-throughput sequencing, characterization and detection of new and conserved cucumber miRNAs. PLoS One 6: e19523.
20. Zeng QY, Yang CY, Ma QB, Li XP, Dong WW, et al. (2012) Identification of wild soybean miRNAs and their target genes responsive to aluminum stress. BMC Plant Biol12: 182.
21. Griffiths-Jones S (2004) The microRNA Registry. NAR 32(Database Issue): D109–D111.
22. Griffiths-Jones S, Grocock RJ, van Dongen S, Bateman A, Enright AJ (2006) miRBase: microRNA sequences, targets and gene nomenclature. NAR (Database Issue): D140–D144.
23. Griffiths-Jones S, Saini HK, van Dongen S, Enright AJ (2008) miRBase: tools for microRNA genomics. NAR 36(Database Issue): D154–D158.
24. Jagadeeswaran G, Zheng Y, Li YF, Shukla LI, Matts J, et al. (2009) Cloning and characterization of small RNAs from *Medicago truncatula* reveals four novel legume-specific microRNA families. New Phytol 184: 85–98.
25. Devers EA, Branscheid A, May P, Krajinski F (2011) Stars and symbiosis: MicroRNA- and MicroRNA*-mediated transcript cleavage involved in arbuscular mycorrhizal symbiosis. Plant Physiol 156: 1990–2010.
26. Mohorianu I, Schwach F, Jing R, Lopez-Gomollon S, Moxon S, et al. (2011) Profiling of short RNAs during fleshy fruit development reveals stage-specific sRNAome expression patterns. Plant J67: 232–246.
27. Shivaprasad PV, Chen HM, Patel K, Bond DM, Santos BA, et al. (2012) microRNA superfamily regulates nucleotide binding site leucine-rich repeats and other mRNAs. Plant Cell 24: 859–874.
28. Garg R, Patel RK, Tyagi AK, Jain M (2011) De novo assembly of chickpea transcriptome using short reads for gene discovery and marker identification. DNA Res 18: 53–63.
29. Meyers BC, Axtell MJ, Bartel B, Bartel DP, Baulcombe D, et al. (2008) Criteria for annotation of plant microRNAs. Plant Cell 20: 3186–3190.
30. Chi XY, Yang QL, Chen XP, Wang JY, Pan LJ, et al. (2011) Identification and characterization of microRNAs from peanut (*Arachis hypogaea* L.) by high-throughput sequencing. PLoS One 6: e27530.
31. Zuker M (2003) Mfold web server for nucleic acid folding and hybridization prediction. Nucleic Acids Res 31: 3406–3415.
32. Sunkar R, Jagadeeswaran G (2008) Insilico identification of conserved microRNAs in large number of diverse plant species. BMC Plant Biol 8: 37.
33. Dai X, Zhao PX (2011) psRNATarget: a plant small RNA target analysis server. Nucleic Acids Res 39: 9–155.
34. Williams L, Grigg SP, Xie M, Christensen S, Fletcher JC (2005) Regulation of Arabidopsis shoot apical meristem and lateral organ formation by microRNA miR166g and its AtHD-ZIP target genes. Development 132: 3657–3668.
35. Kulcheski FR, de Oliveira LF, Molina LG, Almerao MP, Rodrigues FA, et al. (2011) Identification of novel soybean microRNAs involved in abiotic and biotic stresses. BMC Genomics 12: 307.
36. Kim J, Jung JH, Reyes JL, Kim YS, Kim SY, et al. (2005) microRNA-directed cleavage of ATHB15 mRNA regulates vascular development in Arabidopsis inflorescence stems. Plant J 42: 84–94.
37. Byrne ME (2006) Shoot meristem function and leaf polarity: the role of class III HD-ZIP genes. PLoS Genet 2: 785–790.
38. Llave C, Kasschau KD, Rector MA, Carrington JC (2002) Endogenous and silencing-associated small RNAs in plants. Plant Cell 14: 1605–1619.
39. Jones-Rhoades MW, Bartel DP (2004) Computational identification of plant microRNAs and their targets, including a stress-induced miRNA. Mol Cell 14: 787–799.
40. Schwab R, Palatnik JF, Riester M, Schommer C, Schmid M, et al. (2005) Specific effects of microRNAs on the plant transcriptome. Dev Cell 8: 517–527.
41. Song QX, Liu YF, Hu XY, Zhang WK, Ma B, et al. (2011) Identification of miRNAs and their target genes in developing soybean seeds by deep sequencing. BMC Plant Biol 11: 5.
42. Lu S, Sun YH, Amerson H, Chiang VL (2007) MicroRNAs in loblolly pine (*Pinus taeda L.*) and their association with fusiform rust gall development. Plant J 51: 1077–1098.
43. Vaucheret H, Vazquez F, Crete P, Bartel DP (2004) The action of ARGONAUTE1 in the miRNA pathway and its regulation by the miRNA pathway are crucial for plant development. Genes Dev 18: 1187–1197.
44. Huertero CA, Pe'rez B, Rabanal F, Blanco-Melo D, Rosa CD, et al. (2009) Conserved and novel miRNAs in the legume *Phaseolus vulgaris* in response to stress. Plant Mol Biol 70: 385–401.
45. Millan T, Clarke HJ, Siddique KHM, Buhariwalla HK, Gaur PM, et al. (2006) Chickpea molecular breeding: New tools and concepts. Euphytica 147: 81–103.
46. Bartel B, Bartel DP (2003) MicroRNAs: at the root of plant development. Plant Physiol 132: 709–717.
47. Mi S, Cai T, Hu Y, Chen Y, Hodges E, et al. (2008) Sorting of small RNAs into Arabidopsis argonaute complexes is directed by the 59 terminal nucleotide. Cell 133: 116–127.
48. Xue LJ, Zhang JJ, Xue HW (2009) Characterization and expression profiles of miRNAs in rice seeds. Nucleic Acids Res 37: 916–930.
49. Jagadeeswaran G, Zheng Y, Sumathipala N, Jiang HB, Arrese EL, et al. (2010) Deep sequencing of small RNA libraries reveals dynamic regulation of conserved and novel microRNAs and microRNA-stars during silkworm development. BMC Genomics 11: 52.
50. Zhou H, Huang X, Cui H, Luo X, Tang Y, et al. (2010a) miR-155 and its star-form partner miR-155* cooperatively regulate type I interferon production by human plasmacytoid dendritic cells. Blood 116: 5885–5894.
51. Navarro L, Dunoyer P, Jay F, Arnold B, Dharmasiri N, et al. (2006) A plant miRNA contributes to antibacterial resistance by repressing auxin signaling. Science 21: 436–439.

52. Zhang X, Zhao H, Gao S, Wang WC, Katiyar-Agarwal S, et al. (2011) Arabidopsis Argonaute 2 regulates innate immunity via miRNA393(*)-mediated silencing of a Golgi-localized SNARE gene, MEMB12. Mol Cell 42: 356–366.
53. Loudet O, Michael TP, Burger BT, Le Mette C, Mockler TC, et al. (2008) A zinc knuckle protein that negatively controls morning-specific growth in *Arabidopsis thaliana*. Proc Natl Acad SciUSA105: 17193–17198.
54. Lu S, Sun YH, Chiang VL (2008) Stress-responsive microRNAs in Populus. Plant J 55: 131–151.
55. Wong CE, Zhao YT, Wang XJ, Croft L, Wang ZH, et al. (2011) MicroRNAs in the shoot apical meristem of soybean. J Exp Bot 62: 2495–2506.
56. Guzman F, Almerao MP, Korbes AP, Loss-Morais G, Margis R (2012) Identification of MicroRNAs from *Eugenia uniflora* by high-throughput sequencing and bioinformatics analysis. PLoS One 7: e49811.
57. Liu B, Li P, Li X, Liu C, Cao S, et al. (2005) Loss of function of OsDCL1 affects microRNA accumulation and causes developmental defects in rice. Plant Physiol 39: 296–305.
58. Wu F, Shu J, Jin W (2014) Identification and validation of miRNAs associated with the resistance of Maize (*Zea mays L.*) to *Exserohilum turcicum*. PLoS One 9: e87251.
59. Xie DX, Feys BF, James S, Nieto-Rostro M, Turner JG (1998) *COI1*: an *Arabidopsis* gene required for jasmonate regulated defense and fertility. Science 280: 1091–1094.
60. Samach A, Klenz JE, Kohalmi SE, Risseeuw E, Haughn G, et al. (1999) The *UNUSUAL FLORAL ORGANS* gene of *Arabidopsis thaliana* is an F-box protein required for normal patterning and growth in the floral meristem. Plant J 20: 433–445.
61. Gray WM, Estelle M (2000) Function of the ubiquitin proteasome pathway in auxin response. TIBS 25: 133–138.
62. Craig KL, Tyers M (1999) The F-box: a new motif for ubiquitin dependent proteolysis in cell cycle regulation and signal transduction. Prog Biophys Mol Biol 72: 299–328.
63. Gray WM, delPozo JC, Walker L, Hobbie L, Risseeuw E, et al. (1999) Identification of an SCF ubiquitin-ligase complex required for auxin response in Arabidopsis thaliana. Genes Dev 13: 1678–1691.
64. Dharmasiri N, Dharmisiri S, Estelle M (2005) The F-box protein TIR1 is an auxin receptor. Nature 435: 441.
65. Kepinski S, Leyser O (2005) The Arabidopsis F-box protein TIR1 is an auxin receptor. Nature 435: 446–51.
66. Andrade MA, González-Guzmán M, Serrano R, Rodríguez PL (2001) A combination of the F-box motif and kelch repeats defines a large *Arabidopsis* family of F-box proteins. Plant Mol Biol 46: 603.
67. Yin Z, Li Y, Han X, Shen F (2012) Genome-wide profiling of miRNAs and other small non-coding RNAs in the *Verticillium dahliae* inoculated cotton roots. PLoS ONE 7: e35765.
68. Ashraf N, Ghai D, Barman P, Basu S, Gangisetty N, et al. (2009) Comparative analyses of genotype dependent expressed sequence tags and stress-responsive transcriptome of chickpea wilt illustrate predicted and unexpected genes and novel regulators of plant immunity. BMC Genomics 10: 415.
69. Zhu L, Zhang X, Tu L, Zeng F, Nie Y, et al. (2007) Isolation and characterization of two novel dirigent-like genes highly induced in cotton (*Gossypium barbadense* and *G. hirsutum*) after infection by *Verticillium dahliae*. Plant Pathol 89: 41–45.
70. Liu HH,Tian X, Li YJ, Wu CA, Zheng CC (2008) Microarray-based analysis of stress-regulated microRNAs in Arabidopsis thaliana. RNA 14: 836–843.
71. Li WX, Oono Y, Zhu J, He XJ, Wu JM, et al. (2008) The Arabidopsis NFYA5 transcription factor is regulated transcriptionally and post transcriptionally to promote drought resistance. Plant Cell 20: 2238–2251.
72. Zhao B, Ge L, Liang R, LiW, Ruan K, et al. (2009) Members of miR-169 family are induced by high salinity and transiently inhibit the NF-YA transcription factor. BMC Mol Biol 10: 29.
73. Jain M, Nijhawan A, Arora R, Agarwal P, Ray S, et al. (2007) F-box proteins in rice. Genome-wide analysis, classification, temporal and spatial gene expression during panicle and seed development, and regulation by light and abiotic stress. Plant Physiol 143: 1467–1483.
74. Zhang X, Zhen JB, Li ZH, Kang DM, Yang YM, et al. (2011) Expression profile of early responsive genes under salt stress in upland cotton (*Gossypium hirsutum* L.). Plant Mol Biol Rep 29: 626–637.
75. Gao P, Bai X, Yang L, Lv D, Li Y, et al. (2010) Over-expression of osa-MIR396c decreases salt and alkali stress tolerance. Planta 231: 991–1001.
76. Mantri NL, Ford R, Coram TE, Pang ECK (2010a) Evidence of unique and shared responses to major biotic and abiotic stresses in chickpea. Environ Exper Bot 69: 286–292.
77. Colcombet J, Hirt H (2008) Arabidopsis MAPKs: a complex signalling network involved in multiple biological processes. Biochem J 413: 217–226.
78. Rodriguez RE, Mecchia MA, Debernardi JM, Schommer C, Weigel D, et al. (2010) Control of cell proliferation in Arabidopsis thaliana by microRNA miR396. Development 137: 103–112.
79. Winter P, Benko AM, Hüttel B, Ratnaparkhe M, Tullu A, et al. (2000) A linkage map of the chickpea (*Cicer arietinum L.*) genome based on recombinant inbred lines from a *C. arietinum* × *C. reticulatum* cross: localization of resistance genes for fusarium wilt races 4 and 5. Theor Appl Genet 101: 1155–1163.
80. Lu C, Meyers BC, Green PJ (2007) Construction of small RNA cDNA libraries for deep sequencing. Methods 43: 110–117.
81. Stocks MB, Moxon S, Mapleson D, Woolfenden HC, Mohorianu I, et al. (2012) The UEA sRNA workbench: a suite of tools for analysing and visualizing next generation sequencing microRNA and small RNA datasets. Bioinformatics 28: 2059–2061.
82. Moxon S, Schwach F, Dalmay T, MacLean D, Studholme DJ, et al. (2008) A toolkit for analysing large-scale plant small RNA datasets. Bioinformatics 24: 2252–2253.
83. Bonnet E, Wuyts J, Rouzé P, Van de Peer Y (2004) Evidence that microRNA precursors, unlike other non-coding RNAs, have lower folding free energies than random sequences. Bioinformatics 20: 2911–7.
84. Zhang BH, Pan XP, Cox SB, Cobb GP, Anderson TA (2006) Evidence that miRNAs are different from other RNAs. Cell Mol Life Sci 63: 246–254.
85. Hofacker IL (2003) Vienna RNA secondary structure server. Nucleic Acids Res 31: 3429–3431.
86. Edgar R, Domrachey M, Lash AE (2002) Gene Expression Omnibus: NCBI gene expression and hybridization array data repository. Nucleic Acids Res 30: 207–210.
87. Livak KJ, Schmittgen TD (2001) Analysis of relative gene expression data using real-time quantitative PCR and the 2[-Delta DeltaC (T)] Method. Methods 25: 402–408.
88. Conesa A, Gotz S, Garcia-Gomez JM, Terol J, Talon M, et al. (2005) Blast2GO: a universal tool for annotation, visualization and analysis in functional genomics research. Bioinformatics 21: 3674–3676.

15

Biochemical Aspects of the Soybean Response to Herbivory Injury by the Brown Stink Bug *Euschistus heros* (Hemiptera: Pentatomidae)

Renata Velozo Timbó[1,2], Marcelo Hermes-Lima[1], Luciano Paulino Silva[2], Angela Mehta[2], Maria Carolina Blassioli Moraes[2], Débora Pires Paula[2]*

1 Department of Cell Biology, University of Brasília, Campus Universitário Darcy Ribeiro, Brasília, Distrito Federal, Brazil, **2** Department of Biological Control, Embrapa Genetic Resources and Biotechnology, Brasília, Distrito Federal, Brazil

Abstract

Plant defense response is an elaborate biochemical process shown to depend on the plant genetic background and on the biological stressor. This work evaluated the soybean biochemical foliar response to brown stink bug herbivory injury through an analysis of redox metabolism and proteomic 2DE profiles of susceptible (BRS Silvania RR) and resistant (IAC-100) varieties. The activity of lipoxygenase-3, guaiacol peroxidase, catalase and ascorbate peroxidase was monitored every 24 h up to 96 h. In the susceptible variety, injury caused an increase in the activities of lipoxygenase 3 and guaiacol peroxidase, no change in ascorbate peroxidase, and a decrease in catalase. In the resistant variety, injury did not cause an alteration of any of these enzymes. The proteomic profiles were evaluated after 24 h of injury and revealed to have a similar proportion (4–5%) of differential protein expression in both varieties. The differential proteins, identified by mass spectrometry, in the susceptible variety were related to general stress responses, to plant defense, and to fungal infections. However, in the resistant variety, the identified change in protein profile was related to Calvin cycle enzymes. While the susceptible variety showed adaptive changes in redox metabolism and expression of stress-responsive proteins, the resistant showed a defense response to circumvent the biological stressor.

Editor: Joseph Clifton Dickens, United States Department of Agriculture, Beltsville Agricultural Research Center, United States of America

Funding: This work was supported by Embrapa (www.embrapa.br) (DPP AM LPS), CNPq (www.cnpq.br) (LPS), and CAPES (www.capes.gov.br) (RVT). The funders had no role in study design, data collection and analysis, decision to publish, or preparation of the manuscript.

Competing Interests: The authors have declared that no competing interests exist.

* Email: debora.pires@embrapa.br

Introduction

Stink bugs (Hemiptera: Pentatomidae) are one of the most important pests of soybean in Brazil [1]. Pentatomids feed on various structures of the host plant, mainly fruits and immature seeds, affecting their quality, development and maturation [2]. The brown stink bug *Euschistus heros* is considered one of the main soybean pests [3,4]. Information about the soybean defense mechanisms to this herbivore when it colonizes in the late vegetative stages is important to support pest management.

Several varieties of soybean have been studied regarding their defense response to herbivores in order to identify herbivory resistant varieties [5,6,7,8]. These studies have mostly focused on primary and secondary metabolite production as parts of the plant defense strategy aiming to support environmentally friendly soybean production. Plant defense to herbivory injury involves two major systems: (i) the direct defense, involving structural components (e.g. thorns, trichomes), and production of primary metabolites (e.g. proteinase inhibitors, antioxidant enzymes) and non-volatile secondary metabolites (e.g. phenolic acids and isoflavonoids) [9,10]; and (ii) the indirect defense, which is related to the production of volatile organic compounds (VOCs), such as (*E*,*E*)-α-farnesene, methyl salicylate and *cis*-jasmone, which attract natural enemies (predators and parasitoids) of the herbivores that feed on the plants, or in some cases act as herbivore repellents [6,11].

Plant defense response varies according to the plant genetic background, and the type of stressor [10]. Therefore, it is valuable to study the response elicited by the key herbivores in different varieties. Biochemical studies on plant defense have focused on evaluating the activity of enzymes related to the redox metabolism (including antioxidant enzymes) [9,12–14], such as lipoxygenase (LOX), peroxidase (POD), glutathione reductase (GR), ascorbate peroxidase (APX), and catalase (CAT), as well as on the modification of protein profiles [15]. However, despite the significant economic impact of the damage caused by stink bugs to soybean [16,17], there are no reports regarding the soybean redox response and the regulation of protein expression induced by stink bug herbivory.

Plants differ in resistance to their key herbivorous pests according to tolerance, non-preference and antibiosis [18]. In this work we used two soybean varieties with large differences in antibiosis resistance to brown stinkbug: resistant IAC-100 and susceptible BRS Silvania RR. IAC-100 was developed by crossing the parental varieties IAC-12 and IAC 78-2318, showing resistance to chewing and sucking insects [19]. BRS Silvania RR was developed for glyphosate tolerance [20]. The goal was to characterize the biochemical effects of the brown stink bug herbivory injury on the foliar redox response, through the analysis of the activity of lipoxygenase-3 (LOX3), guaiacol peroxidase (GOPx), catalase (CAT) and ascorbate peroxidase (APX), as well as on the foliar protein expression, through two-dimensional electrophoresis and mass spectrometry.

Materials and Methods

Soybean cultivation and stink bug rearing

The seeds of IAC-100 and BRS Silvania RR were germinated in a neutral pH germination paper. Seedlings with equivalent development were transplanted to plastic pots of 350 mL containing mixture of soil and organic substrate (in a proportion of 3:1), maintained in a germination chamber at 25±2°C and 13 h photoperiod, and irrigated every two days. The bioassays were conducted with plants in vegetative stage 3 (V3, first three trifoliolate leaves fully expanded). Stink bugs colonize soybeans during the vegetative stage, during which soybean plants produce higher amount of volatiles compared to the reproductive stage, and are known to have their indirect defense elicited by injury of the brown stink bug [10,17,21]. Understanding the plant defense mechanisms during this stage could help control stink bug pest populations that result in the reproductive stage.

The brown stink bugs were obtained from a laboratory colony at Embrapa Genetic Resources and Biotechnology (Brasília, Brazil), reared according to Moraes *et al.* [18] in plastic cages (26×22 cm) containing cotton fiber soaked in water and fed on peanut seeds [*Arachis hypogaea* (L.)], soybean grains [*Glycine max* (L.) Merrill.], sunflower seeds [*Helianthus annus* (L.)] and green beans [*Phaseolus vulgaris* (L.)], at 26±2°C, 60±10% RH and 14 h photoperiod.

Herbivory bioassays

The herbivory bioassays consisted of no injury (control) or injury to individual V3 plants of the two soybean varieties by three male adult brown stink bugs. Plants were exposed to stink bugs for 24, 48, 72, or 96 h after which the plants were harvested. Simultaneously, plants were harvested from the control. There were six replicate plants per harvest per treatment in a completely randomized design. Only males were used to prevent possible interference of oviposition on the plants, which could trigger other plant responses [6,21]. The stink bugs were deprived of food for 24 h before the bioassays. The plants were covered with transparent plastic bags (with micro holes for aeration) to guarantee the constant presence of the stink bugs during the bioassay. All assays were performed in separate germination chambers (FANEM) for each treatment, so that the VOC released could not interfere with the other treatments. All leaves of the six plants per treatment were collected and transferred to 50 mL tubes and immediately frozen in liquid nitrogen and stored at −80°C for further biochemical analysis.

Quantification of enzyme activities

The enzymatic activity was measured for LOX3, GOPx, APX and CAT in the soybean foliar extracts of three plants from each treatment in the four 24 h sampling intervals. Foliar homogenates were prepared in a mortar containing liquid nitrogen, where the leaves were macerated until the formation of a fine powder. For every 100 mg of macerated leaf, 400 µL of 50 mM potassium phosphate buffer (KPi) containing 10 mM phenylmethanesulfonyl fluoride (PMSF), 5 mM ethylenediaminetetraacetic acid (EDTA) was added. After homogenization, the samples were centrifuged at 14,000×*g* for 15 min at 4°C. The supernatant was quantified by the Bradford method [22] and used in triplicate for each enzyme assay.

The conditions for the enzymatic activity assays are detailed in Table 1. The assays for GOPx, APX and CAT were performed in quartz cuvettes (10×10 mm) by adding 900 µL of KPi pH 7.0 and 1 to 20 µL of foliar supernatant. For LOX3, 2,975 µL Kpi pH 6.5 and 1 to 20 µL of foliar supernatant were used. The reactions were initiated by adding the substrate for each enzyme (100 µL for GOPx, APX and CAT and 25 µL for LOX3) and the absorbance was measured for 150 s at intervals of 10 s with a Smart Plus spectrophotometer (BioRad).

The enzymatic activities ($\mu mol.min^{-1}.mg^{-1}$) were compared by one-way ANOVA, and the paired comparisons among the treatments were performed by Tukey's HSD ($p<0.05$).

Total protein extraction for the proteomic analysis

The total foliar protein profiles of all treatments (three plants/each) were analyzed after 24 h of bioassay by 2-DE. Total protein extraction was performed adapting the TCA/acetone methodology [27]. According to Wildgruber *et al.* [28], this method inhibits the degradation of proteins by minimizing protease activity (due to low pH), and also improves cell lysis and subsequent precipitation of proteins at −20°C. The leaves from the same treatment were pooled and macerated in liquid nitrogen until the formation of a fine powder and chilled cold acetone solution, containing 10% (w/v) TCA and 0.07% (v/v) β-mercaptoethanol, was added in a proportion of 1:2 v/v. After vortex homogenization for 5 min and incubation for 24 h at −20°C, the samples were centrifuged at 10,000×*g* for 20 min at 4°C. The pellets were washed three times in cold acetone containing 0.07% (v/v) β-mercaptoethanol and centrifuged at 10,000×*g* for 10 min at 4°C. To each 12 mg of pellets 1 mL of solubilization buffer [8.5 M urea, 2.5 M thiourea, 2% (w/v) CHAPS, 1% (w/v) DTT, 1% (w/v) Serdolit, 2% (v/v) IPG buffer nonlinear 3–11, 1% (w/v) bromophenol blue] [29] was added and incubated at room temperature for 1 h under gentle stirring. After centrifugation at 14,000×*g* for 1 h at 25°C, the supernatants were collected and the proteins quantified by the Bradford method [22].

Two-dimensional electrophoresis (2-DE) and gel image analysis

The isoelectric focusing (IEF) was carried out in 13 cm Immobiline DryStrips pH 3–11 NL (GE Healthcare). The strips were hydrated overnight with 250 µL of solubilization buffer containing 500 µg of total protein extract (three strips for each treatment). The IEF was conducted at 20°C using the Multiphor II Electrophoresis System (GE Healthcare) coupled to a cooling system bath in one step of 300 V for 1 min and one step of 3,500 V for 5.5 h (all at 2 mA and 5 W). After IEF, the strips were reduced by the addition of 8 mL of equilibration buffer [50 mM Tris-HCl pH 8.8, 6 M urea, 30% (v/v) glycerol, 2% (w/v) SDS] containing 1% (w/v) DTT under gentle agitation for 15 min at room temperature; and alkylated by adding 8 mL of equilibration buffer containing 2.5% (w/v) iodoacetamide under gentle agitation for 15 min. The equilibrated strips were transferred to a 12.5% denaturaing polyacrylamide gel and electrophoresis was performed at 25°C in Tris/glycine/SDS buffer on a multiple Ettan DALTsix (GE Healthcare), coupled to a cooling bath system, according to manufacturer's recommendation. For molecular mass reference, the BenchMark Protein ladder (Invitrogen) was applied on a side of the gel using filter paper. The electrophoresis was carried out in two steps: first at 10 mA/80 V for 1 h and 40 mA/500 V for 4.3–6 h. The 2-DE gels were

Table 1. Enzymatic activity assay conditions in leaves of the soybean varieties BRS Silvania RR and IAC-100 in the V3 stage, with and without (control group) brown stink bug herbivory injury.

Enzyme	Substrate	ε ($mM^{-1}cm^{-1}$)	Wavelength (nm)	Final Volume (μL)	References
LOX3	80 mM sodium linoleate	22.00	234	3,000	[23]
GOPx	10 mM hydrogen peroxide +20 mM guaiacol	26.60	470	1,000	[24]
APX	10 mM hydrogen peroxide +40 mM ascorbic acid	14.50	265	1,000	[25]
CAT	10 mM hydrogen peroxide	0.04	240	1,000	[26]

stained for two days in a solution of 0.8% (w/v) Coomassie Brilliant Blue G-250, 20% (v/v) ethanol, 1.6% (v/v) phosphoric acid, 8% (w/v) ammonium sulfate, and rinsed five times in water to remove excess dye.

The 2-DE gel images were obtained by ImageScanner III (GE Healthcare). Spot detection and measurements, as well as spot matching in the three gel images per treatment, were performed using the ImageMaster 2D Platinum 7.0 software (GE Healthcare). A spot was considered differential when its intensity and area were different in the injured and uninjured gels. The comparison of the total average number of spots between the injured and uninjured treatments was performed by one-away ANOVA ($p<$ 0.05).

Protein identification

For the soybean variety BRS Silvania RR the differential spots were excised and sliced into small fragments. Spot destaining was conducted by successive washes in 30% ethanol and incubation in 50% (v/v) acetonitrile solution containing 25 mM ammonium bicarbonate under agitation for 15 min. The gel fragments were dehydrated with 100% acetonitrile for 10 min under agitation and dried in a centrifugal evaporator for 20 min. The spots were rehydrated with 50 μL of hydrolysis solution consisting of 50 mM Tris-HCl pH 7.6 added to 1 mM calcium chloride and 0.5 μL of trypsin at 1 μg/μL (Sequencing Grade Modified Trypsin, Promega), which was previously heated to 30°C for 15 min, and to 50 μL of 50 mM ammonium bicarbonate. The samples were incubated at 37°C for 14 to 18 h. For the soybean variety IAC-100, trypsin hydrolysis was carried out using the Trypsin Profile IGD Kit (Sigma), according to the manufacturer instructions.

The trypsin-hydrolyzed peptides were analyzed by MALDI-TOF mass spectrometry using an UltraFlex III MALDI TOF/TOF (Bruker Daltonics) controlled by the software FlexControl 3.3.108.0 using the following parameters: mass range 600-3,500 Da; RP_PEPMIX method; 200 automatic shooting at a frequency of 200 Hz, laser intensity between 20–45%; deflection mass of 600 Da; reflective mode; and positive external calibration by peptide calibration standard I. The method to obtain the MS/MS data was the LIFT; AUTO_LIFT method; using the following parameters: 2,000 shots for which, 400 and 1,600 shots for the precursor and the fragments, respectively; laser intensity between 22–55%; and the other parameters were the same as mentioned above. The peaks list files containing the m/z ratios of precursor ions and MS/MS fragmented ions were used to search similar oligopeptides in the National Center for Biotechnology Information (NCBI) database sequences using BLASTp and Mascot software. The parameters used in the Mascot program for MS/MS were: (i) NCBI-nr database, (ii) Viridiplantae taxonomy, (iii) fixed modifications: carbamidomethylation of cysteine, (iv) variable modification: oxidation of methionine, (v) monoisotopic mass spectra data, (vi) mass tolerance of 150 ppm, (vii) fragment tolerance of ±0.05 Da. Only significant hits, as defined by Mascot ($p<0.05$), were accepted.

Results and Discussion

Redox response

The herbivory injury by the stink bug affected the foliar enzymatic response ($p<0.05$) in the susceptible BRS Silvania RR variety (Figure 1). The LOX3 activity was significantly higher at 72 h of herbivory, with a decline at 96 h, but still significantly higher than in the first 48 h of injury and in the uninjured control. The GOPx activity did not differ over the time of injury, but was higher along the time evaluated compared to the uninjured plants. On the other hand, APX activity was not affected during the herbivory period, and CAT activity decreased after 48 h of injury compared to uninjured plants. In contrast, the herbivory injury by the brown stink bug did not affect the foliar activity of any of the enzymes evaluated even after 96 h of injury (Fig. 2) for the resistant IAC-100 variety, suggesting that the redox metabolism was not affected in this variety under *E. heros* feeding injury.

From the enzymes studied, LOX3 and GOPx have a clear relation to the plant defense response and an enhanced activity under herbivory injury has been reported [30–32]. LOX3 catalyzes the production of jasmonic acid (JA) from linolenic acid, which stimulates the expression of genes responsible for producing defense compounds and is related to plant defense against herbivorous insects, and GOPx activates defense genes [30–32]. Chen *et al.* [24] observed that herbivory by *Melanocallis caryaefoliae* (Hemiptera: Aphididae) induced higher activity of LOX3 and GOPx in susceptible pecans, although in moderate and high resistance pecans the GOPx activity did not change. Similarly, Ni *et al.* [33] reported increased activity of GOPx, CAT and polyphenol oxidase in wheat leaves after herbivory by *Diuraphis noxia* (Hemiptera: Aphididae), and that peroxidase activity in barley leaves was significantly higher in susceptible than in resistant plants after herbivory by the same aphid species.

Although uninjured BRS Silvania RR generally had higher activity of three enzymes related to the redox metabolism than IAC-100, LOX3 activity was higher in uninjured IAC-100 than in BRS Silvania RR. This result may be directly related to resistance, because one of the functions of the lipoxygenase pathways the production of JA. The constitutive high activity of lipoxygenase in plants is related to their resistance against insects [34], which may help to explain the resistance of IAC-100 against stink bugs [6,16] and why it produces high level of induced plant volatiles when injured by them, which were not detected in BRS Silvania RR [6].

APX catalyzes the redox reaction between hydrogen peroxide and ascorbate [35]. Bi & Felton [36] reported that APX activity in soybean leaves increased over time after herbivory by *Helicoverpa zea* (Lepidoptera: Noctuidae), with maximum activity at 96 h. In our results, the activity of APX was not affected by stink bug

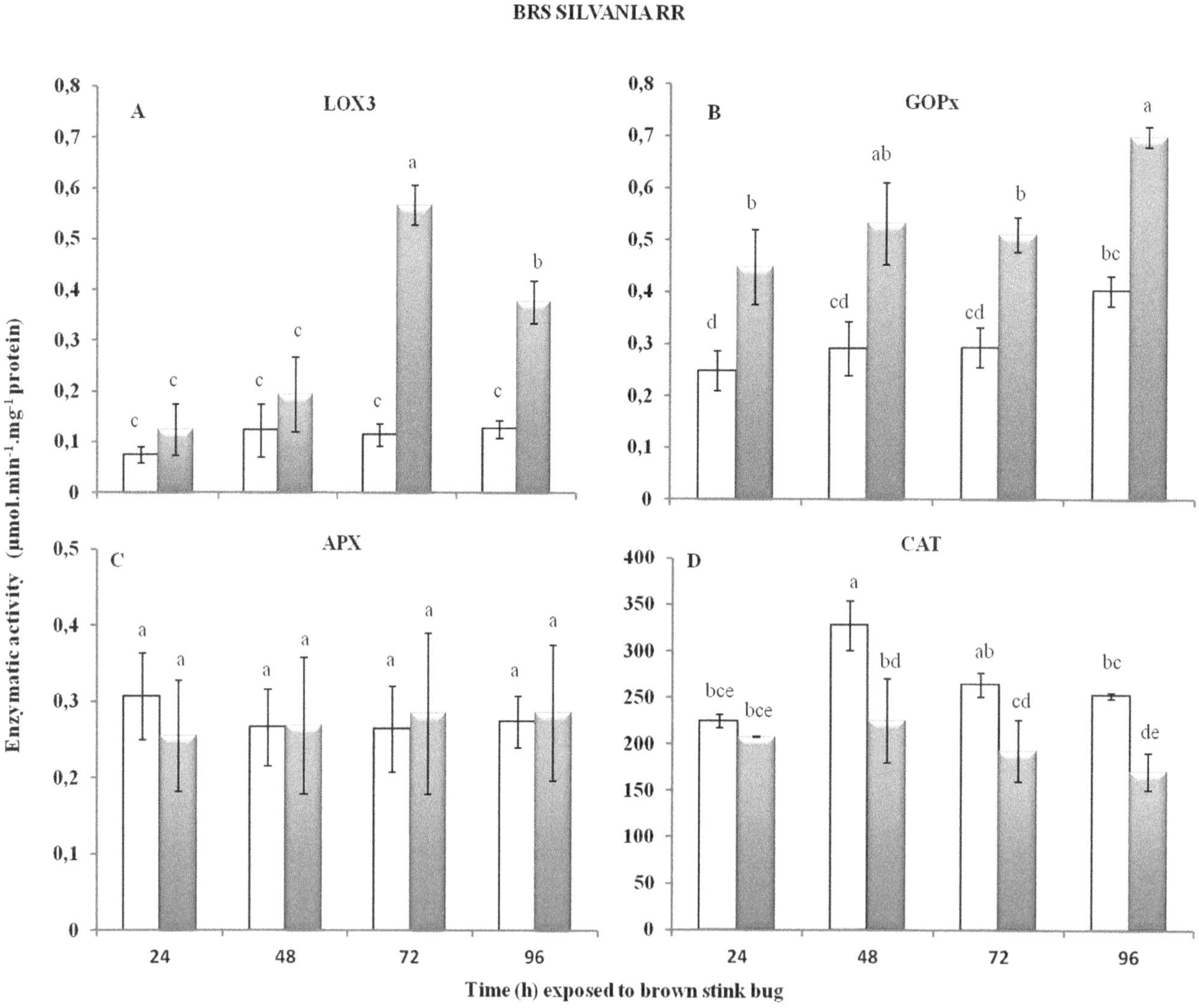

Figure 1. Enzymatic activity for (A) LOX3 ($F_{7,16}=2.32$; $p=0.0779$), (B) GOPx ($F_{7,16}=1.01$; $p=0.4629$), (C) APX ($F_{7,16}=3.35$; $p=0.0213$), and (D) CAT ($F_{7,16}=1.78$; $p=0.1609$) in leaves of BRS Silvania RR soybean variety (V3) with or without brown stink bug herbivory injury along different sampling periods. Injured is represented in grey and not injured in white. Bars with the same letter are not significantly different by Tukey's HSD ($p<0.05$).

herbivory on the susceptible soybean variety. Stink bugs are piercing-sucking insect, therefore provoking less damage to plants compared to chewing insects [37], and possibly inducing lower levels of hydrogen peroxide in the injury process.

CAT performs two major functions: (i) decomposition of hydrogen peroxide (a product of photorespiration and other plant metabolic activities) and (ii) oxidation of hydrogen donors such as methanol, ethanol, formic acid, phenols, with consumption of hydrogen peroxide [25,38]. An increase in CAT activity may be related to the control of hydrogen peroxide levels (which can be stressful at high concentrations) formed in response to herbivory injury [39]. In herbivory studies on various plant species, CAT activity remained unchanged [12,24,40], increased [33] or decreased [36,37]. In our study, a decrease in CAT activity was observed in response to stink bug injury only in the susceptible variety, showing that the varieties are managing the intracellular hydrogen peroxide levels differently. The observed increase in GOPx activity could be compensating the reduction in CAT levels in BRS Silvania RR leaves.

Overall, the susceptible soybean variety (BRS Silvania RR) showed redox gene expression changes, while the resistant variety (IAC-100) did not. The fluctuation in the activity of the redox metabolism enzymes might be associated with redox stress, which could be an indication of herbivory susceptibility. In the resistant variety, the redox genes might be constitutively expressed, so the stink bug injury might not produce enough reactive oxygen species (ROS) to elicit changed expression of these genes. Additional assessment of the redox alterations generated by herbivory across time would include measures such as protein carbonyls, lipid peroxidation and oxidized glutathione (GSSG), which are also classic indicators of redox oxidative stress, as well as other antioxidant enzymes, such as superoxide dismutases and peroxyredoxins [26].

Expression of foliar proteins

In both varieties, the foliar protein profile changed after 24 h of herbivore injury by the brown stink bug. Not all differentially expressed proteins could be identified, even when using the

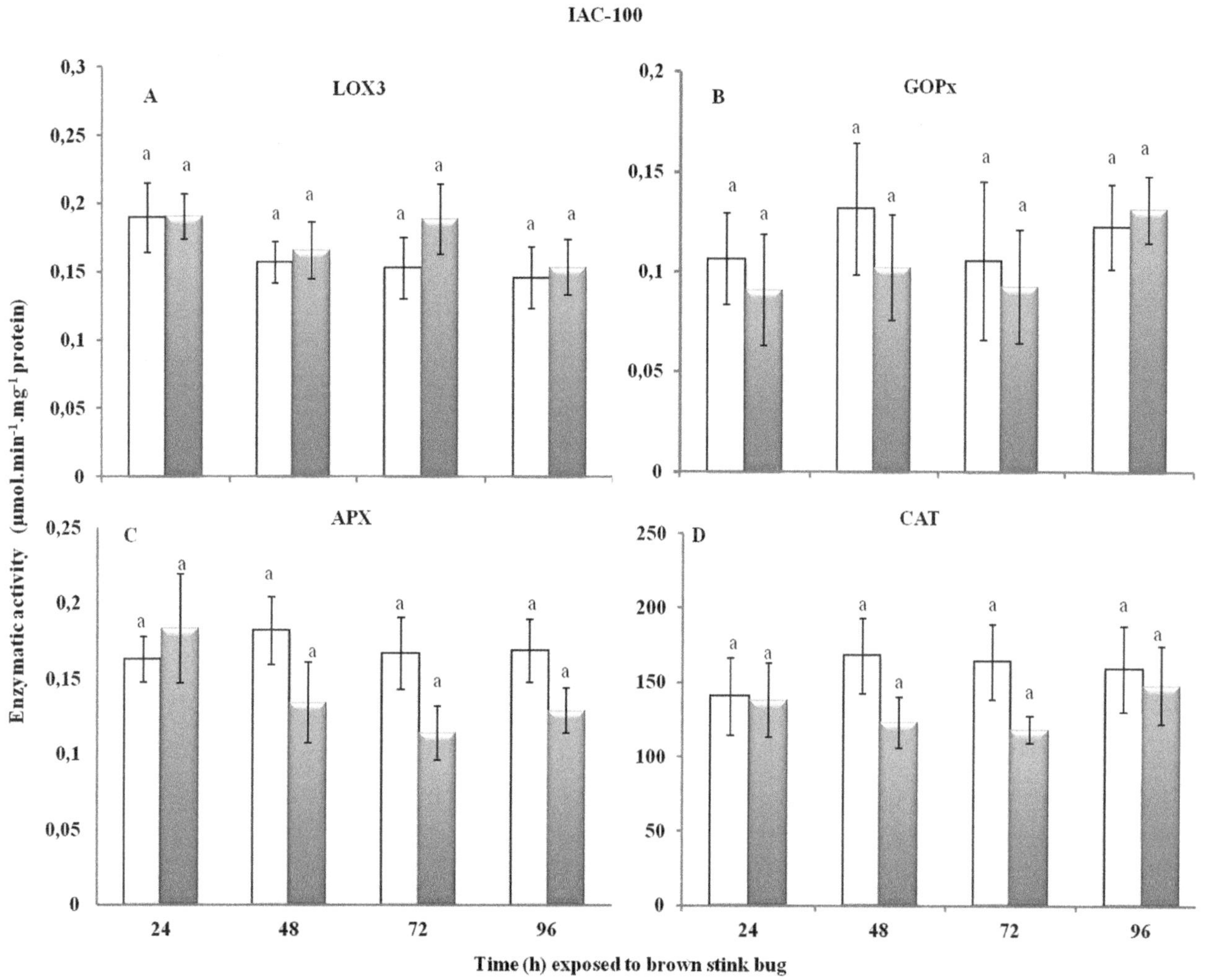

Figure 2. Enzymatic activity for (A) LOX3 ($F_{7,16}=34.52$; $p<0.0001$), (B) GOPx ($F_{7,16}=29.6$; $p<0.0001$), (C) APX ($F_{7,16}=0.16$; $p=0.9907$), and (D) CAT ($F_{7,16}=12.59$; $p<0.0001$) in leaves of IAC-100 soybean variety (V3) with and without brown stink bug herbivory injury along different sampling periods. Injured is represented in grey and not injured in white. Bars with the same letter are not significantly different by Tukey's HSD ($p<0.05$).

soybean genome reference database, possibly due to their low concentration or even absence in the database.

The 2-DE gel analyses of the BRS Silvania RR foliar protein profiles had an average of 386 protein spots per gel, while IAC-100 showed an average of 191 protein spots per gel, and most proteins were observed in the pH range from 4 to 8 and in the molecular mass range from 15 to 100 kDa (Fig. 3). The gel replicates of uninjured and injured plants had a high reproducibility: $r^2=0.95$ and 0.96, respectively for the susceptible; and $r^2=0.93$ and 0.97, respectively for the resistant.

When the foliar protein profile of the injured susceptible plants was compared to the uninjured profile, 17 proteins were differentially expressed, of which eight were up-regulated (Table 2). The identified up-regulated proteins were the chaperonin HSP20, glycine-rich 2b-like protein, both related to biotic and abiotic stress responses [41,42], and one hypothetical protein (gi|147800453).

The identified down-regulated proteins were CP4-EPSPS enzyme, a protease inhibitor, related to plant defense to herbivory, a hypothetical hydrolase function, related to plant defense against fungal infections [43], a heat-shock chaperonin 70 kDa protein (HSP), a glycine-rich RNA binding protein (GRP), related to general plant stress responses, a type 1 glutamine amidotransferase-like (GATase1) and three unknown hypothetical proteins (gi|40538952, gi|242054087, gi|147800453).

CP4-EPSPS participates in the shikimic acid pathway synthetizing the aromatic amino acids phenylalanine, tyrosine and tryptophan, which act as precursors for a wide range of products such as alkaloids, the plant hormone auxin, phytoalexins [44], flavonoids, and phenolic compounds such as lignin [45], and secondary volatile compounds such as methyl salicylate and indole, which are involved in indirect plant defense [6,20,21]. Michereff *et al.* [6] showed that the soybean variety BRS Silvania RR did not release methyl salicylate or indole when injured by the brown stink bug. Therefore the decrease of CP4-EPSPS in BRS Silvania RR might be related to its susceptibility to stink bugs [6,16]. HSPs are not exclusively related to heat stress, and some are induced by stresses such as injury, drought, and salinity [46]. In plants, GRPs are responsible for regulating cellular processes of genic regulation post-transcription, and translation [42,47], as well as cell elongation [48], protoxylem development [49,50], signal transduction [51], RNA chaperone activity [52], plant defense

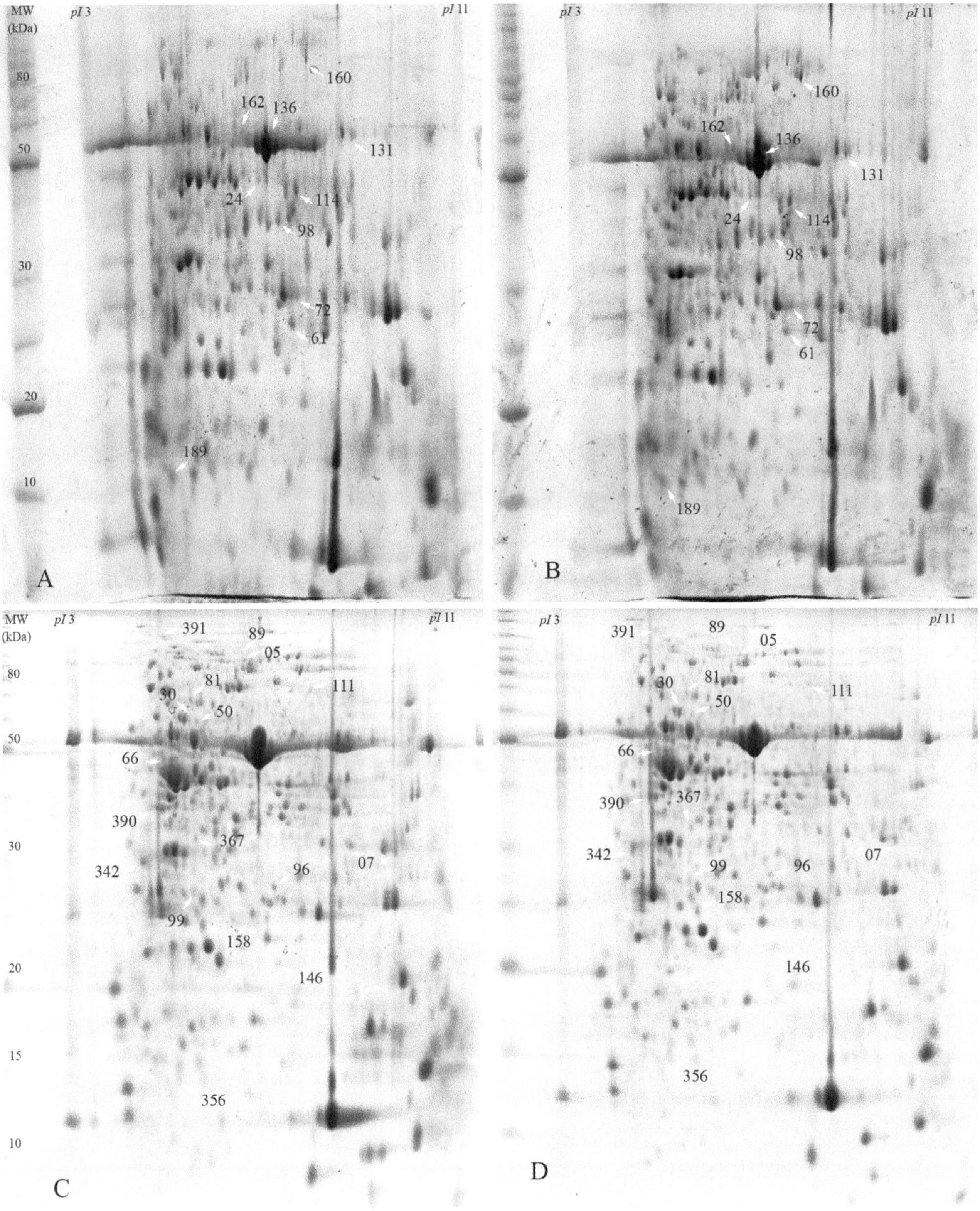

Figure 3. Foliar protein profiles (2-DE, 500 µg) of soybean varieties (V3). Variety IAC-100: (A) injured and (B) not injured, from the total average of 191 (±2) protein spots, 10 were differentially expressed and two identified (Table 2). Variety BRS Silvania RR: (C) injured and (D) not injured, from the total average of 386 (±4) protein spots, 17 were differentially expressed and 11 identified (Table 3). The differentially expressed proteins are indicated with white arrows and numbers, which correspond to identification numbers listed in the respective table.

Table 2. Identification of differentially expressed proteins by MS/MS in BRS Silvania RR soybean leaves (V3) ($p<0.05$).

Spot number	Protein identification	Accession (NCBI)	Molecular mass (kDa) theoretical/experimental	p*I* theoretical/experimental	Peptide sequence	Mascot/Blast Score	Search Type	RMS Error (ppm)	E-value	Expression level (%)
07	GATase1-like domain	gi|255639997	27.0/31.5	5.43/9	LSDFFEILATSEDR	117	MS/MS	44.40	4e-09	↓ 16.7
35*	Unknown	gi|40538952	-/57.8	-/5.6	GFIMKGATNATVDSMV	-	MS/MS	-	6e-13	↑ 30.4
66	CP4-EPSPS	gi|18266432	47.6/47.7	5.13/5	SAVLLAGLNTPGITTVIEPIMTR	57	MS/MS	12.80	2.2e-3	↓ 54.6
81*	70 kDa chaperonin-like	gi|473930207	-/64.6	-/5.6	AVITVPAYFNDAQR	-	MS/MS		1e-06	↓ 56.0
99*	STI-Soybean trypsin inhibitor	gi|351726331	19.8/26.8	-/5.5	GFPITISSPAR	-	MS/MS	-	0.003	↓ 24.0
111*	Unknown	gi|242054087	-/74.6	-/7.7	LSGIIPVHIGKLTK	-	MS/MS	-	5e-06	↓ 15.5
146	Glycine-rich protein 2b-like	gi|356538622	19.3/21.6	6.29/7.09	SLAEGESVEFAIESESDGR	132	MS/MS	11.60	1.6e-10	↑ 250.0
158	20 kDa chaperonin-like	gi|356556406	26.6/21.6	7.79/5.76	DKPSIGTVIAVGPGHLDEEGNR	146	MS/MS	0.86	3.2e-12	↑ 88.0
342	Unknown	gi|1478004532	50.0/31.0	5.43/4.7	SLLYVETADRPGLLVDLVK	90	MS/MS	4.51	1.6e-06	↓ 68.1
356	Glycine-rich RNA-binding	gi|351725469	15.8/15.0	6.58/5.3	AFSQYGEIVETK	77	MS/MS	20.90	2.9e-05	↓ 35.0
367	SGNH plant lipase-like	gi|255646252	40.0/33.4	5.83/5.6	AFFVFGDSLVDNGNNNFLATTAR	85	MS/MS	50.90	4.2e-06	↑ 58.9

* Identification by similarity search with BLASTp using peptide sequence obtained by manual *de novo* sequencing.

against pathogens [53], RNA binding activity [54], among others. Sachetto-Martins *et al.* [55] reported that the expression of GRP is regulated by a number of external stimuli, including stress by cold, water, high salinity, injury, and viral infection. The GATase1 belongs to a large group of enzymes capable of removing ammonia from glutamine and to transfer this group to other substrates to form amino acids, purine, pyrimidine nucleotides, and coenzymes [56]. Further studies are necessary to understand why these proteins were increased or decreased when the plants suffer injury by brown stink bug.

When the protein profiles of the injured and uninjured resistant plants were compared, 10 proteins had quantitative differential expression levels, of which four were increased in the injured plants. Two differentially expressed proteins were identified as ribulose bisphosphate carboxylase oxygenase (RuBisCO), which increased, and glyceraldehyde-3-phosphate dehydrogenase (GAPDH), which decreased (Table 3).

RuBisCO and GAPDH are enzymes in the Calvin cycle [57]. RuBisCO catalyzes the reaction of one carbon dioxide with ribulose-1,5-bisphosphate to form 3-phosphoglycerate (3-PGA), which in turn is catalyzed by phosphoglycerate kinase to 1,3-diphosphoglycerate. This compound, after catalysis by GAPDH produces glyceraldehyde-3-phosphate [43]. The increased expression of RuBisCO, and decreased expression of GAPDH, increases the level of the intermediate 3-phosphoglycerate. 3-Phosphoglycerate is metabolized during glycolysis into phosphoenolpyruvate, which in turn can be used in the shikimic acid pathway. Similar to what occurred in BRS Silvania RR, a protein involved in the shikimic pathway was affected by a piercing-sucking insect on IAC-100 variety. In general, sucking insects act on the shikimic acid pathway, and chewing insects act on the jasmonic acid pathway. Our results corroborate this hypothesis, because the main proteins identified here are involved in the shikimic acid pathway. However in contrast to BRS Silvania RR, the shikimic pathway was up-regulated in IAC-100. Michereff *et al.* [6] showed that IAC 100 injured by brown stink bugs produced higher amounts of methyl salicylate, and this volatile appears to be involved in the indirect defense of soybean.

Overall, the resistant variety (IAC-100) did not exhibit physiological stress to feeding injury by the brown stink bug, and seemed to shift its metabolism to produce secondary compounds for insect defense. In contrast, the susceptible variety (BRS Silvania RR) exhibited signs of physiological stress, both in its enzymes related to redox metabolism and in the protein profile response, which could affect its subsequent development in long term exposure to herbivory injury.

Table 3. Identification of differentially expressed proteins (by MS/MS) in IAC-100 soybean leaves (V3) ($p<0.05$).

<table>
<tr><th>Spot number</th><th>Protein identification</th><th>Accession (NCBI')</th><th>Molecular mass (kDa) theoretical/experimental</th><th>pI theoretical/experimental</th><th>Peptide sequence</th><th>Mascot Score</th><th>Search Type</th><th>RMS Error (ppm)</th><th>E-value</th><th>Expression level (%)</th></tr>
<tr><td>114</td><td>GAPDH</td><td>gi|66026</td><td>36.0/42.7</td><td>6.66/7.2</td><td>K.VIAWYDNEWGYSQR.V</td><td>64</td><td>MS/MS</td><td>61.7</td><td>8.4e-4</td><td>↓ 32.0</td></tr>
<tr><td>136</td><td>RuBisCO large subunit</td><td>gi|131898</td><td>52.0/53.9</td><td>6.23/6.24</td><td>K.TFQGPPHGIQVER.D</td><td>107</td><td>MS/MS</td><td>11.0</td><td>9.1e-08</td><td>↑ 33.0</td></tr>
</table>

Conclusions

This work assessed the foliar redox response and protein expression profile of two soybean varieties (IAC-100, resistant, and BRS Silvania RR, susceptible, at stage V3) under herbivory injury by the brown stink bug. As could be expected from previous reports, the two soybean varieties responded differently at the biochemical level to herbivory by the brown stink bug. The susceptible variety BRS Silvania RR showed adaptive changes in the activity of enzymes related to redox metabolism and in the expression of stress-responsive proteins, probably to minimize biochemical damage caused by herbivory injury. On the other hand, in the resistant variety, the redox genes could be constitutively expressed, and the stink bug injury might not produce enough ROS to elicit changed expression of these genes. Overall, the susceptible variety responded to herbivory in ways that could compromise its development in the long term, while the resistant variety IAC-100 showed a defense response oriented to circumvent the biological stressor. Both varieties were injured by a piercing-sucking insect and the proteins regulated by this stress were linked to the shikimic acid pathway corroborating the hypothesis that sucking insects induce the shikimic acid route of defense more than the jasmonic acid route defense.

Acknowledgments

We would like to thank: Dra. Clara Beatriz Hoffmann-Campo for providing the soybean seeds; Dr. Miguel Borges and Dr. Raul A. Laumann for providing the brown stink bugs; Msc. Démerson André Polli for statistics consulting; MSc. Gabriela Villeth and Dr. Lilian Travassos for supporting the proteomic experiments and analysis; Dr. David A. Andow for English revisions.

Author Contributions

Conceived and designed the experiments: DPP RVT MCBM. Performed the experiments: RVT LPS. Analyzed the data: DPP MHL LPS RVT MCBM AM. Contributed reagents/materials/analysis tools: MHL DPP. Wrote the paper: DPP MHL LPS RVT MCBM AM.

References

1. Corrêa-Ferreira BS, Panizzi AR (1999) Percevejos da soja e seu manejo. Londrina: Embrapa - CNPSo, Circular Técnica 24.
2. Fernandes AP (2000) Análise comparativa da espermogênese normal e diapáusica em percevejos fitófagos (Hemiptera, Pentatomidae) – Revisão Universitas. Biociências (UNICEUB). Brasília-Brasil. 1(1): 87–95.
3. Todd JW, Herzog DC (1980) Sampling phytophagous Pentatomidae on soybean. In: Kogan M, Herzog D C (Ed) Sampling.
4. Companhia Nacional de Abastecimento (CONAB), Acompanhamento da Safra Brasileira de Grãos (2012) http://www.conab.gov.br/OlalaCMS/uploads/arquivos/12_10_17_16_09_58_boletim_graos_-_julho_2012.pdf.
5. Vieira CR, Moraes MCB, Borges M, Sujii ER, Laumann RA (2013) *cis*-Jasmone indirect action on egg parasitoids (Hymenoptera: Scelionidae) and its application in biological control of soybean stink bugs (Hemiptera: Pentatomidae). Biol. Control 64: 75–82.
6. Michereff MFF, Laumann RA, Borges RAM, Michereff-Filho M, Diniz IR, et al. (2011) Volatiles mediating a plant-herbivore-natural enemy interaction in resistant and susceptible soybean cultivars. J. Chem. Ecol. 37: 273–285.
7. Moraes MCB, Laumann RA, Pareja M, Sereno FTPS, Michereff MFF, et al. (2009) Attraction of the stink bug egg parasitoid *Telenomus podisi* to defence signals from soybean activated by treatment with *cis*-jasmone. Entomol. Exp. Appl. 131: 178–188.
8. Vieira SS, Bueno AF, Boff MIC, Bueno RC, Hoffman-Campo CB (2011) Resistance of soybean genotypes to *Bemisia tabaci* (Genn.) biotype B. Neo. Entomol. (Impress) 40: 117–122.
9. Dicke M, Hilker M (2003) Induced plants defence: from molecular biology to evolutionary ecology. Basic Appl. Ecol. 4: 3–14.
10. Courtois EA, Paine CE, Blandinieres PA, Stien D, Bessiere JM, et al. (2009) Diversity of the volatile organic compounds emitted by 55 species of tropical trees: a survey in French Guiana. J. Chem. Ecol. 35: 1349–1362.
11. Colazza S, Wajnberg E (2013) Chemical ecology of insect parasitoids: towards a new era. In: Wajnberg E, Colazza S, Chemical Ecology of Insect Parasitoids, 1st Edition, John Wiley & Sons (2013) 1–8.
12. McPherson RM, Buss GR (2007) Evaluation lepidopteran defoliation resistance in soybean breeding lines containing the stink bug (Hemiptera: Pentatomidae) resistance IAC-100 cultivar in their pedigrees. J. Econ. Entomol. 100: 962–968.
13. Panizzi AR, Mcpherson JE, James JG, Javabery M, McPherson RM (2000) Stink bugs (Pentatomidae). In: Schaefer CW, Panizzi AR (Eds.), Heteroptera of economic importance. Boca Raton: Florida, USA: CRC 432–434 pp.
14. Corrêa-Ferreira BS, Azevedo J (2002) Soybean seed damage by different species of stink bugs. Agr. Forest Entomol. 4: 145–150.
15. Bi JL, Murphy JB, Felton GW (1997) Antinutritive and oxidative components as mechanisms of induced resistance in cotton to *Helicoverpa zea*. J. Chem. Ecol. 23: 97–117.
16. Laumann RA, Farias Neto AL, Moraes MCB, Silva AP, Vieira CR, et al. (2008) Dinâmica populacional de percevejos (Hemiptera: Pentatomidae) em diferentes genótipos de soja. In: IX Simpósio Nacional Cerrado; II Simpósio Internacional Savanas Tropicais, Brasília, DF. Desafios e estratégias para o equilíbrio entre sociedade, agronegócio e recursos naturais: Anais. Planaltina, DF: Embrapa Cerrados, HTTP://www.cpac.embrapa.br/download/747t.
17. Rostas M, Eggert K (2008) Ontogenetic and spatio-temporal patterns of induced volatiles in *Glycine max* in the light of the optimal defense hypothesis. Chemoecol. 18: 29–38.
18. Moraes MCB, Laumann RA, Sujii ES, Pires CSS, Borges M (2005) Induced volatiles in soybean and pigeon pea plants artificially infested with the Neotropical brown stink bug, *Euschistus heros*, and their effect on the egg parasitoid, *Telenomus podisi*. Entomol. Exp. Appl. 115: 227–237.
19. Veiga RFA, Rossetto CJ, Razera LF, Galo PB, Bortoleto M (1999) Caracterização morfológica e agronômica da cultivar de soja 'IAC-100'. Instituto Agronômico de Campinas. Boletim Técnico, 175. 25p.
20. Gris CF, Pinho EVRV, Carvalho MLM, Diniz RP, Andrade T (2013) Lignificação da planta e qualidade de sementes de soja RR pulverizadas com herbicida glifosato. Rev. Ceres. Viçosa 60(2): 286–292.
21. Moraes MCB, Pareja M, Laumann RA, Hoffmann-Campo CB, Borges M (2008) Response of the parasitoid to induced volatiles from soybean damaged by stink bug herbivory and oviposition. J. Plant Interact. 3: 111–118.
22. Bradford MM (1976) A rapid and sensitive method for quantification of microgram quantities of protein utilizing the principle of protein-dye binding. Anal. Biochem. 72: 248–254.
23. Axelrod BT, Cheesbrough TM, Laakso S (1981) Lipoxygenase from soybeans. Methods Enzymol. 71: 441–445.
24. Chen Y, Ni X, Cottrell E, Wood BW, Buntin GD (2009) Changes of oxidase and hydrolase activities in pecan leaves elicited by black pecan aphid (Hemiptera: Aphididae) feeding, J. Econ. Entomol. 102 (3): 1262–1269.
25. Anderson JV, Chevone BI, Hess JL (1992) Seasonal variation in the antioxidant system of Eastern white pine needles. Plant Physiol. 98: 501–508.
26. Hermes-Lima M, Carreiro C, Moreira DC, Polcheira C, Machado DP, et al. (2012) Glutathione status and antioxidant enzymes in a crocodilian species from the swamps of the Brazilian Pantanal. Comp. Biochem. Physiol. Part A. 163: 189–198.
27. Méchin V, Damerval C, Zivy M (2007) Total protein extraction with TCA/acetone. In: Thiellement H, Zivy M, Damerval M, Méchin V (2007) Plant proteomics methods and protocols. Totowa, New Jersey, John M. Walker, 1–8.
28. Wildgruber R, Reil G, Drews O, Parlar H, Görg A (2002) Web-based two dimensional database of *Saccharomyces cerevisae* proteins using immobilized pH gradients from pH 6 to pH 12 and matrix-assisted laser desorption/ionization-time of flight mass spectrometry. Proteomics 2: 727–732.
29. Weiss W, Gorg A (2007) Two-dimensional electrophoresis for plant proteomics. Methods Mol. Biol. 355: 1421–143.
30. Farmer EF, Johnson RR, Ryan CA (1992) Regulation of expression of proteinase inhibitor genes by methyl jasmonate and jasmonic acid. Plant Physiol. 98: 995–1002.
31. Prost I, Dhondt S, Rothe G, Vicente J, Rodriguez MJ, et al. (2005) Evaluation of the antimicrobial activities of plant oxylipins supports their involvement in defense pathogens. Plant Physiol. 139: 1902–1913.
32. Ishige F, Mori H, Yamazaki K, Imaseki H (1993) Identification of a basic glycoprotein induced by ethylene in primary leaves of azuki bean as cationic peroxidase. Plant Physiol. 101: 193–199.
33. Ni X, Quisenberry SS, Heng-Moss T, Markwell J, Sarath G, et al. (2001) Oxidative responses of resistant and susceptible cereal leaves to symptomatic and nonsymptomatic cereal aphid (Hemiptera: Aphididae) feeding. Faculty Publications: Department of Entomology. Paper 113. http://digitalcommons.unl.edu/entomologyfacpub/113.
34. Devi PUM, Reddy PS, Rani NRU, Reddy KJ, Reddy MN, et al. (2000) Lipoxygenase metabolites of alpha-linolenic acid in the development of resistance in pigeonpea, *Cajanus cajan* (L.) Millsp, seedlings against *Fusarium udum* infection. European J. Plant Pathol. 106: 857–865.
35. Pérez FJ, Villegas D, Mejia N (2002) Ascorbic acid and flavonoid-peroxidase reactions as a detoxifying system of H_2O_2 in grapevine leaves. Phytochemistry 60: 573–580.
36. Bi L, Felton GW (1995) Foliar oxidative stress and insect herbivory: primary compounds, secondary metabolites, and reactive oxygen species as components of induced resistance. J. Chem. Ecol. 21: 1511–1530.
37. Ruuhola T, Yang S (2005) Wound-induced oxidative responses in mountain birch leaves. Ann. Botany 97: 29–37.
38. Aebi H (1984) Catalase *in vitro*. B. Chance. Act Chem. Scand. 1: 121–126.
39. Karuppanapandian T, Moon J, Kim C, Manoharan K, Kim W (2011) Reactive oxygen species in plants: their generation, signal transduction, and scavenging mechanisms. Australian J. Crop Sci. 5 (6): 709–725.
40. Gomez SK, Oosterhuis DM, Rajguru SN, Johnson DR (2004) Foliar antioxidant enzyme responses in cotton after aphid herbivory. J. Cotton Sci. 8: 99–104.
41. Li D-C, Yang F, Lu B, Chen DF, Yang WJ (2012) Thermotolerance and molecular chaperone function of the small heat shock protein HSP20 from hyperthermophilic aschaeon, *Sulfolobus solfataricus* P2. Cell Stress and Chaperon. 17: 103–108.
42. Mangeon A, Junqueira RM, Sachetto-Martins G (2010) Functional diversity of the plant glycine-rich proteins superfamily. Plant Signaling Behav. 5 (2): 99–104.
43. Taiz L, Zeiger E (2006) Secondary metabolites and plant defense. In: Taiz L, Zeiger E (Eds.) Plant Physiology, Sinauer Associates Inc., Sunderland, MA, 4th edition. 283–306 pp.
44. Kutchan TM (1995) Alkaloid biosynthesis: the basis for metabolic engineering of medicinal plants. Plant Cell 7: 1059–1070.
45. Vogt T (2010) Phenylpropanoid biosynthesis. Mol. Plant. 3: 2–20.

46. Lindquist S, Craig EA (1998) The heat-shock proteins. Annu. Rev. Genet. 22: 631–677.
47. Burd CG, Dreyfuss G (1994) Conserved structures and diversity of functions of RNA-binding proteins. Science 265: 615–21.
48. Mangeon A, Magioli C, Menezes-Salgueiro AD, Cardeal V, de Oliveira C, et al. (2009) AtGRP5, a vacuole-located glycine-rich protein involved in cell elongation. Planta 230: 253–265.
49. Ringle C, Hauf G, Keller B (2001) Hydrophobic interactions of the structural protein GRP1.8 in the cell wall of protoxylem elements. Plant Physiol. 125: 673–82.
50. Ryser U, Schorderet M, Guyot R, Keller B (2004) A new structural element containing glycine-rich proteins and rhamnogalacturonan I in the protoxylem of seed plants. J. Cell Sci. 117: 1179–90.
51. Park AR, Cho SK, Yun SJ, Jin MY, Lee SH, et al. (2001) Interaction of the arabidopsis receptor protein kinase Wak1 with a glycine-rich protein, AtGRP-3. J. Biol. Chem. 276: 26688–26693.
52. Kim JY, Park SJ, Jang B, Jung CH, Ahn SJ, et al. (2007) Functional characterization of a glycine-rich RNA-binding protein 2 in *Arabidopsis thaliana* under abiotic stress conditions. Plant J. 50: 439–451.
53. Fu ZQ, Guo M, Jeong BR, Tian F, Elthon TE, et al. (2007) A type III effector ADP-ribosylates RNA-binding proteins and quells plant immunity. Nature 447: 284–288.
54. Kim YO, Kim JS, Kang H (2005) Cold-inducible zinc finger-containing glycine-rich RNA-binding protein contributes to the enhancement of freezing tolerance in *Arabidopsis thaliana*. Plant J. 42: 890–900.
55. Sachetto-Martins G, Franco LO, de Oliveira DE (2000) Plant glycine-rich proteins: a family or just proteins with a common motif? Biochim Biophys Acta. 1492: 1–14.
56. Mäntsälä P, Zlakin H (1984) Replacement of the active-site cysteine in glutamine phosphoribosylpyrophosphate amidotranferase by site-directed mutagenesis. J. Soc. Biol. Chem. 259: 14230–14236.
57. Brynat DA, Friggard NU (2006) Porkaryotic photosynthesis and phototrophy illuminate. Trends Microbiol. 14: 488–496.

16

Photosynthetic Diffusional Constraints Affect Yield in Drought Stressed Rice Cultivars during Flowering

Marco Lauteri[1], Matthew Haworth[2*], Rachid Serraj[3,4], Maria Cristina Monteverdi[5], Mauro Centritto[2]

1 Institute of Agro-Environmental and Forest Biology, National Research Council, Porano, Italy, **2** Trees and Timber Institute (IVALSA - Istituto per la valorizzazione del legno e delle specie arboree), National Research Council, Florence, Italy, **3** Consultative Group on International Agricultural Research, Independent Science and Partnership Council Secretariat, Food and Agriculture Organisation of the United Nations, Rome, Italy, **4** International Rice Research Institute, Los Banos, Philippines, **5** Forestry Research Centre, National Research Council, Arezzo, Italy

Abstract

Global production of rice (*Oryza sativa*) grain is limited by water availability and the low 'leaf-level' photosynthetic capacity of many cultivars. *Oryza sativa* is extremely susceptible to water-deficits; therefore, predicted increases in the frequency and duration of drought events, combined with future rises in global temperatures and food demand, necessitate the development of more productive and drought tolerant cultivars. We investigated the underlying physiological, isotopic and morphological responses to water-deficit in seven common varieties of *O. sativa*, subjected to prolonged drought of varying intensities, for phenotyping purposes in open field conditions. Significant variation was observed in leaf-level photosynthesis rates (*A*) under both water treatments. Yield and *A* were influenced by the conductance of the mesophyll layer to CO_2 (g_m) and not by stomatal conductance (g_s). Mesophyll conductance declined during drought to differing extents among the cultivars; those varieties that maintained g_m during water-deficit sustained *A* and yield to a greater extent. However, the variety with the highest g_m and yield under well-watered conditions (IR55419-04) was distinct from the most effective cultivar under drought (Vandana). Mesophyll conductance most effectively characterises the photosynthetic capacity and yield of *O. sativa* cultivars under both well-watered and water-deficit conditions; however, the desired attributes of high g_m during optimal growth conditions and the capacity for g_m to remain constant during water-deficit may be mutually exclusive. Nonetheless, future genetic and physiological studies aimed at enhancing *O. sativa* yield and drought stress tolerance should investigate the biochemistry and morphology of the interface between the sub-stomatal pore and mesophyll layer.

Editor: Wagner L. Araujo, Universidade Federal de Vicosa, Brazil

Funding: This work was funded by: Ministero dell'Istruzione, dell'Università e della Ricerca of Italy (PRIN) grant: "Going to the root of plant productivity: how the rhizosphere interact with the aboveground armament for indirect and direct defense against abiotic and biotic stressors (PRO-ROOT)" URL: http://hubmiur.pubblica.istruzione.it/web/universita/home and Marie Curie Intra European Fellowship (2010-275626) - HAWORTH - Evoutionary adaptation to atmospheric carbon dioxide, URL: http://ec.europa.eu/research/mariecurieactions/about-mca/actions/ief/index_en.htm. The funders had no role in study design, data collection and analysis, decision to publish, or preparation of the manuscript.

Competing Interests: The authors have declared that no competing interests exist.

* Email: matthew.haworth@hotmail.com

Introduction

Global production of rice (*Oryza sativa*) is limited by the availability of freshwater [1] and the tolerance of rice cultivars to drought stress [2]. Nonetheless, *O. sativa* is cultivated in a diverse range of climates and habitats, with lowland varieties generally considered to be more productive, but less tolerant of drought, than their upland counterparts [3]. Selective breeding of *O. sativa* varieties has increased the grain production of individual plants by 70% since 1965 [4]. However, these increases in productivity have been restricted by the comparatively low 'leaf-level' photosynthetic capacity of *O. sativa* relative to other C3 crops [5,6]. Phenotypic screening of the photosynthetic and gas-exchange characteristics of *O. sativa* varieties under water deficit conditions may permit the identification of attributes that confer both drought tolerance and high productivity [7,8].

Yield in varieties of *O. sativa* are closely related to photosynthesis under both well-watered and drought conditions [7,9]. The uptake of CO_2 for photosynthesis involves two major resistance steps at the stomata and at the surface of the mesophyll layer [10,11]. Stomatal (g_s) and mesophyll conductance (g_m) to CO_2 often decrease in unison in response to environmental stresses [7,12–14]. Upland and lowland *O. sativa* cultivars exposed to a soil moisture gradient ranging from high water-availability to severe water-deficit exhibited identical reductions in both g_s and g_m that correlated with the rate of photosynthesis (*A*) and yield. Most notably, while there was no observable difference in *A*, g_s or g_m response to drought between the upland and the lowland varieties, those cultivars that exhibited the highest values of total conductance to CO_2 (g_t: where $g_t = [g_s * g_m]/[g_s + g_m]$) supported higher *A* and yield under all levels of water availability [7]. Those *O. sativa* varieties with higher capacities for CO_2 uptake are therefore the most likely to offer the greatest potential improvement of leaf-level photosynthesis towards achieving enhanced yields [9,15,16]. However, an increase in the capacity for leaf gas exchange may be accompanied by reduced drought tolerance [17,18].

These differences in the photosynthetic and leaf gas-exchange capacities of *O. sativa* varieties have been ascribed to genetic variation [9]. The quantitative trait loci responsible for leaf gas-exchange and photosynthetic parameters in *O. sativa* tend to be found in adjacent regions of the plant genome; indicative of the exertion of selective pressures favouring coordination and high heritability of photosynthetic behaviour [19,20]. Therefore, effective screening and analysis of the photosynthetic and gas exchange characteristics of *O. sativa* varieties under stress conditions has the potential to identify functional traits associated with optimisation of water use efficiency (*WUE*) and the maintenance of yield and *A* under drought [21]. Rates of photosynthesis are not only determined by diffusive CO_2-uptake, but also by the activity of ribulose-1,5-bisphosphate (RubisCO) [22]. However, exposure of thirteen hydroponically grown *O. sativa* varieties to physiological drought did not suggest that the capacity of RubisCO carboxylation (Vc_{max}) or the maximum rate of electron transport required for ribulose-1,5-bisphosphate regeneration (J_{max}) were associated with *A* or grain yield. Rather, *A* and yield were largely determined by g_s and g_m setting the limits for diffusional CO_2 uptake [9], consistent with field based studies [7]. Furthermore, the intrinsic water-use efficiency (A/g_s) of the *O. sativa* varieties was closely related to the ratio of g_m to g_s [9], and this attribute showed the greatest variability between cultivars, and thus potential in terms of possible improvements in yield and growth through maximising *WUE* [5,9,21]. In theory, those varieties with high g_m to g_s ratios should be most suited to maintaining photosynthesis and yield under drought stress conditions, while high values of both g_m and g_s should be favourable for varieties grown under constant high water availability [21].

The observation that *A* and yield of *O. sativa* closely correlate to g_m [7,9,16] has led to a focus on the interface between the leaf mesophyll cells and internal sub-stomatal air space in the improvement of rice productivity [23]. Adachi et al. [16] crossed two varieties of *O. sativa* to produce a new cultivar with higher values of g_m, which were associated with modifications of the mesophyll structure that increased the surface area available for CO_2 uptake; thus demonstrating the potential for enhanced productivity through increased g_m. In addition to morphological adaptations [15], differences in g_m between *O. sativa* varieties are also likely affected by biochemical factors [23] such as the abundance of cooporin proteins responsible for active transport of CO_2 [24,25] (Kaldenhoff, 2012). Furthermore, the abundance and activity of these proteins has been observed to decline under water-stress [26]. Exposure to drought is associated with declines in values of *A*, g_s and g_m [11,27]. Sustained impairment of *A* following the alleviation of drought stress is commonly related with continued biochemical constraints to the uptake and assimilation of CO_2 [28]. However, experimental evidence suggests that *O. sativa* does not display prolonged reductions in *A* following the cessation of drought, and that the observed constraints to *A* under water-stress are largely physical impediments to CO_2 diffusion [7,9,29,30]. This suggests that artificial selection of those *O. sativa* varieties with high g_m would enhance *A* while not negatively affecting drought tolerance or recovery.

Rice production is sensitive to water deficit at different developmental stages of plant growth. For example, early drought stress can affect leaf expansion [3], and later drought stress during flowering or grain-filling stages can disrupt grain formation and development [31,32]. The photosynthetic and yield responses of seven *O. sativa* varieties (five upland and two lowland varieties chosen to represent genetic variability) were analysed in response to sustained drought. As different *O. sativa* varieties have different phenologies (for example Van forms flower inflorescences relatively earlier than Moro), through the use of a drip irrigation system it was possible to apply the stress at the same developmental stage for each variety. This has the advantage over a previous study into the drought tolerance of *O. sativa* varieties conducted at the IRRI in that all varieties experienced drought at the same phenological stage, therefore eliminating a crucial source of error whereby early flowering varieties outperformed late developing cultivars [cf. 7]. Mesophyll conductance to CO_2 is of key interest to the productivity and drought tolerance of *O. sativa* [7,16], however it is not possible to directly measure g_m, and due to the uncertainties associated with its calculation [33] (particularly in the use of the variable J method under stress conditions) it is preferable to simultaneously use two independent methodologies in the estimation of g_m [34]. We therefore employed two independent methodologies in the estimation of g_m: firstly, based upon the synchronous measurement of leaf gas exchange and chlorophyll fluorescence parameters (variable J), and secondly, the analysis of carbon isotope discrimination ($\Delta^{13}C$) in recently formed metabolites to assess their effectiveness in an analysis of *O. sativa* drought tolerance in the field. Further details of the variable J and '$\Delta^{13}C$ of recently synthesised sugars' methods of g_m analysis are presented in the Material and Methods section. This study aimed to: 1) assess the effect of g_s and g_m on *A* and yield in well-watered and drought conditions; 2) analyse differences in g_m calculation between the 'variable J' [35] and '$\Delta^{13}C$ of recently synthesised sugars' [15] methods, and; 3) identify *O. sativa* varieties that exhibit drought tolerance and/or high productivity, and may be suitable for further genomic/biochemical development to further enhance grain production.

Materials and Methods

Plant material and growth conditions

The experiment was conducted at the International Rice Research Institute Experimental Station at Los Banos, Philippines (14° 11′N 121° 15′E, 21 m above sea level) during the dry season (January to May). Rainfall was minimal, however, temperature and leaf to air vapour pressure difference (*VPD*) increased during the experimental period. The soil at the experimental site is a neutral Andaqueptic Haplaquol (pH = 6.5). Seven *O. sativa* cultivars were studied: Apo, IR55419-04 (IR55), IR64, IR71525-19-1-1 (IR71), Moroberekan (Moro), PSBRc80 (PS80) and Vandana (Van). Moroberekan failed to set or produce grains and seemed to suffer from disease and heat sensitivity, particularly under drought conditions and it was not possible to record physiological measurements for this variety under water deficit stress conditions. Seeds were sown in dry soil at a rate of 80 kg ha^{-1} in rows spaced ~25 cm apart. Fertilisation with nitrogen was conducted on two occasions at application rates of 90 and 120 kg ha^{-1}, alongside fertilisation of 30 kg ha^{-1} of phosphate and potassium. Weeds were controlled by the application of standard agricultural herbicides and manual picking. Pesticides were used to control insect pests; in particular stem borers.

Experimental plots for each *O. sativa* variety measured 3 m×2 m, with adjacent irrigated and drought-stressed treatments. The plots were arranged in a randomized block design. A sprinkler system was initially used to provide equal irrigation levels to all plants until flower (panicle) development, when the water deficit treatment was instigated. During the drought stress period, the well-watered treatments were irrigated by use of a drip irrigation system, with soil water potential maintained above -0.04 MPa (monitored using tensiometers placed at soil depths of 15 and 30 cm) – the level at which the more sensitive rice cultivars begin to exhibit a reduction in transpiration [36].

Leaf gas exchange measurements

Leaf gas-exchange and fluorescence of plants grown under both well-watered treatment or stress conditions were simultaneously measured with a LI-6400-40 leaf chamber fluorometer (Li-Cor, Inc., Nebraska, USA) on flag leaves enclosed in a temperature, light and humidity controlled cuvette. The measurements were made *in situ* between 11.00 and 15.00, in saturating PPFD (1400 μmol $m^{-2}s^{-1}$), with relative humidity ranging between 45–55%, and a leaf temperature of 30°C. Instantaneous measurements of steady-state A, g_s, internal [CO_2] (C_i) and the quantum yield of PSII in the light (ΔF/Fm') were made on six to 14 plants per treatment. Measurements of dark respiration (R_d) were also made at ambient CO_2 concentration in the dark on the same leaves. Instantaneous transpiration efficiency was calculated as the ratio of A to g_s.

Carbon isotope discrimination

Three of the same flag leaves used for *in situ* gas exchange measurements were collected in the evening, oven dried at 70°C until their weight remained constant and were then stored until soluble sugar extraction, following the protocol of Richter et al. [37]. Leaves were finely ground in liquid nitrogen and shaken for 60 min in water at room temperature. After centrifugation (15 min at 5000 g), the supernatant was sequentially mixed with cationic (Dowex-50) and anionic (Dowex-1) exchange resins. The residual solution of purified soluble sugars was freeze-dried and carbon isotope composition ($\delta^{13}C$) determined using a mass-spectrometer. Additional isotope ratio mass spectrometry analyses were performed on the pellet samples that remained after the extraction of soluble sugars. Samples of approximately 1 mg were used for $\delta^{13}C$ analysis. Stable isotope ratios were determined using a continuous-flow triple-collector isotope ratio mass spectrometer (ISOPRIME, GV, Manchester, UK). Solid samples were quantitatively combusted in an elemental analyser (Model NA 1500, Carlo Erba, Milan, Italy) and CO_2 was transferred in helium flow to the mass spectrometer. Calculations of carbon isotope discrimination ($\Delta^{13}C$) of solid samples were undertaken following the protocol of Farquhar et al. [38], assuming the carbon isotopic composition of CO_2 in air (δ_{air}) to be −8.0‰.

Estimation of mesophyll conductance

Mesophyll conductance to CO_2 diffusion is the inverse of the total resistance encountered in the transport of CO_2 across the leaf mesophyll. Estimation of g_m is subject to a number of assumptions and sources of error, and it is preferable to employ two independent approaches in its calculation [34,39]. In this study g_m was calculated using the 'variable J' and '$\Delta^{13}C$ of recently synthesised sugars' methods. The calculation of g_m using the variable J method is based on simultaneous measurements of gas-exchange and chlorophyll fluorescence parameters as described by Harley et al. [35] and Loreto et al. [40] (equations 1 and 2):

$$g_m = \frac{A}{Ci - \frac{\Gamma * [J + 8 * (A + R_d)]}{J - 4 * (A + R_d)}} \quad (1)$$

where the electron transport rate (J_F) is calculated from fluorescence [41]:

$$J_F = PPFD * \frac{\Delta F}{F_m} * \alpha * \beta \quad (2)$$

.

Where F_m is the fluorescence maximum and the partitioning factor (β) between photosystems I and II was considered to be 0.5 and leaf absorbance (α) was measured directly [42]. The variable J method is sensitive to the estimation of the CO_2 compensation point to photorespiration (Γ*) and R_d. For the results of this study to be directly comparable with those of Centritto et al. [7], the value of Γ* used in the gas exchange algorithm was calculated using the Rubisco specific factor for an annual herb [43], consistent with previously published *O. sativa* Γ* values [44]. As measurements were conducted under field conditions it was not feasible to produce robust measurements of respiration in the light (R_l); respiration in the dark was therefore used in the calculation of gm, as R_l and R_d have been observed to be broadly consistent when recycling of CO_2 is taken into consideration [45]. Measurements of R_d were performed at ambient [CO_2] concentration in the dark on the same leaves by switching off the light in the leaf cuvette; when CO_2 release from the leaf had become stable for approximately five to ten minutes this was recorded and considered to represent R_d. To calibrate the values of J calculated by fluorescence (J_F) and gas-exchange (J_A), measurements were also conducted under non-photrespiratory conditions using 1% [O_2] [35,40]. Electron transport rate and R_d values are presented in Figures S1 and S2. Total conductance to CO_2 (g_t) was calculated as: $g_t = g_s * g_m/(g_s + g_m)$. The concentration of CO_2 at the chloroplast site (C_c) was then calculated from mesophyll conductance values [46].

The method described by Lauteri et al. [47] for 'on-line' determinations was adapted to calculate g_m by using the $\Delta^{13}C$ of recently synthesized sugars [48]. Discrepancies between $\Delta^{13}C$ of leaf soluble carbohydrates (Δ_{obs}) and $\Delta^{13}C$ expected on the basis of gas-exchange measurements (Δ_{exp}), allow the estimation of g_m utilising the approach of Evans et al. [49] that was subsequently modified by Lloyd et al. [50]:

$$g_m = \frac{(b - b_s - \alpha_1) * {}^{A}/{}_{C_a}}{(\Delta_{exp} - \Delta_{obs}) - (f\,\Gamma * / C_a)} \quad (3)$$

Where: b is the discrimination associated with carboxylation reactions, taken to be 27.5‰; b_s is the fractionation occurring when CO_2 enters solutions (1.1‰ at 25°C); a_1 is the fractionation during diffusion in water (0.7‰); f is the fractionation associated with photorespiration, and; p_a is the partial pressure of CO_2 in air. Two g_m calculations were performed and compared, since two values were taken into account for the fractionation factor f: 7‰ [50] and 0‰ [48,51]. The closest relationship between g_m estimates ($R^2 = 0.911$) using the variable J and $\Delta^{13}C$ of recently synthesized sugars methods in all varieties was found when using a f value of 0‰.

Harvesting

Plants were harvested at physiological maturity by sampling 1 m of the 2 central rows of each plot. Data collected included biomass, plant height, anthesis date, tiller number, straw production, grain yield, and harvest index (*HI*).

Statistical analysis

Analyses of variance were conducted using SAS to generate least squares means for each entry. Leaf gas exchange data were tested using a simple factorial ANOVA (three-way maximum interactions) and, where appropriate, the treatment means of leaf properties and gas exchange parameters were compared using a Tukey *post-hoc* test.

Results

Drought stress induced reductions in yield, *A* and rates of gas-exchange across all the *O. sativa* varieties analysed in this study. The duration of time required for 50% of plants to develop flowers (flower days) did not correlate with yield in well-watered plants, but those varieties with a shorter flowering day period exhibited enhanced yield under drought stress (Fig. 1a). Varieties with shorter flowering days allocated a higher proportion of their total biomass to yield (harvest index: *HI*) under control and drought treatments (Fig. 1b). Positive allometric correlations between plant height and both yield (Fig. 1c) and *HI* (Fig. 1d) were observed, both under watered or drought conditions.

The carbon isotopic composition of the leaves of the *O. sativa* varieties became increasingly enriched in the heavier ^{13}C isotope following water-stress, with the exception of IR64. Changes in the carbon isotopic composition of soluble sugars induced by the drought treatment were less consistent, with three varieties exhibiting increases and four decreases in the ratio of stable carbon isotopes (Table 1). Despite the lack of *genotype x water treatment* interaction, the varieties yielded differences in the 'long-term' isotopic signal of the bulk leaf material; some genotypes showed wide variation (up to −1.5‰ in pellets of Van) while others were more negligible (e.g. 0.2‰ in pellets of IR64). Drought caused also different variation in $\Delta^{13}C$ of soluble sugars: up to −3.5‰ in PS80 but 0.2‰ in IR71. Under both well-watered and rainfed conditions, soluble sugars showed generally lower values of carbon isotopic discrimination when compared with the average $\Delta^{13}C$ values of the flag leaf pellets. Significant differences were observed in the carbon isotopic discrimination values of of leaf pellets between varieties, however this was not replicated in the analysis of soluble sugars (Table 1).

A strong positive correlation was observed between the two estimates of g_m in the *O. sativa* varieties grown under control and drought conditions ($R^2 = 0.930$; $P<0.01$) (Fig. 2), excluding IR64 and PS80. Indeed, the lowland varieties IR64 and PS80 grown under well-watered conditions exhibited high values of g_m derived from the application of the isotopic methodology. It was therefore decided to utilise g_m measurements resulting from the variable J method in the subsequent analyses [an extended discussion of the contrasting merits of the approaches is given in 7]. Rates of *A*, g_s and g_m all showed consistent declines under water-stress conditions (Fig. 3). IR55 showed the highest yield and g_m under well-watered conditions. Variety IR71 exhibited the highest general

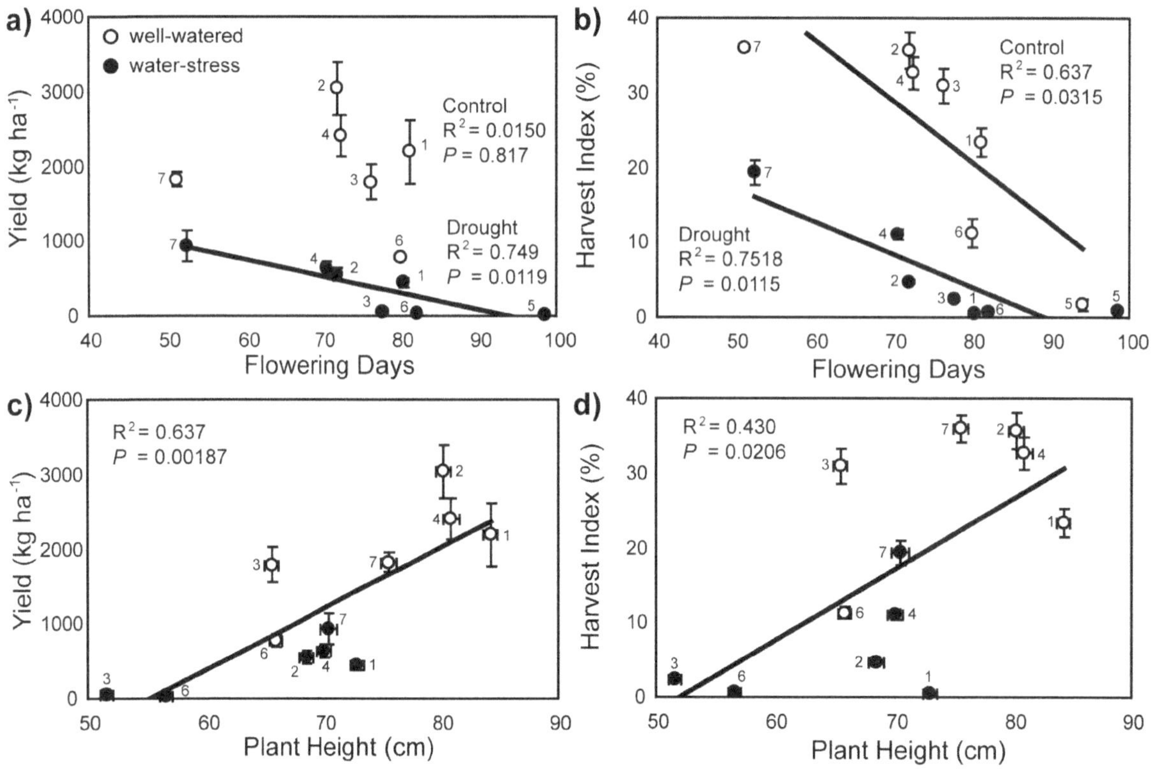

Figure 1. The performance of seven rice varieties under well-watered (open symbols) and drought conditions (closed symbols): a) relationship between yield and flowering days (period of time for 50% of plants to develop flowers) under full-water (linear regression: $R^2 = 0.0150$; $F_{1,4} = 0.0609$; $P = 0.817$) and drought (linear regression: $R^2 = 0.749$; $F_{1,5} = 14.905$; $P = 0.0119$); b) allometric relationship between yield and plant height (linear regression: $R^2 = 0.637$; $F_{1,10} = 17.536$; $P = 0.00187$); c) relationship between harvest index (*HI*: dry weight of grain relative to dry total plant biomass) and flowering days under full-water (linear regression: $R^2 = 0.637$; $F_{1,5} = 8.760$; $P = 0.0315$) and drought (linear regression: $R^2 = 0.752$; $F_{1,5} = 15.144$; $P = 0.0115$), and; d) relationship between *HI* and plant height (linear regression: $R^2 = 0.430$; $F_{1,10} = 7.548$; $P = 0.0206$). Error bars indicate one standard error either side of the mean. Numbers next to data points indicate *Oryza sativa* variety: 1 = Apo; 2 = IR55; 3 = IR64; 4 = IR71; 5 = Moro; 6 = PSBRc80; 7 = Van.

Table 1. Carbon isotope discrimination (‰) of the flag leaf pellet (bulk) and soluble sugars of seven rice cultivars grown under drought and well-watered conditions in ‰.

Genotype	Habitat	Bulk Leaf $\Delta^{13}C$ (‰)			Sugars $\Delta^{13}C$ (‰)		
		Well-watered	Drought	Mean	Well-watered	Drought	Mean
Apo	U	21.6±0.9	20.6±0.3	21.0 ab	18.3±1.3	18.7±0.6	18.5 a
IR55	U	21.4±0.5	20.8±0.0	21.1 a	18.2±0.5	16.2±0.9	17.2 a
IR64	L	20.9±0.2	21.1±0.5	21.0 ab	18.5±0.6	16.6±0.8	17.5 a
IR71	U	21.1±0.2	20.7±0.5	20.9 ab	17.3±1.3	17.5±0.7	17.4 a
Moro	U	19.9±0.4	19.7±0.2	19.8 b	18.3±0.6	19.5±0.2	18.9 a
PS80	L	21.4±0.4	21.2±0.4	21.2 a	19.3±0.3	15.8±0.2	17.2 a
Van	U	21.5±0.3	20.0±0.5	20.7 b	18.1±1.5	15.8±0.4	17.0 a
Mean		21.1 a	20.6 b		18.2 a	17.2 b	

Values indicate the mean of six plants; ± indicates the standard error of mean; different letters indicate significant differences (P≤0.05) among means according to an ANOVA.

values of A, g_s and g_m under both well-watered and drought conditions, while Apo and PS80 showed the lowest values of these photosynthesis and gas-exchange parameters under control and water-stress. The A/g_s ratio of the *O. sativa* varieties either increased (IR71; Van; IR55) or remained constant under drought stress in all varieties with the exception of PS80, where a 42.6% reduction was observed (Fig. 3d). The ratio of [CO_2] within the substomatal pore (C_i) to the external atmospheric [CO_2] (C_a) was largely unaffected by water-stress (Fig. 4a). In contrast, the ratio of the [CO_2] within the chloroplast envelope (C_c) to C_a was diminished in all *O. sativa* varieties during drought treatment (Fig. 4b), indicative of transport limitations of CO_2 in the mesophyll for photosynthesis. *Oryza sativa* variety IR71 showed the smallest proportional reduction of 16.8% in the C_c/C_a ratio, while Apo exhibited the largest reduction of 33.8%.

Yield and A exhibited significant positive correlations with g_s (Fig. 5a and 5b). However, the positive correlations between yield and A with g_m (Fig. 5c and 5d) were stronger than those observed with g_s or total conductance (g_{tot}); suggesting that g_m is the major determinant of yield and A under both well watered and water-stressed conditions. The yield of the *O. sativa* varieties was strongly related to A under both control and drought treatments (Fig. 5g). In order to gauge the response of these diffusive limitations to CO_2 transport and their effect on yield and A, the proportional change (Δ) in response to drought was calculated. The proportional reduction in yield (Δyield) was not related to Δg_s, Δg_m or Δg_{tot} (Fig. 6a; 6b; 6c). However, ΔA was significantly correlated to Δg_m and Δg_{tot}. IR71 exhibited the highest levels of A and g_m (Fig. 3a and 3c), but also the lowest proportional reduction in g_m and A (Fig. 6d). The highest yields observed under control conditions were found in IR55 (3039.6±359.1 kg ha^{-1}) and IR71 (2407.4±284.4 kg ha^{-1}); under water-stress the yield of IR55 fell to 552.7±80.3 kg ha^{-1} and IR71 to 640.3±93.0 kg ha^{-1}. The lowest Δyield reduction under drought conditions was observed in variety Van (Fig. 6), which also produced the highest yield under

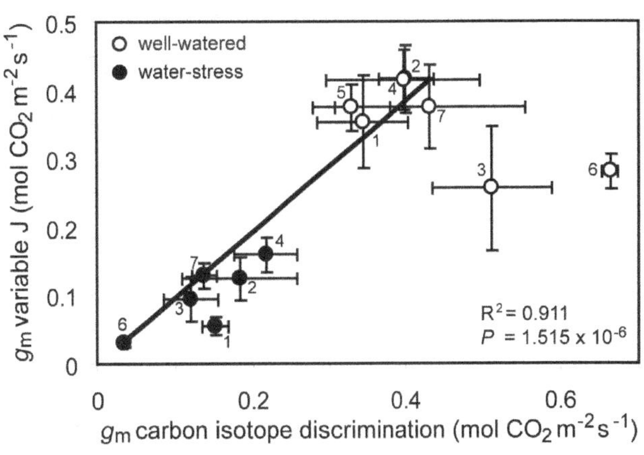

Figure 2. Comparison between the estimates of mesophyll conductance of CO_2 (g_m) obtained by applying two independent methods: the variable *J* method and the 'δ13C of recently synthesised sugars' method (linear regression: $R^2=0.932$; $F_{1,9}=122.750$; $P=1.515\times10^{-6}$). Each data point represents the average value of three observations based upon the Δ13C of recently synthesised sugars and six to fourteen gas-exchange measurements utilising the variable J method. Error bars as in Figure 1. The regression line excludes the two data points on the right of the graph (IR64 and PS80) with anomalously high g_m derived from the δ13C of recently synthesised sugars. Numbers next to data points indicate *Oryza sativa* variety as in Figure 1.

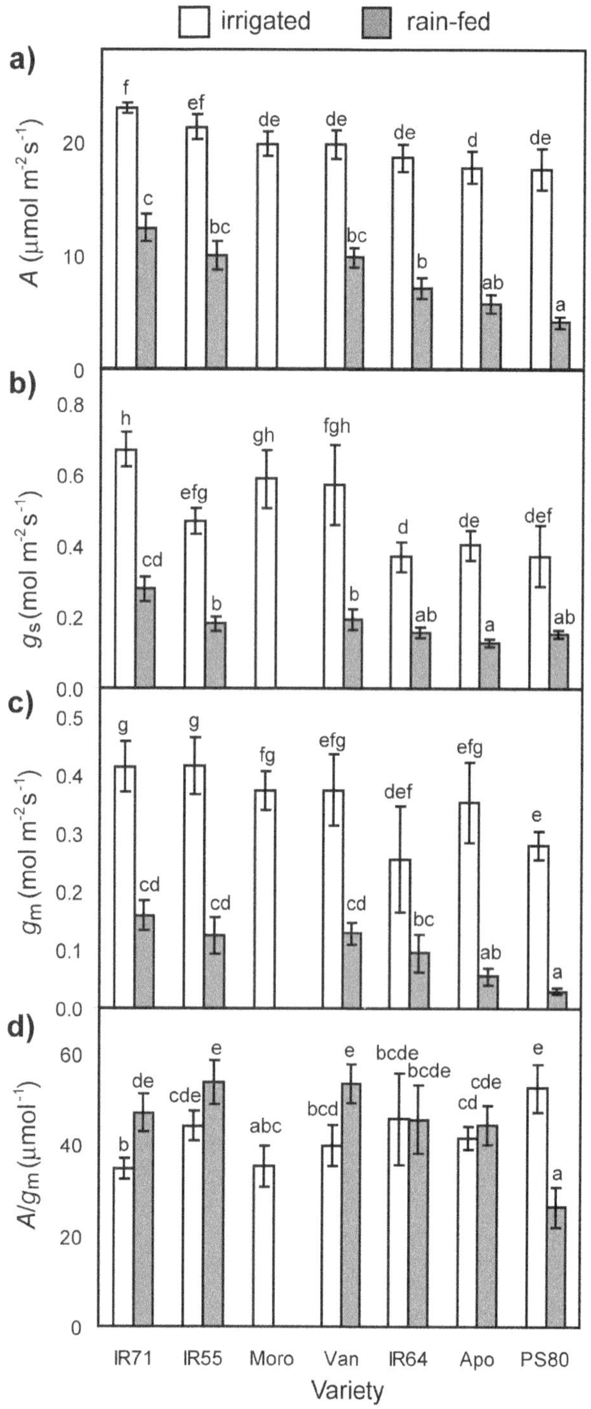

Figure 3. Measurements of (a) photosynthesis rate (A), (b) stomatal conductance (g_s), (c) mesophyll conductance (g_m), and (d) intrinsic transpiration efficiency (A/g_s) in control and water-stressed leaves of the seven *Oryza sativa* genotypes. The measurements were made on the flag leaf in saturating PPFD (1400 µmol $m^{-2}s^{-1}$), with relative humidity ranging between 45–55%, and a leaf temperature of 30°C. Data are means of 4 to 7 plants per treatment. Error bars as in Figure 1. Different letters denote significant differences among means derived using a factorial ANOVA and Tukey *post-hoc* test.

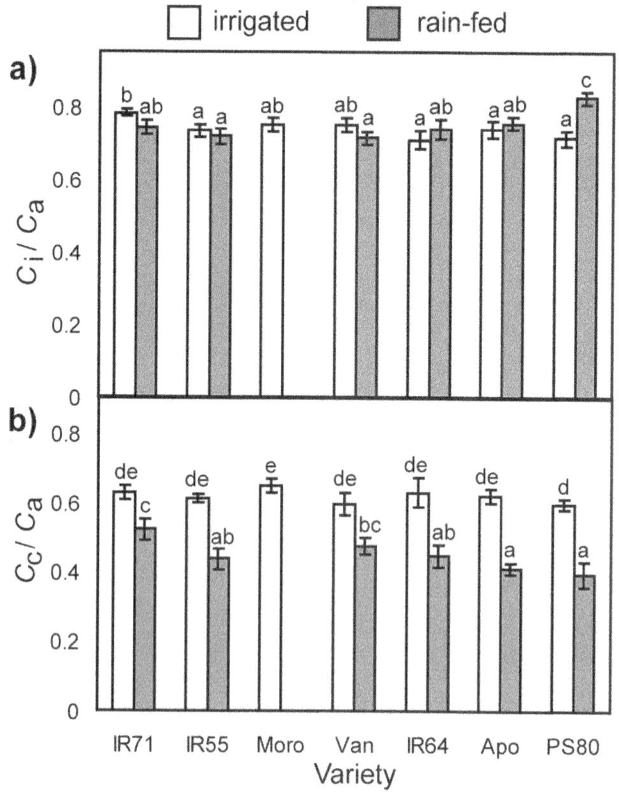

Figure 4. Measurements of (a) the intercellular [CO_2] (C_i) to the ambient [CO_2] (C_a) ratio (C_i/C_a), and (b) the chloroplastic [CO_2] (C_c) to the ambient [CO_2] ratio (C_c/C_a) in control and water-stressed leaves of the seven *Oryza sativa* genotypes. The measurements were made on the flag leaf in saturating PPFD (1400 µmol $m^{-2}s^{-1}$), with relative humidity ranging between 45–55%, and a leaf temperature of 30°C. Data are means of 4 to 7 plants per treatment. Error bars as in Figure 1. Different letters denote significant differences among means derived using a factorial ANOVA and Tukey *post-hoc* test.

water-stress (935.3±206.0 kg ha^{-1}). IR71 exhibited the second lowest Δyield and produced second largest yield during water-stress of the varieties analysed (Fig. 6g). The A/g_s ratio was not significantly related to either yield or *HI* under both well-watered or drought conditions (Fig. 7a and 7c). However, the g_m:g_s ratio was significantly correlated with both yield and *HI* (Fig. 7b and 7d), suggesting that those *O. sativa* varieties with greater g_m/g_s ratio will exhibit enhanced productivity under both well-watered and water-stressed conditions.

Discussion

The seven *O. sativa* varieties exhibited a diverse range of morphological, isotopic, photosynthetic and leaf gas-exchange responses to drought stress. A number of factors determined the yield of the *O. sativa* cultivars analysed in this study. Despite water deficit being instigated at the same developmental stage for all varieties, phenology had an effect on yield under drought as rapidly developing plants maintained yield to a greater extent than slower developers (Fig. 1a) [52]. Despite the type and phenolgical timing of the drought stress being different to that utilised in the investigation of Centritto et al. [7], the effect on *A* and yield observed in the two studies are broadly consistent. Those plants that were able to accumulate biomass rapidly produced higher

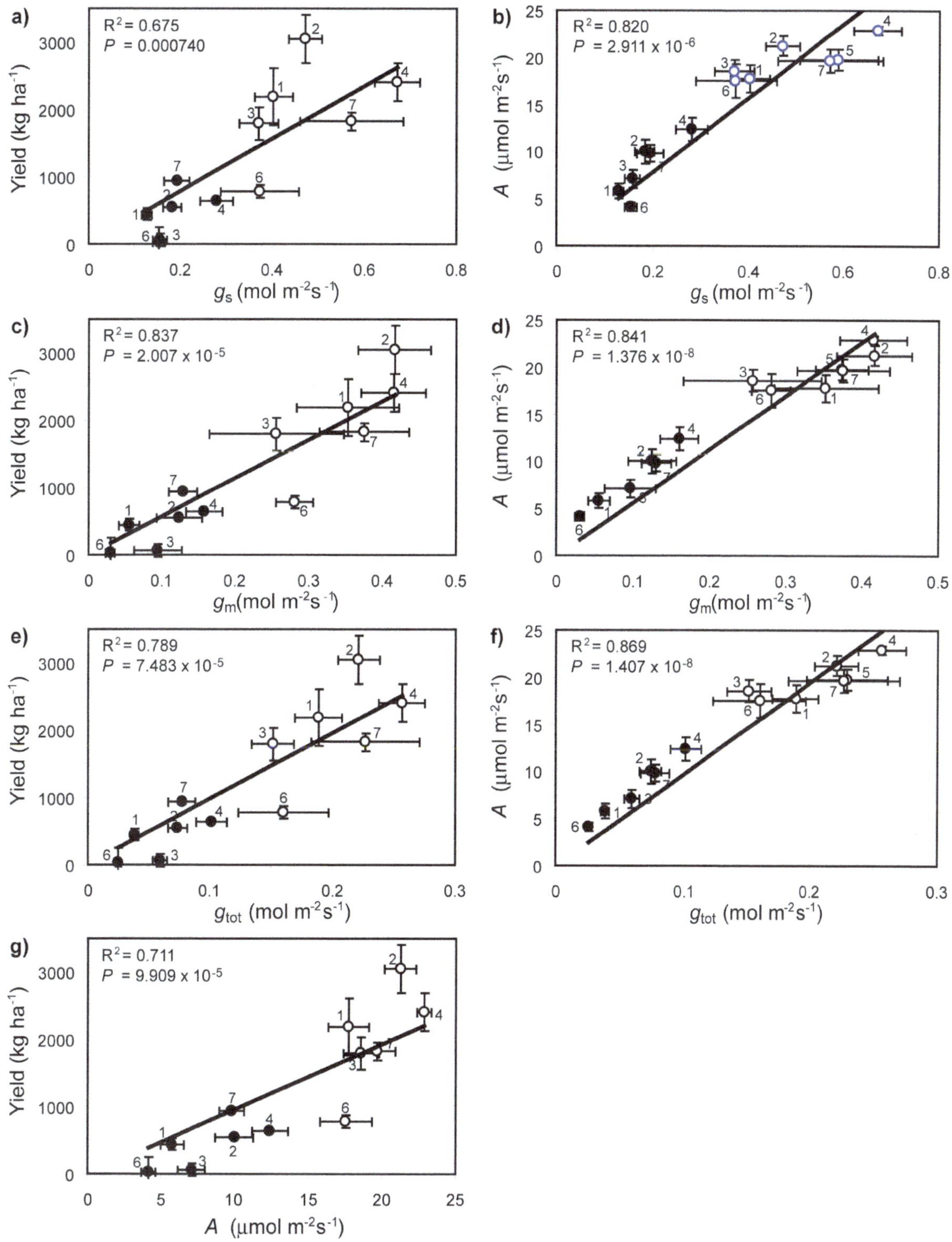

Figure 5. Interaction of diffusive conductance parameters to CO_2 uptake with yield and photosynthesis (*A*) under well-watered (open symbols) and drought conditions (closed symbols): a) relationship between yield and stomatal conductance (g_s) (linear regression: $R^2=0.696$; $F_{1,10}=22.900$; $P=0.000740$); b) relationship between *A* and g_s (linear regression: $R^2=0.873$; $F_{1,11}=75.721$; $P=2.911\times10^{-6}$); c) relationship between yield and mesophyll conductance (g_m) (linear regression: $R^2=0.850$; $F_{1,10}=56.611$; $P=2.911\times10^{-5}$); d) relationship between *A* and g_m (linear regression: $R^2=0.952$; $F_{1,11}=217.071$; $P=1.376\times10^{-8}$); e) relationship between yield and total conductance (g_{tot}) (linear regression: $R^2=0.806$; $F_{1,10}=41.418$; $P=7.483\times10^{-5}$); f) relationship between *A* and g_{tot} (linear regression: $R^2=0.952$; $F_{1,11}=216.173$; $P=1.407\times10^{-8}$), and; g) relationship between yield and *A* (linear regression: $R^2=0.795$; $F_{1,10}=38.665$; $P=9.909\times10^{-5}$). Error bars as in Figure 1. Numbers next to data points indicate *Oryza sativa* variety as in Figure 1.

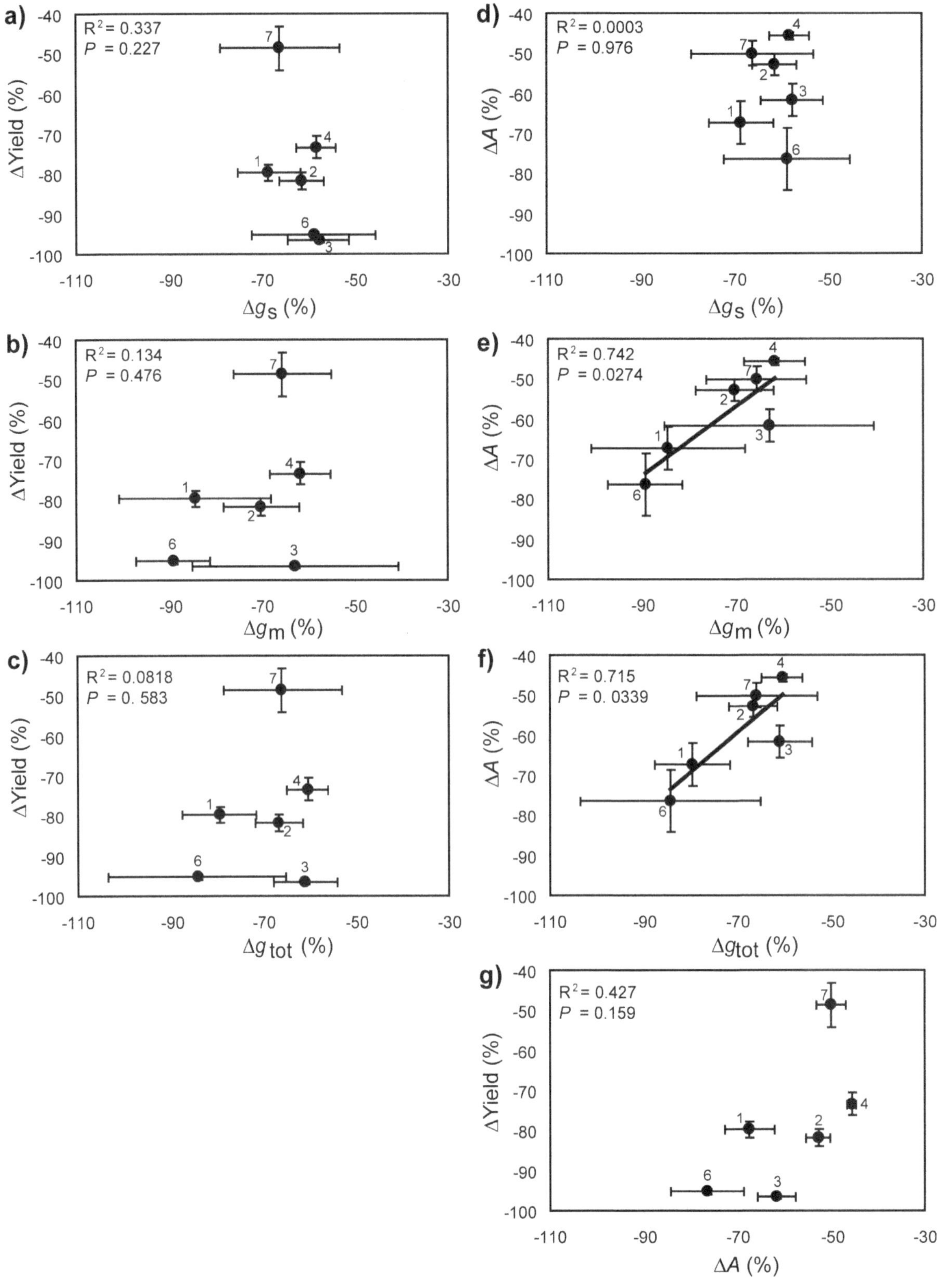

Figure 6. Changes in yield and photosynthesis in relation to modification of diffusive resistances to CO_2 uptake following water-stress. Those varieties that experienced smaller reductions in parameters were more tolerant of drought. a) relationship between Δyield and Δg_s (linear regression: $R^2 = 0.337$; $F_{1,4} = 2.032$; $P = 0.227$); b) relationship between Δyield and Δg_m (linear regression: $R^2 = 0.134$; $F_{1,4} = 0.618$; $P = 0.476$); c) relationship between Δyield and Δg_{tot} (linear regression: $R^2 = 0.0818$; $F_{1,4} = 0.356$; $P = 0.583$); d) relationship between ΔA and Δg_s (linear regression: $R^2 = 0.0003$; $F_{1,4} = 0.00106$; $P = 0.976$); e) relationship between ΔA and Δg_m (linear regression: $R^2 = 0.742$; $F_{1,4} = 11.527$; $P = 0.0274$); f) relationship between ΔA and Δg_{tot} (linear regression: $R^2 = 0.715$; $F_{1,4} = 10.042$; $P = 0.0339$), and; g) relationship between Δyield and ΔA (linear regression: $R^2 = 0.427$; $F_{1,4} = 2.979$; $P = 0.159$). Error bars as in Figure 1. Numbers next to data points indicate *Oryza sativa* variety as in Figure 1.

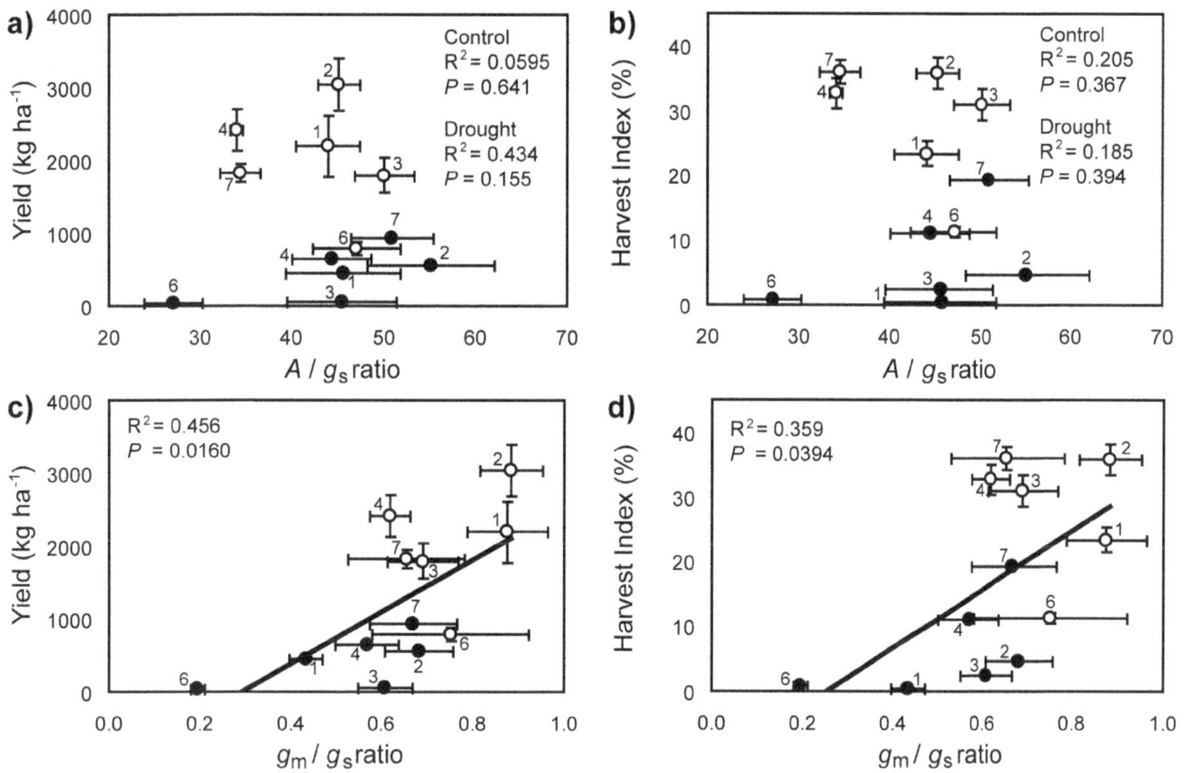

Figure 7. Interaction of yield and A with transpiration efficiency (A/g_s) and the ratio of g_m to g_s in well-watered (open symbols) and drought conditions (closed symbols): a) relationship between yield and A/g_s under full (linear regression: $R^2=0.0595$; $F_{1,4}=0.253$; $P=0.641$) and water-stressed (linear regression: $R^2=0.434$; $F_{1,4}=3.072$; $P=0.155$) conditions; b) relationship between harvest index (*HI*) and A/g_s under full (linear regression: $R^2=0.205$; $F_{1,4}=1.032$; $P=0.367$) and water-stressed (linear regression: $R^2=0.185$; $F_{1,4}=0.909$; $P=0.394$) conditions; c) relationship between yield and g_m:g_s (linear regression: $R^2=0.456$; $F_{1,10}=8.379$; $P=0.0160$), and; d) relationship between *HI* and g_m:g_s (linear regression: $R^2=0.359$; $F_{1,10}=5.610$; $P=0.0394$). Error bars as in Figure 1. Numbers next to data points indicate *Oryza sativa* variety as in Figure 1.

grain yields than those that accumulated biomass less efficiently under both well-watered and drought conditions. This may be due to allometric relationships between biomass and yield [53], or the positive influence of increased photosynthetic area on 'whole-plant' carbon uptake [54]. Leaf-level photosynthetic rates were also closely related to yield in all varieties under both well-watered and drought conditions (Fig. 5g). Photosynthesis rates were in turn determined by diffusive limitations to CO_2-uptake (Fig. 5b; 5d; 5f). The most significant of these resistances to CO_2-uptake in *O. sativa* occurs at the interface between the internal sub-stomatal air-space and the mesophyll layer [15,55]. Consistent with other studies, g_m in *O. sativa* correlated most closely with A and yield (Fig. 5d) [7,9], and represents a critical 'rate-limiting' step in the determination of grain productivity and assessment of the photosynthetic response to drought stress.

The 'bulk' carbon isotope discrimination values of the *O. sativa* varieties decreased following drought treatment; indicative of water-deficits over the life-span of the flag-leaf reducing discrimination against the heavier ^{13}C isotope [56,57]. However, this pattern of reduced carbon isotope discrimination is not apparent in the carbon isotopic composition of recently synthesised sugars (Table 1). These findings may reflect 'shorter-term' variations in *WUE* in the plants experiencing drought stress, a modification of the photosynthetic pathway affecting sugar metabolism [28] or the breakdown of previously stored carbohydrates with a different isotopic signature into more simple sugars over the course of the stress treatment [58,59]. Differences in the diffusional uptake of CO_2 likely affected carbon isotopic composition [60]; however, this was not apparent from our g_s, g_m (variable J) and g_{tot} datasets (Fig. 5), suggesting that other factors affected carbon isotopic discrimination and possibly the effectiveness of the Δ13C of recently synthesised sugars method of estimating g_m in varieties IR64 and PS80 under control conditions (Fig. 2). It is noteworthy that under water-deficit conditions, g_m estimates based on the variable J method closely correlated with those derived from the Δ13C of recently synthesised sugars (Fig. 2). These findings do not support the suggestion that reductions in C_i may affect estimation of the total photosynthetic electron transport rate associated with enhanced photorespiration, thus rendering the variable J method unreliable in the investigation of g_m in plant responses to environmental stress [34].

Selective breeding has improved the productivity of *O. sativa* varieties through improved growth form and the increased proportional allocation of biomass as grain [4]; yet these increases in yield have been constrained by the relatively low leaf-level photosynthetic rate of *O. sativa* in comparison to other C3 crops [5,61]. The results of this study indicate that under well-watered and water-deficit conditions, A is an important determinant of yield in *O. sativa* and is largely related to the rate of transport of CO_2 across the mesophyll (Fig. 5d). Further improvements in *O. sativa* grain yield may likely be achieved through the enhancement of g_m to CO_2 inducing enhanced A [6,9,16,62]. Those *O. sativa* varieties that showed the greatest photosynthetic performance and yield under control conditions were not necessarily the most effective varieties under drought conditions. Varieties that maintained transport of CO_2 during water-stress sustained *A* and yield to a greater extent than varieties that exhibited more pronounced diffusive limitations to photosynthetic CO_2-uptake (Fig. 5). Drought stress induced a relatively consistent mean reduction in g_s in the *O. sativa* varieties (Fig. 3), yet the C_i:C_a ratio was unaffected (Fig. 4), indicating that g_s was not limiting to *A* under water-deficit, or that the gas-exchange data were possibly biased by the occurrence of non-uniform stomatal closure over the leaf lamina and consequent overestimation of C_i [27,63,64]. However, the C_i/C_a ratio in the *O. sativa* varieties was reduced by 16.8 to 33.8% after drought, consistent with the impairment of *A* through mesophyll diffusive limitations [7,11,12,65,66]. Mesophyll conductance to CO_2 declined during the drought treatment, constraining *A* and growth (Fig. 5; 6). However, significant variability in the extent of reductions in g_m were observed in the *O. sativa* cultivars ranging from −61.7% (IR71) to −89.3% (PS80). Furthermore, those varieties that showed the smallest reductions in g_m also exhibited greater *A*; demonstrating that g_m largely determines *A* in *O. sativa* under differing degrees of water-deficit [7,9]. These differences in the ability to sustain g_m values under drought stress are likely related to the maintenance of biochemical function in the mesophyll such as the activity of cooporin and carbonic anhydrase [eg. 25,67].

The conductance of the mesophyll layer to the transport of CO_2 determined *A* and yield in the seven *O. sativa* varieties studied. Under conditions of water-stress it has been suggested that a low g_s (to reduce transpirative water-loss) and a high g_m (to maintain the uptake of CO_2) are beneficial [11,68]. However, in the case of *O. sativa*, yield and *HI* were only weakly related to the ratio of g_m:g_s ($R^2<0.460$) (Fig. 7). The reduction in g_s under water-deficit is consistent with active stomatal control in the optimisation of *WUE* [69,70]; nevertheless, the lack of a strong correlation to the g_m:g_s ratio may be indicative of the lack of a stomatal limitation to *A* under drought stress in the *O. sativa* varieties apparent in the C_i:C_a ratios (Fig. 4a). The photosynthetic and yield responses of *O. sativa* under well-watered and drought conditions were more dependent upon g_m than g_s (Fig. 5b and 5d). This is consistent with earlier studies reporting that modification and improvement of g_m is more effective in the enhancement of *O. sativa* productivity [16] and likely drought tolerance [7,9]. The strong dependence of *A* and yield on g_m is also evident in the lack of a relationship between A/g_s and yield or *HI* (Fig. 7a and 7c). This may suggest that stomatal control is either less effective [70,71], or plays a reduced role, in respect to mesophyll limitations to CO_2 in a drought intolerant crop such as *O. sativa* in comparison to crops adapted to more water-limited environments [72].

The disparity in *A* values observed between the cultivars raises the possibility of crossing different varieties to enhance leaf-level *A* and yield in subsequent generations [6,8,16]. Selection of drought tolerance on the basis of the ability of a particular genotype to maintain g_m function under drought stress would be viable [11,73]. However, genetic sequencing and identification of quantitative trait loci permit the possibility of extracting genetic material from a *O. sativa* cultivar with a valuable characteristic (such as high g_m and/or stable g_m under water deficit) and in the creation of a new variety with a suite of desired attributes [17,19,74–76]. For example, from the results of this study it would be desirable to combine the high g_m and yield of IR55 under well-watered conditions with the capacity to sustain g_m under water-stress exhibited by IR71 and water-stressed yield of Van (Fig. 3). In this context it may be possible to phenotypically screen a large number of *O. sativa* genotypes to identify quantitative trait loci responsible for g_m values and g_m response to drought. Those varieties developed to be resistant to drought may be suited to more marginal land-types or in scenarios of insufficient water availability. Nonetheless, given the sporadic and unpredictable nature of sustained and persistent drought events it is unlikely that large numbers of farmers would select a drought tolerant variety such as Van over a more productive cultivar such as IR55, and thus limit their harvest to accommodate the risk of interruption to water supplies [73,77]. An ideal *O. sativa* variety would therefore possess both values of g_m capable of sustaining high *A* and the ability to maintain g_m in the event of water-deficits [8,78]. However, the results of this study suggest that those attributes may be mutually exclusive in the species analysed. Furthermore, no single parameter analysed in this study exclusively accounted for yield under either well-watered or drought conditions; therefore attempts to improve the productivity of *O. sativa* cultivars may also involve selection for a suite of characteristics that improve productivity under specific conditions of water availability. Nonetheless, future efforts to improve *O. sativa* productivity should focus on biochemical [eg. 25] and physical modification [eg. 16] of the interface between the mesophyll layer and sub-stomatal air-space to enhance CO_2 uptake and *A*.

Conclusions

Physiological analysis of *O. sativa* has great potential in identifying those photosynthetic and leaf gas-exchange characteristics that confer attributes such as high yield and tolerance to drought. The analysis of the seven *O. sativa* varieties indicates that under both well-watered and drought stressed conditions yield and *A* are determined by diffusional constraints to CO_2 uptake and not by biochemical limitations. The varieties showed significant variation in leaf-level photosynthetic rates, symptomatic of the photosynthetic constraints to *O. sativa* production that have previously been recorded. Selection of those *O. sativa* genotypes that exhibit higher rates of leaf-level *A* would likely be beneficial for grain yields. Photosynthesis was most closely related to g_m, and the tolerance of the *O. sativa* varieties to drought was largely determined by the ability to maintain g_m values under water-deficits. However, the highest g_m and yield under well-watered conditions (IR55) and the most pronounced ability to sustain g_m under water-deficit (IR71) were observed in different varieties. This may suggest that the desired attributes for enhanced yield during high water availability and the ability to maintain g_m, and thus sustain grain production under water-deficits, may be mutually exclusive. Nonetheless, the results of studies such as this could make it possible through selective breeding, or the identification of a particular genetic expression associated with a desired attribute, to increase leaf-level photosynthetic rates and drought tolerance in the same variety of *O. sativa*. These findings indicate that the interface between the internal sub-stomatal air-space and the mesophyll layer offers the greatest potential in terms of possible modification to enhance the performance of *O. sativa* under optimal and stressed growth conditions in the future.

Supporting Information

Figure S1 Dark respiration (R_d) values of seven *Oryza sativa* varieties under irrigated (white) and rain-fed (grey) conditions used in the calculation of g_m using the variable J method.

Figure S2 Electron transport rate (ETR) values of seven *Oryza sativa* varieties under irrigated (white) and rain-

fed (grey) conditions used in the calculation of g_m using the variable J method.

Acknowledgments

The following are thanked for scientific discussion and technical assistance: L. Spaccino for IRMS analyses (CNR-IBAF) and R. Torres G. Dimayuga and the drought physiology team at IRRI for technical support (IRRI, Philippines). The comments of two anonymous referees significantly improved this manuscript.

Author Contributions

Conceived and designed the experiments: MC ML RS. Performed the experiments: MC ML MCM. Analyzed the data: MH MC. Contributed reagents/materials/analysis tools: MC ML RS. Wrote the paper: MH MC ML.

References

1. Pandey S, Bhandari H, Ding S, Prapertchob P, Sharan R, et al. (2007) Coping with drought in rice farming in Asia: insights from a cross-country comparative study. Agricultural Economics 37: 213–224.
2. Serraj R, McNally KL, Slamet-Loedin I, Kohli A, Haefele SM, et al. (2011) Drought resistance improvement in rice: an integrated genetic and resource management strategy. Plant Production Science 14: 1–14.
3. Parent B, Suard B, Serraj R, Tardieu F (2010) Rice leaf growth and water potential are resilient to evaporative demand and soil water deficit once the effects of root system are neutralized. Plant, Cell & Environment 33: 1256–1267.
4. Peng S, Cassman KG, Virmani SS, Sheehy J, Khush GS (1999) Yield potential trends of tropical rice since the release of IR8 and the challenge of increasing rice yield potential. Crop Science 39: 1552–1559.
5. Hubbart S, Peng S, Horton P, Chen Y, Murchie EH (2007) Trends in leaf photosynthesis in historical rice varieties developed in the Philippines since 1966. Journal of Experimental Botany 58: 3429–3438.
6. Gu J, Yin X, Stomph TJ, Struik PC (2014) Can exploiting natural genetic variation in leaf photosynthesis contribute to increasing rice productivity? A simulation analysis. Plant, cell & environment 37: 22–34.
7. Centritto M, Lauteri M, Monteverdi MC, Serraj R (2009) Leaf gas exchange, carbon isotope discrimination, and grain yield in contrasting rice genotypes subjected to water deficits during the reproductive stage. Journal of Experimental Botany 60: 2325–2339.
8. Long SP (2014) We need winners in the race to increase photosynthesis in rice, whether from conventional breeding, biotechnology or both. Plant, Cell & Environment 37: 19–21.
9. Gu J, Yin X, Stomph T-J, Wang H, Struik PC (2012) Physiological basis of genetic variation in leaf photosynthesis among rice (*Oryza sativa* L.) introgression lines under drought and well-watered conditions. Journal of Experimental Botany 63: 5137–5153.
10. Niinemets Ü, Diaz-Espejo A, Flexas J, Galmes J, Warren CR (2009) Role of mesophyll diffusion conductance in constraining potential photosynthetic productivity in the field. Journal of Experimental Botany 60: 2249–2270.
11. Flexas J, Ribas-Carbó M, Diaz-Espejo A, Galmés J, Medrano H (2008) Mesophyll conductance to CO_2: current knowledge and future prospects. Plant, Cell & Environment 31: 602–621.
12. Centritto M, Loreto F, Chartzoulakis K (2003) The use of low [CO_2] to estimate diffusional and non-diffusional limitations of photosynthetic capacity of salt-stressed olive saplings. Plant, Cell & Environment 26: 585–594.
13. Pallozzi E, Fortunati A, Marino G, Loreto F, Agati G, et al. (2013) BVOC emission from *Populus × canadensis* saplings in response to acute UV-A radiation. Physiologia Plantarum 148: 51–61.
14. Hu L, Wang Z, Huang B (2010) Diffusion limitations and metabolic factors associated with inhibition and recovery of photosynthesis from drought stress in a C3 perennial grass species. Physiologia Plantarum 139: 93–106.
15. Scafaro AP, Von Caemmerer S, Evans JR, Atwell BJ (2011) Temperature response of mesophyll conductance in cultivated and wild *Oryza* species with contrasting mesophyll cell wall thickness. Plant Cell and Environment 34: 1999–2008.
16. Adachi S, Nakae T, Uchida M, Soda K, Takai T, et al. (2013) The mesophyll anatomy enhancing CO_2 diffusion is a key trait for improving rice photosynthesis. Journal of Experimental Botany 64: 1061–1072.
17. Manickavelu A, Nadarajan N, Ganesh SK, Gnanamalar RP, Chandra Babu R (2006) Drought tolerance in rice: morphological and molecular genetic consideration. Plant Growth Regulation 50: 121–138.
18. Huang X-Y, Chao D-Y, Gao J-P, Zhu M-Z, Shi M, et al. (2009) A previously unknown zinc finger protein, DST, regulates drought and salt tolerance in rice via stomatal aperture control. Genes & Development 23: 1805–1817.
19. Hu SP, Zhou Y, Zhang L, Zhu XD, Li L, et al. (2009) Correlation and quantitative trait loci analyses of total chlorophyll content and photosynthetic rate of rice (*Oryza sativa*) under water stress and well-watered conditions. Journal of Integrative Plant Biology 51: 879–888.
20. Gu J, Yin X, Struik PC, Stomph TJ, Wang H (2012) Using chromosome introgression lines to map quantitative trait loci for photosynthesis parameters in rice (*Oryza sativa* L.) leaves under drought and well-watered field conditions. Journal of Experimental Botany 63: 455–469.
21. Flexas J, Niinemets Ü, Gallé A, Barbour M, Centritto M, et al. (2013) Diffusional conductances to CO_2 as a target for increasing photosynthesis and photosynthetic water-use efficiency. Photosynthesis Research 117: 45–59.
22. Farquhar GD, Caemmerer S, Berry JA (1980) A biochemical model of photosynthetic CO_2 assimilation in leaves of C3 species. Planta 149: 78–90.
23. Giuliani R, Koteyeva N, Voznesenskaya E, Evans MA, Cousins AB, et al. (2013) Coordination of leaf photosynthesis, transpiration, and structural traits in rice and wild relatives (Genus *Oryza*). Plant Physiology 162: 1632–1651.
24. Kaldenhoff R (2012) Mechanisms underlying CO_2 diffusion in leaves. Current Opinion in Plant Biology 15: 276–281.
25. Hanba YT, Shibasaka M, Hayashi Y, Hayakawa T, Kasamo K, et al. (2004) Overexpression of the barley aquaporin HvPIP2;1 increases internal CO_2 conductance and CO_2 assimilation in the leaves of transgenic rice plants. Plant and Cell Physiology 45: 521–529.
26. Tyerman SD, Niemietz CM, Bramley H (2002) Plant aquaporins: multifunctional water and solute channels with expanding roles. Plant, Cell & Environment 25: 173–194.
27. Loreto F, Centritto M (2008) Leaf carbon assimilation in a water-limited world. Plant Biosystems 142: 154–161.
28. Chaves M, Flexas J, Pinheiro C (2009) Photosynthesis under drought and salt stress: regulation mechanisms from whole plant to cell. Annals of Botany 103: 551–560.
29. Cornic G (2000) Drought stress inhibits photosynthesis by decreasing stomatal aperture – not by affecting ATP synthesis. Trends in Plant Science 5: 187–188.
30. Pinheiro C, Chaves MM (2011) Photosynthesis and drought: can we make metabolic connections from available data? Journal of Experimental Botany 62: 869–882.
31. Kamoshita A, Rodriguez R, Yamauchi A, Wade LJ (2004) Genotypic variation in response of rainfed lowland rice to prolonged drought and rewatering. Plant Production Science 7: 406–420.
32. Boonjung H, Fukai S (1996) Effects of soil water deficit at different growth stages on rice growth and yield under upland conditions. 2. Phenology, biomass production and yield. Field Crops Research 48: 47–55.
33. Pons TL, Flexas J, von Caemmerer S, Evans JR, Genty B, et al. (2009) Estimating mesophyll conductance to CO_2: methodology, potential errors, and recommendations. Journal of Experimental Botany 60: 2217–2234.
34. Gilbert ME, Pou A, Zwieniecki MA, Holbrook NM (2012) On measuring the response of mesophyll conductance to carbon dioxide with the variable J method. Journal of Experimental Botany 63: 413–425.
35. Harley PC, Loreto F, Dimarco G, Sharkey TD (1992) Theoretical considerations when estimating the mesophyll conductance to CO_2 flux by analysis of the response of photosynthesis to CO_2. Plant Physiology 98: 1429–1436.
36. Bois JF, Couchat PH, Lasceve G (1985) Relationships between transpiration and photosynthesis during a water stress. Acta Horticulturae 171: 297–314.
37. Richter A, Wanek W, Werner RA, Ghashghaie J, Jaeggi M, et al. (2009) Preparation of starch and soluble sugars of plant material for the analysis of carbon isotope composition: a comparison of methods. Rapid Communications in Mass Spectrometry 23: 2476–2488.
38. Farquhar GD, Ehleringer JR, Hubick KT (1989) Carbon isotope discrimination and photosynthesis. Annual Review of Plant Physiology and Plant Molecular Biology 40: 503–537.
39. Rodeghiero M, Niinemets Ü, Cescatti A (2007) Major diffusion leaks of clamp-on leaf cuvettes still unaccounted: how erroneous are the estimates of Farquhar et al. model parameters? Plant, cell & environment 30: 1006–1022.
40. Loreto F, Harley PC, Dimarco G, Sharkey TD (1992) Estimation of mesophyll conductance to CO_2 flux by three different methods. Plant Physiology 98: 1437–1443.
41. Genty B, Briantais J-M, Baker NR (1989) The relationship between the quantum yield of photosynthetic electron transport and quenching of chlorophyll fluorescence. Biochimica et Biophysica Acta (BBA) - General Subjects 990: 87–92.
42. Laisk A, Loreto F (1996) Determining photosynthetic parameters from leaf CO_2 exchange and chlorophyll fluorescence (ribulose-1, 5-bisphosphate carboxylase/oxygenase specificity factor, dark respiration in the light, excitation distribution between photosystems, alternative electron transport rate, and mesophyll diffusion resistance. Plant Physiology 110: 903–912.
43. Galmes J, Flexas J, Keys AJ, Cifre J, Mitchell RAC, et al. (2005) Rubisco specificity factor tends to be larger in plant species from drier habitats and in species with persistent leaves. Plant Cell and Environment 28: 571–579.
44. Furbank RT, von Caemmerer S, Sheehy J, Edwards G (2009) C4 rice: a challenge for plant phenomics. Functional Plant Biology 36: 845–856.
45. Pinelli P, Loreto F (2003) $^{12}CO_2$ emission from different metabolic pathways measured in illuminated and darkened C-3 and C-4 leaves at low, atmospheric and elevated CO_2 concentration. Journal of Experimental Botany 54: 1761–1769.

46. Loreto F, Dimarco G, Tricoli D, Sharkey TD (1994) Measurements of mesophyll conductance, photosynthetic electron transport and alternative electron sinks of field grown wheat leaves. Photosynthesis Research 41: 397–403.
47. Lauteri M, Scartazza A, Guido MC, Brugnoli E (1997) Genetic variation in photosynthetic capacity, carbon isotope discrimination and mesophyll conductance in provenances of *Castanea sativa* adapted to different environments. Functional Ecology 11: 675–683.
48. Scartazza A, Lauteri M, Guido MC, Brugnoli E (1998) Carbon isotope discrimination in leaf and stem sugars, water-use efficiency and mesophyll conductance during different developmental stages in rice subjected to drought. Australian Journal of Plant Physiology 25: 489–498.
49. Evans JR, Sharkey TD, Berry JA, Farquhar GD (1986) Carbon isotope discrimination measured concurrently with gas exchange to investigate CO_2 diffusion in leaves of higher plants. Australian Journal of Plant Physiology 13: 281–292.
50. Lloyd J, Syvertsen JP, Kriedemann PE, Farquhar GD (1992) Low conductances for CO_2 diffusion from stomata to the sites of carboxylation in leaves of woody species. Plant Cell and Environment 15: 873–899.
51. von Caemmerer S, Evans JR (1991) Determination of the average partial pressure of CO_2 in chloroplasts from leaves of several C3 plants. Australian Journal of Plant Physiology 18: 287–305.
52. Pantuwan G, Fukai S, Cooper M, Rajatasereekul S, O'Toole JC (2002) Yield response of rice (*Oryza sativa* L.) genotypes to different types of drought under rainfed lowlands: Part 1. Grain yield and yield components. Field Crops Research 73: 153–168.
53. Siddique K, Kirby E, Perry M (1989) Ear: stem ratio in old and modern wheat varieties; relationship with improvement in number of grains per ear and yield. Field Crops Research 21: 59–78.
54. Koyama K, Kikuzawa K (2009) Is whole-plant photosynthetic rate proportional to leaf area? A test of scalings and a logistic equation by leaf demography census. The American Naturalist 173: 640–649.
55. Tomás M, Flexas J, Copolovici L, Galmés J, Hallik L, et al. (2013) Importance of leaf anatomy in determining mesophyll diffusion conductance to CO_2 across species: quantitative limitations and scaling up by models. Journal of experimental botany 64: 2269–2281.
56. Zhao B, Kondo M, Maeda M, Ozaki Y, Zhang J (2004) Water-use efficiency and carbon isotope discrimination in two cultivars of upland rice during different developmental stages under three water regimes. Plant and Soil 261: 61–75.
57. Tipple BJ, Berke MA, Doman CE, Khachaturyan S, Ehleringer JR (2013) Leaf-wax n-alkanes record the plant–water environment at leaf flush. Proceedings of the National Academy of Sciences 110: 2659–2664.
58. Quick W, Chaves M, Wendler R, David M, Rodrigues M, et al. (1992) The effect of water stress on photosynthetic carbon metabolism in four species grown under field conditions. Plant, Cell and Environment 15: 25–35.
59. Yang J, Zhang J, Wang Z, Zhu Q (2001) Activities of starch hydrolytic enzymes and sucrose-phosphate synthase in the stems of rice subjected to water stress during grain filling. Journal of Experimental Botany 52: 2169–2179.
60. Scheidegger Y, Saurer M, Bahn M, Siegwolf R (2000) Linking stable oxygen and carbon isotopes with stomatal conductance and photosynthetic capacity: a conceptual model. Oecologia 125: 350–357.
61. Schmitt MR, Edwards GE (1981) Photosynthetic capacity and nitrogen use efficiency of maize, wheat, and rice - a comparison between C-3 and C-4 Photosynthesis. Journal of Experimental Botany 32: 459–466.
62. Kiran TV, Rao YV, Subrahmanyam D, Rani NS, Bhadana VP, et al. (2013) Variation in leaf photosynthetic characteristics in wild rice species. Photosynthetica 51: 350–358.
63. Terashima I (1992) Anatomy of non-uniform leaf photosynthesis. Photosynthesis research 31: 195–212.
64. Mott KA, Buckley TN (2000) Patchy stomatal conductance: emergent collective behaviour of stomata. Trends in Plant Science 5: 258–262.
65. Sun P, Wahbi S, Tsonev T, Haworth M, Liu S, et al. (2014) On the use of leaf spectral indices to assess water status and photosynthetic limitations in *Olea europaea* L. during water-stress and recovery. Plos One 9: e105165.
66. Tsonev T, Wahbi S, Sun P, Sorrentino G, Centritto M (2014) Gas exchange, water relations and their relationships with photochemical reflectance index in *Quercus ilex* plants during water stress and recovery. International Journal of Agriculture and Biology 16: 335–341.
67. Warren CR. Stand aside stomata, another actor deserves centre stage: the forgotten role of the internal conductance to CO_2 transfer; 2007 Jul 22–27; Glasgow, Scotland. Oxford University Press. pp.1475–1487.
68. Centritto M, Tognetti R, Leitgeb E, Střelcová K, Cohen S (2011) Above ground processes - anticipating climate change influences. In: Bredemeier M, Cohen S, Godbold DL, Lode E, Pichler V et al., editors. Forest Management and the Water Cycle: An Ecosystem-Based Approach. London: Springer. pp.31–64.
69. Haworth M, Elliott-Kingston C, McElwain J (2013) Co-ordination of physiological and morphological responses of stomata to elevated [CO_2] in vascular plants. Oecologia 171: 71–82.
70. Haworth M, Elliott-Kingston C, McElwain JC (2011) Stomatal control as a driver of plant evolution. Journal of Experimental Botany 62: 2419–2423.
71. Haworth M, Gallagher A, Elliott-Kingston C, Raschi A, Marandola D, et al. (2010) Stomatal index responses of *Agrostis canina* to carbon dioxide and sulphur dioxide: implications for palaeo-[CO_2] using the stomatal proxy. New Phytologist 188: 845–855.
72. Kubiskem ME, Abrams MD (1993) Stomatal and nonstomatal limitations of photosynthesis in 19 temperate tree species on contrasting sites during wet and dry years. Plant, Cell and Environment 16: 1123–1129.
73. Venuprasad R, Lafitte HR, Atlin GN (2007) Response to direct selection for grain yield under drought stress in rice. Crop Sci 47: 285–293.
74. Karaba A, Dixit S, Greco R, Aharoni A, Trijatmiko KR, et al. (2007) Improvement of water use efficiency in rice by expression of HARDY, an *Arabidopsis* drought and salt tolerance gene. Proceedings of the National Academy of Sciences of the United States of America 104: 15270–15275.
75. Lanceras JC, Pantuwan G, Jongdee B, Toojinda T (2004) Quantitative trait loci associated with drought tolerance at reproductive stage in rice. Plant Physiology 135: 384–399.
76. Xu Y, This D, Pausch RC, Vonhof WM, Coburn JR, et al. (2009) Leaf-level water use efficiency determined by carbon isotope discrimination in rice seedlings: genetic variation associated with population structure and QTL mapping. Theoretical and Applied Genetics 118: 1065–1081.
77. Wassmann R, Jagadish SVK, Heuer S, Ismail A, Redona E, et al. (2009) Climate change affecting rice production: the physiological and agronomic basis for possible adaptation strategies. In: Sparks DL, editor. Advances in Agronomy, Vol 101. pp.59–122.
78. Blum A (2005) Drought resistance, water-use efficiency, and yield potential–are they compatible, dissonant, or mutually exclusive? Australian Journal of Agricultural Research 56: 1159–1168.

17

Cloning of *Gossypium hirsutum* Sucrose Non-Fermenting 1-Related Protein Kinase 2 Gene (*GhSnRK2*) and Its Overexpression in Transgenic *Arabidopsis* Escalates Drought and Low Temperature Tolerance

Babatunde Bello, Xueyan Zhang, Chuanliang Liu, Zhaoen Yang, Zuoren Yang, Qianhua Wang, Ge Zhao, Fuguang Li*

State Key Laboratory of Cotton Biology, Cotton Research Institute, Chinese Academy of Agricultural Sciences, Beijing, China

Abstract

The molecular mechanisms of stress tolerance and the use of modern genetics approaches for the improvement of drought stress tolerance have been major focuses of plant molecular biologists. In the present study, we cloned the *Gossypium hirsutum* sucrose non-fermenting 1-related protein kinase 2 (*GhSnRK2*) gene and investigated its functions in transgenic Arabidopsis. We further elucidated the function of this gene in transgenic cotton using virus-induced gene silencing (VIGS) techniques. We hypothesized that *GhSnRK2* participates in the stress signaling pathway and elucidated its role in enhancing stress tolerance in plants via various stress-related pathways and stress-responsive genes. We determined that the subcellular localization of the *GhSnRK2*-green fluorescent protein (GFP) was localized in the nuclei and cytoplasm. In contrast to wild-type plants, transgenic plants overexpressing *GhSnRK2* exhibited increased tolerance to drought, cold, abscisic acid and salt stresses, suggesting that *GhSnRK2* acts as a positive regulator in response to cold and drought stresses. Plants overexpressing *GhSnRK2* displayed evidence of reduced water loss, turgor regulation, elevated relative water content, biomass, and proline accumulation. qRT-PCR analysis of *GhSnRK2* expression suggested that this gene may function in diverse tissues. Under normal and stress conditions, the expression levels of stress-inducible genes, such as *AtRD29A, AtRD29B, AtP5CS1, AtABI3, AtCBF1*, and *AtABI5*, were increased in the *GhSnRK2*-overexpressing plants compared to the wild-type plants. *GhSnRK2* gene silencing alleviated drought tolerance in cotton plants, indicating that VIGS technique can certainly be used as an effective means to examine gene function by knocking down the expression of distinctly expressed genes. The results of this study suggested that the *GhSnRK2* gene, when incorporated into Arabidopsis, functions in positive responses to drought stress and in low temperature tolerance.

Editor: Girdhar K. Pandey, University of Delhi South Campus, India

Funding: This work was supported by Major Program of Joint Funds (Sinkiang) of the National Natural Science Foundation of China (Grant No. U1303282). The funders had no role in study design, data collection and analysis, decision to publish, or preparation of the manuscript.

Competing Interests: The authors have declared that no competing interests exist.

* Email: aylifug@163.com

Introduction

Plants have developed complex signaling pathways in response to various environmental stresses, such as salt, drought, and cold, and have acquired metabolic functions and developmental methods to survive changing environmental conditions [1]. Improving crop resistance to drought stress would be the most valuable means to improve agricultural productivity and to reduce crop loss caused by environmental stress. As a result, understanding the mechanisms of drought tolerance and developing drought-resistant crop plants have been major targets of plant molecular biologists and biotechnologists. Low-temperature constraints have been progressively overcome by the recognition of cold-tolerant genes for applications in transgenic plants. Transgenic approach has reveal many possibilities to improve cold stress in plants by incorporating or deleting genes that regulate a particular characteristic [2]. These approaches also provide unique opportunities to improve the genetic quality of plants via the development of particular crop varieties that exhibit enhanced resistance to biotic and cold stresses and improved nutritional quality. The plant response to salt stress typically results in osmotic alterations, which play a major role in ensuring osmotic balance in plant cells. During plant stress responses, the regulation of gene expression involves both universal and unique changes in the transcript levels of certain plant genes [3]. Plants directly or indirectly respond to stresses by initiating signal transduction pathways. Various abiotic stresses result in both general and specific effects on plant growth and development. For example, drought limits plant growth due to difficulties in maintaining turgor pressure, photosynthetic decline, osmotic stress-imposed constraints on plant processes and interference with nutrient availability as the soil dries [4].

Protein kinases and phosphatases are major elements of stress signals which are transmitted to different cellular regions via specialized signaling pathways. Some of the protein kinases involved in stress signal transduction in plants, such as mitogen-activated protein kinases (*MAPKs*) [5,6,7,8], glycogen synthase kinase 3 (*GSK3*) [9,10], and S6 kinase (*S6K*) [11], are similar among all eukaryotic organisms, whereas others, including calcium-dependent protein kinases (*CDPKs*) [12,13,14] and *SNF1*-related kinases (*SnRKs*), are plant-specific.

SnRK2 is an important stress-related protein kinase in plants that has been implicated in stress and abscisic acid-mediated signaling pathways [15]. Previously, it was reported that *SRK2C/SnRK2.8*, a subclass II member, was strongly activated by drought stress and that plants overexpressing *SRK2C* exhibited improved drought tolerance as a result of the up-regulation of many stress-inducible genes [16]. *SnRK2s* have a molecular weight of approximately 40 kDa and are monomeric serine/threonine protein kinases [17,18]. Based on phylogenetic analysis, three groups of *SnRK2* family members have been identified. Group 1 consists of kinases not activated by ABA, group 2 consists of kinases not activated or activated very weakly by ABA, and group 3 consists of kinases strongly activated by ABA. The amino acid sequences of all *SnRK2s* can be separated into two regions, the highly conserved N-terminal kinase domain and the regulatory C-terminal domain, which contains stretches of acidic amino acids. Furthermore, the C-terminal domain consists of two subdomains, Domain I and Domain II. Domain I is characteristic of all *SnRK2* family members and is required for activation by osmotic stress. Based on our investigation, we found that the expression of *GhSnRK2* was induced by PEG. PEG was quite commonly used in physiological experiments to induce controlled drought stress. In the present study, we generated a gene construct containing *GhSnRK2* driven by the constitutive cauliflower mosaic virus (CaMV) 35S promoter and transformed this construct into *Arabidopsis* in order to investigate the functional analysis of *GhSnRK2* gene. We monitored the activities of this gene with respect to drought and low temperature tolerance in transgenic plants. We further elucidated the function of this gene in transgenic cotton using virus-induced gene silencing (VIGS) techniques. We showed that *GhSnRK2*, the cotton *SnRK2* gene, is involved in multi-stress responses. Our results promote the analysis of gene function in *G. hirsutum* to facilitate the exploitation of desirable genes from this species. The elucidation of *GhSnRK2* gene function contributes to our understanding of the mechanism by which this plant adapts to abiotic stress and provides a valuable gene resource for plant breeders.

Materials and Methods

Cloning of theGhSnRK2 gene

Total RNA was extracted from CCRI24 cotton tissues using Trizol (Sigma-Aldrich) according to the manufacturer's instructions. Reverse transcription (RT) was performed using total RNA extracted from seedlings oligo(dT)16 primer, and SuperScript II reverse transcriptase (Promega). The RT product was used in PCRs to amplify the predicted *GhSnRK2* open reading frame using primer star polymerase (enzymes). The cDNA regions of *GhSnRK2* were cloned into the T-simple vector. All of the clones were confirmed via sequencing. Primers specific to the sequence of *GhSnRK2* were designed, synthesized, and used to clone the *SnRK2* gene. For plant transformation, *GhSnRK2* cDNAs were introduced into the modified pCAMBIA2301 plant transformation vector under the control of the CaMV 35S promoter.

Localization of the *GhSnRK2*-GFP fusion protein

The method of [19] was adopted to perform the subcellular localization assay. The *GhSnRK2* coding sequence was cloned and ligated into the XbaI and SpeI sites of the PCAMBIA2301-GFP vector to generate PCAMBIA2301-*GhSnRK2*-GFP, which expressed the *GhSnRK2*-GFP fusion protein under the control of the CaMV35S promoter. The construct was used for transient transformation of onion. Onion epidermal peels were bombarded with DNA-coated gold particles, and GFP expression was visualized 24 h later. Transformed onion cells were observed under a confocal microscope.

Effect of polyethylene glycol (PEG) treatment on *GhSnRK2* gene expression

The effect of PEG on the expression level of the *GhSnRK2* gene was evaluated via qRT- PCR. The root of a three-week-old upland cotton plant was submerged in 10% PEG solution, and samples were collected from the root at one hour intervals for 6 hours. RNA was extracted from the samples, and RT was performed as described above to generate cDNA for qRT- PCR analysis.

Arabidopsis transformation and screening of transgenic plants

pCAMBIA2301 carrying *GhSnRK2* was introduced into *Agrobacterium tumefaciens* strain GV3101. The transgenic *Arabidopsis* plants were generated using the flower dipping method [20], and transgenic plants were selected based on their growth in 0.8% agar containing half-strength MS salts and kanamycin. The transformants were transferred to soil and allowed to set seed. The T3 generation of the transgenic plants was used for all experiments.

Plant materials and growth conditions

Arabidopsis thaliana seeds and ecotype Col-0 wild-type, mutant, and transgenic seeds were surface-sterilized using bleach (5% Sodium hypochlorite) and 0.1% Triton X-100. After cold exposure at 4°C for 2 d, the seeds were germinated and cultured in plates containing 0.5 × MS medium, 0.8% agar, and 1% sucrose under continuous light at 22°C. For plants grown in soil, 7-day-old seedlings were transferred from the MS plates to soil and cultured under 80–100 μmol $m^{-2}s^{-1}$ photoperiodic cycles of 16 h light and 8 h dark at 22°C in a growth chamber under fluorescent light using one cool-light and one warm-light tube, each of which was suspended several inches above the plants. Mature seeds obtained from transformed plants (T_0 generation) were cultured in MS medium containing the antibiotic kanamycin in order to screen for transformed plants. The seeds were allowed to grow in the medium for two weeks before transfer to soil for growth continuation. Cotton seeds were planted in moist soil and grown in a growth chamber with a 14 h photoperiod at a 20/30°C night/day temperature cycle, with a light intensity of 400 $\mu molm^{-2}s^{-2}$ and at 60% relative humidity.

Drought tolerance assay

Drought stress tolerance was measured by transferring 2-week-old plants cultured in Petri dishes to pots (10-cm diameter) filled with a 1:1 (v/v) vermiculite:perlite mixture. The seedlings were cultured for 2 weeks with constant watering before the water was withheld. After 9 days without water, all of the pots were re-watered simultaneously, and plant re-growth was scored 4 days later. Thirty-five plants for each individual transgenic line were used in each repeated experiment, and one representative image is shown. Plants were scored as survivors if there were healthy green young leaves after re-watering treatment. The survival rate was calculated as the ratio of number of surviving plants to the total number of treated plants in the flower pot. Representative transgenic and WT plants of were photographed before and after drought treatment. Each stress assay was repeated at least three times. The water use efficiency (WUE) of the plants was calculated as the amount of water needed to maintain the weight of each experimental pot containing the plants. To determine the soil water content (SWC), the pots were immediately weighed after saturation with water prior to the initiation of drought stress and then periodically weighed during the drought stress period. The SWC was calculated as (final fresh weight-dry weight)/(initial weight–dry weight)x100. The plants were exposed to water stress by reducing the water content to 25 to 30% of the SWC. For plant tugor pressure assay, drought stressed plants were removed from the growth chamber and transferred to a dark laboratory cupboard for 4 h before turgor measurements using the pressure chamber [21]. To measure the drought tolerance of *GhSnRK2* gene silenced and non-silenced plants, water was withheld for 5 days, approximately two weeks post-inoculation, and the survival rate was recorded as the percentage of plants that survived after re-watering for 3 days. Each of the ten treated groups consisted of five plants.

Freezing and cold stress treatment

Transgenic plants overexpressing the *GhSnRK2* gene and wild-type *Arabidopsis* plants were cultured in pots filled with a 1:1 mixture of perlite and vermiculite under a long-day photoperiod at 22°C. Freezing stress was performed by transferring 4-week-old plants to a chamber in which the temperature was adjusted to −4°C and −8°C for 10 h, respectively, and then returned to the typical standard growth conditions. For cold stress, 4-week-old plants were transferred to 4°C for three weeks. The plants were analyzed after recovery for 7 days under normal growth conditions. The survival rate (%) under freezing and cold conditions was calculated as the percentage of clearly green plants after returning to normal conditions, and plants exhibiting clear sign of wilting were denoted as dead. Thirty-five plants for each individual line were used in each repeated experiment, and one representative image is shown.

Transpirational water loss assay

The leaves of mutant and wild-type seedlings at the rosette stage were detached and placed in a weighing boat, and the changes in fresh weight over time were monitored using an electronic balance. The rate of water loss was calculated as the loss in fresh weight of the samples. Six plants of each transgenic and WT line were analyzed in this assay, and this experiment was replicated three times [22].

Relative water content

The relative water content of the leaves was measured according to the method of [23]. Fully expanded leaves were cut from the plants, and the fresh weight (FW) was recorded immediately. Then, the fresh portions were immersed in distilled water for 4 h and the turgid weight (TW) was recorded. Finally, the dry weight (DW) was recorded after drying for 48 h at 80°C in an oven. The RWC was calculated according to the following formula: $RWC (\%) = (FW-DW)/(TW-DW) \times 100$.

Measurement of stomatal closure in response to ABA treatment

Stomatal closure assays were conducted as described by [24]. Leaves from *GhSnRK2* transgenic and WT plants were immersed in a solution containing 50 μM $CaCl_2$, 10 mM KCl, and 10 mM 2-(N-morpholino) ethanesulfonic acid (MES)-Tris, pH 6.15, and were exposed to light for 2 h. Subsequently, various concentrations of ABA were added to the solution. The stomatal apertures were measured after 2 h of ABA treatment.

Determination of proline content

The free proline content was measured using the method described by [25]. Leaf segments were homogenized in 3% sulfosalicylic acid, and the homogenates were centrifuged at 3000 g for 20 min. Mixtures containing 2 ml of sample supernatant, 2 ml of acetic acid, and 2 ml of 2.5% acid ninhydrin solution were boiled for 30 min, and the absorbance at 520 nm (A520) was measured.

Chlorophyll content assay

The determination of chlorophyll content was performed according to the method of [26]. Extracts were obtained from 0.1 g (fresh weight) leaf samples from 4-week-old plants and were homogenized in 1 ml of 80% acetone to quantify the chlorophyll content via spectrophotometric analysis.

Seedling growth in response to ABA and NaCl treatment

The sensitivity of seedling germination to ABA and NaCl was assessed on MS agar plates containing various concentrations of ABA and NaCl solution [27]. Seedlings from WT and transgenic plants were placed on MS agar plates supplemented with distilled water or different concentrations of ABA or NaCl and were placed in the growth chamber at 22°C.

Germination assay

To evaluate seed germination, the method of [27] was employed. Seeds from wild-type and transgenic plants were surface-sterilized using sodium hypochlorite and were placed on Murashige and Skoog (MS) solid medium containing various concentrations of ABA (0 μm, 0.3 μm, or 0.5 μm). For the evaluation of seed germination under salt stress, the MS medium was supplemented with various concentrations of NaCl (0 mM, 50 mM, or 100 mM). The percentage of germinating seeds was recorded after 7 days.

Biomass accumulation

Plants were harvested for biomass measurements after 4 weeks of germination. The fresh weight of each individual plant shoot and root was measured immediately after harvesting. For fresh weight biomass, the fresh weight of the root and the shoot was measured immediately after harvesting, and the dry weight was recorded in (g) after drying in an oven to a constant weight at 70° for 48 h.

qRT-PCR

Total RNA was extracted from *Arabidopsis* seedlings using Trizol reagent (Takara) and was treated with RNase-free DNase (Promega). cDNA was synthesized using a Promega kit according to the manufacturer's protocol. qPCR was performed using a SYBR green PCR master mix kit. After PCR, the data was quantified using the comparative CT method ($2^{-}\Delta\Delta CT$ method) based on the CT values [28]. *AtACT2* (gene accession At3G18780) and cotton *histone3* (gene accession AF024716) were used as internal controls, and the relative expression level of each target gene was quantified. Each RT-PCR measurement was performed in triplicate. All RT-PCR experiments were reproduced at least three times using independent cDNA samples.

Silencing construct development and cotton transformation

The method described by [29] was employed to construct the VIGS vector. Specific fragments (approximately 300 bp) were amplified via PCR and cloned into the T-simple vector according to the manufacturer's specifications (Promega). The resulting clones were sequenced. The plasmids were digested using the restriction enzymes Xba1 and BamH1 and were subsequently ligated into the VIGS vector (*PYL156:(pTRV-RNA2)-GhSnRK2*). The resulting constructs were named according to the putative function of the stress-related gene. For transformation, *Agrobacterium* (GV3101) carrying *GhSnRK2* derivatives were cultured at 28°C in LB medium containing appropriate antibiotics. Cells were harvested from cultures grown overnight, re-suspended in inoculation buffer (200 μM acetosyringone, 10 mM MES, pH 5.5), and incubated for 2 h at room temperature in a shaker. *Agrobacterium* strains containing the *GhSnRK2* derivative *PYL156:(pTRV-RNA2)-GhSnRk2*, *PYL192:(pTRV-RNA1)*, *PYL156:(pTRV-RNA2)-GrCLA1*(positive control), or*PYL156:(pTRV-RNA2)* vector (OD 1.5) were mixed at a 1 : 1 ratio in 5 mM MES buffer (pH 5.5) and inoculated into *Gossypium hirsutum* leaves using a needleless syringe. The inoculated plants were maintained in a green house at 23±2°C for effective viral infection and spreading.

Experimental design and statistical analysis

The data obtained were subjected to statistical analysis and were expressed as the means and standard error (SE) of at least three replicates. All experimental data are presented as the means of at least three independent replicates, and the analysis for significance was performed using Student's T-test.

The sequences of primers used in this study are listed in Table S2.

Results

Phylogenetic analysis, subcellular localization, and expression pattern of *GhSnRK2* in cotton plant

The phylogenetic tree constructed using the full length amino acid sequence of selected *SnRK2* genes to analyze the evolutionary relationship between *GhSnRK2* and other *SnRK2* family genes is shown in the neighbor-joining tree developed based on an alignment of the complete protein sequences. The bootstrap values are shown on the branches. *GhSnRK2* clustered with the known stress-related genes *AtSnRK2.1*, AED91326.1; *AtSnRK2.4*, AEE28666.1; *AtSnRK*2.5, AED97781.1; *AtSnRK2.10*, AEE33751.1; and *Oryza sativa* (*RK1*), ABB89146.1. *SnRK2.10* and *SnRK2.4* are closely related to *GhSnRK2*, suggesting that *GhSnRK2* belongs to the *SnRK2* family (Figure 1A). Alignment of the *GhSnRK2* amino acid sequence with that of other *SnRK2s* revealed that the *GhSnRK2* protein is highly similar to other *SnRK2s*; the relatively conserved motif is underlined. The deduced amino acid sequence displays relatively high homology with the monocot *SnRK2* family members *Oryza sativa* (*RK1*), ABB89146 and with the dicot species *AtSnRK2.10*, AEE33751.1 (Figure S1A). The *GhSnRK2*-GFP fusion protein driven by the CaMV 35S promoter was transiently expressed in onion epidermal cells, and the green fluorescent *GhSnRK2* protein signals were localized to the nuclei and the cytoplasm, whereas GFP alone was detected throughout the cell (Figure1B). qRT-PCR was performed to quantify the expression level of the *GhSnRK2* gene. Treatment of 3-week-old CCRI24 upland cotton cultivar plants with 10% PEG for different periods induced the expression of the *GhSnRK2* gene for 3 h, after which the expression level declined (n = 3) (Figure 1C). To investigate the distribution of *GhSnRK2* in different tissues, samples from root (RT), stem (ST), cauline leaves (CL), rosette leaves (RL) and flowers (FL) were analyzed. *GhSnRK2* gene expression was detected in all of the examined tissues in distinct expression patterns and was highest in the root and lowest in the flower, suggesting that the *GhSnRK2* gene may most actively function in the root (Figure 1D). Cotton *histone3* (gene accession AF024716) was used as internal control to normalize expression data. The values are presented as the means of three replicates, and the error bars denote the SE. Different letters denote the means ± standard deviation displaying significant difference at P≤0.05.

Plant transformation vector and the expression pattern of the *GhSnRK2* gene in the transgenic lines

The schematic representation of the 35S-*GhSnRK2* construct used for *Arabidopsis* transformation (Figure 2A). The full-length *GhSnRK2* cDNA was introduced into the *Arabidopsis* genome via the floral dip method using *Agrobacterium* GV3101; the XbaI–BamHI region of *GhSnRK2* was inserted between the CaMV 35S promoter and the nopaline synthase gene terminator in the pCAMBIA2301 vector to generate the recombinant plasmid *35S-GhSnRK2-NOST*. The neomycin phosphotransferase II (*NPTII*) gene, driven by the nopaline synthase gene promoter (NOSP), was carried by the *pCAMBIA2301* vector for transgenic cell selection in kanamycin-containing LB medium. *GhSnRK2* in transgenic *Arabidopsis* plants was confirmed by PCR analysis. Genomic DNA from the first generation of the plants (T1) was extracted and used as a template for gene-specific primers. Of the 43 plants evaluated, 35 were positive for *GhSnRK2* amplification. No product was formed by the amplification reaction using untransformed plant DNA as the template (Figure S2B). At T1 generation, the ratio of dead to surviving plants was approximately 1:3 in kanamycin-containing LB medium. However, at T3 generation, all of the plants became homozygous, as their survival rate was 100% in kanamycin-containing LB medium (Figure S2A, Table S1). The values are expressed as the mean germination rate (%) of approximately 200 seeds. qRT-PCR analysis was performed to evaluate the differences in transgene expression between all of the independently generated transgenic lines. Various transgene expression patterns were detected in the transgenic lines overexpressing the *GhSnRK2* gene, but no expression was detected in the WT line, indicating that the *GhSnRK2* genes were successfully introduced into the transformed plants (Figure 2B). The values are presented as the means of three experimental replicates; the error bars indicate the standard deviations. *AtACT2* (gene accession At3G18780) was used as internal control to normalize expression data. Three transgenic lines (L1, L2, and L4) were selected for further investigation based on their seed availability and relative expression pattern.

Overexpression of *GhSnRK2* in transgenic plants improves their tolerance to drought and cold stresses

To investigate the roles of *GhSnRK2* in stress response pathways, the responses of the three selected transgenic *GhSnRK2*-overexpressing *Arabidopsis* lines to drought and low temperature stresses were analyzed. Constitutive overexpression of

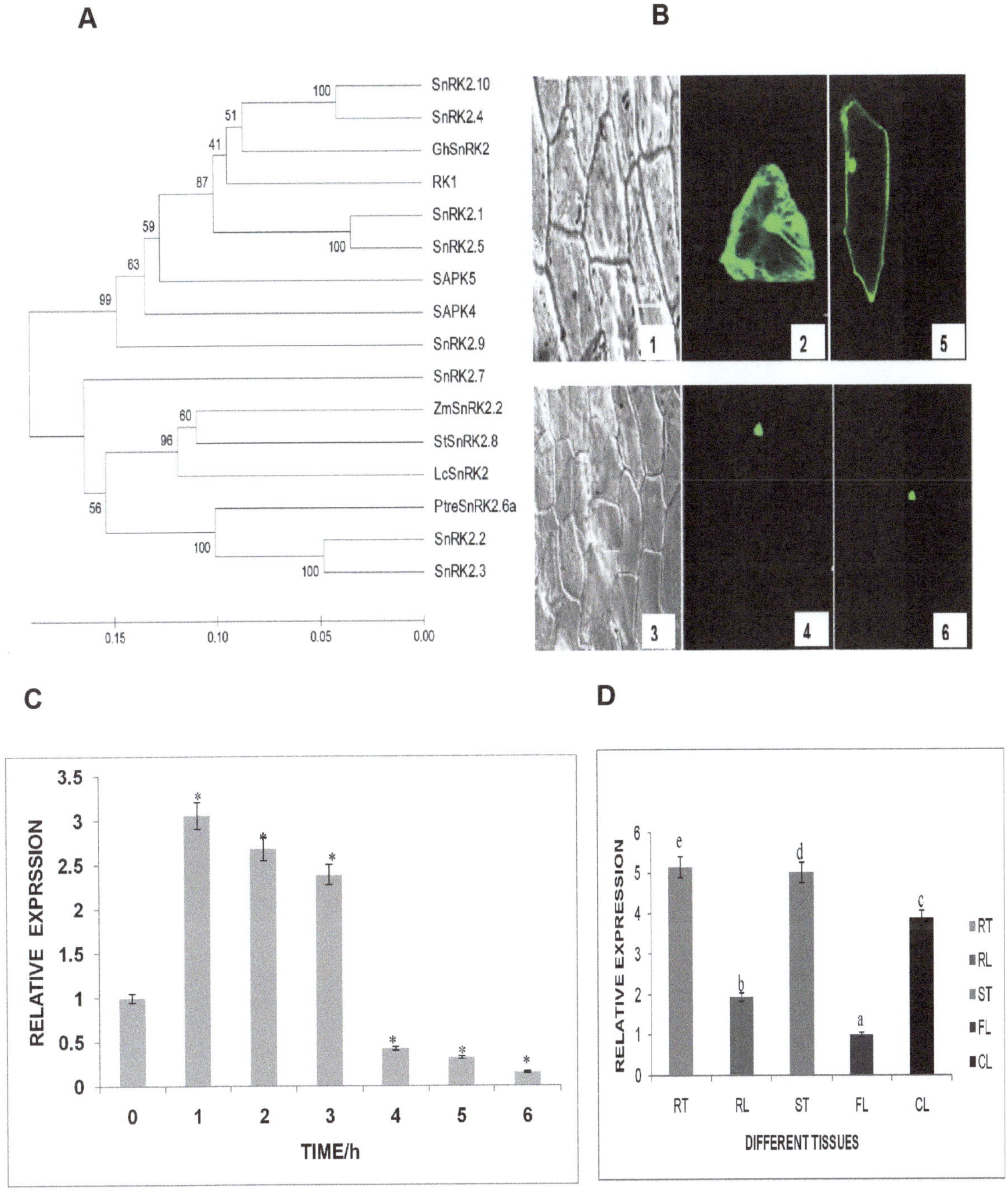

Figure 1. Phylogenetic analysis, subcellular localization, and expression pattern of *GhSnRK2* in cotton plant. (A) A phylogenetic tree of *GhSnRK2* and other *SnRK2* proteins from different plants was constructed using the neighbor-joining method with MEGA 5. The sequences used for analysis are listed by accession number: *Litchi chinensis* (*LcSnRK2*), AFX72761.1; *Arabidopsis thaliana* (*AtSNRK2.2*), CP002686.1; *Arabidopsis thaliana* (*AtSnRK2.3*), AED98274.1; *AtSnRK2.1*, AED91326.1; *AtSnRK2.9*, AEC07398.1; *AtSnRK2.10*, AEE33751.1; *AtSnRK2.4*, AEE28666.1; *AtSnRK2.5*, AED97781.1; *AtSnRK2.7*, AEE87152.1; *Populus tremula* (*PtreSnRK2.6a*), AGW51610.1; *Zea mays* (*ZmSnRK2.2*), NM_001137717.1; *Solanum tuberosum* (*StSnRK2.8*), AFR68945.1; *Oryza sativa* (*RK1*), ABB89146.1; *Oryza sativa* (*SAPK4*), BAD18000.1; *Sorghum bicolor* (*SAPK4*), AGM39623.1; and *Zea mays* (*SAPK5*), ACG42286.1. The bootstrap values are shown on the tree branches. (B) Subcellular localization of the *GhSnRK2-GFP* protein. (2 and 5) GFP alone; (4 and 6) *GhSnRK2-GFP* in onion epidermal cells; (1 and 3) corresponding bright-field images. (C) The expression pattern of the *GhSnRK2* gene in cotton plants subjected to 10% PEG stress. The gene expression data were normalized to that of the cotton histone 3 gene. The values are presented as the means of three experimental replicates. The vertical axis represents the relative expression level. The values from1 to 6 indicate the time (h) of PEG treatment. Asterisk denotes a significant difference ($P<0.05$) compared with the control (0 h). (D) Relative expression levels of the *GhSnRK2* gene in various cotton plant tissues. Samples from root (RT) stem (ST), cauline leaves (CL), rosettes leaves (RL) and flowers (FL) were analyzed. The vertical axis represents the relative expression level. The letters denote significant differences ($P<0.05$) based on Duncan's multiple range tests. The cotton histone 3 gene was used as an internal control for normalization of the gene expression data.

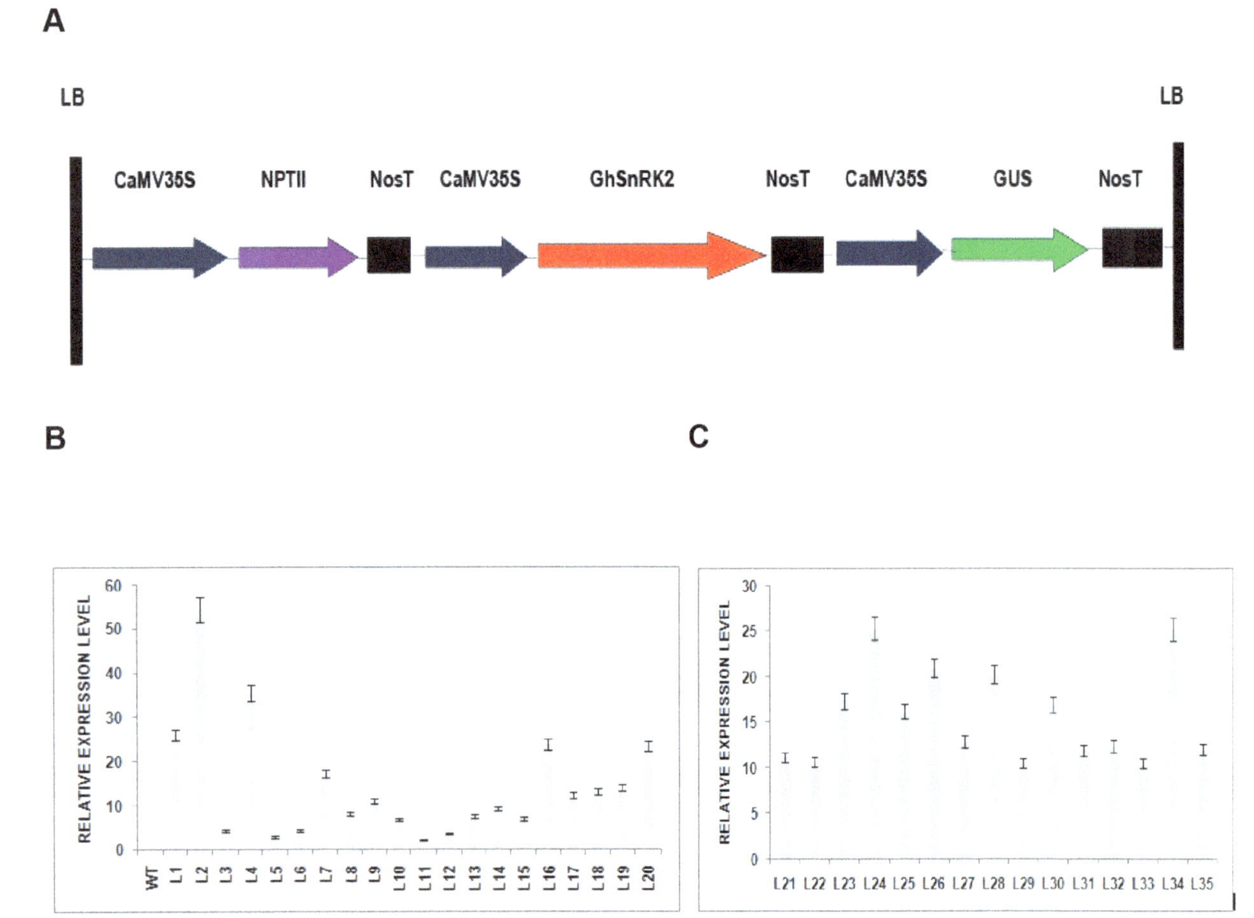

Figure 2. Plant transformation vector and the expression pattern of the *GhSnRK2* gene in the transgenic lines. (A) Schematic representation of the T-DNA region of the binary vector *pCAMBIA2301-GhSnRK2*. (B) Expression pattern of the *GhSnRK2* gene in the transgenic plants. Various upregulated expression patterns of the *GhSnRK2* gene in transgenic lines were detected, as indicated by the vertical axis. The values are presented as the means of three experimental replicates; the error bars indicate the standard deviations. The *AtACT2* gene was used as an internal control for normalization of gene expression.

the 35S-*GhSnRK2* gene resulted in an increase in the drought tolerance of the transgenic plants. After withholding water for 9 days, some of the WT plants grew slowly, wilted severely and died, whereas few of the *GhSnRK2* transgenic plants wilted. The recovery rate of the transgenic plants was higher than that of the WT plants after re-watering. Indeed, the survival rate was significantly different between the *GhSnRK2* transgenic and wild-type plants: L2 (100%), L4 (94.4%), L1 (88.9%), and WT (4.4%); Student's T-test ($n = 3$) ($p < 0.05$) (Figure 3A, B). These results suggested that the *GhSnRK2* gene is involved in the response to drought tolerance.

Under low temperature stress conditions, the survival rate of the transgenic lines was significantly higher than that of the WT line. Most of the transgenic plants were intact and recovered from this stress, whereas most WT plants were found to have died and could not recover after transferring them to the normal growth conditions. The survival rates at 4°C were: WT (16.67%), L2 (80.55%), L1 (77.78%), and L4 (78.33%). The survival rates at −4°C were: WT (5.56%), L1 (54.44%), L2 (58.33%), and L4 (56.11%). At −8°C, only 3.3% of the WT plants survived, whereas the survival rates of transgenic lines ranged between 30% and 33.33% after 7 days under normal growth conditions (Figure 3C, D). However, the survival rate under different low temperatures clearly indicates the difference between the transgenic and WT plants, suggesting that the *GhSnRK2* overexpression alleviates low temperature stress in transgenic Arabidopsis. Our findings revealed that the water use efficiency was higher in *GhSnRK2*-overexpressing plants compared to the corresponding WT plants, suggesting that *GhSnRK2*-overexpressing plants exhibit greater photosynthetic potential during drought stress treatment (Figure 3E). Normal turgor pressure which is regulated by the amount of water in the plant's cells is required for healthy growth of a plant and is a powerful determinant of the plant's drought tolerance. We determined the turgor of the *GhSnRK2* transgenic plant to understand the role of turgor pressure in cellular signaling during water deficit condition. Turgor values of *GhSnRK2* transgenic plant was notably higher compared to the corresponding WT plant and this may be attributable to osmotic adjustment, like stomatal closure, and the maintenance of water content which prevent plants from desiccation and turgor loss. Student's T-test ($n = 3$) ($p < 0.05$) (Figure 3F).

Biochemical and physiological assays of plants overexpressing the *GhSnRK2* gene

The rate of water loss and the RWC were investigated to further understand the tolerance of *GhSnRK2*-overexpressing plants to

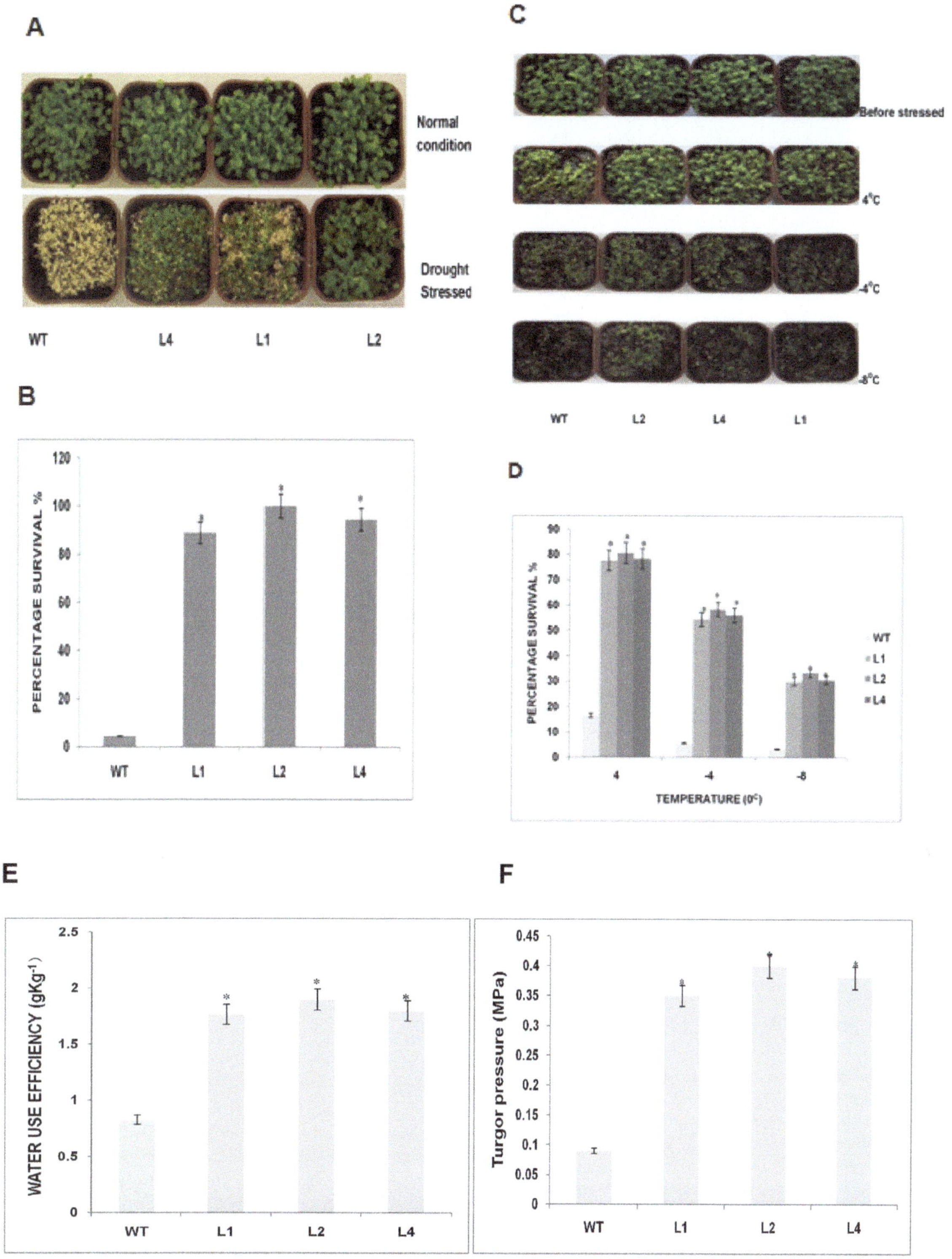

Figure 3. Survival rates of *GhSnRK2* transgenic plants under drought and low temperature stresses. (**A**) WT and *GhSnRK2* transgenic plants before and after drought stress. (**B**) Seedlings were cultured for 2 weeks with constant watering before the withholding of water. After 9 days without water, all of the plants were irrigated, and plant re-growth was scored 4 days later. The plants were scored as survivors if there were healthy green young leaves after re-watering. The survival rate was calculated as the ratio of the number of surviving plants to the total number of treated plants in the flower pot. Asterisk denotes a significant difference ($P<0.05$). (**C**) WT and *GhSnRK2* transgenic plants before and after low temperature stress. (**D**) Survival rates under low temperature stress conditions. The survival rate after transferring 4-week-old *GhSnRK2* transgenic and WT plants to a low temperature chamber at 4°C, −4°C or −8°C for 10 h, followed by returning the plants to normal growth conditions. Clearly green plants after returning to the normal growth condition were scored as survivors, and plants exhibiting clear signs of wilting were denoted as dead. The mean survival rates of the WT line were compared with those of the transgenic lines using Student's T-test. (E) The water use efficiency of the WT and *GhSnRK2* transgenic plants. Asterisk denotes a significant difference ($P<0.05$).

water stress via the maintenance of a higher RWC and a reduced rate of water loss. Our results revealed that due to their smaller stomata aperture, the transgenic *Arabidopsis* plants overexpressing the *GhSnRK2* gene lost water more slowly than the WT plant in the same period under normal conditions, which may underlie the capacity of the transgenic plants to maintain a higher leaf RWC and to tolerate water stress conditions; "tstat">t Critical two-tail (Figure 4A). The *GhSnRK2*-overexpressing plants exhibited a higher RWC than the WT plants, which is most likely due to the reduction in transpiration-mediated water loss. The RWCs in each line were L2(74.56%), L4(72.09%), L1 (70.04%), and WT (53.98%); Student's T-test (n = 3) ($p<0.05$) (Figure 4B). It is known that during the plant response to abiotic stress, stomata typically close to reduce the rate of water loss due to transpiration. Moreover, abscisic acid (ABA) regulates the stomatal aperture under water-deficient conditions. We found that the stomata apertures of the *GhSnRK2* transgenic plants in response to exogenous ABA treatment were smaller than those of the WT plants, indicating the protective effect of the *GhSnRK2* protein in response to ABA. At 0 µM ABA, the stomata aperture of the WT plants was observed slightly wider than that of the *GhSnRK2* transgenic plants. However, when the concentration of ABA was increased to 1 µM, the aperture size of the transgenic plants was significantly reduced. A similar pattern was detected at 5 µM ABA (n = 5) (Figure 4C, D). Free proline is an osmoprotective molecule that accumulates under stress conditions. Our results revealed that the proline contents of the *GhSnRK2* transgenic lines were significantly higher than those of the WT line, indicating that the transgenic plants accumulate a higher amount of proline than the WT plants. Moreover, the upregulation of Arabidopsis pyrroline-5-carboxylate synthetase 1, a proline biosynthesis gene, in *GhSnRK2*-overexpressing plants suggested that proline production may be improved by *GhSnRK2* via the regulation of proline biosynthesis genes; Student's T-test (n = 3) ($p<0.05$) (Figure 4E). Greater leaf chlorophyll content at all stages of plant development has been associated with improved transpiration efficiency under drought stress, and this trait may indicate the presence of drought avoidance mechanisms. Based on our results, the chlorophyll content of the *GhSnRK2* transgenic lines was higher than that of the WT line, suggesting that *GhSnRK2*-overexpressing plants exhibit greater photosynthetic potential ($p<0.05$) (Figure 4F).

Overexpression of *GhSnRK2* in transgenic plants enhances seedling growth in response to NaCl and exogenous ABA treatment

To investigate the effect of *GhSnRK2* overexpression in response to NaCl and exogenous ABA treatment, a root growth experiment was conducted to elucidate the physiological differences between the transgenic and WT plants. The transgenic plants overexpressing *GhSnRK2* exhibited enhanced seedling growth under NaCl and ABA stress treatments, suggesting that *GhSnRK2* may be involved in the oxidative stress response pathway. Seedlings of WT and *GhSnRK2* transgenic lines grew normally in 0 mM NaCl. In 100 mM and 150 mMNaCl, the transgenic seedlings formed longer roots and displayed significantly larger growth than the WT seedlings. When the NaCl concentration was increased to 200 mM, the growth of the WT seedlings was completely inhibited, and the color of these seedlings was found to have turned brownish, whereas the transgenic seedlings remained green and continued to grow, although at a slower rate (n = 3) (Figure 5A, B).

The seedlings from the WT and *GhSnRK2*-overexpressing lines cultured under normal conditions displayed no significant difference. However, when various concentrations of ABA (0.5 µM, 1 µM, or 2 µM) were introduced to the MS medium and the plants were allowed to grow vertically, the root growth of the *GhSnRK2*-overexpressing lines, although slightly inhibited, displayed greater elongation than the WT line, the root growth of which was severely inhibited (n = 3). These results indicated that *GhSnRK2*-overexpressing plants were more tolerant to ABA than the WT plants (Figure 5C, D).

GhSnRK2 overexpression in transgenic plants enhances seed germination in response to NaCl and exogenous ABA treatment

We investigated the germination of *GhSnRK2*-overexpressing plants under exogenous ABA and salt stresses to determine whether *GhSnRK2* is involved in stress response pathways. The germination rates of WT and *GhSnRK2*-overexpressing seeds were similar at 0 mM NaCl and ABA. However, when different concentrations of ABA or NaCl were introduced, the germination of the WT and *GhSnRK2*-overexpressing seeds was inhibited. When 50 mM NaCl was introduced to the "MS" medium, the germination rate was 47% for the WT line and 85%, 89%, and 86% for the L1, L2, and L4 lines, respectively. At100 mM NaCl, the germination rates were: WT (20.5%), L1(52%), L2 (56%), and L4 (54%)(Figure 6A, B). These results suggested that the *GhSnRK2*-overexpressing plants were more tolerant to salt stress than the corresponding WT plants. ABA plays a prominent role in the regulation of germinating and post-germinating growth arrest and mediates the adaptation of the plant to stress. To investigate this hypothesis, we measured the response of *GhSnRK2*-overexpressing plants to different concentrations of ABA during the germination stage. At 0.3 µm ABA the seed germination rates were WT (45%), L1 (69%), L2 (73%), and L4 (70%), respectively. When the concentration of ABA was increased to 0.5 µm, the germination rate of the WT line was clearly reduced (27%), whereas the germination rate of the transgenic lines ranged from 43% to 48% (Figure 6A, C). In summary, our findings revealed that the *GhSnRK2*-overexpressing mutants were more tolerant to NaCl and exogenous ABA stresses than the corresponding WT plants. Student's t-test was used to compare the mean germination rates. Asterisks denote a significant difference compared to the control ($P\leq 0.05$).

We further investigated the biomass accumulation of *GhSnRK2* transgenic and WT plants. The transgenic plants exhibited highly significant increases in both the fresh and dry weight biomasses compared with the corresponding WT plants (Figure 6D, E). These increases were all statistically significant; suggesting that overexpression of the *GhSnRk2* gene increases biomass accumulation in plants. The increased biomass accumulation may be attributable to the increased chlorophyll content in *GhSnRK2* overexpressing plant, as chlorophyll in plant cells carries out the bulk of energy fixation in the process of photosynthesis and is probably the most-often used estimator of plant biomass.

Expression analysis of stress-responsive marker genes

To further elucidate the biological function and molecular mechanisms of the *GhSnRK2* gene, we determined the transcript levels of several stress-associated genes in *GhSnRK2*-overexpressing lines. Seedlings from 10-day-old *GhSnRK2* transgenic and WT *Arabidopsis* lines cultured in MS medium were in the presence or absence of NaCl (250 mM) for 2 h. qRT-PCR was performed as described in the Materials and Methods section. Our findings revealed that *GhSnRK2* regulates the expression levels of ABA and stress-responsive marker genes, suggesting that *GhSnRK2* may positively affect ABA signaling and plant stress

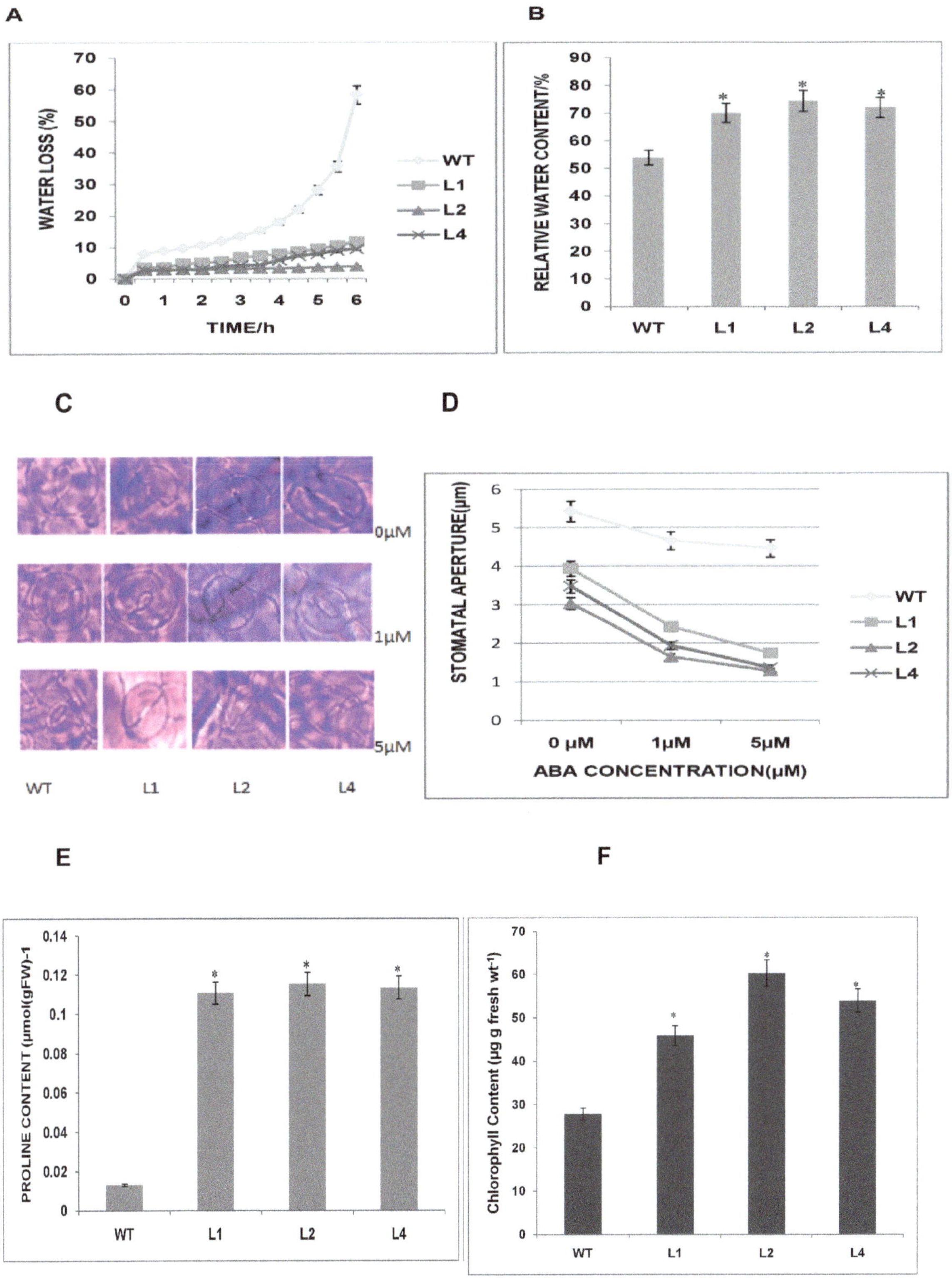

Figure 4. Biochemical and physiological assays of plants overexpressing the *GhSnRK2* gene. (A) Rate of water loss from *GhSnRK2* transgenic plants. Six plants of each transgenic and WT line were analyzed. Three biological replicates produced similar results. Asterisk denotes a significant difference ($P<0.05$). (B) The RWC of *GhSnRK2* transgenic plants. (C) Stomata aperture size of WT and *GhSnRK2* transgenic plants treated with different concentrations of ABA (D) The stomata apertures were measured after 2 h of treatment with different concentrations of ABA, and the mean values of the WT and transgenic lines at each ABA concentration were compared using Student's T-test ($P<0.05$). Asterisk denotes a significant difference ($P<0.05$). (E) Proline accumulation in WT and *GhSnRK2* transgenic plants. The proline content of *GhSnRK2* transgenic lines was consistently higher than that of the WT line. Student's T-test revealed a significant difference ($p<0.05$) between the transgenic and WT lines ($n=3$). (F) The chlorophyll content of the WT and *GhSnRK2* transgenic plants. The mean values were compared using Student's "t-test". Asterisk denotes a significant difference ($P<0.05$).

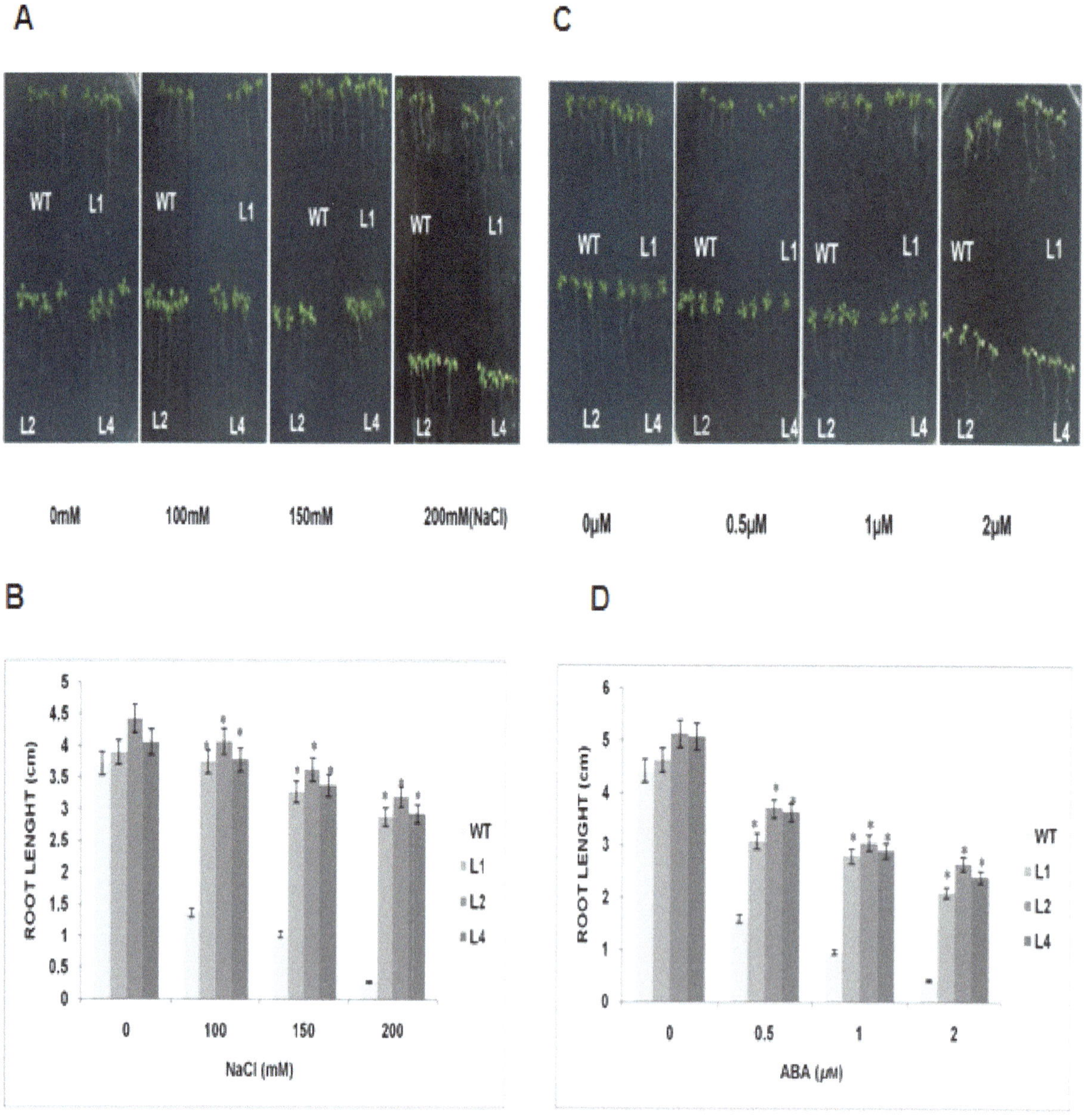

Figure 5. Seedling growth of the WT and *GhSnRK2*-overexpressing lines in response to NaCl and ABA treatment. (A) WT and *GhSnRK2* transgenic plants subjected to different concentrations of NaCl. (B) Approximately one-week-old seedlings were transferred to 1/2MS medium supplemented with different concentrations of NaCl; the root length was measured after 7 days. Each of the three biological replicates consisted of 16 plants. The mean values were compared using Student's T-test ($p<0.05$). (C) The root growth of the WT and *GhSnRK2* transgenic plants treated with different concentrations of ABA. (D) The plants were cultured vertically on MS-containing medium treated with different concentrations of ABA (0.5 μM, 1 μM, or 2 μM); the root length was measured after 7 days. Each of the three biological replicates consisted of 16 plants. Student's T-test was used to compare the mean values of the WT and transgenic lines. Asterisk denotes a significant difference ($P<0.05$).

responses. The expression levels of C-repeat binding factor-1 (*AtCBF1*; accession: ABV27087), delta-1-pyrroline-5-carboxylate-synthase-1 (*AtP5CS1*; accession:NP001189715), protein abscisic acid-insensitive-5 (*AtABI5*; accession:NP565840), protein abscisic acid-insensitive-3 (*AtABI3*; accession:CAA48241), desiccation-responsive genes (*Atrd29B*; accession: BAA02375 and *Atrd29A*; accession:BAA02376) in the *GhSnRK2*-overexpressing lines were significantly higher than those of the WT lines ($p\leq0.05$), revealing that *GhSnRK2* is actively involved in stress signaling pathways (Figure 7). Three experimental replicates produced similar results ($n = 3$); Student's T-test.

VIGS efficiency and the transcript level of *GhSnRK2* in gene silenced plants

We employed the VIGS technique to further dissect the function of the *GhSnRK2* gene in transgenic cotton. The TRV VIGS vectors were modified based on a pTRV1 containing RNA-dependent RNA polymerase (RdRp), movement protein (MP), a 16 kDa cysteine-rich protein (16K), CaMV 35S promoters (2X35S) and a NOS terminator (NOSt) T-DNA vector. pTRV2 contains the coat protein (CP), multiple cloning sites (MCSs), CaMV 35S promoters (2X35S) and a NOS terminator (NOSt) T-DNA vector. Both vectors contain Rz, which is designated as a

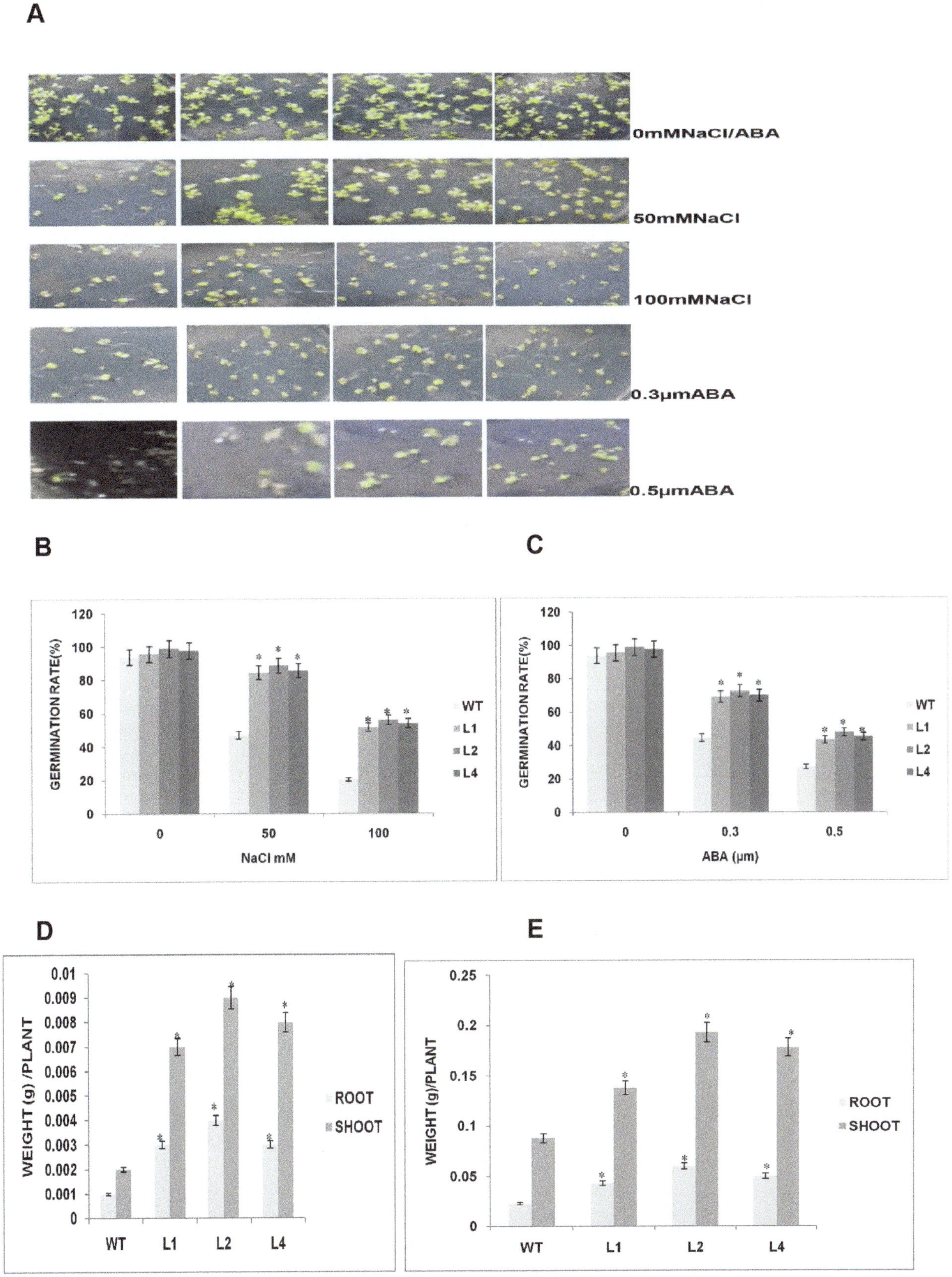

Figure 6. Seed germination of the WT and *GhSnRK2* plants subjected to NaCl and exogenous ABA treatment and biomass accumulation of these plants. (A) Seed germination frequency of the WT and *GhSnRK2* transgenic plants cultured on MS medium supplemented with different concentrations of NaCl (50 mM or 100 mM) or ABA (0 μM, 0.3 μM, or 0.5 μM). (B) The survival rate of the WT and *GhSnRK2* transgenic plants cultured in MS medium containing 50 mM or100 mMNaCl. The mean values were compared using Student's T-test ($p<0.05$). (C) The

germination rate in MS medium supplemented with 0.3 μM or 0.5 μM ABA. The values are presented as the mean germination rates (%) of approximately 200 seeds. Asterisk denotes a significant difference ($P<0.05$). (D) Biomass accumulation of the *GhSnRK2* transgenic and WT plants. For dry weight biomass, the dry weight in the roots and shoots was recorded after drying in an oven to a constant weight at 70°C for 48 h. (E) Fresh weight biomass of *GhSnRK2* transgenic plants and corresponding WT plants. The fresh weight of the roots and shoots was measured immediately after harvesting. Each of the three biological replicates consisted of 12 plants. Student's T-test was performed. Asterisk denotes a significant difference ($P<0.05$).

self-cleaving ribozyme, LB (left border) and RB (right border) of the T-DNA (Figure S1B). After inoculating *Agrobacterium* strains containing *GhSnRK2* derivatives (PYL156: (pTRV-RNA2)-*GhSnRk2,* PYL192:(pTRV-RNA1), PYL156:(pTRV-RNA2)-*GhCLA1* (positive control), or PYL156:(pTRV-RNA2) vector) into *Gossypium hirsutum* leaves, the phenotype of both silenced and non-silenced plants was monitored daily for gene silencing efficiency. Loss of normal green coloration in plant leaves and albino phenotype was detected in PYL156:(pTRV-RNA2)-*GrCLA1* silenced plants, which served as the positive control indicating gene silencing efficiency. The *GrCLA1* gene represents the ideal visual marker for gene silencing efficiency due to its involvement in chloroplast development, as its loss of function results in an albino phenotype in true leaves (Figure 8A). The transcript levels of *GhSnRK2* in gene silenced and non-silenced plants were analyzed via qRT-PCR. The transcript level of *GhSnRK2* in the gene silenced two plant cultivars were downregulated compared to the non-silence plants inoculated with empty vector and WT plants, which exhibited a higher expression level. The downregulation of *GhSnRK2* indicated that the gene was successfully knocked down in the silenced plants (Figure 8B). qRT-PCR analysis of the distribution of the TRV-construct in the silenced plants revealed that its infection results in gene-specific transcript degradation, that pTRV-*GhSnRK2* functions in diverse tissues, and that pTRV-*GhSnRK2* is abundantly expressed in the root. The roots of gene silenced plants exhibited further down regulated expression levels compared to the stems and the leaves. The TRV-construct displays increased infectivity and meristem invasion, both of which are key requirements for efficient VIGS-based functional characterization of genes in root tissues (Figure 8C).

Physiological assay and responses of *GhSnRK2* gene silenced plants to various stresses

To further elucidate the function of the *GhSnRK2* gene in stress tolerance, we investigated the response of *GhSnRK2*gene silenced plants to various stresses. Our findings validated the importance of this technique for stress tolerance studies. Under water-deficient conditions, the silenced plants wilted and drooped, regardless of the cultivar (Figure 9A). The phenotype of the non-silenced plants inoculated with empty vector (TRV) was indistinguishable from that of the WT plants. The two cotton cultivars (CRI99668 and CRI409) inoculated with the target gene displayed similar phenotypes and symptoms. The WT and control vector-treated plant survival rates were 80% and 77.3%, respectively, whereas the survival rate of the two gene silenced cotton cultivars (CRI99668 and CRI409) were 34.66% and 38.67%, respectively (Figure 9B). Student's T-test revealed a significant difference ($p< 0.05$) between the survival rate of the gene silenced and non-silenced plants. The rate of water loss from the detached leaves was determined in both the gene silenced and non-silenced plants (Figure 9C). After monitoring the water loss for 6-h, we found that the *GhSnRK2* gene silenced plants inoculated lost more water than the non-silenced plants inoculated with empty vector and the WT plants. The rate of water loss was slightly higher in the CRI99668 cultivars than in the CRI409 cultivars. Under similar conditions, the gene silenced plants, regardless of the cultivar, exhibited a substantially reduced water content compared with the non-silenced plants inoculated with empty vector and the WT plants (Figure 9D). Oxidative stress, including that mediated by salinity, alters the physiological and morphological responses of plants. To investigate the effect of salt stress on *GhSnRK2* gene silenced plants, the gene silenced and non-silenced plants were treated with 150 mM NaCl for 7 days, and the effect of NaCl was measured following this stress. Our findings revealed that the chlorophyll content of the non-silenced plants was higher than that of the gene silenced plants, indicating that the *GhSnRK2* gene may be involved in the oxidative stress response (Figure 9E). Growth retardation was detected in the *GhSnRK2* silenced plants, and the effect of salt accumulation in the plant cells was detected as elevated blisters, which were visibly detectable on the leaf surface of *GhSnRK2* gene silenced plants (Figure 9F).

Discussion

The ability of plants to withstand water shortage while sustaining proper physiological activities can be associated with drought tolerance. In this study, we generated a gene construct containing *GhSnRK2* driven by the CaMV 35S promoter and functionally characterized it via gene overexpression in Arabidopsis. We further elucidated the function of this gene in transgenic cotton plants using VIGS techniques. We found that the amino acid sequence of *GhSnRK2* is highly similar to that of other *SnRK2* proteins, and a neighbor-joining tree developed based on an alignment of the complete protein sequences revealed that *GhSnRK2* clustered with known stress-related genes from other plants. The conserved motif detected demonstrated that *GhSnRK2* is a functional protein. The conserved domain (CD) region in these proteins plays a crucial role in protein interactions, DNA binding, enzyme activity, and other important cellular processes. Furthermore, we found that *GhSnRK2* was localized to the cytoplasm and the nucleus, suggesting that *GhSnRK2* might perform diverse functions in cotton plants. Our findings revealed that *GhSnRK2* is abundantly expressed in the root and is widely distributed throughout plant tissues, suggesting that it may function in diverse tissues. Transgenic *Arabidopsis* plants overexpressing *GhSnRK2* exhibited enhanced tolerance to various abiotic stresses. Consistent with previous studies, *SnRK2* genes have been implicated in the response to multi-environmental stresses [16,30,31]. The improved drought tolerance of plants overexpressing the *GhSnRK2* gene may be due to the abundance of gene expression in roots. In addition, protein kinases that are specifically found in roots, such as *SnRK2C*, may perform certain important functions in root tissues as a sensor of water and nutrients in soil. The root tip plays a vital role in the response to water and nutrient detection via appropriate signal transduction [32]. Turgor regulation of the *GhSnRK2* transgenic plant under water deficit condition may be attributable to the dynamic process of cell wall adjustment of the transgenic plant due to their enhanced drought-tolerant. Plants can maintain turgor by solute accumulation, i.e. by osmotic adjustment, and possibly by elastic adjustment of their cell walls [33].

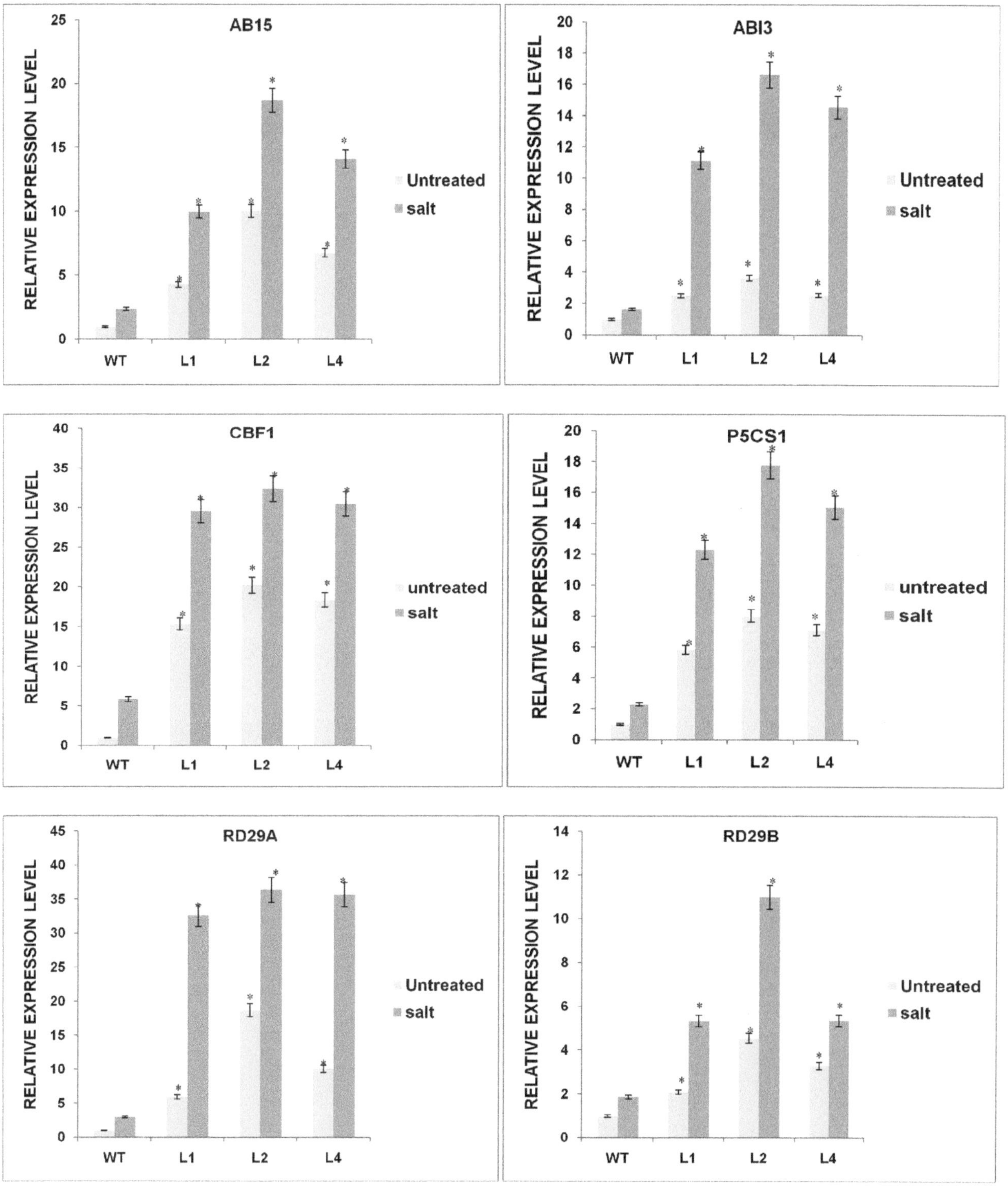

Figure 7. Expression analysis of stress-responsive marker genes. The relative transcript levels of the stress-responsive genes *AtABI5, AtABI3, AtP5CS1, AtRD29A, AtCBF1*, and *AtRD29B* in the *GhSnRK2*-overexpressing and WT lines. qRTPCR was performed for gene expression analysis. The vertical axis displays the expression pattern. Three biological replicates produced similar results.

Based on our results, the transgenic plants overexpressing *GhSnRK2* exhibited enhanced low temperature tolerance compared to WT plants. This result may be attributable to the resistance of the membrane system of the transgenic plant cells to cold stress. These findings corroborated the results of [30], who reported enhanced multi-stress tolerance in *Arabidopsis* plants overexpressing *TaSnRK2.8*. Numerous studies have demonstrated that the membrane systems of the cell are the primary site of low temperature contusion in plants [34,35]. Indeed, it is well confirmed that low temperature-mediated membrane injury

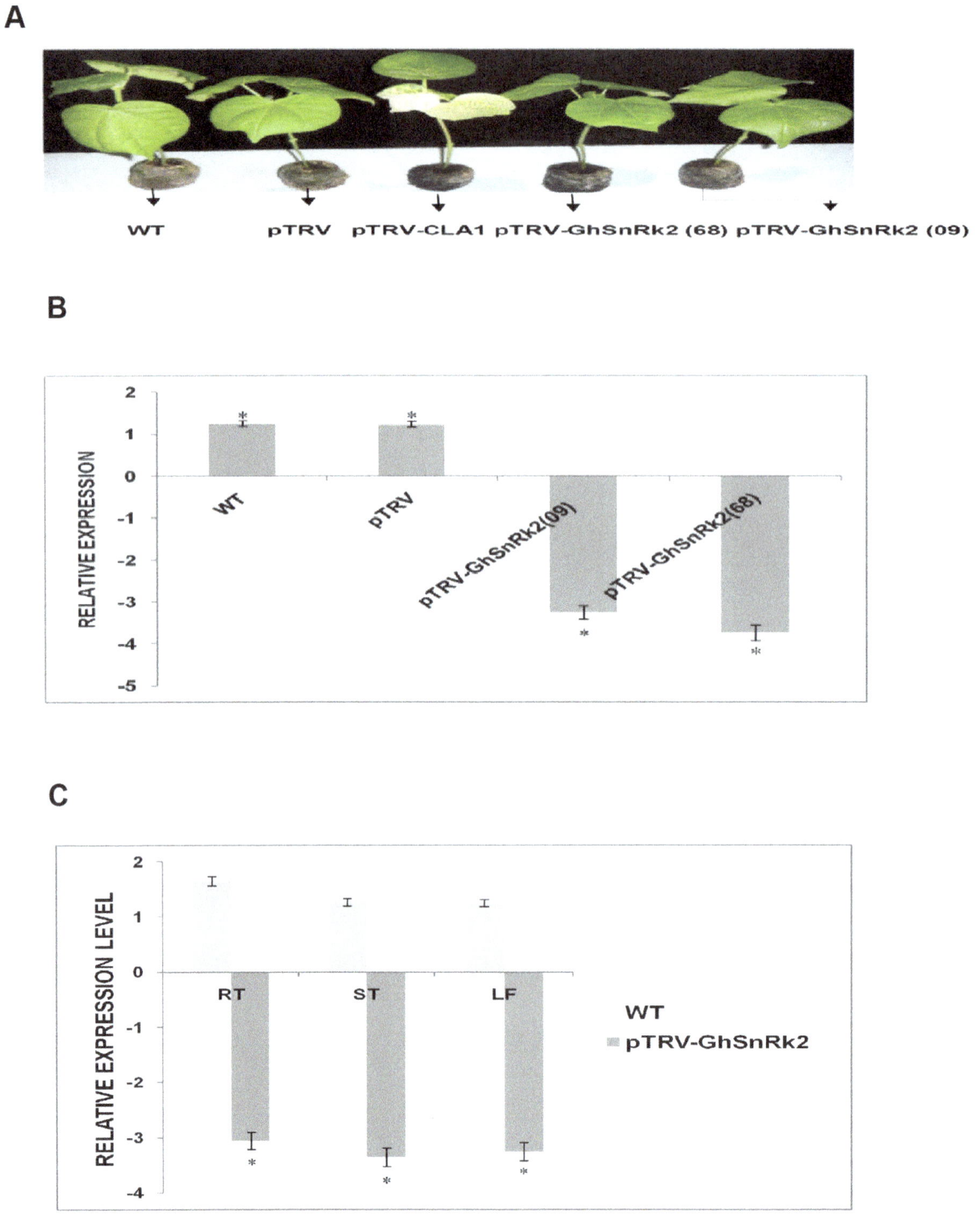

Figure 8. Silencing efficiency and transcript level of the *GhSnRK2* gene silenced plants. (A) Phenotype of gene silenced and non-silenced plants; 14 dpi. Wild-type (WT), negative control Empty vector (pTRV), positive control (pTRV-GrCLA1), gene silenced CRI409 cotton cultivars [pTRV-GhSnRk2 (09)], and gene silenced CRI99668 cotton cultivars [pTRV-*GhSnRK2* (68)]. (B) The silencing efficiency as determined by the expression pattern of WT, vector control (pTRV) and *GhSnRK2* gene silenced plants based on qRT-PCR. (C) The distribution of the TRV- construct in the gene silenced plants. Samples from the root, the stem and the leaves of gene silenced and non-silenced plants were analyzed via qRT-PCR. The values are presented as the means of three biological replicates. Asterisk denotes a significant difference ($P<0.05$).

occurs primarily as a result of acute dehydration associated with low temperature [35,36].

Our findings revealed the accumulation of compatible osmolytes, such as free proline, in the *GhSnRK2*-overexpressing plants; therefore, we speculate that the *GhSnRK2* transgenic plants exhibit increased stress tolerance by regulating downstream gene expression and accumulating a larger amount of compatible osmolytes, which may account for the increase in stress tolerance

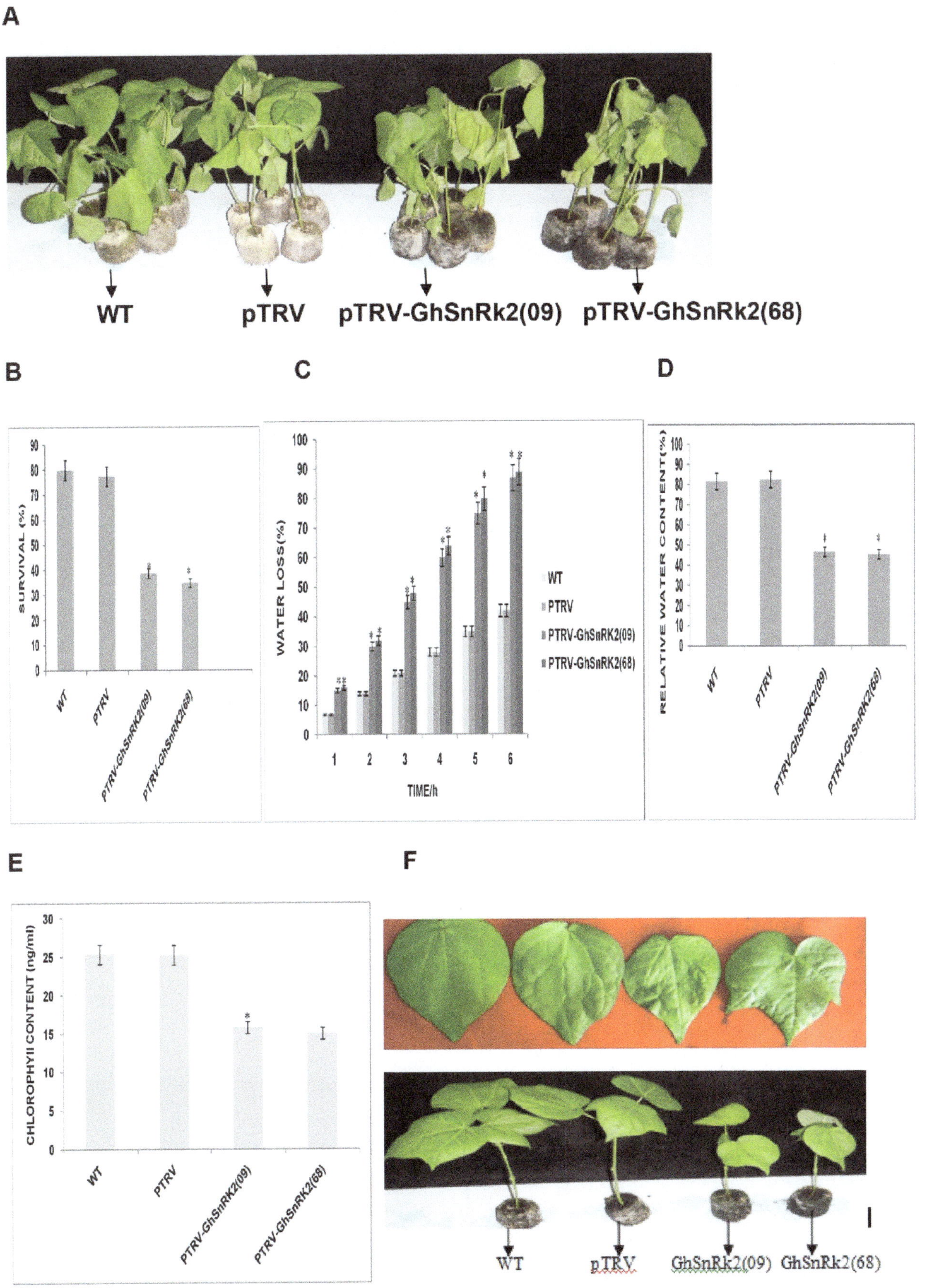
A
WT
pTRV
pTRV-GhSnRk2(09)
pTRV-GhSnRk2(68)
B
SURVIVAL (%)
WT
PTRV
PTRV-GhSnRK2(09)
PTRV-GhSnRK2(68)
C
WATER LOSS(%)
TIME/h
WT
PTRV
PTRV-GhSnRK2(09)
PTRV-GhSnRK2(68)
D
RELATIVE WATER CONTENT(%)
WT
PTRV
PTRV-GhSnRK2(09)
PTRV-GhSnRK2(68)
E
CHLOROPHYII CONTENT (ng/ml)
WT
PTRV
PTRV-GhSnRK2(09)
PTRV-GhSnRK2(68)
F
WT
pTRV
GhSnRk2(09)
GhSnRk2(68)

Figure 9. Physiological assay of *GhSnRK2* gene silenced plants. (A) Phenotype of drought stressed plants. Water was withheld from *GhSnRK2* gene silenced and non-silenced plants for 5 days approximately two weeks post-inoculation. Each of the ten treated groups consisted of five plants. A photograph of each group representative was captured. (B) The survival rate was determined by withholding water from *GhSnRK2* gene silenced and non-silenced plants for 5 days approximately two weeks post-inoculation, and the survival rate was recorded as the percentage of plants that survived after re-watering for 3 days. (C) The relative water loss was determined two weeks after inoculation. The reduction in the fresh weight from the initial weight was determined at the indicated time and represented as the percentage of water loss. The experiment was replicated three times. (D) The mean RWC was analyzed after immersing the fresh detached leaves in distilled water for 4 h and oven drying at 80°C for 48 h. The values are presented as the means of three biological replicates. Asterisk denotes a significant difference ($P<0.05$). (E) The change in the chlorophyll content of *GhSnRK2* gene silenced and non-silenced plants under salt stress. (F) The phenotype of *GhSnRK2* gene silenced and non-silenced plants under salt stress.

of *GhSnRK2* plants. Proline is an osmoprotective molecule that accumulates in response to water stress and salinity [37,38]. Proline is considered as a typical physiological parameter for evaluating abiotic stress tolerance and resistance in plants. Many plants reduce their cellular osmotic potential via the accumulation of intracellular organic osmolytes, such as proline, to maintain a stable intracellular environment when subjected to external environmental stresses [39,40,41].

We detected decreased stomatal apertures in the *GhSnRK2* transgenic plants compared with the corresponding WT plants. When exposed to adverse environmental conditions, the stomata must close for the plant to survive [42]. Stomatal closure reduces water loss, particularly in plants that have been exposed to water stress conditions caused by high solute concentrations in the nutrient medium, such that minimizing the rate of water loss due to transpiration reduces the accumulation of toxic ions [43]. The reduced rate of water loss from *GhSnRK2* transgenic plants may be due to the decrease in the size of the stomata aperture. ABA is involved in stomata movement, enhanced drought tolerance and plant osmoregulation. Under water stress conditions, ABA is released into leaves, inducing the release of potassium salts from the guard cells, resulting in stomatal closure. Plant guard cells modulate the opening and closure of the stomata in response to phytohormones and various environmental signals, such as light and temperature, thereby regulating gas exchange for photosynthesis and water status via transpiration [44]. The *SnRK2* gene *SRK2E*, or *SnRK2.6*, plays a significant role in stomatal closure in *Arabidopsis* leaves [45,46].

The *GhSnRK2* gene was upregulated after treatment with 10% PEG, which may suggest that the *GhSnRK2* gene sufficiently confers drought tolerance, as one strategy that can be employed to mimic the effect of drought on plants is treatment with PEG. PEG was quite commonly used in physiological experiments to induce controlled drought stress.

We speculate that the *GhSnRK2*-overexpressing plants were more tolerant to NaCl than the WT plants, which may be due to the improved accumulation of Na+ in the vacuoles, increased sequestration of Na+ into the vacuoles, improved cellular Na+ exclusion, or their greater capacity for vacuolar osmoregulation. High salt can result in oxidative damage to cell membranes. The ability to avert such damage is consistent with the degree of tolerance exhibited by the plant [47]. Under salt stress, this damage is less extensive in transgenic plants overexpressing the *GhSnRK2* gene than in WT plants because *GhSnRK2* overexpression can tolerate the accumulation of Na+ in the vacuoles, thus preventing the toxicity of excess Na+. Hypersensitivity to ABA treatment during seed germination and early seedling development is typically followed by improved drought tolerance [48,49]. Our finding reveals that the transgenic plants overexpressing *GhSnRK2* exhibited enhanced seedling growth under ABA stress treatment. This may be attributable to ABA-independent signaling pathway involving SnRK2 in root growth. This finding reveals the complexity in ABA-signaling, as different signaling components may function in different cells or tissues in ABA signal transduction. *SnRK2.2* and *SnRK2.3* double mutant showed strong ABA-insensitive phenotypes in seed germination and root growth inhibition.[50]. The decreased stomatal apertures detected in the *GhSnRK2* transgenic plants may be attributable to the role of *SnRK2* in the regulation of ABA-induced activation of plasma membrane anion channels in guard cells and concomitant response to their closure. These results suggest that *SnRK2* functions in ABA signaling of stomatal closure. *SRK2E* knockout mutant lost the ABA-dependent stomatal closure. SRK2E affects ABA-signaling in stomatal closure, but not in germination stage [51]. ABA and PYR/PYLs inhibit protein phosphatase2C (PP2C), which in turn relieves the repression of positive factors, such as SnRK2s [52].

We analyzed the expression of the stress-inducible marker genes *RD29A, RD29B, P5CS1, ABI3, CBF1*, and *ABI5* in *GhSnRK2*-overexpressing plants, and our results revealed increases in the transcript levels of these marker genes. Consistent with a previous study, the upregulation of stress response genes, such as *P5CS1*, can contribute to enhanced salt stress tolerance in plants [53]. *ABI3* and *ABI5* activation is necessary to sustain the germination of seedlings during intense drought stress [54]. *ABI5* and *ABI3* regulate numerous ABA responses, such as osmotic water permeability of the plasma membrane, stomatal closure, drought-induced resistance, germination, and inhibition of vegetative growth. We speculated that *GhSnRK2* functions in the transcriptional regulation of ABA-inducible genes in seedlings, suggesting that *SnRK2* protein kinases are involved in several processes of ABA signal transduction, including transcriptional as well as post-transcriptional regulation pathways. Therefore, we suggest that stress signaling pathways may be involved in *GhSnRK2*-mediated stress tolerance.

Our findings revealed the downregulation of *GhSnRK2* in gene silenced plants, indicating that this gene was knocked down in the gene silenced plants. TRV spread vigorously throughout the entire plant, including meristem tissue, and the symptoms induced by TRV are not severe compared with those induced by other viruses [55]. The albino phenotype observed in the *GhCLA1*gene silenced plants in this study is an indicator of effective VIGS function in inoculated plants, which is consistent with previous findings [56]. The results of this study revealed that *GhSnRK2* gene silencing in *Gossypium hirsutum* greatly reduced its tolerance to drought stress, corroborating the findings of [57], who reported that silencing of the function of the *SpMPK1*, *SpMPK2*, and *SpMPK3* genes in tomato plants alleviates their tolerance to drought stress. The decrease in the chlorophyll content of silenced plants in response to salt stress indicates that *GhSnRK2* may be involved in oxidative stress. High-concentration salt stress inhibits plant biochemical processes. Decreased chlorophyll content in salt-stressed pumpkin plants was detected by [58]. The stunted growth observed in salt-stressed *GhSnRK2* gene silenced plants corroborate the findings of [59], who reported that the typical effect of salt stress in plants is growth retardation as a result of cell elongation inhibition.

Conclusions

Despite the significant innovations that have been made in elucidating the genetic mechanisms underlying drought tolerance, considerable challenges remain. The present study revealed that the genetic manipulation of the *GhSnRK2* gene from cotton using a transgenic technique results in enhanced drought and cold stress tolerance in plants. Thus, *GhSnRK2* is hypothesized to participate in the stress signaling pathway, and therefore, overexpression of *GhSnRK2* may alleviate abiotic stress by regulating stress-responsive genes. Moreover, the findings in this study have helped to elucidate that *GhSnRK2* enhances stress tolerance in plants by affecting various stress-related pathways. Thus, the *GhSnRK2* gene represents a candidate gene for future research of abiotic stress signaling pathways and the genetic modification of novel *Gossypium hirsutum* varieties.

Supporting Information

Figure S1 Multiple sequence alignment of *GhSnRK2* and closely related *SnRK2s* from other plants species and VIGS construct. (A) Alignment of *GhSnRK2* and closely related *SnRK2s*. The relatively conserved motif is underlined. The deduced amino acid sequence displays relatively high homology with the monocot *SnRK2* family members *Oryza sativa* (*RK1*), ABB89146 and with the dicot species *AtSnRK2.10*, AEE33751.1. (B) Virus-induced gene silencing construct. The TRV VIGS vectors were modified based on a pTRV1 containing RNA-dependent RNA polymerase (RdRp), movement protein (MP), a 16 kDa cysteine-rich protein (16K), CaMV 35S promoters (2X35S) and a NOS terminator (NOSt) T-DNA vector. pTRV2 contains the coat protein (CP), multiple cloning sites (MCSs), CaMV 35S promoters (2X35S) and a NOS terminator (NOSt) T-DNA vector. Both vectors contain Rz, which is designated as a self-cleaving ribozyme, LB (left border) and RB (right border) of the T-DNA.

Figure S2 Survival rate in kanamycin-containing medium and PCR confirmation of *GhSnRK2* gene expression in transgenic Arabidopsis. (A) The survival rate of WT and *GhSnRK2* transgenic plants in MS medium supplemented with the antibiotic kanamycin. Photograph of a representative plant was captured after 9 days of germination in kanamycin-containing medium. (B) Confirmation of *GhSnRK2* gene in transgenic Arabidopsis. Genomic DNA from the first generation of the plants (T1) was extracted and used as a template for gene-specific primers. Lane 1: DNA molecular marker III standard; lane 2: negative control.

Table S1 Survival (%) of *GhSnRK2* transgenic plants in kanamycin-containing medium. At T1 generation, the ratio of dead to surviving plant was approximately 1:3 in kanamycin-containing LB medium. The survival rate at T3 generation was 100% in kanamycin-containing LB medium. The values are expressed as the mean germination rate (%) of approximately 200 seeds.

Table S2 The lists of primers sequences used in this study.

Acknowledgments

Our appreciation goes to the staff and students of the State Key Laboratory of Cotton Biology, Chinese Academy of Agricultural Sciences, Cotton Research Institute, Anyang, China, for their support during the course of this research. We also thank Jiahe Wu for VIGS technical support and suggestions during manuscript preparation.

Author Contributions

Conceived and designed the experiments: BB FL XZ Zhaoen Yang CL QW GZ Zuoren Yang. Performed the experiments: BB FL XZ Zhaoen Yang CL QW GZ Zuoren Yang. Analyzed the data: BB FL XZ Zhaoen Yang CL QW GZ Zuoren Yang. Contributed reagents/materials/analysis tools: BB FL XZ Zhaoen Yang CL QW GZ Zuoren Yang. Wrote the paper: BB FL XZ Zhaoen Yang CL QW GZ Zuoren Yang.

References

1. Genoud T, Metraux JP (1999) Crosstalk in plant cell signaling: structure and function of the genetic network. Trends in Plant Science 4: 503–507.
2. Kumar N, Bhatt RP (2006) Transgenics: An emerging approach for cold tolerance to enhance vegetables production in high altitude areas. Indian J. Crop Sci 1: 8–12.
3. Shinozaki K, Yamaguchi-Shinozaki K (2000) Molecular responses to dehydration and low temperature: differences and cross-talk between two stress signaling pathways. Curr Opin Plant Biol 3: 217–223
4. Guo-Tao H, Shi-Liang M, Li-Ping B, Li Z, Hui M, et al. (2000) The role of root border cells in plant defense. Trends Plant Sci 5: 128–133
5. Cho K, Agrawal GK, Jwa NS, Kubo A, Rakwal R (2009) Rice OsSIPK and its orthologs: a "central master switch" for stress responses. Biochem Biophys Res Commun 379: 649–653.
6. Mishra NS, Tuteja R, Tuteja N (2006) Signaling through MAP kinase networks in plants. Arch Biochem Biophys 452: 55–68.
7. Pitzschke A, Schikora A, Hirt H (2009) MAPK cascade signalling networks in plant defence. Curr Opin Plant Biol 12: 421–426.
8. Rodriguez MC, Petersen M, Mundy J (2010) Mitogen-activated protein kinase signaling in plants. Annu Rev Plant Biol 61: 621–649.
9. Jonak C, Hirt H (2002) Glycogen synthase kinase 3/SHAGGY-like kinases in plants: an emerging family with novel functions. Trends Plant Sci 7: 457–461.
10. Koh S, Lee SC, Kim MK, Koh JH, Lee S, et al. (2007) T-DNA tagged knockout mutation of rice OsGSK1, an orthologue of Arabidopsis BIN2, with enhanced tolerance to various abiotic stresses. Plant Mol Biol 65: 453–466.
11. Mahfouz MM, Kim S, Delauney AJ, Verma DP (2006) Arabidopsis TARGET OF RAPAMYCIN interacts with RAPTOR, which regulates the activity of S6 kinase in response to osmotic stress signals. Plant Cell 18: 477–490.
12. Das R, Pandey GK (2010) Expressional analysis and role of calcium regulated kinases in abiotic stress signaling. Curr Genomics 11: 2–13.
13. Hrabak EM, Chan CWM, Gribskov M, Harper JF, Choi JH, et al. (2003) The Arabidopsis CDPK-SnRK superfamily of protein kinases. Plant Physiol 132: 666–680.
14. Wurzinger B, Mair A, Pfister B, Teige M (2011) Cross-talk of calcium-dependent protein kinase and MAP kinase signaling. Plant Signal Behav 6: 8–12.
15. Coello P, Hirano E, Hey SJ, Muttucumaru N, Martinez-Barajas E, et al. (2012) Evidence that abscisic acid promotes degradation of SNF1-related protein kinase (SnRK) 1 in wheat and activation of a putative calcium-dependent SnRK2. J Exp Bot 63: 913–924
16. Umezawa T, Yoshida R, Maruyama K, Yamaguchi-Shinozaki K, Shinozaki K (2004) SRK2C, a SNF1-related protein kinase 2,improves drought tolerance by controlling stress-responsive gene expression in Arabidopsis thaliana. Proc Natl Acad Sci. USA 101: 17306–173011.
17. Hardie DG (1999) Plant protein serine/threonine kinases: classification and functions. Annu Rev Plant Physiol Plant Mol Biol 50: 97–131.
18. Mikołajczyk M, Awotunde OS, Muszyńska G, Klessi DF, Dobrowolska G (2000) Osmotic stress induces rapid activation of a salicylic acid-induced protein kinase and a homolog of protein kinase ASK1 in tobacco cells. Plant cell 12: 165–178
19. Wang X, Xu W, Xu Y, Chong K, Xu Z, et al (2004) Wheat RAN1, a nuclear small G protein, is involved in regulation of cell division in yeast. Plant Sci 167: 1183–1190.
20. Clough SJ, Bent AF (1998) Floral dip: a simplified method for Agrobacterium-mediated transformation of Arabidopsis thaliana. Plant J 16(6):735–743
21. Tyree MT, Hammel HT (1972) The measurement of the turgor pressure and the water relations of plants by the pressure-bomb technique. J Exp Bot 23: 267–282
22. Duan J, Zhang M, Zhang H, Xiong H, Liu P, et al. (2012) OsMIOX, a myoinositol oxygenase gene, improves drought tolerance through scavenging of reactive oxygen species in rice (Oryza sativa L.). Plant Sci 196: 143–151.

23. Parida AK, Dasgaonkar VS, Phalak MS, Umalkar GV, Aurangabadkar LP (2007) Alterations in photosynthetic pigments, protein, and osmotic components in cotton genotypes subjected to short-term drought stress followed by recovery. Plant Biotechnology Reports 1: 37–48.
24. Pei ZM, Kuchitsu K, Ward JM, Schwarz M, Schroeder JI (1997 Mar) Differential abscisic acid regulation of guard cell slow anion channels in Arabidopsis wild-type and abi1 and abi2 mutants. Plant cell 9(3):409–423.
25. Bates L, Waldren RP, Teare ID (1973) Rapid determination of free proline for water stress studies. Plant and Soil 39: 205–207
26. Arnon DI (1949) Copper enzymes in isolated chloroplasts, polyphenoxidase in beta vulgaris. Plant physiology 24: 1–15.
27. Xiong L, Ishitani M, Lee H, and Zhu JK (2001b) The *Arabidopsis LOS5/ABA3* locus encodes a molybdenum cofactor sulfurase and modulates cold and osmotic stress responsive gene expression. Plant cell 13: 2063–2083.
28. Livak KJ, Schmittgen TD (2001) Analysis of Relative Gene Expression Data Using Real-Time Quantitative PCR and the 2DDCT Method. METHODS 25: 402–408
29. Gao X, Britt RC, Shan L, He P (2011) Agrobacterium-Mediated Virus-Induced Gene Silencing Assay In Cotton. J. Vis. Exp 54:e2938
30. Kobayashi Y, Murata M, Minami H, Yamamoto S, Kagaya Y (2005) Abscisic acid-activated SNRK2 protein kinases function in the gene-regulation pathway of ABA signal transduction by phosphorylating ABA response element binding factors. Plant J 44: 939–949.
31. Zhang H, Mao X, Wang C, Jing R (2010) Overexpression of a common wheat gene *TaSnRK2.8* enhances tolerance to drought, salt and low temperature in *Arabidopsis*. PLoS ONE 5: e16041.
32. Hawes MC, Gunawardena U, Miyasaka S, Zhao X (2000) The role of root border cells in plant defense. Trends Plant Sci 5: 128–133.
33. Dainty J (1976) Water relations of plant cells. *In* AP Gottingen, MH Zimmermann, eds, Encyclopedia of Plant Physiology, Vol 2. New Series, Part A. Springer-Verlag, Berlin, pp 12–35.
34. Levitt J (1980) Response of plant to Environmental Stress, water, radiation, salt and other stresses. Academic press, New York.
35. Steponkus PL (1984) Role of the plasma membrane in freezing injury and cold acclimation. Annu. Rev. Plant Physiol 35: 543–584.
36. Steponkus PL, Uemura M, Webb MS (1993) A contrast of the cryostability of the plasma membrane of winter rye and spring oat-two species that widely differ in their freezing tolerance and plasma membrane lipid composition. In: Steponkus P L, editor. Adv. Low-Temperature Biol. Vol. 2. London: JAI Press pp. 211–312.
37. Claussen W (2005) Proline as a measure of stress in tomato plants. Plant Sci 168: 241–248.
38. Younis ME, Hasaneen MNA, Tourky MNS (2009) Plant growth, metabolism and adaptation in relation to stress conditions. XXIV. Salinity biofertility interactive effects on proline, glycine and various antioxidants in Lactuca sativa. Plant Omics J 2: 197–205.
39. Zhu JK (2002) Salt and drought stress signal transduction in plants. Annu Rev Plant Biol 53: 247–273
40. Granier C, Tardieu F (1999) Water deficit and spatial pattern of leaf development variability in responses can be simulated using a simple model of leaf development. Plant Physiol 119: 609–620.
41. Wang ZQ, Yuan YZ, Ou JQ, Lin QH, Zhang CF (2007) Glutamine synthetase and glutamate dehydrogenase contribute differentially to proline accumulation in leaves of wheat (*Triticum aestivum*) seedlings exposed to different salinity. Plant Physiol 164: 695–701.
42. Pareek A, Sopory SK, Bohnert HJ, Govindjee EDS (2010) Abiotic Stress Adaptation in Plants: Physiological, Molecular and Genomic Foundation, Springer, Dordrecht, pp. 283e305.
43. Everard JD, Gucci R, Kahn JA, Flore WH (1994). Gas exchange and carbon partitioning in the leaves of celery (Apium graveolens L.) at various levels of root zone salinity, Plant Physiol 106: 281e292.
44. Schroeder JI, Kwak JM, Allen GJ (2001) Guard cell abscisic acid signalling and engineering drought hardiness in plants. Nature 410: 327–330.
45. Yoshida R, Hobo T, Ichimura K, Mizoguchi T, Takahashi F (2002) ABA-activated SnRK2 protein kinase is required for dehydration stress signaling in Arabidopsis Plant Cell Physiol 43: 1473–1483.
46. Mustilli AC, Merlot S, Vavasseur A, Fenzi F, Giraudat J (2002) Arabidopsis OST1 Protein Kinase Mediates the Regulation of Stomatal Aperture by Abscisic Acid and Acts Upstream of Reactive Oxygen Species Production. Plant Cell 14: 3089–3099.
47. Chinnusamy V, Jagendorf A, Zhu JK (2005) Understanding and improving salt tolerance in plants, Crop Sci 45: 437e448.
48. Hu H, Dai M, Yao J, Xiao B, Li X, et al. (2006) Overexpressing a NAM, ATAF, and CUC (NAC) transcription factor enhances drought resistance and salt tolerance in rice. Proc Natl Acad Sci USA 103: 12987–12992.
49. Ko JH, Yang SH, Han KH (2006) Upregulation of an Arabidopsis RING-H2 gene, *XERICO*, confers drought tolerance through increased abscisic acid biosynthesis. Plant J 47 343–355.
50. Hiroaki F, Paul E, Jian-Kang Z (2007) Identification of Two Protein Kinases Required for Abscisic Acid Regulation of Seed Germination, Root Growth, and Gene Expression in *Arabidopsis*. The Plant Cell 19 (2):485–494.
51. Riichiro Y, Tokunori H, Kazuya I, Tsuyoshi M, Fuminori T, et al. (2002) ABA-Activated SnRK2 Protein Kinase is Required for Dehydration Stress Signaling in *Arabidopsis*. *Plant Cell Physiol* 43 (12): 1473–1483.
52. Sang-Youl P, Pauline F, Noriyuki N, Davin RJ, Hiroaki F, et al. (2009) Abscisic Acid Inhibits Type 2C Protein Phosphatases via the PYR/PYL Family of START Proteins. Science 324: 1068
53. Kavi PB, Zonglie H, Cuo-Hua M, Chein-An AH, DeshPal SV (1995) Overexpression of A1-Pyrroline-5-Carboxylate Synthetase Increases Proline Production and Confers Osmotolerance in Transgenic Plants Plant Physiol 108: 1387–1394
54. Lopez-Molina L, Mongrand S, McLachlin DT, Chait BT, Chua NH (2002) ABI5 acts downstream of ABI3 to execute an ABA-dependent growth arrest during germination. Plant J 32: 317–328.
55. Ratcliff F, Martin-Hernandez AM, Baulcombe DC (2001) Tobacco rattle virus as a vector for analysis of gene functions by silencing. Plant J 25: 237–245
56. Liu Y, Schiff M, Dinesh-Kumar S (2002) Virus-induced gene silencing in tomato. Plant J 31: 777–786.
57. Cui L, Jian-Min Y, Yun-Zhou L, Zhen-Cai Z, Qiao-Li W, et al. (2013) Silencing the *SpMPK1*, *SpMPK2*, and *SpMPK3* Genes in Tomato Reduces Abscisic Acid—Mediated Drought Tolerance Int. J. Mol. Sci 14: 21983–21996
58. Senay S, Fikret Y, Sebnem K, Sebnem E (2011) The effect of salt stress on growth, chlorophyll content, lipid peroxidation and antioxidative enzymes of pumpkin seedling. African Journal of Agricultural Research 6(21): 4920–4924
59. Yasar F, Ellialtioglu S, Yildiz K (2008) Effect of salt stress on antioxidant defense systems, lipid peroxidation, and chlorophyll content in green bean. Russian J. Plant Physiol 55(6): 782–786.

18

Comparative Transcriptomic Analysis of the Response to Cold Acclimation in *Eucalyptus dunnii*

Yiqing Liu[1,2]*, Yusong Jiang[2], Jianbin Lan[2], Yong Zou[2], Junping Gao[1]

1 Department of Ornamental Horticulture, China Agricultural University, Beijing 100193, China, **2** College of Life Science & Forestry, Chongqing University of Art & Science, Yongchuan 402160, China

Abstract

Eucalyptus dunnii is an important macrophanerophyte with high economic value. However, low temperature stress limits its productivity and distribution. To study the cold response mechanisms of *E. dunnii*, 5 cDNA libraries were constructed from mRNA extracted from leaves exposed to cold stress for varying lengths of time and were evaluated by RNA-Seq analysis. The assembly of the Illumina datasets was optimized using various assembly programs and parameters. The final optimized assembly generated 205,325 transcripts with an average length of 1,701 bp and N50 of 2,627 bp, representing 349.38 Mb of the *E. dunnii* transcriptome. Among these transcripts, 134,358 transcripts (65.4%) were annotated in the Nr database. According to the differential analysis results, most transcripts were up-regulated as the cold stress prolonging, suggesting that these transcripts may be involved in the response to cold stress. In addition, the cold-relevant GO categories, such as 'response to stress' and 'translational initiation', were the markedly enriched GO terms. The assembly of the *E. dunnii* gene index and the GO classification performed in this study will serve as useful genomic resources for the genetic improvement of *E. dunnii* and also provide insights into the molecular mechanisms of cold acclimation in *E. dunnii*.

Editor: Turgay Unver, Cankiri Karatekin University, Turkey

Funding: 1. Natural Science Foundation of Chongqing Province, China (Grant NO. cstc2013jcyjA80035), 2. Key Program for forestry of Chongqing University of Art and Science (Grant NO. 201302), 3. National Science Foundation for Young Scientists of China (Grant NO. 31340016). The funders had no role in study design, data collection and analysis, decision to publish, or preparation of the manuscript.

Competing Interests: The authors have declared that no competing interests exist.

* Email: liung906@163.com

Introduction

Rapid population increase and the consequent increase in the requirement for different types of paper products, as well as the emphasis on paper as an environmentally friendly packaging material, have led to an increased demand for wood [1]. The imbalance between the supply and demand for forest products is growing. Eucalyptus is an economically important forest tree that grows in tropical and subtropical regions [2,3]. Eucalyptus trees can be highly productive over a short rotation period, tolerate a wide range of soils and commonly exhibit a straight stem form in those species utilized in production forestry. Furthermore, eucalypts, unlike many trees, do not have a true dormant period and retain their foliage, which enables growth during warm winter periods [4]. Nevertheless, in Eucalyptus plantations, low temperature stress limits their productivity and distribution. When the temperature drops to 8°C or below, Eucalyptus trees would exhibit various symptoms of cold injury due to their inability to adapt to the low temperature [5]. Cold stress also alters the physiological status, such as transient increases in hormone levels (e.g., ABA), changes in the membrane lipid composition, accumulates of compatible osmolytes (such as soluble sugars, betaine, and proline) and increases in antioxidant levels [6,7]. In contrast, temperate plants can withstand freezing temperatures following a period of low, but non-freezing temperatures, a process called cold acclimation. The mechanisms of cold acclimation have been extensively investigated in *Arabidopsis thaliana* [8] and other important crop species such as maize and barley [9,10]. Cold stress has been shown to induce changes in physiology and gene expression, and hundreds of cold-responsive genes have been identified so far [11]. However, in tropical and subtropical plants, especially *E. dunnii*, the molecular mechanisms of the cold response are not clear.

The physiological and biochemical changes that occur during plant cold acclimation result primarily from changes in the expression of cold-responsive genes. In general, Cold-responsive genes could be classified into two groups: 1) functional proteins, which directly protect plants against environmental stresses, and 2) regulatory proteins, which regulate the expression of downstream target genes in the stress response [4]. The first group mainly comprises enzymes involved in the biosynthesis of various osmoprotectants, such as late embryo genes is abundant (LEA) proteins, antifreeze proteins, chaperones, and detoxification enzymes[8,12]. The second group mainly includes transcription factors and protein kinases [12]. The best-characterized transcription factors (TFs) involved in the plant cold response are the class of AP2/ERF (APETALA2/ethylene-responsive element binding proteins), one kind of subfamily was known as CBF/DREB(C-repeat binding factor/dehydration resistance element binding protein), which regulate cold-responsive gene expression by binding to DRE/CRT cis-elements in the promoter region of cold-responsive genes [6,13]. Changes in the expression of cold-responsive contribute to the differences in plant cold tolerance. For example, *Solanum commersonii* and *S. tuberosum*, which are closely related species that differ in their cold acclimation abilities, exhibit considerable differences in the expression levels of cold-responsive genes [6,14]. Chen *et al.* found that the activities of some detoxification enzymes, such as catalase (CAT), superoxide dismutase (SOD), peroxidase (POD) and esterase (EST) are increased in response to cold stress, whereas the plant's metabolic activity is decreased [15–17]. Some cold-induced genes have been cloned from Eucalyptus plants. For example, four CBF paralogs were previously isolated from *E. gunnii*, and qRT-PCR analysis demonstrated that they exhibited complementary expression profiles in a range of natural standard and cold conditions [18]. Navarro *et al.*found overexpression of *EguCBF1a* or *EguCBF1b* in the cold-sensitive *E. urophylla·E. grandis* hybrid could enhance its freezing tolerance [19].

Given the importance of cold-responsive genes in plant cold tolerance, studying the cold response at the transcription level may be a key step in identifying specific tolerance mechanisms. Next generation sequencing (NGS) provides a high throughput approach for analyzing genes involved a particular process at transcription level. Compared to the traditional sequencing techniques, NGS is more robust and demonstrates greater resolution and inter-lab portability compared to several microarray platforms. NGS could detect millions of transcripts and is beneficial to explore new genes and their expression profiling independent of a reference genome [6,20,21]. For example, cDNA libraries for *E. gunnii* have been constructed to identify genes involved in cell protection (such as PCP, Lti6b and metallothionein), LEA/dehydrin accumulation, and cryoprotection [22,23]. Despite its obvious potential, these next generation sequencing methods have not been applied for *E. dunnii* yet.

The goal of this study was to construct a comprehensive transcriptome to investigate the molecular mechanism of cold tolerance in *E. dunnii*. The plants were exposed to low temperature (4°C) for 0, 3, 6, 12, and 24 h, and the first two expanded leaves below apical bud of *E. dunnii* were collected for high throughput RNA-Seq analysis. Paired-end (PE) reads from the RNA-Seq output were then assembled *de novo* to build an *E. dunnii* transcriptome, which was subjected to a comparative analysis. This analysis provides preliminary global insight into the molecular mechanism of cold tolerance and a good base for future basic research in *E. dunnii*.

Results

Physiological changes in *E. dunnii* in response to cold stress

Firstly, we detected the concentration of proline during the cold treatment (at 4°C) from 0 to 48 h. The concentration of proline decreased slightly from 0 to 3 h, but it increased rapidly as the cold stress prolonging (Fig. 1A). The decrease in proline content at 0 to 3 h might be caused by a transient stress response of *E. dunnii* to the low temperature shock. However, prolonged exposure to low temperature (24 h) resulted in proline accumulation.

Plant cells could accumulate amounts of reactive oxygen species under environmental stress, which result in severe damage of proteins, membrane lipid, DNA and other cellular components [15]. CAT could catalyze the decomposition of hydrogen peroxide to water and oxygen, and it is important in protecting the cell from oxidative damage by reactive oxygen species (ROS) [16–17]. The activity level of CAT changed during cold acclimation in *E. dunnii*. We observed an almost 25% increase in CAT activity after 3 h, and a nearly two-fold change after 24 h of cold stress (Fig. 1B). H_2O_2 is one kind of ROS molecule. The H_2O_2 concentration increased nearly 50% after 3 h, and then continued to increase at a more moderate rate, remaining at high levels until 24 h (Fig. 1C). These results indicated that *E. dunnii* plants are sensitive to the cold stress.

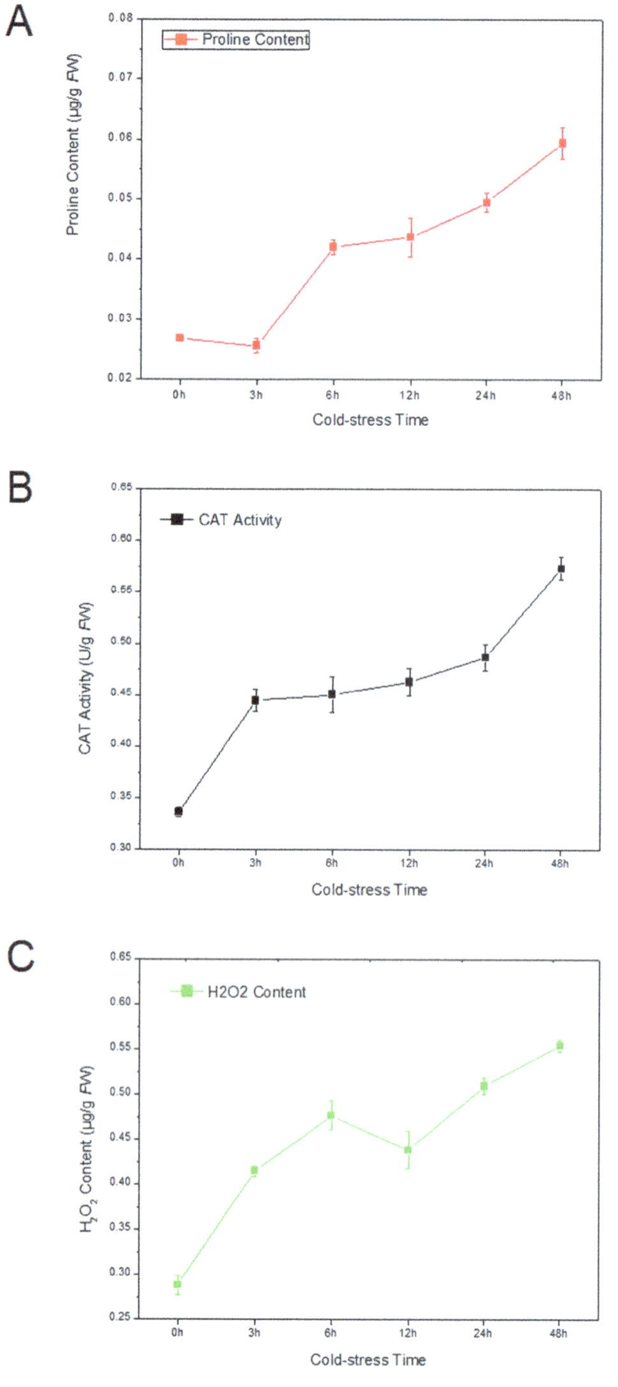

Figure 1. Changes in proline content (A), CAT activity (B) and H_2O_2 content (C) under low temperature (4°C) treatment over time.

Table 1. Total number of reads for each treatment sample, as obtained by Illumina sequencing.

Duration of low temperature (4°C) treatment	Paired-end reads	Total length	Total number of contigs	Average length	N50 of contigs	Alignment rate (%)
0 h	25,407,247	120,616,917	118,761	1,015	1,817	90.3
3 h	24,817,373	179,471,852	149,467	1,200	2,136	91.7
6 h	25,703,824	215,673,491	161,627	1,334	2,343	90.6
12 h	34,870,702	302,614,952	195,733	1,546	2,630	92.4
24 h	33,846,411	217,801,712	160,461	1,357	2,367	91.2
TOTAL	–	349,381,021	205,325	1,701	2,827	94.5

RNA Sequencing, *de novo* assembly and functional annotation

To study the *E. dunnii* transcriptome in response to cold stress, we transferred plantlets with 10 leaves to a climate-chamber (4°C) and collected the first two expand leaves below apical bud at 0, 3, 6, 12, and 24 h time points, respectively. For the RNA-Seq analysis, we obtained 25,407,247, 24,817,373, 25,703,824, 34,870,702, and 33,846,411 clean paired-end reads, respectively (data not shown).

To obtain a more reliable and comprehensive transcriptome database, these five libraries were pooled together and then performed the *de novo* assembly. The pipeline for the bioinformatics analysis of the RNA-Seq data is shown in Fig. 2. The parameters of the contig databases assembled by each individual assembler, such as the alignment rate, sensitivity, accuracy and length distribution, were significantly different. Overall, the contig database produced by Trinity was significantly better than those from the other assemblers (Table S1). The optimal contig database contained 205,325 contigs ≥300 bp in length. The average length of these contigs was 1,701.6 bp, the N50 number was 2,827 bp, and the maximum length was 15,965 bp (Table 1). Additionally, there were 148,151 contigs with a length≥600 bp, 105,494 contigs with a length ≥1,200 bp, and 33,700 contigs with a length ≥ 3,000 bp (Fig. 3). The assembled contigs (≥300 bp) were deposited in the NCBI Transcriptome Shotgun Assembly (TSA) database under the accession number PRJNA208093.

Sequence similarity search against the NCBI non-redundant protein database (NR) was conducted using a locally installed BLAST program for functional annotation. Among all the assembled contigs (≥300 bp), 134,358 (65.4%) were annotated with BLASTx hits, matching 80,578 unique protein accessions (Table S2). For contigs longer than 600 bp, 80.9% had BLASTx hits, and for longer than 900 bp, the percentage increased to 88.8% (Fig. 3), indicating that most contigs, particularly the longer contigs, represent protein-encoding transcripts. As the completed genome information of *E. dunnii* was not available at this time, 70,967 contigs (34.6%) had no hits to any known proteins in the Nr database (Fig. 4A), suggesting that these contigs might be non-coding regions or potentially new genes [24].

In addition, among 134,358 contigs with BLASTx results, 52,265 (38.9%), 26,602 (19.8%), and 24,721 (18.4%) showed high sequence similarity to *Vitis vinifera*, *Populus* and *Glycine max*, respectively, but only 265 contigs (0.2%) shared homology with Eucalyptus (Fig. 4B). Alternatively, our results could indicate that *E. dunnii* is more closely to *V. vinifera* than *G. max* or *Arabidopsis* evolutionarily. Interestingly, some other plant transcriptomes, such as *Craterostigma plantagineum* [25] and *Fraxinus* spp. [26],

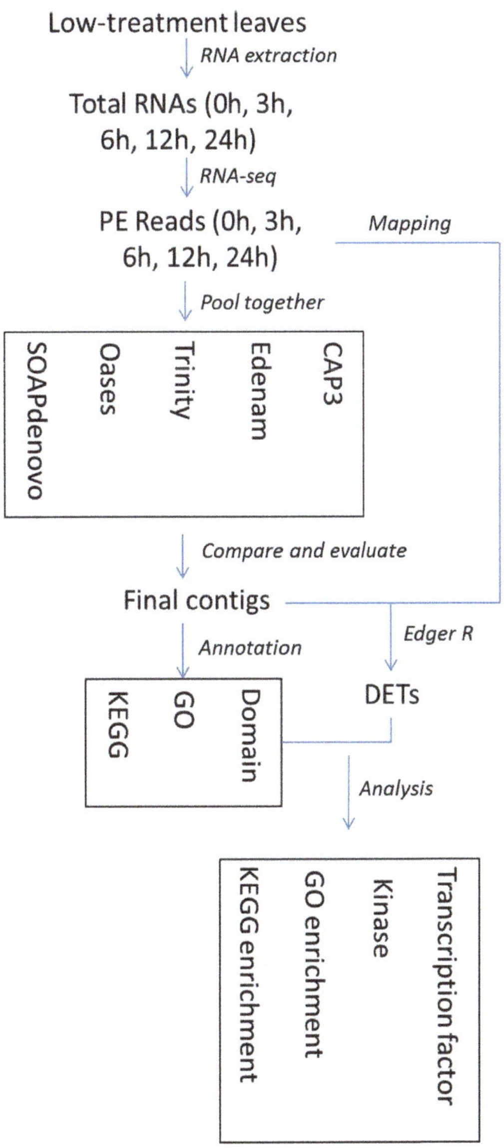

Figure 2. The pipeline for the bioinformatics analysis of the deep sequencing data.

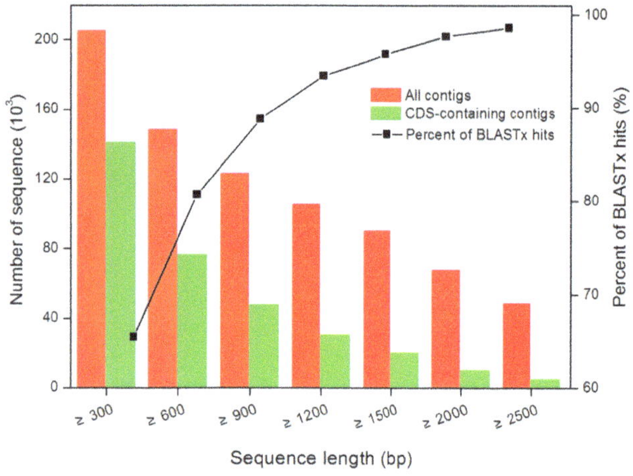

Figure 3. Annotation rate and proportion of long-CDS-containing sequences. A total of 205,325 contigs were used for the BLASTx search. The contig length is indicated on the X-axis. The size distributions of the final assembled contigs (red) and the number of long-CDS-containing contigs (green) are indicated on the left Y-axis. The percentage of BLASTx hits to size-grouped contigs is indicated by the diamond symbol.

display the same distribution pattern of top-hit species. These results could be simply explained by the number of genes deposited in Nr database. For example, by November 2013, the NCBI database contained 78,045 *V. vinifera* transcripts, 11,4590 *P. trichocarpa* transcripts, and 81,270 *G. max* transcripts, but only 7,146 Eucalyptus transcripts.

Differential expression between the groups and qRT-PCR validation

To characterize the digital gene expression profiles of the *E. dunnii* in response to low temperature, we performed a short-read alignment of each library using Perl script provided by the Trinity software package. For samples treated at 4°C for 0, 3, 6, 12, and 24 h, a total of 90.3%, 91.6%, 92.1%, 91.2%, and 91.5% of the reads could be aligned back to the contigs, and 64.2%, 63.8%, 65.1%, 62.7% and 61.5% aligned concordantly exactly once. To eliminate the effect of library size, edgeR (empirical analysis of digital gene expression in R) was used to create an effective library size for each sample. The number of aligned reads per transcript was normalized to FPKM based on an RESM-based algorithm. Differentially expressed transcripts (DETs) with FDR $\leq$0.05 and $\log_2$ fold-change ($\log_2$FC) $\geq$1 between pairs of samples were identified by edgeR [27]. The edgeR analysis generated 10 DET sets (0 h vs. 3 h, 0 h vs. 6 h, 0 h vs. 12 h, 0 h vs. 24, 3 h vs. 6 h, 3 h vs. 12 h, 3 h vs. 24 h, 6 h vs. 12 h, 6 h vs. 24 h, and 12 h vs. 24 h) with 11,395, 11,908, 11,901, 12,671, 8,935, 10,641, 11,843, 9,531, 11,489, and 10,230 DETs, respectively (Fig. 5). We also found that 7,059, 7,348, 7,479, and 7,636 DETs were up-regulated in the 0 h vs. 3 h, 0 h vs. 6 h, 0 h vs. 12 h, and 0 h vs. 24 h comparison sets, respectively. These results demonstrated that the number of up-regulated DETs increased as the duration of cold stress prolonged.

To validate the expression patterns of each DET obtained from the comparative RNA-Seq studies, we randomly selected 31 transcripts from the annotated DETs for qRT-PCR analysis. Noteworthy, qRT-PCR results are often affected by the choice of reference genes. Previously, a report explored the expression stability of reference genes which are using in gene expression test in Eucalyptus in response to various abiotic stresses by qRT-PCR [28]. The authors found that expression of some genes, such as *PP2A-3/SAND*, *UPL7*, *UBC2* and *GAPDH*, are stable enough in all tested samples, while *ACT2* gene was not stable in response to environmental stimuli as expected. As mentioned in the paper of Cassan-Wang et al., GAPDH is a good choice as reference gene in qRT-PCR assay in Eucalyptus [28]. Therefore,we selected two most commonly used reference genes, beta-actin and GAPDH,- because these two genes could be mutual support, mutual correction, and minimize the experimental errors. The results showed that the expression patterns of 25 DETs were compatible with the RNA-Seq analysis (Table S3), suggesting that the differential expression analysis based on high-throughput RNA sequencing produced reliable expression data.

Gene ontology (GO) and Kyoto Encyclopedia of Genes and Genomes (KEGG) enrichment analysis of DETs

GO (Gene Ontology) and KEGG (Kyoto Encyclopedia of Genes and Genomes) annotation were applied to the BLASTx results to provide comprehensive functional information for each transcript. In total, we obtained 198,528 GO annotations for 62,965 transcripts and 966 unique Enzyme Codes (ECs) for 28,295 transcripts (Table S2). Among the 62,965 transcripts with GO terms, 34,064 (54.1%) were assigned to the Biological process category, 19,959 (31.7%) to the Molecular function category, and 28,965 (46.0%) to the Cellular component category. In addition, 20,025 (31.8%) unique transcripts were assigned GO terms from all three categories (Fig. 6 and Table S4). To understand the mechanism of the cold stress response in *E. dunnii*, the DETs were subjected to GO and KEGG enrichment analysis. Under the GO category 'Biological process', the 'response to stress' and 'translational initiation' were the most highly enriched terms, with P.ad-values of 0 and 0.02, respectively. Under the category 'Molecular function', the 'quinolinate synthetase A activity' were the most highly enriched term, with a P.ad-values of 0.04. Under the category 'Cellular component', the 'cell part' was the most highly enriched term, with a P-value of 8.5E-11 (Table S4). KEGG analysis identified 27,688 contigs with pathway information were involved in 137 KEGG pathways. Among these 137 KEGG pathways, 'arginine and proline metabolism' and 'tropane, piperidine and pyridine alkaloid biosynthesis' were the two most significantly enriched KEGG pathways (Table S5).

Cold-responsive transcription factors and protein kinases in *E. dunnii*

Transcription factors and protein kinases are crucial upstream regulators that respond to various biotic and abiotic stresses in plants [29,30]. In this study, we identified a total of 586 contigs involving in transcription factor activity, which were classified into 65 types of transcription factors, including AP2, bZIP, JmjC, and SRF-TF. In order to verify the expression pattern of these transcription factors, an additional 5 transcripts were selected to carry out qRT-PCR analysis. The results displayed that the expression trend of these transcripts agreed with the results of RNA-seq analysis (Table S3). The START and bzip domains transcription factor families were the largest groups represented in the cold-responsive transcription factors, containing 64 (39 up-regulated and 25 down-regulated) and 62 (43 up-regulated and 19 down-regulated) unique transcripts, respectively. The next largest groups were the UDF (31 up-regulated and 24 down-regulated), AP2 (27 up-regulated and 24 down-regulated) and Sigma70 (24 up-regulated and 9 down-regulated) families (Table S6). In

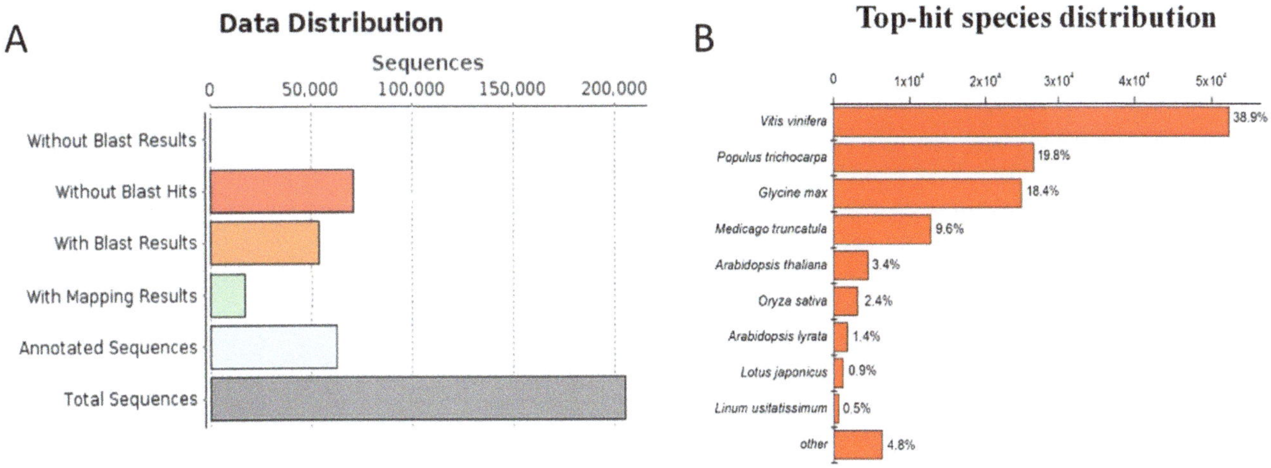

Figure 4. Distribution of the BLASTx results (A) and the top-hit species distribution of the *E. dunnii* transcriptome (B). A total of 205,325 contigs ≥300 bp in length were used for the sequence similarity searches, and 134,358 contigs produced BLASTx results. All of the contigs with BLASTx results were used for the species distribution analysis. Overall, 52,265 (38.9%), 26,602 (19.8%) and 24,721 (18.4%) contigs showed strong similarity to *Vitis vinifera*, *Populus* and *Glycine max*, but only 265 contigs (0.2%) shared the highest homology with Eucalyptus.

addition, we identified 169 contigs related to protein kinase activity, which were classified into 8 types of protein kinases based on their domains (Table S6).

Discussion

Improving *de novo* transcriptome assembly

The most critical step of an RNA-Seq study is the *de novo* assembly, especially for species without genome information [31–

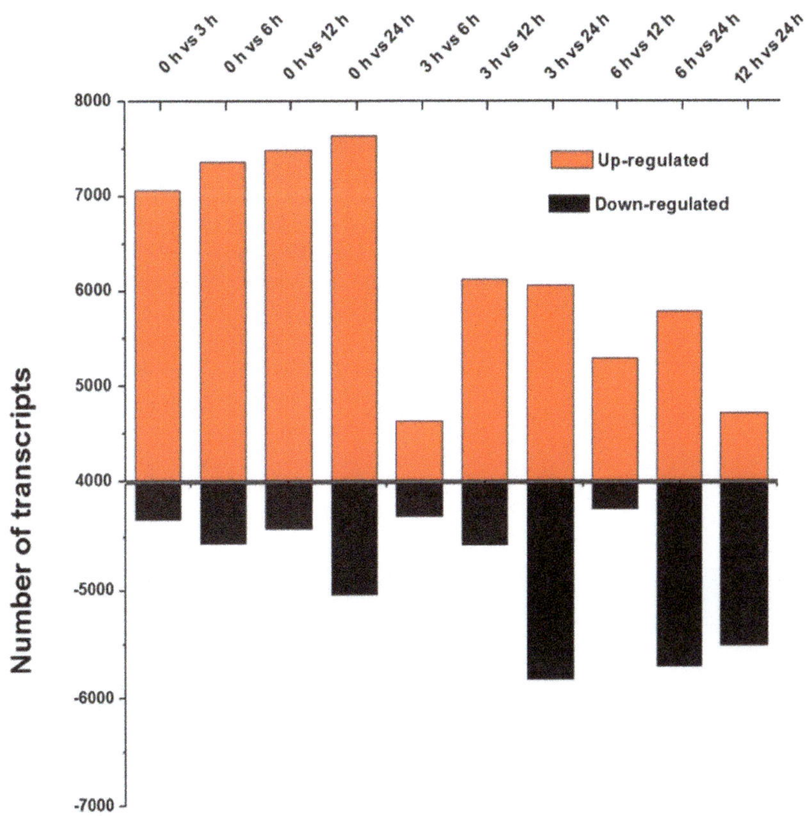

Figure 5. Transcripts that exhibited differential expression pattern. In total, 20,5325 contigs were used for the differential expression analysis, and the differential transcripts were identified by edgeR using the following parameters: FDR ≤0.05 and $\log_2$fold-change ≥1.

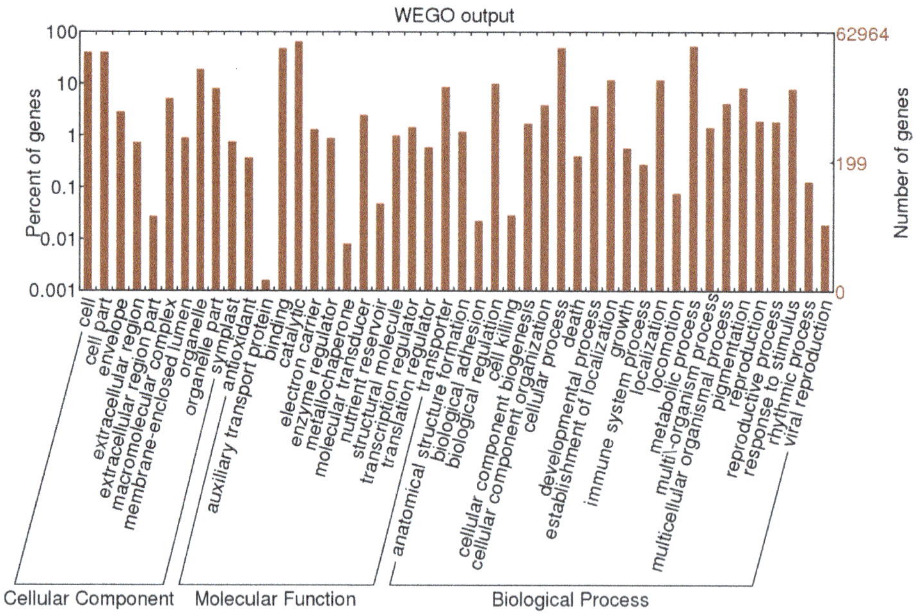

Figure 6. GO assignment of all contigs in the *E. dunnii* transcriptome. The contigs mapped to three main categories: Biological process, Cellular component and Molecular function. The right-hand y-axis indicates the number of annotated contigs.

34]. More and more genomes and/or transcriptomes sequences have been completed due to the development of high-throughput sequencing technologies. However, major published studies on transcriptome *de novo* assembly have typically used a single assembly program [35–37]. In this study, we compared the quality of 5 assemblers (Trinity, Osease, SOAP*denovo*, Edena, and Cap3) and then used the optimal combined strategy to construct the *E. dunnii* transcriptome database. When the reads were assembled using Trinity, Cap3, Edena, Oases, and SOAP *de novo*, the N50 (contig ≥300 bp) values were 2,827 bp, 2,551 bp, 1,368 bp, 1,838 bp, and 1,336 bp, respectively (Table S1). Although the accuracy and sensitivity of the contigs assembled by Trinity were the highest compared to the other assemblers, the assembly strategy still needs further optimization to obtain higher accuracy and sensitivity (Table S1). Different assembly software programs used different algorithms, such as the traditional OLC approach of the Edena assembler and the de Bruijn graph approach of the Oases and SOAP *de novo* assemblers [38]. For a particular species, these different algorithms have multiple advantages and disadvantages, which should be taken into account when selecting the most suitable assembler to complete the process of *de novo* assembly in different species. However, neither Trinity nor any other assembler is individually capable of assembling the results satisfactorily. When assembling the sweet potato transcriptome, Tao *et al.* [31] found that only 80% of the reads mapped back to contigs assembled by Trinity, implying that approximately 20% of the reads were not used effectively in the assembly process. In addition, sequencing quality, which is the foundation for obtaining an ideal assembly, should be improved. Xiao *et al.* [38] found that trimming all raw read sequences at the 3′-end and merging the assemblies from different assemblers significantly improved assembly outcome. Some researchers have also suggested that combining data produced by two or more sequencing methods, such as Illumina sequencing and 454 sequencing, could generate a more satisfactory assembly [39]. Combined assemblies use different assembling software and/or different assembling parameters, which means they benefit from the advantages of different software packages.

To date, there have been no standard criteria to evaluate the quality of transcriptome assemblies. Researchers appraise the quality of an assembly mainly by examining the data distribution of the assembly [40,41]. Besides the data distribution, we assessed the assembly quality using numerous metrics. Due to the lack of genomic resources for *E. dunnii*, we downloaded Eucalyptus genes with full-length from GenBank to use as reference sequences. The overlapping high-scoring segment pairs (HSPs) were only calculated once to determine the sensitivity. For each individual assembly, Trinity achieved higher sensitivity than Cap3, Oases and SOAP *de novo*. However, the final assembly generated by Edenam exhibited the highest sensitivity, which was slightly higher than that of Trinity (Table S1). To calculate the accuracy, we considered all unmatched components to be false positives, and Trinity exhibited a greatest accuracy. Taken together, the results from the above metrics indicate that our final assembly quality is optimal.

DETs involving in proline metabolism and quinoline alkaloid biosynthesis

Free proline in plant cells can significantly improve cold resistance [42], as it acts as a type of osmotic adjuster that can reduce the cell freezing point and stabilize intracellular water. Furthermore, free proline can also protect the cell from excessive dehydration and lipid peroxidation [42]. The accumulation of proline is frequently associated with whole plant tolerance to chilling and other stresses [6]. In this study, we observed that the free proline content was increased more than two-fold after 48 h cold treatment (Fig. 1), which were consistent with that accumulation pattern in Arabidopsis and cassava [6,43].

KEGG analysis showed that the 'arginine and proline metabolism' pathway was significantly enriched (Table S5) in *E. dunnii* during cold-stress. A total of 576 transcripts were involved in the 'arginine and proline metabolism' pathway, with 79 transcripts being up-regulated in response to cold stress at 24 h (Tables S6, S2).

In higher plants, proline can be synthesized via the glutamate (Glu) pathway or the ornithine (Orn) pathway, depending on the initial substrate [44,45]. P5CS (delta 1-pyrroline-5-carboxylate synthetase), a key enzyme in the Glu pathway, functions as a bifunctional enzyme to transform Glu to GSA [46]. The accumulation of free proline could improve the ability of stress resistance in many plants, which regulated by the expression of *p5cs* [47,48]. In *E. dunnii*, 3 transcripts were annotated as *p5cs*, and all three transcripts were up-regulated, particularly contig_6788, whose expression increased more than 10-fold when the plants were exposure to low temperature (Fig. 7 and Table S2). This transcriptome result correlated well with the change in free proline content, suggesting that at 4°C, the Glu pathway was activated to increase the free proline content to protect the plant against cold stress. δ-OAT (ornithine-oxo-acid transaminase) is a key enzyme in the Orn pathway that catalyzes the transformation of L-Orn into GSA. Because δ-OAT can catalyze arginine to glutamate, it could be involved in proline synthesis and accumulation [49]. However, we only identified two contigs (contig_6006 and contig_60065) annotated as δ-OATin in the *E. dunnii* transcriptome, and neither was up- or down-regulated in response to cold stress (Fig. 7 and Table S2). Based on the expression profiles of these transcripts, we hypothesize that the Orn pathway may play a less important role than the Glu pathway during cold acclimation or that it may represent an alternative pathway for cold acclimation in *E. dunnii*.

Free proline accumulation is affected not only by the proline biosynthesis pathway but also by the proline degradation pathway. Under normal conditions, free proline functions as a feedback regulator to inhibit *p5cs* expression and concurrently induce ProDH (Proline dehydrogenase) gene expression. In contrast to the normal condition, *p5cs* expression is hyperactive during cold acclimation, whereas ProDH expression is inhibited, resulting in the accumulation of more and more free proline in plant cells. In *Arabidopsis* and other plants, proline levels are mainly determined by balance of biosynthetic and catabolic pathways, controlled by P5CS and ProDH genes, respectively [6]. Nanjo *et al.* found that proline degradation was inhibited in *Arabidopsis* transformed with *At*ProDH [43], suggesting that free proline levels increased in leaves.

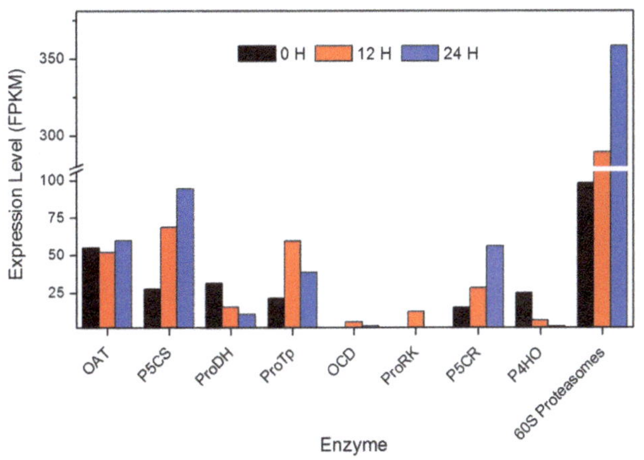

Figure 7. The expression level of some key enzymes involved in the 'arginine and proline metabolism' pathway during cold acclimation. Both the up-regulated expression of OAT (ornithine-oxo-acid transaminase), P5CS (pyrroline-5-carboxylatesynthase), ProTp (proline transporter), ProRK (proline-rich receptor protein kinase), P5CR (pyrroline-5-carboxylate reductase), and the down-regulated expression of ProDH (proline dehydrogenase), P4HO (prolyl 4-hydroxylase), OCD (ornithine cyclodeaminase) could result in proline accumulation.

Secondary metabolism and its products are also involved in the response to various stresses in plants, representing a process that formed over a long evolutionary period [50–52]. There is some evidence that secondary metabolic products and environmental factors (biotic and abiotic) are closely linked, as in the case of alkaloids, which play an important role in resisting insects and herbivores via chemical defense mechanism [53]. In addition to 'arginine and proline metabolism', the DETs were significantly enriched in 'quinoline alkaloid biosynthesis' pathway during cold acclimation, based on KEGG pathway analysis (Table S5). Early in the cold stress period (0–6 h), 40% of transcripts related to quinoline alkaloid metabolism were up-regulated more than 2-fold compared to the 0 h time point (Table S2), including contig_65006 and contig_65485. This suggests that the up-regulation of transcripts in response to low temperatures may play a crucial role in plant stress tolerance. However, when the duration of cold stress exceeded 6 h, the expression levels of these up-regulated transcripts decreased gradually, dropping to their initial levels(i.e., comparable to their expression at 0 h) by 24 h (Table S2). This suggests that there may be a relationship between quinoline alkaloid biosynthesis and abiotic factors, although this relationship may not be as simple and direct as the relationship between the biological environment and chemical defense [54,55]. Further research is needed to explore this relationship in depth. Many researchers believe that plants produce secondary metabolites such as alkaloids at the cost of slower growth [56,57]. However, when biotic and abiotic stresses become severe enough to affect their survival, the plants have no choice but to produce some secondary metabolites for protection against such rigorous stress conditions.

'Response to stress' and 'translational initiation' response to low temperature

Under the GO category 'Biological process', the terms 'response to stress' and 'translational initiation' accounted for1.76% of the total 198,528 GOs, but the DETs accounted for 15.4% of the transcripts involved in these GO terms. Additionally, both of two GO terms were significantly enriched in four comparison sets (0 h vs.3 h, 0 h vs. 6 h, 0 h vs. 12 h, and 0 h vs. 24 h) according to the GO enrichment analysis (Table S4). The largest proportion of the 'Biological process' terms included the 'metabolic process' (30.07%), 'cellular process' (27.99%), and 'biological regulation' (5.75%) (Table S4), indicating comprehensive changes in *E. dunnii* gene expression before and after the cold stress. However, although only a few transcripts were identified as belonging to 'response to cold stress', as up-term of 'response to stress', these transcripts represented the most important components that are directly involved in protecting plants from cold stress. A total of 50 transcripts were annotated under this term based on GO categorization, and most were up-regulated in response to low temperature treatment. In particular, 26 transcripts involved in the 'response to cold stress' were not expressed under normal conditions but were induced by exposure to low temperature (Table S7 and Fig. S1). ROS scavenging enzymes, including catalase (CAT), superoxide dismutase (SOD), and glutathione transferase (GST), have been demonstrated to play key roles in the removal of ROS [17,58–61]. During exposure to low temperature, the CAT activity was increased (from 0.34 to 0.56 U/g *Fw*), which was in accordance with the expression level of the corresponding transcripts in the *E. dunnii* transcriptome (Figs. 1, and 7).

Although the expression of peroxidases such as CAT and SOD increased significantly as the duration of cold exposure increased, these enzymes were still unable to completely clear the increased levels of H_2O_2, resulting in a significant increase in the amount of H_2O_2 during cold acclimation (Fig. 1). In this study, we found that some genes (e.g., MAP kinase and TCH2; Table S2) that are known to be involved in the response to other stresses (including salinity, heat and drought) in other plants are also involved in cold acclimation, which could support the hypothesis that the same gene have different functions in different plants.

The GO term 'translational initiation' was enriched in response to cold acclimation. A total of 254 transcripts were annotated under in this term, and 89 exhibited a greater than two-fold change in expression during the low temperature treatment (Table S8). Translation initiation in eukaryotes depends on many eukaryotic initiation factors (eIFs) that stimulate both the recruitment of the initiator tRNA, Met-tRNAiMet, and mRNA to the 40S ribosomal subunit and the subsequent scanning of the mRNA for the AUG start codon [62–64]. The largest of these initiation factors, the eIF-3 complex, organizes a web of interactions among several eIFs that assemble on the 40S subunit and participate in the different reactions involved in translation [62,65]. In plants, eIF-3plays the role of the central protein and interacts with many other translation initiation factors, such as eIF-4F, eIF-4G, eIF-4B, and eIF-1A [66]. Among the 89 contigs we identified that were annotated as 'translational initiation', 18 containedeIF-3 (eukaryotic translation initiation factor 3), and almost all were up-regulated during cold acclimation (Table S8). Daniel *et al.* [67] found that the expression level and phosphorylation state of these factors described above is subject to alteration during development, environmental stress (e.g., heat shock, and starvation), or viral infection. Tuteja [68] evaluated the roles of translation initiation, transcription factors, protein kinases, free proline, and reactive oxygen species in plant stress tolerance and found that these factors typically have synergistic effects in response to stress in plants. We also found that some transcripts encoding transcription factors, protein kinases (Table S6), translation initiation factors and antioxidant enzymes were up- or down-regulated in *E. dunnii* during cold acclimation, suggesting that the plant response to cold acclimation is a complex and global process.

Cold-responsive transcription factor genes in *E. dunnii*

In *Arabidopsis*, at least 5 transcription factor families have been reported to be involved in the cold stress response process, including AP2-EREBP, MYB, NAC, bHLH and WRKY family [29]. Wang *et al.* found there were many families of transcription factor, such as bHLH family, MYB family, WRKY family, NAC family and so on, responding to cold acclimation in *C. sinensis* [16]. Meanwhile, An *et al.* identified 6 AP2-EREBP and 5 Myb transcription factors participated in the process of cold stress in treated cassava [6]. In our study, many transcripts were annotated as AP2 transcription factor based on the domain analysis. Among theses, 27 transcripts were up-regulated and 24 down-regulated under cold stress (Table S6). In present work, we tested 5 AP2 TF genes by qRT-PCR and found four were up-regulated, one was down-regulated during cold-stress, and the changing trend of the two methods was accordant (Table S3). The AP2-EREBP family plays a major role in the early stages of the cold response and is the major regulator that functions in activating cold-regulated effectors in Arabidopsis and other plants [69,6]. In Eucalyptus plant, the CBF proteins, belonging to A-1 subfamily of ERF/AP2 TF family has been reported involved in response to cold stress in *E. gunnii* and *E. globules* [19,70–72].

Besides the AP2 family, the bZIP family has also been demonstrated to be involved in the cold response in *Arabidopsis* and *C. sinensis* [69,16]. In this study, we found that bZIP family was the most enriched TF family, containing 62 genes (43 up-regulated and 9 down-regulated). Differential expression of bZIP TFs implies that other environmental or hormonal pathways may be involved in cold response in *E. dunnii*.

In addition, four novel transcription factor families (JmjC, SRF-TF, and Sigma70-like) were also identified. Although their homologous genes in other plant species have not yet been reported in response to cold stress, the expression level of these genes were markedly changed before and after cold stress, suggesting they might be specific to *E. dunnii* or attractive targets for further functional characterization in plant.

Materials and Methods

Plant materials

Eucalyptus dunnii was used in this study. The plantlets of *E. dunnii* with 10 leaves were grew in a climate-chamber, with a temperature of 25°C, 200 $\mu Em^{-2}s^{-1}$ illumination and a 14/10 h light/dark photoperiod. After eight weeks, the plants were moved into another climate-chamber with a temperature of 4°C and 200 $\mu Em^{-2}s^{-1}$ continuous illumination for cold-stress. For physiological measurement, we harvested the first two expanded leaves of these plantlets at 0, 3, 6, 12, 24 and 48 h after cold treatment, respectively. For RNA-seq, leaves from 6 plants treated by 0, 3, 6, 12, and 24 h were mixed for RNA isolation and sequencing. For test of physiological changes, leaves of plants treated by all time points were used. The harvested leaves were immediately frozen in liquid nitrogen for use.

Analysis of physiological parameters

The proline content of the leaves was analyzed using a free proline ELISA kit (Omega, Georgia, USA) according to the manufacturer's instructions. The CAT activity and H_2O_2 content were measured using a CAT ELISA kit (Omega, Georgia, USA) and a H_2O_2 ELISA kit (Omega, Georgia, USA), respectively. All measurements were performed on the platform of Epoch-ELIASA (Shmadzu, Tokyo, Japan), and all analysis were repeated three times in this study.

RNA extraction, library construction and RNA sequencing

Total RNA was isolated from the leaves by using Trizol reagent (Invitrogen, CA, USA) according to the manufacturer's instructions, and the RNAwas treated with RNase-free DNase I (TaKaRa, Dalian, China). The purity, concentration and RNA integrity number (RIN) were determined using a SMA3000 and/or Agilent 2100 Bioanalyzer. The total RNA was then sent to Beijing Genomics Institute (BGI) -Shenzhen (Shenzhen, China) for RNA sequencing.

More than 20 µg of total RNA extracted from each group of plants exposed to low temperatures (n>3) was used to construct the cDNA libraries. First, the polyadenylated RNAs (mRNAs) were purified and retrieved using magnetic beads coated with a poly-T oligo. These mRNAs were then mixed with fragmentation media and fragmented. The fragmented mRNAs were subjected to reverse transcription using reverse transcriptase and random primers. The second-strand cDNA synthesis was performed using DNA polymerase I and RNase H. Finally, the resulting dscDNAs were repaired by adding a single 'A' base, and specific Illumina adapters were ligated to the repaired ends. Fragments of approximately 200 bp in size were purified and retrieved from the gels. To construct the fragmented cDNA library, these fragments, which served as the template, were enriched by PCR using two primers that annealed to the ends of the adapters. The cDNA libraries constructed above were sequenced using an Illumina Hiseq2000. The PE read information and quality values were generated using the Illumina sequencing-by-synthesis, image analysis and base-calling procedures.

Denovo assembly and functional annotation

Sequencing quality was assessed using fastQC software [http://www.bioinformatics.bbsrc.ac.uk/projects/fastqc/], and the PE reads were *de novo* assembled by five different assemblers: the Trinity software package (v2013-02-25) [73] with default parameters, the Oases software package (v0.1.21) [74] with a different K-value, the Edenam software package (v2013-07-15) [75] with a different M-value, the SOAP *de novo* software package (v2013-07-15) [76] with different K- and P-values, and the Cap3 software package (v12.07.21) [77] with default parameters. To evaluate the quality of the assemblies produced by the different assemblers, the PE reads were aligned back to the contigs assembled by a different assembler using Bowtie2 software (v2.0.0) [78], and the alignment rate was calculated. Subsequently, we analyzed the length distribution information of these contigs, such as the N50 number, average length, max length and total contig number, using common Perl scripts. Due to the lack of genomic information for Eucalyptus, 535 Eucalyptus sequences containing complete CDSs were downloaded from GenBank [http://www.ncbi.nlm.nih.gov/] and used as reference sequences to calculate the sensitivity and accuracy. Furthermore, we analyzed the best candidate coding sequence (CDS) for each contig from different assemblers and obtained the ratios of long CDS-containing transcripts to contigs with corresponding lengths.

All of the contigs ($\geq$300 bp) produced by the Trinity software package were subjected to a similarity search against the NR database downloaded from GenBank utilizing local NCBI-BLAST software (v2.2.28+). The BLASTx searches were performed using a threshold E-value of $<10^{-3}$, max_target_seqs of 5, and an xml output file format. The BLASTx results were imported into Blast2GO software (v2.6.7) [79], and local functional annotation was performed. Enzyme codes, gene ontology (GO), and Kyoto Encyclopedia of Genes and Genomes (KEGG) pathways were retrieved from the KEGG web server (http://www.genome.jp/kegg/) [80]. GO classification [81] was performed using the WEGO program (http://wego.genomics.org.cn/cgibin/wego/index.pl) [82].

Differential expression profiling and enrichment

To investigate the expression level of each transcript at the five treatment time points, the PE reads for each sample were aligned back to the optimal assembly result (assembled by the Trinity assembler) using Perl scripts provided by the Trinity software package. Using these scripts, we obtained the digital expression levels of each transcript and normalized these data with a RESM-based algorithm to obtain the FPKM (Fragments per Kilobase per Million Mapped Fragments) values of each transcript. Based on the normalized expression profiles, the effect and bias introduced by library size and/or RNA composition were eliminated using edgeR [83], and significant differentially expressed transcripts (DETs) were identified with a P.ad-value $\leq$0.05 and $\log_2$ fold-change ($\log_2$ FC) $\geq$1.

The DET enrichment analysis was performed using the common Perl and R scripts. We first counted the number of transcripts involved in each KEGG pathway from the Trinity assembled contigs and/or DETs. Based on the transcript numbers in the contig database and DETs, we determined the enriched KEGG pathway. Then, the P-value was adjusted by the Bernoulli equation, and a P.ad-value<0.05 was the threshold value for significant enrichment results. We applied a similar approach for the GO enrichment analysis.

Expression level verification

To verify the reliability and accuracy of the NGS-based expression level analysis, we randomly selected 31 transcripts from the contig database and evaluated the expression profiles among the five samples using quantitative real-time PCR. The primers for these transcripts are listed in Table S5. The first-strand cDNA was synthesized from 500 ng of total RNA using oligo (dT), random hexamers, and Moloney murine leukemia virus (M-MLV) reverse transcriptase (Invitrogen, CA, USA) according to the manufacturer's instructions. The real-time PCR was performed using the IQ5 Real-Time PCR System (Bio-Rad, CA, USA) in a total volume of 20 μL containing 100 ng of cDNA template, 1× SYBR *Premix Ex Taq*™II (Perfect Real Time, TaKaRa), and 400 nM of each primer. Serial dilutions of each cDNA were used to generate a quantitative PCR standard curve to calculate the corresponding PCR efficiencies. The PCR conditions were as follows: initial denaturation at 95°C for 30 s, followed by 40 cycles of denaturation at 95°C for 5 s, primer annealing at 60°C for 30 s, and DNA extension at 72°C for 30 s. Two most commonly used reference genes, beta-actin and GAPDH, were selected for internal controls. Three biological replicates were used, and melting curve analysis was performed to check the amplification specificity. The relative expression levels were calculated using the BIO-RAD IQ5 standard edition Optical System software (version 2.1) and a normalized expression (ddCt) model.

Supporting Information

Figure S1 Differences of 'response to stimulus' (A), 'response to cold' (B), 'transcription factor activity' (C) and 'kinase regulator activity' (D) between each pair of samples. Overlap examinations were performed based on the resulting gene lists of four comparisons by VENNY. Overlap among four groups, D0 vs D3 (blue), D0 vs D6 (yellow), D0 vs D12 (yellow) and D0 vs D24 (red), were shown here.

Table S1 The characteristics of contig databases assembled by different assembler.

Table S2 Sequence annotations of *E. dunnii* transcripts and gene expression profiling of five samples.

Table S3 Comparison of expression patterns between RNA-Seq expression and qRT-PCR.

Table S4 GO classification of *E. dunnii* teanscriptome and differentially.

Table S5 KEGG classification of *E. dunnii* teanscriptome and differentially expressed transcripts indentified between each pairs comparisons.

Table S6 Transcription factor and kinase of *E. dunnii* response to low-temperature stress.

Table S7 Expression patterns of some transcripts involved in 'Response to cold'.

Table S8 Expression patterns of some transcripts involved in 'translational initiation'.

Acknowledgments

The authors are extremely grateful to Professor Yizheng Zhang (Sichuan University) for technical advice and assistance in data processing.

Author Contributions

Conceived and designed the experiments: YQL JPG. Performed the experiments: YQL YSJ JBL YZ. Analyzed the data: YSJ YQL JPG. Contributed reagents/materials/analysis tools: YQL YSJ. Wrote the paper: YSJ.

References

1. Leslie AD, Mencuccini M, Perks M (2012) The potential for Eucalyptus as a wood fuel in the UK. Appl Energ 89: 176–182.
2. Munoz F, Valenzuela P, Gacitua W (2012) Eucalyptus nitens: nanomechanical properties of bark and wood fibers. Appl Phys A-Mater 108: 1007–1014.
3. Brawner JT, Lee DJ, Meder R, Almeida AC, Dieters MJ (2013) Classifying genotype by environment interactions for targeted germplasm deployment with a focus on Eucalyptus. Euphytica 191: 403–414.
4. Gomat HY, Deleporte P, Moukini R, Mialounguila G, Ognouabi N, et al. (2011) What factors influence the stem taper of Eucalyptus: growth, environmental conditions, or genetics? Ann For Sci 68: 109–120.
5. Sands PJ, Landsberg JJ (2002) Parameterisation of 3-PG for plantation grownEucalyptus *globulus*. Forest Ecol Manage 163: 273–292.
6. An D, Yang J, Zhang P (2012) Transcriptome profiling of low temperature-treated cassava apical shoots showed dynamic responses of tropical plant to cold stress. BMC Genomics 13: 64.
7. Pennycooke JC, Cox S, Stushnoff C (2005) Relationship of cold acclimation, total phenolic content and antioxidant capacity with chilling tolerance in petunia (*Petuniax hybrida*). Environ Exp Bot 53: 225–232.
8. Zhao ZG, Tan LL, Dang CY, Zhang H, Wu QB, et al. (2012) Deep-sequencing transcriptome analysis of chilling tolerance mechanisms of a subnival alpine plant, Chorispora bungeana. BMC Plant Biol 12: 222.
9. Fernandes J, Morrow DJ, Casati P, Walbot V (2008) Distinctive transcriptome responses to adverse environmental conditions in *Zea mays* L. Plant Biotechnol J 6: 782–798.
10. Ziemann M, Kamboj A, Hove RM, Loveridge S, El-Osta A, et al. (2013) Analysis of the barley leaf transcriptome under salinity stress using mRNA-Seq. Acta Physiol Plant 35: 1915–1924.
11. He XD, Li FG, Li M, Weng QJ, Shi JS, et al. (2012) Quantitative genetics of cold hardiness and growth in Eucalyptus as estimated from *E. urophylla* x *E. tereticornis* hybrids. New Forest 43: 383–394.
12. Yang QS, Wu JH, Li CY, Wei YR, Sheng O, et al. (2012) quantitative proteomic analysis reveals that antioxidation mechanisms contribute to cold tolerance in plantain (*Musa paradisiaca* L.; ABB Group) seedlings. Mol Cell Proteomics 11: 1853–1869.
13. Tatusov RL, Koonin EV, Lipman DJ (1997) A genomic perspective on protein families. Science 278: 631–637.
14. O'Rourke JA, Yang SS, Miller SS, Bucciarelli B, Liu JQ, et al. (2013) An RNA-Seq transcriptome analysis of orthophosphate-deficient white lupin reveals novel insights into phosphorus acclimation in plants. Plant Physiol 161: 705–724.
15. Wang XC, Yang YJ (2003) Research progress on resistance breeding of tea plant. J Tea Sci 23: 94–98.
16. Wang XC, Zhao QY, Ma CL, Zhang ZH, Cao HL, et al. (2013) Global transcriptome profiles of *Camellia sinensis* during cold acclimation. BMC Genomics 14: 415.
17. Torres R, Teixidó N, Usall J, Abadias M, Mir N, et al. (2011) Anti-oxidant activity of oranges after infection with the pathogen Penicillium digitatum or treatment with the biocontrol agent Pantoea agglomerans CPA-2. Biol Control 57: 103–109.
18. Fernandez M, Villarroel C, Balbontin C, Valenzuela S (2010) Validation of reference genes for real-time qRT-PCR normalization during cold acclimation in *Eucalyptus globulus*. Trees-Struct Funct 24: 1109–1116.
19. Navarro M, Ayax C, Martinez Y, Laur J, El Kayal W, et al. (2011) Two EguCBF1 genes overexpressed in Eucalyptus display a different impact on stress tolerance and plant development. Plant Biotechnol J 9: 50–63.
20. Ponciano G, McMahan CM, Xie WS, Lazo GR, Coffelt TA, et al. (2012) Transcriptome and gene expression analysis in cold-acclimated guayule (*Parthenium argentatum*) rubber-producing tissue. Phytochemistry 79: 57–66.
21. Liu SH, Wang NF, Zhang PY, Cong BL, Lin XZ, et al. (2013) Next-generation sequencing-based transcriptome profiling analysis of *Pohlia nutans* reveals insight into the stress-relevant genes in Antarctic moss. Extremophiles 17: 391–403.
22. Fernandez M, Aguila SV, Arora R, Chen KT (2012) Isolation and characterization of three cold acclimation-responsive dehydrin genes from *Eucalyptus globulus*. Tree Genet Genomes 8: 149–162.
23. Fernandez M, Valenzuela S, Barraza H, Latorre J, Neira V (2012) Photoperiod, temperature and water deficit differentially regulate the expression of four dehydrin genes from *Eucalyptus globulus*. Trees-Struct Funct 26: 1483–1493.
24. Li XY, Sun HY, Pei JB, Dong YY, Wang FW, et al. (2012) *De novo* sequencing and comparative analysis of the blueberry transcriptome to discover putative genes related to antioxidants. Gene 511: 54–61.
25. Rodriguez MC, Edsgard D, Hussain SS, Alquezar D, Rasmussen M, et al. (2010) Transcriptomes of the desiccation-tolerant resurrection plant *Craterostigma plantagineum*. Plant J 63: 212–228.
26. Bai X, Rivera-Vega L, Mamidala P, Bonello P, Herms D A, et al. (2011) Transcriptomic signatures of ash (*Fraxinus spp.*) phloem. PLoS One 6: e16368.
27. Dussert S, Guerin C, Andersson M, Joet T, Tranbarger TJ, et al. (2013) Comparative transcriptome analysis of three oil palm fruit and seed tissues that differ in oil content and fatty acid composition. Plant Physiol 162: 1337–1358.
28. Hua CW, Marcal S, Hong Y, Eduardo L, Victor C, et al. (2012) Reference Genes for High-Throughput Quantitative ReverseTranscription-PCR Analysis of Gene Expression in Organs andTissues of Eucalyptus Grown in Various EnvironmentalConditions. Plant Cell Physiology 53(12): 2101–2116.
29. Feng BM, Lu DH, Ma X, Peng YB, Sun J, et al. (2012) Regulation of the Arabidopsis anther transcriptome by DYT1 for pollen development. Plant J 72: 612–624.
30. Ragusa M, Statello L, Maugeri M, Majorana A, Barbagallo D, et al. (2012) Specific alterations of the microRNA transcriptome and global network structure in colorectal cancer after treatment with MAPK/ERK inhibitors. J Mol Med-JMM 90: 1421–1438.
31. Tao X, Gu YH, Wang HY, Zheng W, Li X, et al. (2012) Digital gene expression analysis based on integrated *de novo* transcriptome assembly of sweet potato [*Ipomoea batatas* (L.) Lam.]. PLoS One 7: e36234.
32. Birol I, Jackman SD, Nielsen CB, Qian JQ, Varhol R, et al. (2009) *De novo* transcriptome assembly with ABySS. Bioinformatics 25: 2872–2877.
33. Wang L, Li PH, Brutnell TP (2010) Exploring plant transcriptomes using ultra high-throughput sequencing. Brief Funct Genomics 9: 118–128.
34. Wu HL, Chen D, Li JX, Yu B, Xiao XY, et al. (2013) *De Novo* characterization of leaf transcriptome using 454 sequencing and development of EST-SSR markers in tea (*Camellia sinensis*). Plant Mol Biol Rep 31: 524–538.
35. Schafleitner R, Tincopa LR, Palomino O, Rossel G, Robles RF, et al. (2010) A sweet potato gene index established by *de novo* assembly of pyrosequencing and Sanger sequences and mining for gene-based microsatellite markers. BMC Genomics 11: 604.
36. Shi CY, Yang H, Wei CL, Yu O, Zhang ZZ, et al. (2011) Deep sequencing of the Camellia sinensis transcriptome revealed candidate genes for major metabolic pathways of tea-specific compounds. BMC Genomics 12: 131.
37. Duan J, Xia C, Zhao G, Jia J, Kong X (2012) Optimizing *de novo* common wheat transcriptome assembly using short-read RNA-Seq data. BMC Genomics 13: 392.
38. Xiao M, Zhang Y, Chen X, Lee EJ, Barber CJS, et al. (2013) Transcriptome analysis based on next-generation sequencing of non-model plants producing specialized metabolites of biotechnological interest. J Biotechnol 166: 122–134.
39. Ong WD, Voo LYC, Kumar VS (2012) *De novo* assembly, characterization and functional annotation of pineapple fruit transcriptome through massively parallel sequencing. PLoS One 7: e46937.
40. Paszkiewicz K, Studholme DJ (2010) *De novo* assembly of short sequence reads. Brief Bioinform 11: 457–472.
41. Verma P, Shah N, Bhatia S (2013) Development of an expressed gene catalogue and molecular markers from the *de novo* assembly of short sequence reads of the lentil (*Lens culinaris Medik.*) transcriptome. Plant Biotechnol J 11: 894–905.
42. Bates LS, Waldren RP, Teare ID (1973) Rapid determination of free proline for water-stress studies. Plant Soil 39: 205–207.
43. Nanjo T, Kobayashi M, Yoshiba Y, Kakubari Y, Yamaguchi-Shinozaki K, et al. (1999) Antisense suppression of proline degradation improves tolerance to freezing and salinity in *Arabidopsis thaliana*. Febs Letters 461: 205–210.
44. Kishor P, Hong Z, Miao GH, Hu C, Verma D (1995) Overexpression of [delta]-pyrroline-5-carboxylate synthetase increases proline production and confers osmotolerance in transgenic plants. Plant Physiol 108: 1387–1394.
45. Delauney AJ, Verma DPS (2002) Proline biosynthesis and osmoregulation in plants. Plant J 4: 215–223.
46. Forlani G, Scainelli D, Nielsen E (1997) [delta]1-pyrroline-5-carboxylate dehydrogenase from cultured cells of potato (purification and properties). Plant Physiol 113: 1413–1418.
47. Verslues PE, Sharma S (2010) Proline metabolism and its implications for plant-environment interaction. Arabidopsis Book 8: e0140.

48. Ábrahám E, Rigó G, Székely G, Nagy R, Koncz C, et al. (2003) Light-dependent induction of proline biosynthesis by abscisic acid and salt stress is inhibited by brassinosteroid in Arabidopsis. Plant Mol Biol 51: 363–372.
49. Delauney AJ, Hu CA, Kishor PB, Verma DP (1993) Cloning of ornithine delta-aminotransferase cDNA from Vigna aconitifolia by trans-complementation in *Escherichia coli* and regulation of proline biosynthesis. J Biol Chem 268: 18673–18678.
50. Tena G, Boudsocq M, Sheen J (2011) Protein kinase signaling networks in plant innate immunity. Curr Opin Plant Biol 14: 519–529.
51. Goossens A, Hakkinen ST, Laakso I, Seppanen-Laakso T, Biondi S, et al. (2003) A functional genomics approach toward the understanding of secondary metabolism in plant cells. Proc Natl Acad Sci U S A 100: 8595–8600.
52. Vom Endt D, Kijne JW, Memelink J (2002) Transcription factors controlling plant secondary metabolism: what regulates the regulators? Phytochemistry 61: 107–114.
53. Constabel CP, Lindroth RL (2010) The impact of genomics on advances in herbivore defense and secondary metabolism in Populus. In: Jansson S, Bhalerao R, Groover A, editors. Genetics and genomics of Populus, vol 8. London: Springer. pp. 279–305.
54. Mithofer A, Boland W (2012) Plant defense against herbivores: chemical aspects. Annu Rev Plant Biol 63: 431–450.
55. Osbourn A (2010) Secondary metabolic gene clusters: evolutionary toolkits for chemical innovation. Trends Genet 26: 449–457.
56. Dixon DP, Skipsey M, Edwards R (2010) Roles for glutathione transferases in plant secondary metabolism. Phytochemistry 71: 338–350.
57. Wu B, Li Y, Yan HX, Ma YM, Luo HM, et al. (2012) Comprehensive transcriptome analysis reveals novel genes involved in cardiac glycoside biosynthesis and mlncRNAs associated with secondary metabolism and stress response in *Digitalis purpurea*. BMC Genomics 13: 15.
58. Heller J, Tudzynski P (2011) Reactive oxygen species in phytopathogenic fungi: signaling, development, and disease. Annu Rev Phytopathol 49: 369–390.
59. Reape TJ, McCabe PF (2010) Apoptotic-like regulation of programmed cell death in plants. Apoptosis 15: 249–256.
60. Baxter A, Mittler R, Suzuki N (2013) ROS as key players in plant stress signalling. J Exp Bot 65: 1229–1240.
61. Dey S, Ghose K, Basu D (2010) Fusarium elicitor-dependent calcium influx and associated ros generation in tomato is independent of cell death. Eur J Plant Pathol 126: 217–228.
62. Hinnebusch AG (2006) eIF3: a versatile scaffold for translation initiation complexes. Trends Biochem Sci 31: 553–562.
63. Sizova DV, Kolupaeva VG, Pestova TV, Shatsky IN, Hellen CU (1998) Specific interaction of eukaryotic translation initiation factor 3 with the 5' nontranslated regions of hepatitis C virus and classical swine fever virus RNAs. J Virol 72: 4775–4782.
64. Dever TE (2002) Gene-specific regulation by general translation factors. Cell 108: 545–556.
65. Maquat LE, Tarn WY, Isken O (2010) The pioneer round of translation: features and functions. Cell 142: 368–374.
66. Preiss T, Hentze MW (2003) Starting the protein synthesis machine: eukaryotic translation initiation. Bioessays 25: 1201–1211.
67. Gallie DR, Le H, Caldwell C, Robert LT, Nam XH, et al. (1997) The phosphorylation state of translation initiation factors is regulated developmentally and following heat shock in wheat. J Biol Chem 272: 1046–1053.
68. Tuteja N (2007) Mechanisms of high salinity tolerance in plants. Methods Enzymol 428: 419–438.
69. Lee BH, Henderson DA, Zhu J-K (2005) The Arabidopsis Cold-Responsive Transcriptome and Its Regulation by ICE1. Plant Cell17: 3155–3175.
70. Kayal WE, Navarro M, Marque G (2006) Expression profile of CBF-like transcriptional factor genes from Eucalyptus in response to cold. Journal Experiment Botany57: 2455–2469.
71. Navarro M, Marque G, Ayax C (2009) Complementary regulation of four Eucalyptus CBF genes under various cold conditions. Journal Experiment Botany 60: 2713–2724.
72. Fernandez M, Aguila SV, Arora R (2012) Isolation and characterization of three cold acclimation-responsive dehydrin genes from Eucalyptus globulus. TREE GENETICS & GENOMES, 8: 149–155.
73. Grabherr MG, Haas BJ, Yassour M, Levin JZ, Thompson DA, et al. (2011) Full-length transcriptome assembly from RNA-Seq data without a reference genome. Nature Biotechnol 29: 644–652.
74. Schulz MH, Zerbino DR, Vingron M, Birney E (2012) Oases: robust *de novo* RNA-seq assembly across the dynamic range of expression levels. Bioinformatics 28: 1086–1092.
75. Hernandez D, Tewhey R, Veyrieras JB, Farinelli L, Østerås M, et al. (2014) *De novo* finished 2.8 Mbp Staphylococcus aureus genome assembly from 100 bp short and long range paired-end reads. Bioinformatics 30: 40–49.
76. Luo R, Liu B, Xie Y, Li Z, Huang W, et al. (2012) SOAP*denovo*2: an empirically improved memory-efficient short-read *de novo* assembler. Gigascience 1: 18.
77. Huang X, Madan A (1999) CAP3: a DNA sequence assembly program. Genome Res 9: 868–877.
78. Langmead B, Trapnell C, Pop M, Salzberg SL (2009) Ultrafast and memory-efficient alignment of short DNA sequences to the human genome. Genome Biol 10: R25.
79. Conesa A, Gotz S (2008) Blast2GO: a comprehensive suite for functional analysis in plant genomics. Int J Plant Genomics 2008. doi: 10.1155/2008/619832
80. Kanehisa M, Goto S (2000) KEGG: kyoto encyclopedia of genes and genomes. Nucleic Acids Res 28: 27–30.
81. Arasan SK, Park JI, Ahmed NU, Jung HJ, Lee IH, et al. (2013) Gene ontology based characterization of expressed sequence tags (ESTs) of *Brassica rapa cv. Osome*. Indian J Exp Biol 51: 522–530.
82. Ye J, Fang L, Zheng H, Zhang Y, Chen J, et al. (2006) WEGO: a web tool for plotting GO annotations. Nucleic Acids Res 34: W293–297.
83. Robinson MD, McCarthy DJ, Smyth GK (2010) edgeR: a Bioconductor package for differential expression analysis of digital gene expression data. Bioinformatics 26: 139–140.

19

Functional Potential of Soil Microbial Communities in the Maize Rhizosphere

Xiangzhen Li[1,2]❡, Junpeng Rui[3]❡, Jingbo Xiong[4], Jiabao Li[2], Zhili He[5], Jizhong Zhou[5], Anthony C. Yannarell[6], Roderick I. Mackie[1,7]*

1 Energy Biosciences Institute, University of Illinois at Urbana-Champaign, Urbana, IL, United States of America, **2** Key Laboratory of Environmental and Applied Microbiology, CAS, Environmental Microbiology Key Laboratory of Sichuan Province, Chengdu Institute of Biology, Chinese Academy of Sciences, Chengdu, Sichuan, China, **3** Key Laboratory of Mountain Ecological Restoration and Bioresource Utilization, CAS, Ecological Restoration Biodiversity Conservation Key Laboratory of Sichuan Province, Chengdu Institute of Biology, Chinese Academy of Sciences, Sichuan, PR China, **4** School of Marine Sciences, Ningbo, Zhejiang, China, **5** Institute for Environmental Genomics and Department of Microbiology and Plant Biology, University of Oklahoma, Norman, OK, United States of America, **6** Department of Natural Resources and Environmental Sciences, University of Illinois at Urbana-Champaign, Urbana, IL, United States of America, **7** Department of Animal Sciences, University of Illinois at Urbana-Champaign, Urbana, IL, United States of America

Abstract

Microbial communities in the rhizosphere make significant contributions to crop health and nutrient cycling. However, their ability to perform important biogeochemical processes remains uncharacterized. Here, we identified important functional genes that characterize the rhizosphere microbial community to understand metabolic capabilities in the maize rhizosphere using the GeoChip-based functional gene array method. Significant differences in functional gene structure were apparent between rhizosphere and bulk soil microbial communities. Approximately half of the detected gene families were significantly ($p<0.05$) increased in the rhizosphere. Based on the detected *gyrB* genes, Gammaproteobacteria, Betaproteobacteria, Firmicutes, Bacteroidetes and Cyanobacteria were most enriched in the rhizosphere compared to those in the bulk soil. The rhizosphere niche also supported greater functional diversity in catabolic pathways. The maize rhizosphere had significantly enriched genes involved in carbon fixation and degradation (especially for hemicelluloses, aromatics and lignin), nitrogen fixation, ammonification, denitrification, polyphosphate biosynthesis and degradation, sulfur reduction and oxidation. This research demonstrates that the maize rhizosphere is a hotspot of genes, mostly originating from dominant soil microbial groups such as Proteobacteria, providing functional capacity for the transformation of labile and recalcitrant organic C, N, P and S compounds.

Editor: Shiping Wang, Institute of Tibetan Plateau Research, China

Funding: This work was funded by the Energy Biosciences Institute, Environmental Impact and Sustainability of Feedstock Production Program at the University of Illinois, Urbana (http://www.energybiosciencesinstitute.org/program_project/environmental-impact-and-sustainability-feedstock-production). The author RJ was also supported from the National Natural Science Foundation of China (41371268, 41301272) (http://www.nsfc.gov.cn/publish/portal1/), Strategic Priority Research Program of the Chinese Academy of Sciences (XDB15010000) (http://english.cas.cn/). The funders had no role in study design, data collection and analysis, decision to publish, or preparation of the manuscript.

Competing Interests: Xiangzhen Li is a PLOS ONE Editorial Board member. This does not alter the authors' adherence to PLOS ONE Editorial policies and criteria.

* Email: r-mackie@illinois.edu

❡ These authors contributed equally to this work.

Introduction

The rhizosphere is a unique ecological niche that shapes microbial community structure through the interactions of plant species, root exudates, soil properties, and many other factors [1,2]. Organic substances released from roots may support high microbial biomass and metabolic activity and thus the assembly of more active and distinct microbial communities in the rhizosphere as compared to those in the bulk soil. Interactions between plants and microorganisms in the rhizosphere control important biogeochemical processes, such as nutrient cycling and greenhouse gas emissions. While some pathogenic microbes undermine plant health, many beneficial microbes in the rhizosphere provide plants with mineral nutrients, phytohormones, and also help to protect the plant against phytopathogens [1,3]. Thus, characterization of the functional potential of rhizosphere microbial communities is important in order to link soil community activity to plant growth and health.

Previous studies show that the activities of many enzymes are higher in the rhizosphere than in bulk soil [4], due to higher microbial activity and enzymes secreted from roots and microorganisms. Biochemical measurements of soil enzyme activities provide general information, but little is known about the origin of various types of enzyme. With high throughput pyrosequencing-based approaches [5] and microarray-based metagenomic technologies, e.g., GeoChip [6,7], the composition, structure, and metabolic potential of microbial communities can be rapidly analyzed. For example, GeoChip 3.0 covers ca. 57,000 gene sequences in >292 gene families, allowing quantification of hundreds and thousands of functional genes from known microbes in each sample [7]. Thus, the phylogenetic nature of the data acquisition enables the assignment of metabolic capabilities to specific microbial groups in each niche. Such studies bridge the gap between microbial diversity and potential enzyme activities, which have both been previously investigated separately in the rhizosphere [1–3].

Since maize (*Zea mays*) is an important crop, comprehensive phylogenetic investigations of its rhizosphere microbiome have been conducted using high throughput pyrosequencing techniques [5,8] and other methods [9,10]. Previous research shows that the rhizosphere of maize is preferentially colonized by Proteobacteria, Bacteroidetes, and Actinobacteria [5], and these organisms may be active in utilizing plant exudates and organic polymers. However, it is important to determine which microorganisms are active in the rhizosphere and what functional genes and pathways are present.

The objective of this study, was to identify important functional genes characterizing the rhizosphere microbial community using a GeoChip-based functional gene array to understand metabolic capabilities in the maize rhizosphere. We hypothesize that the rhizosphere of maize is a hotspot of genes, mostly originating from dominant microbial groups, and providing functional capacity for the transformation of labile and recalcitrant organic C, N, P and S compounds.

Materials and Methods

Field description and sampling

Maize (*Zea mays*) was grown near the University of Illinois (Champaign, IL, USA, 40°03′N, 88°12W, 230 m elevation) in a soil classified as a fine-silty, mesic Typic Endoaquoll formed from loess and silt parent material. The field site is a typical arable soil in the U.S. Corn Belt, and it has been in continuous cultivation for over 100 years. The *Z. mays* cultivar cv 34B43 (Pioneer Hi-Bred International, Des Moines, IA, USA) was planted in May following standard agricultural practices in this region with annual applications of 20 g m^{-2} of mixed urea, ammonia, and nitrate fertilizer at planting. No specific permissions were required for research activities in this location, and the field studies did not involve endangered or protected species.

Samples were collected in August 4, 2009, which represents the early reproductive (R) growth stages (the R1–R2 or blister growth stage defined by Hanway [11]). Our rationale for sampling at this time is that it allows us to assess the cumulative growing season influence of the plant on its rhizosphere community at an important early stage in the development of the cereal corn kernels. Thus, this growth period should provide the clearest link between rhizosphere community composition/activity and yield-relevant plant health. However, it should be remembered that the microbial communities are dynamic over the growing season, and this sampling only provides a snapshot view of rhizosphere communities at this important growth stage.

To collect rhizosphere samples, plants were carefully excavated using a shovel, and the soil loosely attached to the root was removed. Fresh roots were collected into a 50 ml centrifuge tube, rinsed in phosphate buffer solution, shaken by hand and centrifuged at 8000 rpm for 10 min. Bulk soil cores of 0–30 cm depth were also collected between sampled plants for paired comparison to the rhizosphere samples. Triplicate samples were taken for both rhizosphere and bulk soil, in which each individual sample was a pool of four plants or soil cores, and each plant or soil core was on average 1 m apart. All the samples were transferred to the lab on dry ice, pre-processed and stored at −80°C for downstream applications. Soil total organic carbon and nitrogen were measured using an element analyzer.

GeoChip analysis

Soil DNA was extracted by freeze-grinding mechanical lysis as described previously [12]. To determine the abundance of functional genes in the rhizosphere and bulk soils, GeoChip 3.0 was used. This gene array covers ca. 57,000 gene sequences in > 292 gene families [7]. For GeoChip analysis, genomic DNA was first amplified using the Templiphi kit (GE Healthcare, NJ) with 100 ng community DNA as the template. Three micrograms of amplified DNA were used for each hybridization. Three technical replicates of each rhizosphere and bulk soil sample were used for microarray hybridization, for a total of 18 hybridizations. Details for template amplification, labeling and hybridization, image processing is described in detail by He et al. [7].

Microarray data were analyzed using the data analysis pipeline developed at the University of Oklahoma (http://ieg.ou.edu/microarray/). Briefly, signal intensity of good quality microarray spots was extracted. Normalized intensity of each gene was calculated by dividing the signal intensity by the mean intensity of the microarray. The matrices of microarray data resulting from our pipeline were considered as 'species' abundance in statistical analyses. Microbial diversity indices were analyzed with the Diversity Index Calculator (http://www.reading.ac.uk/ssc/software/diversity/diversity.html). Differences in signal intensity between bulk soil and rhizosphere were tested with two-tailed paired t-tests. All of the original microarray signal data were deposited at NCBI Gene Expression Omnibus (GEO) under accession number GSE61338.

Results

Overall differences of microbial communities between rhizosphere and bulk soil

A total of 6201 genes were detected by GeoChip in all samples, with 5777 genes (93.2% of total 6201 genes) enumerated in the rhizosphere and 1983 genes (32.0%) enumerated in the bulk soils. In the rhizosphere, 5390 genes were from bacteria, 103 from archaea and 246 from fungi. In the bulk soil, 1849 genes were from bacteria, 38 from archaea and 84 from fungi. The detected genes were distributed across 248 gene families, of which 53 gene families were only detected in the rhizosphere. Among them, 38 gene families were related to organic compound utilization and degradation. In addition, two gene families were only detected in the bulk soil, and both were related to organic transformation. Gene abundances of 140 gene families were significantly ($p<0.05$) increased in the rhizosphere (Table S1).

GeoChip 3.0 includes the DNA gyrase subunit B gene (*gyrB*) [7] as a phylogenetic marker to characterize the diversity, composition and structure of microbial communities. A total of 164 *gyrB* probes showed positive signals in both bulk soil and rhizosphere samples. These *gyrB* probes detected were primarily related to Gammaproteobacteria, Betaproteobacteria, Firmicutes, Bacteroidetes and Cyanobacterial phyla (Fig. 1). Signal intensities were higher in the rhizosphere than in the bulk soil. Also, the diversity of functional genes calculated from GeoChip data was significantly higher in the rhizosphere ($p<0.05$) (Table 1). This is in contrast to the microbial species diversity based on pyrosequencing of the 16S rRNA gene, which was lower in the rhizosphere than that in the bulk soil [5].

Even though the GeoChip array may have some bias in probe selection, the above results showed the maize rhizosphere

Table 1. Soil carbon and nitrogen content (%) and microbial diversity indices based on 454-pyrosequencing and GeoChip data in the rhizosphere and bulk soils.

	Bulk soil	Rhizosphere	Significant test
Total carbon	2.26±0.11	2.43±0.08	*
Total nitrogen	0.18±0.01	0.19±0.01	
454 Pyrosequencing data			
Shannon's diversity index	6.79±0.05	6.29±0.11	**
Shannon's evenness	0.94±0.01	0.90±0.01	**
GeoChip data			
Shannon's diversity index	6.56±0.45	7.86±0.30	*
Shannon's evenness	0.95±0.01	0.96±0.004	

a) Pyrosequencing data are from Li et al. [5]. Samples were collected from the same field site in different years.
b) The value is the average of triplicate samples ± standard deviation.
*p<0.05,
**p<0.01.

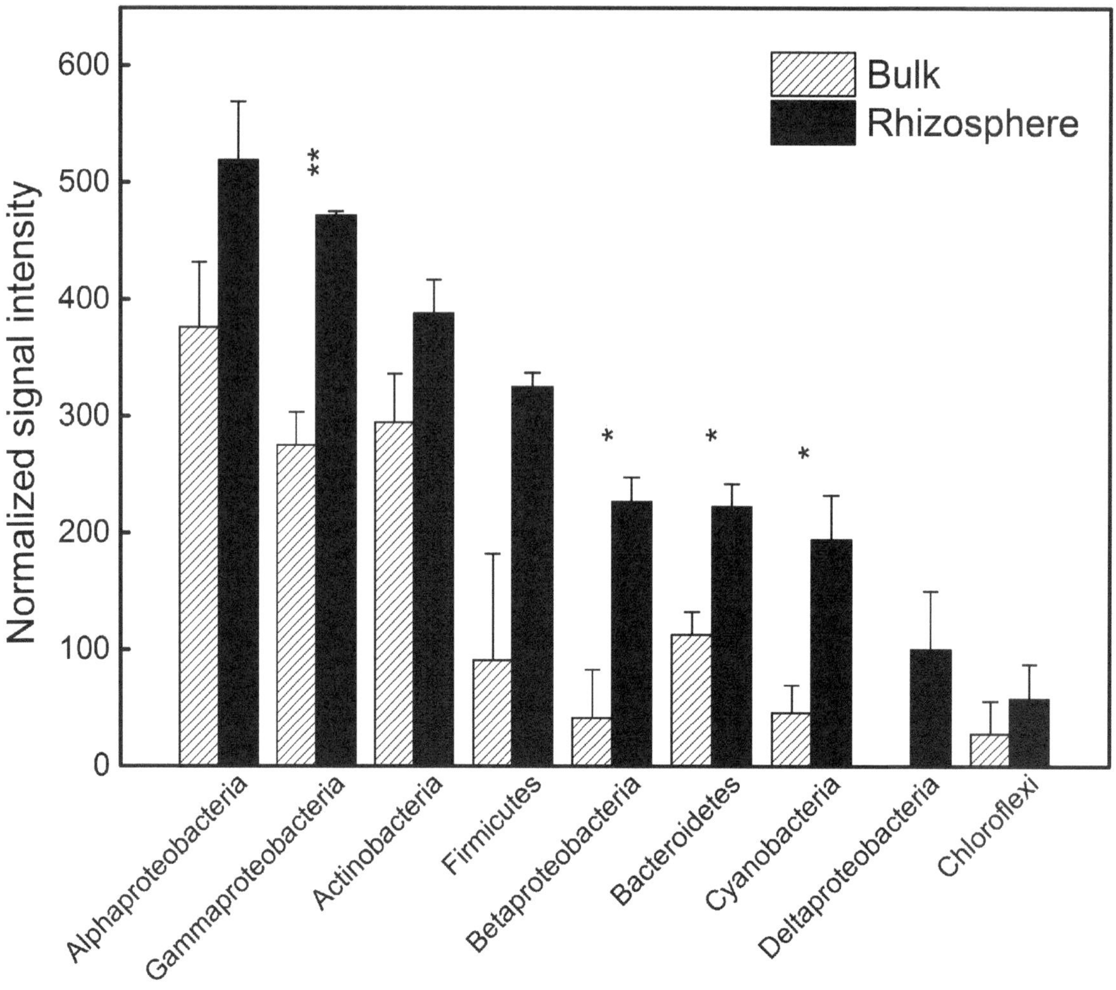

Figure 1. Normalized signal intensity of *gyrB* genes derived from different phylogenetic groups in the rhizosphere and bulk soil. All data are presented as means ± standard errors (n = 3). *p<0.05, and **p<0.01.

microbial community had distinct structure and metabolic potential compared to those from the bulk soil. Though many different types of genes were detected, we were especially interested in microbial genes important to nutrient cycling and microbial interactions, such as antibiotic genes and their phylogenetic origins, which were further examined below.

Autotrophic CO_2 fixation

GeoChip 3.0 contains probes for detecting genes encoding key enzymes involved in autotrophic carbon fixation pathways: ribulose-l,5-bisphosphate carboxylase/oxygenase (Rubisco) for Calvin cycle, propionyl-CoA/acetyl-CoA carboxylase (Pcc/Acc) for 3-hydroxypropionate/malyl-CoA cycle, carbon monoxide dehydrogenase (CODH) for reductive acetyl-CoA pathway, and ATP citrate lyase (AclB) for reductive acetyl-CoA pathway [7]. The Pcc/Acc, Rubisco and CODH pathways were identified to be dominant in both the bulk soil and rhizosphere samples (Fig. 2), and their relative intensity was higher ($p<0.05$) in the rhizosphere than that in bulk soil, suggesting higher capacity in CO_2 fixation in the rhizosphere. The AclB pathway was not detected in any of the samples. The CODH genes were mainly distributed in Alphaproteobacteria phylum in both rhizosphere and bulk soil, especially the genera *Rhodopseudomonas* and *Stappia* (Table S2). Also, Rubisco genes were mainly found in Proteobacteria in both rhizosphere and bulk soil, especially the genera *Burkholderia* and *Magnetospirillum*. The *pcc/acc* genes for autotrophic carbon fixation were identified mainly in Proteobacteria and Actinobacteria in both rhizosphere and bulk soil, especially the genera *Verminephrobacter*, and *Mycobacterium*, respectively. In contrast, the *pcc/acc* genes related to genera *Bradyrhizobium* and *Caulobacter* were only detected in the rhizosphere (Table S3).

Carbon degradation

Rhizosphere soil had more organic C than bulk soil (Table 1). Among 248 detected gene families, 176 gene families were involved in C degradation and organic matter remediation. These genes included those for the degradation of labile C, e.g., starch (alpha-amylase), hemicelluloses (arabinofuranosidase and xylanase), cellulose (cellobiase, cellobiose dehydrogenase and endoglucanase, mainly in Actinobacteria, Alphaproteobacteria and fungi), aromatics (limonene epoxide hydrolase, vanillate monooxygenase and vanillin dehydrogenase, mainly in Proteobacteria, e.g. *Sphingomonas*, *Bradyrhizobium* and *Pseudomonas*), chitin (endochitinase, mainly in fungi and Bacteroidetes), and lignin (manganese peroxidase and glyoxal oxidase, mainly in fungi, especially the class Basidiomycetes) (Fig. 2).

Nitrogen cycling

Total soil N levels were similar in the maize rhizosphere and in the bulk soil (Table 1). GeoChip detected genes critical to all the major pathways in the N cycle. Most of genes were more abundant in the rhizosphere than bulk soil (Fig. 3).

Firstly, *nifH* for N fixation was significantly ($p<0.05$) enriched in the rhizosphere compared to the bulk soil (Table S2 and S3). The most abundant *nifH* genes detected in all samples were from uncultured microorganisms (80–87% of total *nifH* hybridization signal intensities), which suggests that many N_2-fixing microorganisms are still unknown. Among four defined clusters of *nifH* genes, cluster I genes were most represented (6% of total *nifH* signal in rhizosphere and 12% in bulk soil), especially those from *Rhodobacter* and *Rhodopseudomonas*.

Secondly, two genes encoding key enzymes (glutamate dehydrogenase/*gdh* and urease/*ureC*) for ammonification (N mineralization) were significantly ($p<0.05$) enriched in the maize rhizosphere compared to bulk soil (Table S2 and S3). *Gdh* genes were only detected in the rhizosphere, mainly distributed in the bacterial phylum Actinobacteria and archaeal genus *Thermoplasma*, while *ureC* genes were mainly distributed in the bacterial phyla Proteobacteria and Actinobacteria.

Thirdly, two genes encoding key enzymes (ammonia monooxygenase/*amoA* and hydroxylamine oxidoreductase/*hao*) for nitrification were detected. Both of the bacterial and archaeal *amoA* genes were dominant in the rhizosphere. The *hao* genes were only detected in the rhizosphere and largely derived from Rhodobacteraceae.

Fourthly, two key genes (nitrate reductase/*napA* and c-type cytochrome nitrite reductase/*nrfA*) for dissimilatory N reduction to ammonium (DNRA) were significantly ($p<0.05$) overrepresented in the rhizosphere compared to bulk soil, mainly derived from *Desulfitobacterium* and *Syntrophobacter*. Also, three genes (*nasA*, *nirA* and *nirB*) for assimilatory N reduction to ammonium were significantly ($p<0.05$) enriched in the rhizosphere compared to bulk soil. The *nasA* gene was mainly found in the uncultured bacteria, the *nirA* gene in uncultured archaea, and the *nirB* gene in Verrucomicrobiae, Actinomycetales, and genus *Pseudomonas*.

In addition, the abundances of five key genes (*narG*, *nirS/nirK*, *norB* and *nosZ*) involved in the denitrification process were significantly higher ($p<0.05$) in the rhizosphere than in the bulk soil. Most of these genes were found in uncultured bacteria. Some of *narG* genes detected in the maize rhizosphere were indicated in Actinobacteria and Proteobacteria, which is consistent with a previous report [13].

Phosphorus utilization

GeoChip 3.0 contains probes targeting phosphorus utilization, exopolyphosphatase (Ppx) for inorganic polyphosphate degradation, polyphosphate kinase (Ppk) for polyphosphate biosynthesis in prokaryotes, and phytase for phytate degradation [7]. The total signal intensity of *ppx* and *ppk* genes was significantly ($p<0.05$) increased in the rhizosphere (Fig. 4), suggesting a possible enhancement of polyphosphate transformation and the availability of inorganic P in the maize rhizosphere. These genes were mainly distributed in Proteobacteria and Actinobacteria. The abundance of phytase genes, mainly derived from fungi *Ceriporia* sp. in the rhizosphere and *Obesumbacterium proteus* in bulk soil, was not significantly different between rhizosphere and bulk soil, indicating that organic P degradation capability may be not significantly different.

Sulfur cycling

Four enzymes involved in the sulfur cycling of microbial communities are included in GeoChip 3.0: sulfite reductase (DsrA/DsrB) for sulfur reduction, sulfite oxidase (Sox) for sulfur oxidation, dissimilatory adenosine-5′-phosphosulfate reductase (AprA) for both microbial sulfate reduction and sulfur oxidation processes [7]. The total signal intensity of *dsrA* and *sox* genes was significantly ($p<0.05$) increased in the rhizosphere (Fig. 4), suggesting a possible increase in the sulfur transformation rate. *sox* genes were mainly derived from Alphaproteobacteria, especially the order Rhizobiales, while the other three genes were mainly affiliated with Deltaproteobacteria.

Antibiotic resistance

Eleven genes for antibiotic resistance are included in GeoChip 3.0: five for transporters, four for β-lactamase genes, and two for tetracycline and vancomycin resistance [7]. We found most of these genes to be significantly more abundant in the rhizosphere than in bulk soil (Fig. 5; $p<0.05$). The small multidrug resistance

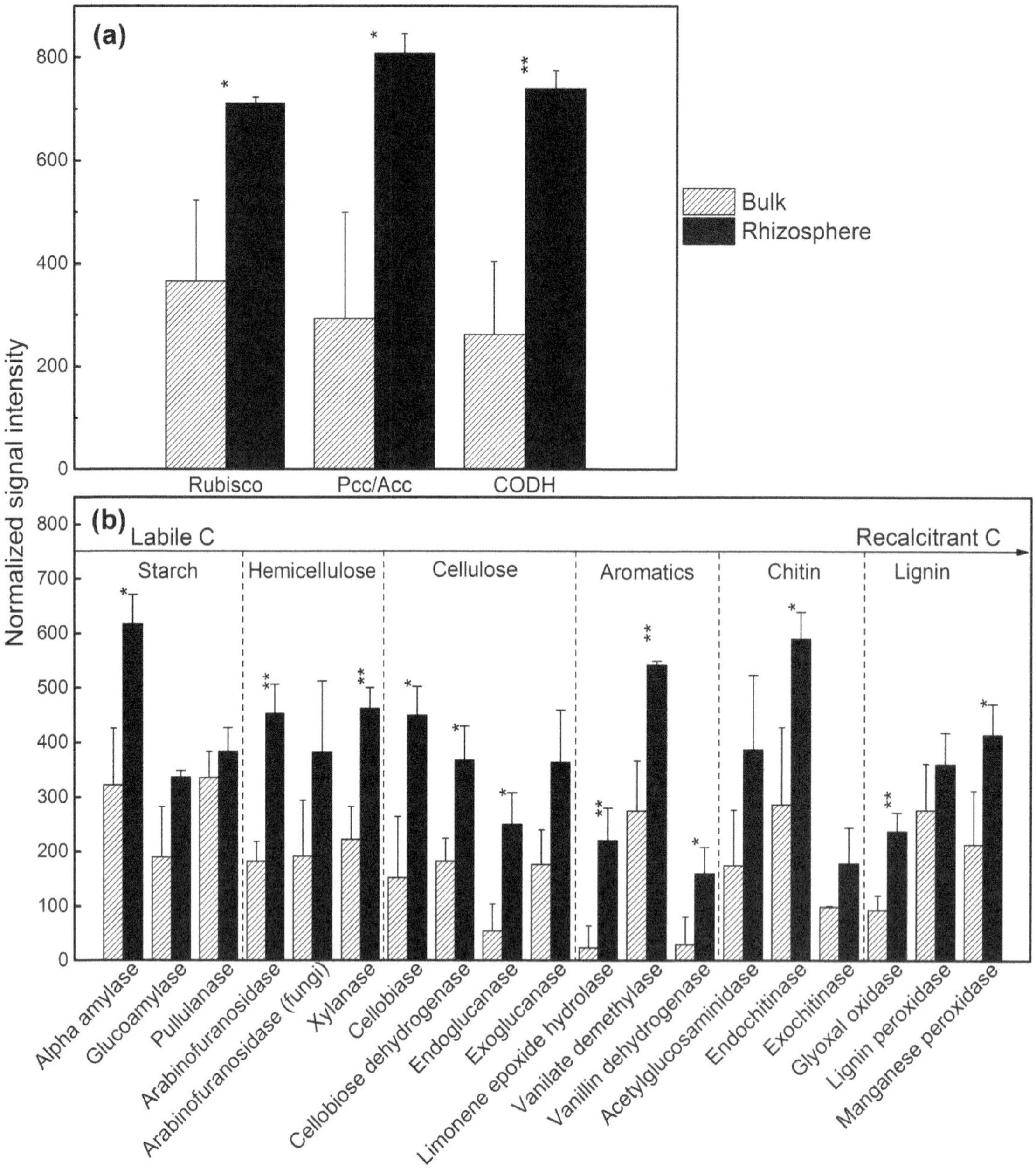

Figure 2. Normalized signal intensity of genes involved in (a) carbon fixation, (b) carbon degradation pathways in the rhizosphere and bulk soil. All data are presented as means ± standard errors (n = 3). *$p<0.05$, and **$p<0.01$.

gene family (SMR) showed the highest signal intensity among these five transporter genes, which were mainly affiliated with genera *Mycobacterium*, *Bacillus* and *Yersinia* in both rhizosphere and bulk soil. Major facilitator super gene family (MFS) also showed high signal intensity after SMR, and they were mainly derived from the genus *Burkholderia*. The ABC antibiotic transporter was only detected in the rhizosphere.

Discussion

GeoChip-based analysis in this study reveals the assemblage of dominant microbiomes from an explicitly functional perspective rather than from taxonomic classification only. Our previous study showed that the maize rhizosphere had significantly reduced bacterial species diversity in comparison to bulk soil [5]. In contrast, GeoChip analysis revealed that functional diversity increased in the rhizosphere (Table 1). Root exudates are driving forces to determine the assemblage of rhizosphere microbial community [14,15]. There are a wide variety of compounds, such as organic acids, sugars, amino acids, lipids, enzymes and aromatics, in the root exudates [1]. Half of the detected gene families were significantly more abundant in the rhizosphere than those in the bulk soil, suggesting that a broad diversity of metabolic activities is enhanced in the rhizosphere. Rhizosecretion (exudate) may activate the growth of previously dormant microbes and their metabolic potentials, including fast-growing r-strategist microbes [14]. The r-strategists may respond rapidly to the freshly input

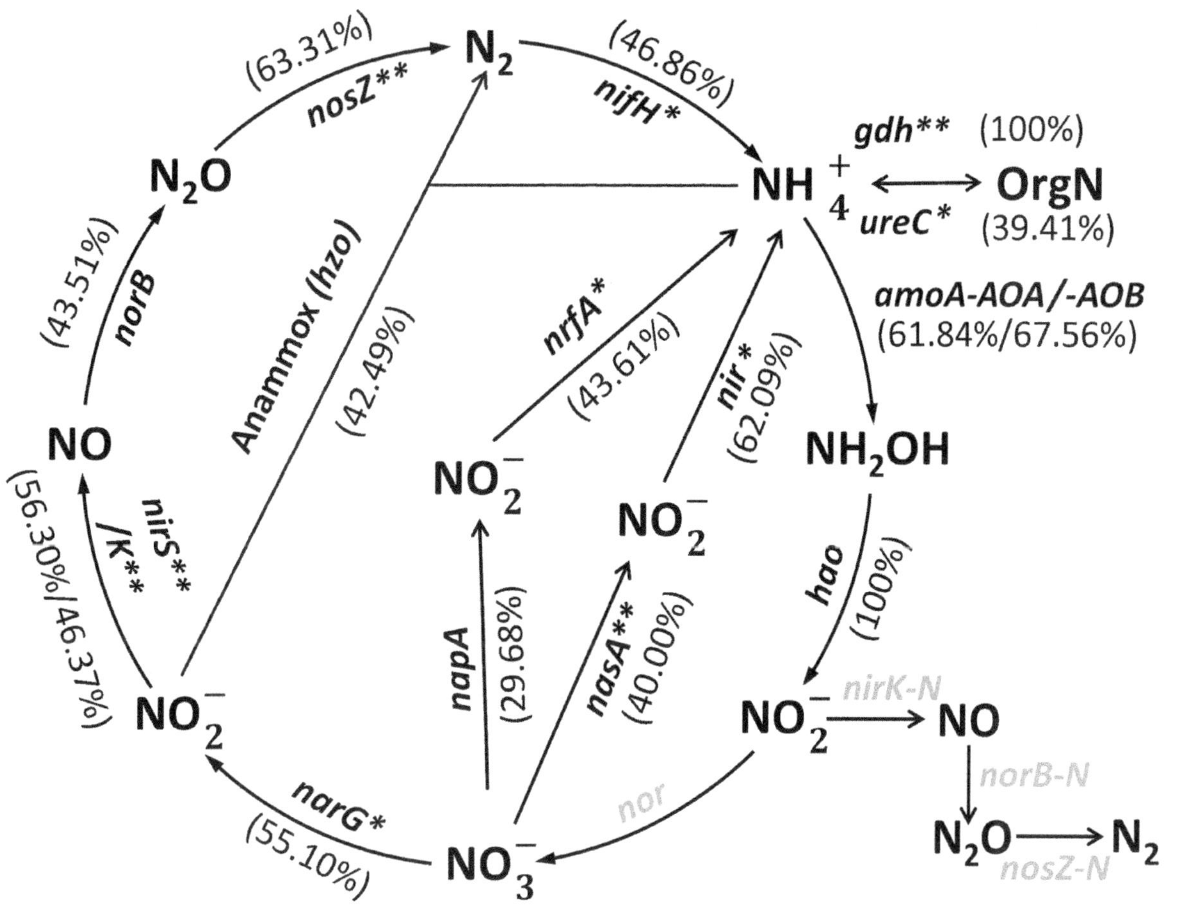

Figure 3. Differences in the abundance of N cycling genes in the rhizosphere and bulk soil. The numbers in brackets indicate the percentage difference of a functional gene signal intensity between rhizosphere and bulk soil samples relative to the normalized signal intensity in the bulk soil sample. The gray-colored genes were not detected by GeoChip 3.0. $^{*}p<0.05$, and $^{**}p<0.01$.

organic matter by increasing their growth, functional gene expression and decomposition rates [16].

Enzyme activities involved in carbon and nitrogen cycling are influenced by plant-microorganism-fauna interactions [17]. Geochip data in this study showed that the changes in the structure and metabolic potential of rhizosphere microbial communities were reflected in a significant increase in the abundance of many functional genes compared to the bulk soil. The higher abundances of functional genes (e.g., CODH, xylanase, glyoxal oxidase and vanilate demethylase) in the rhizosphere suggests that plants select functional groups with a strong capacity for C source utilization. We speculate that the rhizosphere may favor those decomposers for labile C compounds and plant polymers since the gene abundances responsible for the degradation of these organic substrates were significantly over-represented in the rhizosphere. The ability to carry out both simple and complex carbohydrate catabolism was mainly detected in Proteobacteria, Actinobacteria, Firmicutes, and fungi (Table S3), whereas the ability to catabolize complex aromatic compounds was indicated mainly by Proteobacteria and fungi. The increased potential to transform lignin-like compounds and other naturally occurring polyaromatics in the rhizosphere might be related to manganese peroxidase found in Basidiomycetes [18].

A previous study showed that increased microbial activity and nitrogen mineralization were coupled to the changes in microbial community structure in the rhizosphere of Bt corn [19]. The stimulatory effect of maize roots on denitrification was observed at both high and low soil nitrate levels [20]. Our GeoChip results suggested that N fixation, ammonification, denitrification, and N reduction were significantly higher in the maize rhizosphere compared to the bulk soil, especially by members of Proteobacteria (Table S3), possibly leading to more active microbially mediated N transformation in the rhizosphere. GeoChip analysis further showed that the genes involved in C, N, P, S cycling stimulated by the maize rhizosphere were mainly found in organisms belonging to the phyla Proteobacteria and Actinobacteria. This is also supported by the finding that the maize rhizosphere was preferentially colonized by Proteobacteria and Actinobacteria [5]. Within these phyla, the genera *Bradyrhizobium*, *Pseudomonas*, *Rhodopseudomonas* and *Mycobacterium* were overrepresented in the rhizosphere, and these genera harbored diverse functional genes for nutrient transformation. GeoChip data showed that *Bradyrhizobium* played important roles in autotrophic carbon fixation and aromatic degradation; *Pseudomonas* involved in aromatics degradation and assimilatory N reduction to ammonium; *Rhodopseudomonas* involved in autotrophic carbon fixation and N fixation; *Mycobacterium* involved in autotrophic carbon fixation and antibiotic resistance. A metaproteomics study also revealed that bacterial proteins were mostly linked to the Proteobacteria and Actinobacteria in the rhizosphere of rice [21], suggesting that these two bacterial phyla constitute important parts of 'core rhizosphere microbiome' that maintains various

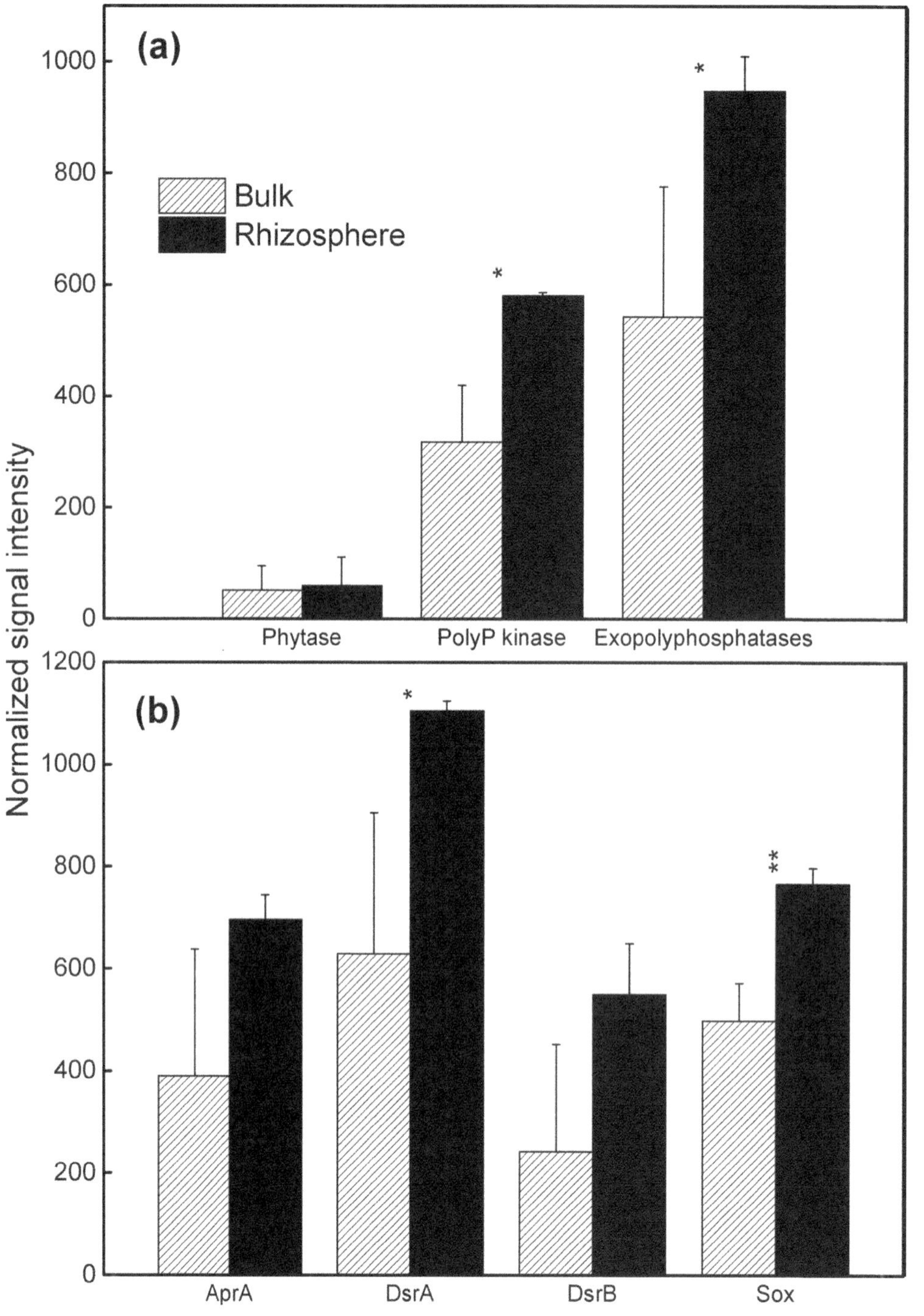

Figure 4. Normalized signal intensity of genes involved in (a) phosphorus utilization, (b) sulfur cycling in the rhizosphere and bulk soil. All data are presented as means ± standard errors (n = 3). *$p<0.05$, and **$p<0.01$.

normal functions and are also effective in controlling soil-borne pathogens.

Since the rhizosphere processes are driven by complex interactions between roots, micro-organisms and soil fauna [17], higher competition for mineral elements in the rhizosphere than bulk soil is expected. GeoChip data supported the hypothesis that microbial competition in the rhizosphere might be greater than that found in bulk soil, as indicated by the higher abundance of antibiotic resistance genes in the rhizosphere. Some microorganisms produce unique antimicrobial metabolites toxic to other microorganisms, allowing them to colonize and proliferate on plant surfaces in the presence of other microbial communities. The genera *Burkholderia, Mycobacterium*, *Bacillus* and *Yersinia* were found to be important antibiotic producers in this study, corroborating their abundant distributions in the rhizosphere of many crops, such as maize [1,5]. The higher occurrence of

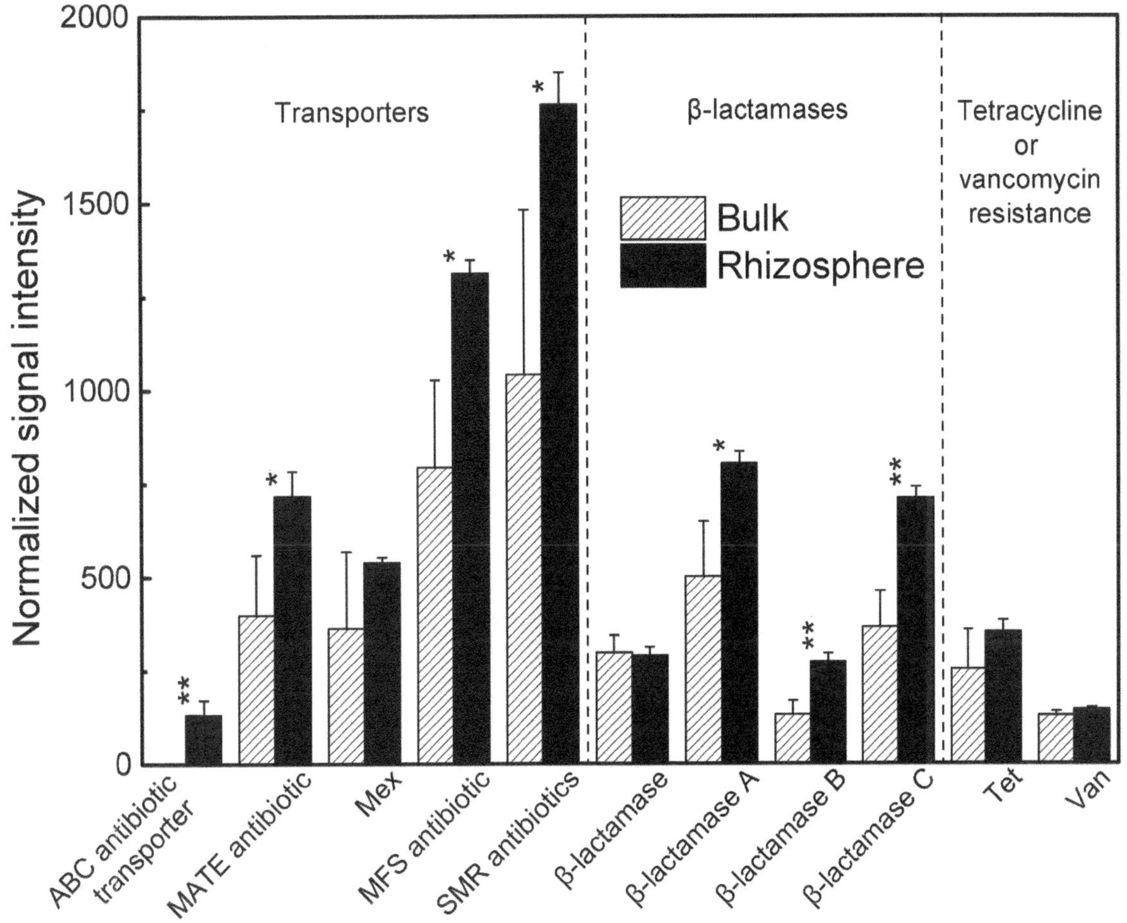

Figure 5. Normalized signal intensity of antibiotic resistance genes in the rhizosphere and bulk soil. All data are presented as means ± standard errors (n = 3). *$p<0.05$, and **$p<0.01$.

antibiotic resistance genes also provide the frontline defense for plant roots against attack by soilborne pathogens [22]. Microbes that can degrade or detoxify these metabolites via specific functional genes gain a competitive advantage [1].

With GeoChip-based functional gene arrays, we have been able to identify functional potentials with the most relevant focus on functional ecology in the rhizosphere of maize. The data supports the concept that the rhizosphere of maize is a hotspot of genes for the transformation of labile and recalcitrant organic C, N, P and S compounds. These functional genes mostly originate from dominant microbial groups, such as Proteobacteria. This study provides comprehensive baseline information on functional capacity in the maize rhizosphere. Specific plant-microbe interactions should be further defined by linking the plant C metabolism by microbial species with a particular functional role in the rhizosphere [2].

Supporting Information

Table S1 Normalized signal intensity of gene families significantly different in the rhizosphere and bulk soil ($p<0.05$).

Table S2 The dominant genes affiliated to different genera.

Table S3 Log_{10} values of normalized signal intensity of dominant gene categories affiliated to each phylum.

Author Contributions

Conceived and designed the experiments: XL RIM. Performed the experiments: XL JX. Analyzed the data: JR JX XL. Contributed reagents/materials/analysis tools: XL JR JZ JX RIM. Contributed to the writing of the manuscript: JR XL JL ACY ZH JX RIM.

References

1. Mendes R, Garbeva P, Raaijmakers JM (2013) The rhizosphere microbiome: significance of plant beneficial, plant pathogenic, and human pathogenic microorganisms. FEMS Microbiol Rev 37: 634–663.
2. Singh BK, Millard P, Whiteley AS, Murrell JC (2004) Unravelling rhizosphere-microbial interactions: opportunities and limitations. Trends Microbiol 12: 386–393.
3. Marschner P, Crowley D, Yang CH (2004) Development of specific rhizosphere bacterial communities in relation to plant species, nutrition and soil type. Plant Soil 261: 199–208.
4. Asmar F, Singh T, Gahoonia, Nielsen NE (1995) Barley genotypes differ in activity of soluble extracellular phosphatase and depletion of organic phosphorus in the rhizosphere soil. Plant Soil 172: 117–122.

5. Li X, Rui J, Mao Y, Yannarell A, Mackie R (2014) Dynamics of the bacterial community structure in the rhizosphere of a maize cultivar. Soil Biol Biochem 68: 392–401.
6. He ZL, Van Nostrand JD, Zhou JZ (2012) Applications of functional gene microarrays for profiling microbial communities. Curr Opin Biotechnol 23: 460–466.
7. He ZL, Deng Y, Van Nostrand JD, Tu QC, Xu MY, et al. (2010) GeoChip 3.0 as a high-throughput tool for analyzing microbial community composition, structure and functional activity. ISME J 4: 1167–1179.
8. Dohrmann AB, Kuting M, Junemann S, Jaenicke S, Schluter A, et al. (2013) Importance of rare taxa for bacterial diversity in the rhizosphere of Bt- and conventional maize varieties. ISME J 7: 37–49.
9. Aira M, Gomez-Brandon M, Lazcano C, Baath E, Dominguez J (2010) Plant genotype strongly modifies the structure and growth of maize rhizosphere microbial communities. Soil Biol Biochem 42: 2276–2281.
10. Chelius MK, Triplett EW (2001) The diversity of archaea and bacteria in association with the roots of *Zea mays* L. Microb Ecol 41: 252–263.
11. Hanway JJ (1963) Growth Stages of Corn (*Zea mays*, L.). Agron J 55: 487–492.
12. Zhou J, Bruns MA, Tiedje JM (1996) DNA recovery from soils of diverse composition. Appl Environ Microbiol 62: 316–322.
13. Philippot L, Piutti S, Martin-Laurent F, Hallet S, Germon JC (2002) Molecular analysis of the nitrate-reducing community from unplanted and maize-planted soils. Appl Environ Microbiol 68: 6121–6128.
14. Fierer N, Bradford MA, Jackson RB (2007) Toward an ecological classification of soil bacteria. Ecology 88: 1354–1364.
15. Mendes LW, Kuramae EE, Navarrete AA, van Veen JA, Tsai SM (2014) Taxonomical and functional microbial community selection in soybean rhizosphere. ISME J 8: 1577–1587.
16. Goldfarb KC, Karaoz U, Hanson CA, Santee CA, Bradford MA, et al. (2011) Differential growth responses of soil bacterial taxa to carbon substrates of varying chemical recalcitrance. Front Microbiol 2: doi:10.3389/fmicb.2011.00094.
17. Bonkowski M, Cheng WX, Griffiths BS, Alphei G, Scheu S (2000) Microbial-faunal interactions in the rhizosphere and effects on plant growth. Eur J Soil Biol 36: 135–147.
18. Kersten P, Cullen D (2007) Extracellular oxidative systems of the lignin-degrading Basidiomycete *Phanerochaete chrysosporium*. Fungal Genet Biol 44: 77–87.
19. Velasco AGV, Kowalchuk GA, Manero FJG, Ramos B, Yergeau E, et al. (2013) Increased microbial activity and nitrogen mineralization coupled to changes in microbial community structure in the rhizosphere of Bt corn. Appl Soil Ecol 68: 46–56.
20. Mahmood T, Ali R, Malik KA, Shamsi SRA (1997) Denitrification with and without maize plants (*Zea mays* L.) under irrigated field conditions. Biol Fertil Soils 24: 323–328.
21. Wang H-B, Zhang Z-X, Li H, He H-B, Fang C-X, et al. (2010) Characterization of metaproteomics in crop rhizospheric soil. J Proteome Res 10: 932–940.
22. Cook RJ, Thomashow LS, Weller DM, Fujimoto D, Mazzola M, et al. (1995) Molecular mechanisms of defense by rhizobacteria against root disease. Proc Natl Acad Sci U S A 92: 4197–4201.

20

Proteomic Analysis of the Salt-Responsive Leaf and Root Proteins in the Anticancer Plant *Andrographis paniculata* Nees

Daryush Talei[1,2]*, Alireza Valdiani[3], Mohd Yusop Rafii[4], Mahmood Maziah[3,4,5]*

1 Medicinal Plants Research Center, Shahed University, Tehran, Iran, 2 Department of Cell and Molecular Biology, Faculty of Biotechnology and Biomolecular Sciences, Universiti Putra Malaysia, UPM Serdang, Selangor DE, Malaysia, 3 Department of Biochemistry, Faculty of Biotechnology and Biomolecular Sciences, Universiti Putra Malaysia, UPM Serdang, Selangor DE, Malaysia, 4 Institute of Tropical Agriculture, Universiti Putra Malaysia, UPM Serdang, Selangor DE, Malaysia, 5 Institute of Bioscience, Universiti Putra Malaysia, UPM Serdang, Selangor DE, Malaysia

Abstract

Separation of proteins based on the physicochemical properties with different molecular weight and isoelectric points would be more accurate. In the current research, the 45-day-old seedlings were treated with 0 (control) and 12 dS m^{-1} of sodium chloride in the hydroponic system. After 15 days of salt exposure, the total protein of the fresh leaves and roots was extracted and analyzed using two-dimensional electrophoresis system (2-DE). The analysis led to the detection of 32 induced proteins (19 proteins in leaf and 13 proteins in the root) as well as 12 upregulated proteins (four proteins in leaf and eight proteins in the root) in the salt-treated plants. Of the 44 detected proteins, 12 were sequenced, and three of them matched with superoxide dismutase, ascorbate peroxidase and ribulose-1, 5-bisphosphate oxygenase whereas the rest remained unknown. The three known proteins associate with plants response to environmental stresses and could represent the general stress proteins in the present study too. In addition, the proteomic feedback of different accessions of *A. paniculata* to salt stress can potentially be used to breed salt-tolerant varieties of the herb.

Editor: Keqiang Wu, National Taiwan University, Taiwan

Funding: Funding provided by Universiti Putra Malaysia, Andrographis grant (MYR). The funders had no role in study design, data collection and analysis, decision to publish, or preparation of the manuscript.

Competing Interests: The authors have declared that no competing interests exist.

* Email: d.talei@shahed.ac.ir (DT); maziahm@upm.edu.my (MM)

These authors contributed equally to this work.

Introduction

Andrographis paniculata is a medicinal herb from the family Acanthaceae. The plant extract contains diterpene compounds with a broad scope of pharmaceutical properties such as anticancer, antibacterial, antivirus and anti-hepatitis [1]. Salinity stress alters various biochemical and physiological responses in plants and causes adverse effects on various physiological processes such as photosynthesis, growth and development [2,3]. From a molecular perspective, salt stress comprises many factors, including oxidative stress, osmotic stress, ion stress, nutritional imbalance or a combination of these factors [4,5].

Plants produce antioxidants and protective enzymes such as superoxide dismutase, catalase, peroxidase, glutathione reductase, polyphenol oxidase that scavenge the reactive oxygens or prevent their formation, to reduce oxidative damage [6]. In response to salinity, plants accumulate organic compounds (proline, glycine-betaine, malate, and polyol) in the cytoplasm, which function as osmolytes preventing conformational changes of macromolecules especially proteins [7].

Under certain level of salinity, some of those major macromolecules involved with various cellular processes are prone to the conformational changes of their native structures affecting the number and level of proteomes in the affected tissues. These changes should be monitored through profiling the proteomes of the affected tissues/organs, after being exposed to salt, regularly. The changes could include protein modifications, proteolysis, subcellular localizations, and interaction with other proteins [8]. Many proteins undergo post-translational modifications including phosphorylation, which play an important role in subcellular localization [8,9].

With respect to salinity, several salt-responsive proteins have been reported in rice (*Oryza sativa*) [10,11], *Arabidopsis thaliana* [12], and wheat (*Triticum aestivum*) [13]. These proteins associate with major cellular processes including photosynthesis, photorespiration, metabolic regulation, signal transduction, control of ion channels, oxidative stress defence and protein folding.

An enormous lack of information related to the salt responsive proteins of *A. paniculata* is tangible. Therefore, the present study emphasized on the analysis and identification of the salt stress responsive proteins of this herb under salinity stress conditions using two-dimensional gel electrophoresis.

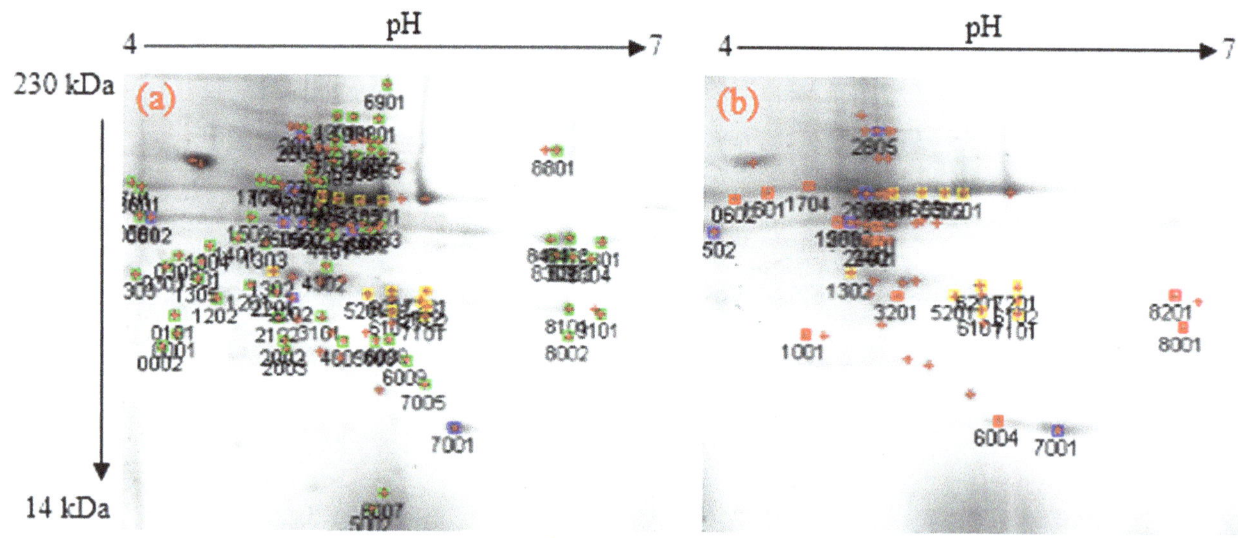

Figure 1. The super composite images of 2-DE gels representing the repressed leaf proteins of *A. paniculata* in control (a) and salt-treated samples (b).

Results and Discussion

Analysis of protein spots separated by 2-DE PAGE using the PDQuest software

A total of 162 induced upregulated and downregulated proteins were detected in the control and salt-treated leaf and root samples of *A. paniculata* seedlings. Dissimilar expression patterns of the expressed proteins were observed in both of the control and salt-treated plants. Figures 1 and 2 represent the typical super composite images of the leaf and root samples. The analysis revealed a total of 15 upregulated and four downregulated leaf proteins in the control and salt-treated samples, respectively as shown in Table 1. Furthermore, 12 upregulated and eight downregulated root proteins were detected in the control and salt-treated samples, respectively (Table 1). Comparison of the total number of representing proteins (repressed, induced, upregulated and downregulated proteins) in the control and salt-treated samples using independent samples t-test showed non-significant differences in the studied organs such as leaves, roots, as well as the whole plant (Table 2).

Several proteins were differentially expressed (repressed, induced, upregulated and downregulated) in the seedlings exposed to high salinity (12 dS m^{-1}). Previous studies on different plant species under salinity stress have discovered a number of differently expressed proteins. These proteins link to several functional categories such as transcription factors, solute accumulation enzymes and stress tolerance proteins e.g. superoxide dismutase (SOD) [14–18], ascorbate peroxidases [19–22], transcription factors [23,24], protein kinases [25], protein phosphatases [26], ATP generation [24] and calcium related signalling molecules [27]. As a matter of fact, the role of the expressed proteins of *A. paniculata* remained unidentified in this study. For

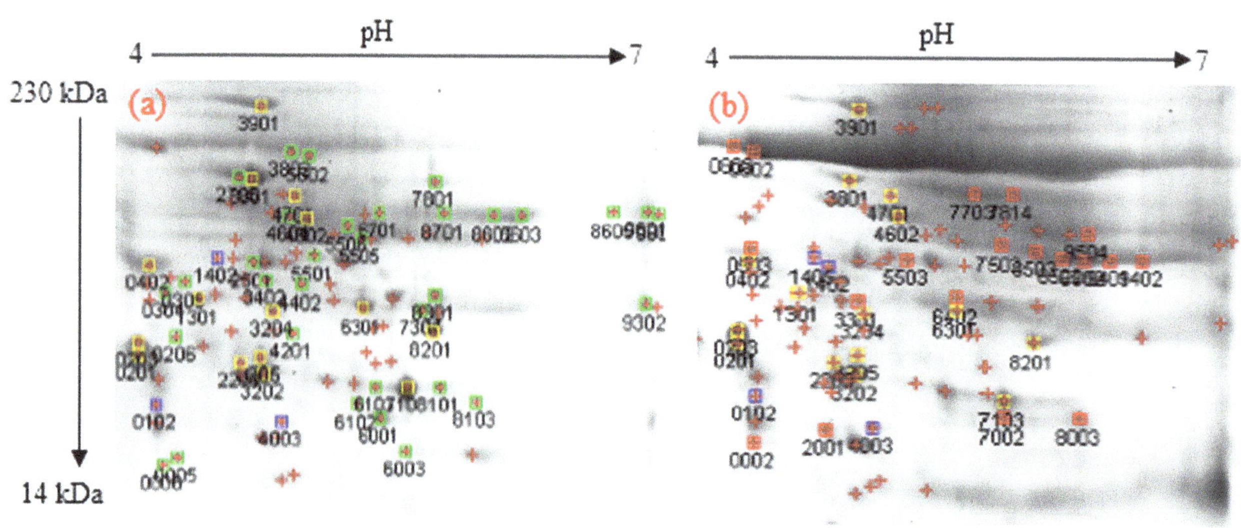

Figure 2. The super composite images of 2-DE gels representing the repressed root proteins of *A. paniculata* in control (a) and salt-treated samples (b).

Table 1. The number of unique/common expressed proteins of the leaf and root samples of *A. paniculata* under control and salinity condition.

Protein types	Leaf		Root	
	Control	Salt-treated	Control	Salt-treated
Repressed	85	71	77	17
Induced	33	19	63	13
Upregulated	15	4	12	8
Downregulated	4	15	8	12

Repressed proteins: The protein spots absent in the salt-treated samples and present in the controls, Induced proteins: The protein spots present in the salt-treated samples and absent in the controls, Downregulated proteins: The protein spots that their size and intensity was decreased in the salt-treated samples as compared to the control samples, Upregulated proteins: The protein spots that their size and intensity was increased in the salt-treated samples as compared to the control samples.

this reason, further functional analyses of the proteins will be useful to identify their roles in salt tolerance mechanism, in the next studies.

As was anticipated, the protein expression patterns of the leaves and roots were different in the control and slat-treated accessions. In this study, the protein was considered as "induced", when it was only present in the treated samples, but not in the control. The number of the unique proteins of the leaf and root samples is available in Table 1. There were altogether 19 and 13 induced protein spots in the treated leaf and root samples, respectively, four of which were "commonly" induced in both of the control and salt-treated samples. The uniquely induced proteins might be related to the specific genes involved with the plant's response to salinity stress whereas the "commonly" induced proteins could represent a general stress response at various degrees of salt stress. Apart from the induced proteins, the software also facilitated the detection of repressed protein, i.e. the protein spots which are absent on the salt-treated samples, but are present in the control samples. Table 3 represents the number of the repressed proteins.

Salt stress affects almost every aspect of the plant physiology and metabolism. Since, root is the primary target of many abiotic stresses; therefore, the current research considered it as the first responding organ of the plant to the salt stress. The response of *A. paniculata* roots to salinity stress was more complex, in which the number of the induced and upregulated proteins in the root were higher than the number of these proteins in the leaf samples. Almost all of the induced and upregulated proteins in the leaf and root samples were different in terms of the estimated isoelectric points (pI), and molecular weights (MW) with the exception of three detected identical proteins (Table 3). The higher number of the repressed proteins in the root compared to the leaf in response to salt treatment indicated that roots are more prone to the effects of salt stress and are severely affected by salinity. The independent samples t-test results showed non-significant differences between the induced proteins of the salt-treated leaf and root samples, as well as between the upregulated proteins of the salt-treated leaf and root samples in terms of the molecular weights (Table 4). However, the average molecular weights of the induced and upregulated proteins in the salt-treated root were higher than the leaf (Table 4).

Overall, the total number of expressed proteins in the salt-treated leaves and roots was decreased, which might reflect the adverse effects of salinity on growth and development of the plant. Many cellular and metabolic processes of plants are known to be affected by salinity, including the reductions in the stromal volume of chloroplast, generation of reactive oxygen species (ROS), photosynthesis, respiration, biosynthesis of protein, nucleic acid, lipids, and pigments [28]. Reportedly, in some plant species such as *Oryza sativa* [29] and *Bruguiera parviflora* [30], decrease in the number of proteins happened in response to salinity.

The magnitude of upregulated proteins in the root and leaf samples of the treated plants varied from spot to spot. The magnitude of expression or intensity reflects the level of induction of the responsive proteins, which in turn indicates the pathways responding the salt stress. The expression intensities of the four upregulated proteins varied in the leaves and roots of the treated plants as shown by the histograms in Figure 3 and 4, respectively.

Unfortunately, the whole genome of the plant is still unsequenced; hence, the online search of the database with emphasis on molecular weight and isoelectric point information

Table 2. Comparison of the total number of representing proteins (repressed, induced, upregulated and downregulated proteins) in the control and salt-treated samples using independent samples t-test.

Organ	Group	N	Mean	SD	T	df	Sig.	SE Difference
Leaf	Control	4	34.25	35.88	0.3^{ns}	6	0.774	23.34
	Salt-treated	4	27.25	29.85				
Root	Control	4	40.00	35.15	1.56^{ns}	6	0.171	17.67
	Salt-treated	4	12.50	3.70				
Whole plant	Control	8	37.13	33.03	1.24^{ns}	14	0.243	13.88
	Salt-treated	8	19.88	21.21				

N: Number of samples, SD: Standard deviation, T: T-test value, df: Degree of freedom, Sig: Statistical significance, SE: Standard error, ns: Non-significant.

Table 3. List of the induced and upregulated protein spots detected in the salt-treated leaf and root samples of *A. paniculata* at seedling stage.

Salt-treated leaf				Salt-treated root			
Spot No.	**pI (pH)**	**Molecular weight (kDa)**	**Regulation trend**	**Spot No.**	**pI (pH)**	**Molecular weight (kDa)**	**Regulation trend**
L-0002	4.35	21.28	**	R-0602	4.38	61.33	**
L-0503	4.48	42.3	**	R-1001	4.65	28.42	**
L-0802	4.38	78.41	**	R-1601	4.42	63.4	**
L-0803	4.44	72.47	**	R-1704	4.78	66.27	**
L-2001	4.78	18.8	**	R-1501	4.85	50.14	**
L-3301	4.95	35.74	**	R-2402	4.98	47.83	**
L-5503	5.28	41.37	**	R-2401	5.12	46.78	**
L-6402	5.35	38.48	**	R-2501	5.04	48.61	**
L-7002	5.65	20.72	**	R-3201	5.25	38.26	**
L-7502	5.62	47.31	**	R-3601	5.14	63.73	**
L-7703	5.45	65.42	**	R-6004	5.52	21.64	**
L-7814	5.65	64.73	**	R-8001	6.41	32.72	**
L-8003	6.15	20.4	**	R-8201	6.35	38.47	**
L-8502	5.82	37.42	**	R-0502	4.28	47.23	*
L-8503	5.98	32.75	**	R-2202	5.05	37.82	*
L-9401	6.15	31.34	**	R-2502	4.12	46.88	*
L-9402	6.22	30.87	**	R-2503	4.92	48.74	*
L-9502	6.12	29.83	**	R-2602	5.04	56.73	*
L-9504	6.18	48.54	**	R-2805	5.11	87.82	*
L-0102	4.46	24.84	*	R-5403	5.34	47.36	*
L-1402	4.82	41.83	*	R-7001	5.87	19.62	*
L-1405	4.74	43.18	*	-	-	-	-
L-4003	5.12	18.35	*	-	-	-	-

**Induced proteins and * Upregulated proteins based on two fold. pI: Isoelectric point.

Table 4. Comparison of the induced and upregulated proteins in the salt-treated leaf and root samples of *A. paniculata* on the basis of molecular weight using independent samples t-test.

Protein	Group	N	Mean	SD	T	Df	Sig.	SE Difference
Induced	Leaf	19	40.96	17.95	0.967^{ns}	30	0.341	5.98
	Root	13	46.74	14.37				
Upregulated	Leaf	4	32.05	12.37	1.594^{ns}	10	0.142	10.65
	Root	8	49.03	19.14				

N: Number of samples, SD: Standard deviation, T: T-test value, df: Degree of freedom, Sig: Statistical significance, SE: Standard error, ns: Non-significant.

sole, yielded little success, as experienced during this study. Consequently, identification of the proteins became impossible in this manner.

Classification of the Identified Functional Proteins

In total, 12 induced, upregulated and downregulated leaf and root proteins of the plant were identified using mass spectrometry technique. Tables 5 and 6 show the identity of the protein spots as determined by the peptide mass fingerprinting method and Table 7 demonstrates the math rate and correlation coefficient of protein spots among three replicate of the leaf and root samples at control and salinity levels using the PDQuest software.

The sequencing result revealed the two salt stress responsive enzymes known as superoxide dismutase (SOD) and ascorbate peroxides (APX), which have a central role in oxidative plant defense mechanisms. Plants possess a number of antioxidant enzymes such as SOD, APX and glutathione reductase (GR) for protecting themselves against the injurious effects of reactive oxygen species (ROS) [31]. Salinity stress induces production of ROS in plants. Major scavenging mechanisms of ROS connect to the production of SOD and APX enzymes. Antioxidant enzymes are essential elements in the defense mechanisms. Therefore, the inability of plants in tolerating salt stress is attributed to a decrease in the activity of the mentioned antioxidant enzymes [16]. In accordance with similar studies SOD and APX were induced, under salinity condition in the present experiment as well. Increment in the activity of SOD correlates to the increased level of protection against the damages associated with oxidative stress [32,33]. In agreement with the present results, the increased activity of antioxidant enzymes under salt stress has been reported in some salt tolerant plants such as cotton [34], cucumber [35], pea [36], *Solanum tuberosum* [37], *Pisum sativum* [38], wild tomato species *Lycopersicon pennellii* [16], rice (*Oryza sativa*) [17], and soybean (*Glycine max*) [39].

Ascorbate peroxidase is a hydrogen peroxide-scavenging enzyme that is essential for plants to protect chloroplasts and the other cell components from damages driven by hydrogen peroxide and its hydroxyl radicals [40]. The enzyme can improve the levels of salt and drought tolerance by scavenging H_2O_2 that is unreachable for catalase because of their high similarity with H_2O_2 and also their presence in different subcellular locations [41]. Analysis of the APX activity in some plant species such as *Arabidopsis thaliana* [21], *Oryza sativa*, *Hordeum vulgare*, *Triticum aestivum*, *Lolium perenne*, and *Zea mays* showed a significant high resolution [14,19,20].

Another upregulated enzyme was ribulose-1, 5-bisphosphate oxygenase also known as Rubisco. This enzyme is a critical factor in the process of photosynthesis. During salinity stress, the available water capacity (AWC) is limited in soil, and since water is extremely crucial for photosynthesis, the photosynthetic rate of the plants should be lower under such a condition compared to the standard situation. However, leaf analysis showed that Rubisco was upregulated in the salt-treated plants, but it is unclear whether the increment of Rubisco was due to an increase in photosynthetic activity of the plant or caused by another reason. Nevertheless, according to the previous findings, Rubisco is also associated with proline accumulation [42] that the same happened to the salt-treated plants of the present study.

The intensive response of the plants to salinity in the root compared to leaf, suggesting roots as a principal organ for mediating the plant's response to salinity. Most of the induced and upregulated proteins differed in their estimated isoelectric points (pI) and molecular weights (MW) in the leaf and root of the salt-treated plants, which might suggest that different metabolic

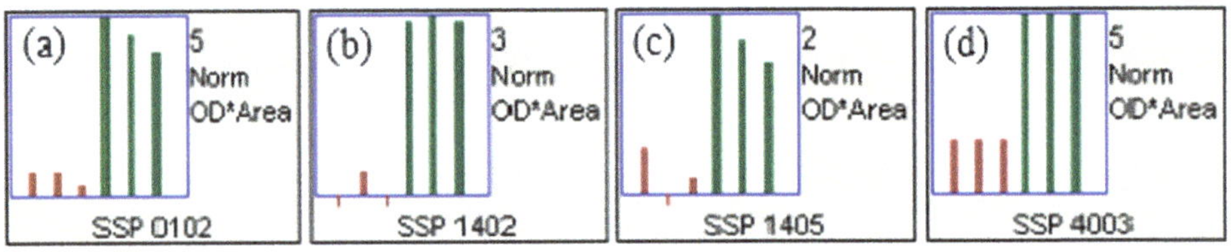

Figure 3. Histogram representing the four upregulated leaf proteins of *A. paniculata* in the salt-treated plants. The red color refers to the quantity of leaf proteins in control samples (downregulated) and the green color shows the quantity of leaf proteins in salt-treated samples (upregulated). Spot No. 0102 (a), spot No. 1402 (b), spot No. 1405 (c) and spot No. 4003 (d).

pathways be functional in the leaf and root of the plant during the stress. A few of these responsive proteins were identified, and their functions are related to plant responses to the stress conditions.

Despite the lack of applicable information on the genome of *A. paniculata*, protein profiling of this plant will serve as an introduction to its proteomics. The current results can also provide the researchers with a reliable background based on protein function in different parts of *A. paniculata* and their effects on biological phenomena of the herb. However, a pure focus on the performance of the proteins of this plant species will release a huge volume of useful information in the future. Undoubtedly, two dimensional proteomics analysis is one of the most efficient technologies to acquire cutting-edge knowledge on the genetics and physiology of plants, including *A. paniculata*.

The systematic proteomic analysis of *A. paniculata* under salinity stress led to the identification of upregulated, downregulated, repressed and induced proteins in control and salt-treated plants. The combination of two-dimensional electrophoresis (2-DE) and mass spectrometry revealed the plasticity of protein expression during salinity stress. These proteins belong to several functional groups such as osmotic stress related proteins and general stress proteins. Induction of some of these proteins was observed in the roots of salt-treated *A. paniculata* plants, and this might be connected to the plant response to salt stress. The proteins might either be directly involved in the protection against environmental stresses or play a role in stress regulation and signal transduction. Overexpression of such proteins should make the plant more resistant to the stress if the limiting factor in resistance is either the timing or level of expression of the gene in question. Functional testing by overexpression could initiate developing new plant varieties with a higher capability of resistance.

Application of proteomic data in *A. paniculata* breeding

As a promising horizon in the breeding of *A. paniculata*, intraspecific hybridization has led to considerable results by increment in the contents of anticancer diterpenes such as andrographolide (AG) [43], and neoandrographolide (NAG), and 14-deoxy-11,12-didehydroandrographolide (DDAG) [44]. Furthermore, recently conducted genetic investigations have revealed that morphological- and DNA-based genetic distance could affect the heterosis of phytochemical as well as Agro-morphological characteristics in *A. paniculata* [45,46]. In this regard, proteomic data can be effectively employed for breeding purposes [47] and assessing the impact of outcrossing on diversification of protein pattern of the herb through the recombinant proteins. A prospective study should tend to quantify the proteomic data and link them to the breeding approaches of this precious herb, in the future.

Conclusion

Proteins are the final production of genes and one of the basic adaptive strategies in plants under salinity stress. Overexpression and identification of the salt-induced proteins can be potentially used in genetic engineering of *A. paniculata* and improving its salt tolerance capacity. Since, cells have the ability to monitor the severity of stress and the degree of stress-induced damage, the

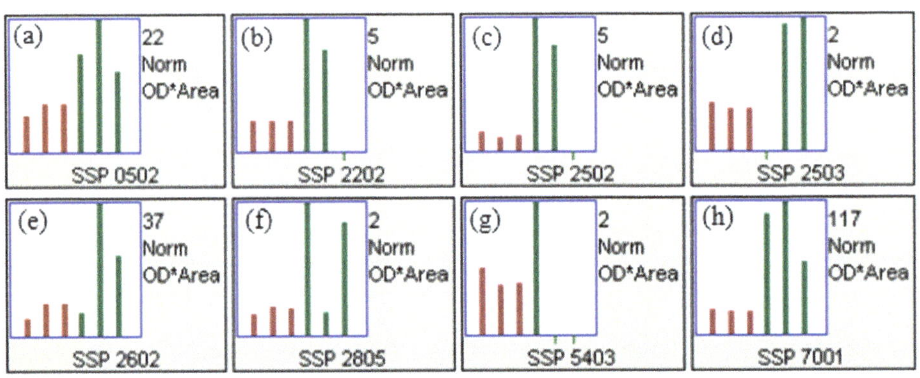

Figure 4. Histogram representing the eight upregulated root proteins of *A. paniculata* in the salt-treated plants. The red color refers to the quantity of root proteins in control samples (downregulated) and the green color shows the quantity of root proteins in salt-treated samples (upregulated). Spot No. 0502 (a), spot No. 2202 (b), spot No. 2502 (c) and spot No. 2503 (d), spot No. 2602 (e), spot No. 2805 (f) and spot No. 4503 (g) and spot No. 7001 (h).

Table 5. Identification of the 12 induced and upregulated protein spots in the salt-treated leaf and root samples of *A. paniculata* as determined by 2-DE analysis, MALDI-TOF-MS, and MASCOT analysis.

Spot No.	Spot ID	Regulation trend	Protein name	Sequence	Mass (Da)	Mascot score	MPN
L-0802	167858	**	Ribulose-1,5-bisphosphate carboxylase/oxygenase	K.D¯DILAAFR.V	49310	427	6(6)
L-0402	167859	***	(Unknown protein)	VNFPEVPR	Unknown	Unknown	Unknown
L-4003	167860	*	Putative uncharacterized protein	K. ALFSQITTRF.-	35649	432	3(3)
L-3202	167861	***	Putative uncharacterized protein	R.SVDETLR.T	30801	545	5(4)
L-7002	167862	**	Superoxide dismutase (Cu-Zn)	K.GGHELSLSTGNAGGR.L	20883	372	3(3)
L-2001	167863	**	(Unknown protein)	LITFEGEK	Unknown	Unknown	Unknown
L-5503	167867	**	Os04g0416400 protein (Fragment)	R.GLFIIDKEGVIQHSTINNLAIGR.S	15378	57	1(1)
L-8003	167869	**	Putative uncharacterized protein	R.GQQLLDTFR.V	12438	44	1(1)
R-6501	167864	*	Q0VYC8|Peroxidase 1	R.SPNVFDNR.Y	39905	66	1(1)
R-7001	167865	*	Superoxide dismutase (Cu-Zn)	K.EHGAPEDETR.H	15217	262	3(2)
R-2401	167866	**	Putative uncharacterized protein	K.TLVFQFSVK.H	49129	300	4(4)
R-2501	167857	**	Putative salt-inducible protein	RM¯EFGKIGMLMTLFKQLK	18737	372	4(4)

***Downregulated proteins, ** Induced proteins and * Upregulated proteins. MPN: Matched peptide number. Mascot score: Total ion score for the entire protein and for ions complemented by 100% of the confidence index (C.I).

identified proteins in this study could represent a so-called "stress proteome" in response to salinity for *A. paniculata*.

Materials and Methods

Chemicals and Reagents

All the chemicals, including HEPES [4-(2-hydroxyethyl)-1-piperazineethanesulfonic acid], thiourea, bovine serum albumin (BSA), iodoacetamide, and ethylenediaminetetraacetic acid (EDTA), ammonium acetate and Phenylmethanesolfonyl fluoride (PMSF) and the other chemicals used for protein extraction and separation were supplied with analytical grade. Double distilled water (ddH$_2$O) was used for preparing all solutions and buffers.

Plant Material and Growth Conditions

According to Talei et al. [48], the seeds of the salt-tolerant accession of *A. paniculata* (11329) were collected from the Agro Gene Bank, Universiti Putra Malaysia. The seeds were germinated as described by Talei et al. [49,50] and then incubated in a growth chamber under a light/dark regime of 14/10 h at 28–30°C, and relative humidity between 60–75%. The germinated seeds at initial leaf stage were transferred into the hydroponic system (Hoagland nutrient solution) to provide the plants with continual nutrient supplies. The seedlings were grown in a greenhouse at a controlled temperature of 20°C during the night and 28°C during the day. The 45-day-old seedlings were then treated with 0 (control) and 12 dS m^{-1} of sodium chloride solution. The sodium chloride was increased with a daily increment to accomplish the final concentrations of NaCl (12 dS m^{-1}). Salt treatments above 12 dS m^{-1} would be lethal to the seedlings within a short period [51]; therefore, the 12 dS m^{-1} salt treatment was the maximum salinity used in this study. After 15 days of salinity exposure, fresh leaves and roots were collected and frozen immediately in liquid nitrogen to prevent protein degradation.

Protein Sample Preparation and Quantification for 2-DE

Five grams of the fresh and fully expanded leaves from 5–10 healthy 45-day-old plants of each accession were collected and thoroughly washed with deionised water, and were then frozen in liquid nitrogen. Two grams of the frozen leaf and root tissue samples of each control and salt-treated plants (accession 11329) were ground into a fine powder using the autoclaved and pre-chilled mortar and pestle, and then homogenized with 10 mL of extraction buffer (20 mM HEPES/KOH pH 7.5, 40 mM KCl, 1 mM EDTA, 10% (v/v) Glycerol and 1 mM PMSF) as described by Talei et al. [52].

The pellet was air dried and solubilised in a solubilization buffer containing 9.8 M Urea, 4% CHAPS, 65 mM DTT and 200 mM Tris base. The solubilised samples were dissolved in 500 μL rehydration buffer. The Bradford method [53] determined the total protein concentration by employing bovine serum albumin as a standard. The measurements were in triplicate, at 595 nm, using a Perkin Elmer (Lambda 25 UV/Vis) spectrophotometer.

First and Second Dimension Protein Separation

The electrophoresis techniques used for proteome analysis were including the isoelectric focusing (IEF) and sodium dodecyl sulphate polyacrylamide gel electrophoresis (SDS-PAGE). Two-dimensional gel electrophoresis was carried out to resolve the protein mixtures as described by O'Farrell [54]. Approximately, 200 μg of the total proteins were extracted from the leaves and roots of *A. paniculata* seedlings, and then were solubilised in 20 μL solubilization buffer. The solubilised samples were placed in the separate tubes and incubated at room temperature for 30 min,

Table 6. Physicochemical characteristics of the 12 induced and upregulated protein spots in the salt-treated leaf and root samples of *A. paniculata* as determined by 2-DE analysis, MALDI-TOF-MS, and MASCOT analysis.

Spot No.	Spot ID	TMM (kDa)	EMM (kDa)	EpI (pH)	TpI (pH)	Protein coverage (%)	Function	Taxonomy identifier
L-0802	167858	4.93	1.02	6.43	4.38	18	Salt-inducible protein	883475 [*Barleria* sp.]
L-0402	167859	-	0.96	-	4.26	-	Unknown	NSHR [*Populus trichocarpa*]
L-4003	167860	3.56	1.18	8.91	5.12	5	Probable salt-inducible protein	3847 [*Glycine max*]
L-3202	167861	3.08	0.82	8.56	4.82	22	Probable salt-inducible protein	3332 [*Picea sitchensis*]
L-7002	167862	2.09	1.41	5.45	5.65	15	Antioxidant (salt induced protein)	4577 [*Zea mays*]
L-2001	167863	-	0.88	-	4.78	Unknown	Unknown	NSHR [*Oryza punctate*]
L-5503	167867	1.54	2.51	5.87	5.28	14	Antioxidant	39947 [*Oryza sativa* subsp. *Japonica*]
L-8003	167869	1.24	1.08	10.11	6.15	-	Probable salt-inducible protein	29760 [*Vitis vinifera*]
R-6501	167864	3.99	0.95	7.63	6.02	2	Detoxification (salt induced protein)	4058 [*Catharanthus roseus*]
R-7001	167865	1.52	1.14	5.45	5.87	21	Antioxidant(salt induced protein)	3880 [*Medicago truncatula*]
R-2401	167866	4.91	1.07	4.45	5.12	8	Probable salt-inducible protein	81972 [*Arabidopsis lyrata* subsp. *Lyrata*]
R-2501	167857	1.87	1.47	5.87	5.04	15	Probable salt-inducible protein	NSHR [*Arabidopsis*]

EpI: Experimental isoelectric point. TpI: Theoretical isoelectric point. TMM: Theoretical molecular mass, EMM: Experimental molecular mass, Protein coverage: Percentage of protein coverage. Mr: The nominal mass of an ion or molecule that is calculated using the mass of the most abundant isotope of each element rounded to the nearest integer (Da). NSHR: No significant hits to report.

Table 7. The results of automath settings of protein spots in the leaf and root sample of *Andrographis paniculata*.

Replicates	Gel name	Spot No.	Matched	Match rate 1 (%)	Match rate 2 (%)	Correlation coefficient
1	LC1	96	85	88	57	1.00
2	LC1	96	87	90	59	0.98
3	LC1	96	87	90	59	0.96
1	LC4	95	71	74	48	0.50
2	LC4	95	71	74	48	0.59
3	LC4	95	71	74	48	0.60
1	RC1	114	112	98	82	0.92
2	RC1	114	114	100	84	1.00
3	RC1	114	112	98	82	0.99
1	RC4	57	47	82	34	0.50
2	RC4	57	50	87	37	0.58
3	RC4	58	49	84	36	0.47

LC1: Leaf sample at control level, LC4: Leaf sample at 12 dS m^{-1} salinity level, RC1: Root sample at control level, RC4: Root sample at 12 dS m^{-1} salinity level. High correlation coefficients indicate high degree of confidence of the matched spots.

and then mixed with 250 µL of rehydration buffer (9.8 M Urea, 2% CHAPS, 0.5% IPG buffer, 65 mM DTT and 0.1% Bromophenol blue). The total content of each tube was pipetted in a separate channel of the rehydration tray through an IPG strip gel (pH 4–7), in a bead-like manner. An IPG strip was located on each channel gel side down, ensuring that no bubbles were trapped underside the plastic side of the strip.

The strips were rehydrated at room temperature for 16 hours, and the strips were then transferred to an IEF focusing tray for the isoelectric focusing. The strip was placed on an IEF Mini-Protean II electrophoresis cell, and the first dimension (IEF) was performed with the focusing program of 250 V for two min, 500 V for 30 min, 1000 V for 1 h, 4000 V for 2 h, and a final focusing of 14000 volt-hour focusing step was at 4000 V Equilibration of the IPG strips were firstly performed with an equilibration buffer (6 M Urea, 50 mM Tris- HCl, pH 8.8, 2% SDS, 30% glycerol and 2% DTT (w/v), for 15 min. Immediately after the first equilibration step, the alkylation procedure was carried out by incubating each strip in the second equilibration buffer with 2.5% of iodoacetamide (IAA).

The second dimensional electrophoresis was implemented using the Laemmli method [55] on the 12% polyacrylamide gels using the Mini-Protean II electrophoresis cell. The electrophoretic separation was carried out using the running buffer (3% Tris base, 14.4% glycine, 1% SDS), for 90 minutes at 100 V.

Staining and digitization of protein pattern

Prior to staining, the gels were fixed overnight in a fixative solution (30% Ethanol/10% Acetic acid) on a benchtop shaker SASTEC; ST-344 model, at 100 rpm. The gels were stained with 0.25% Coomassie Brilliant Blue R-250 in 40% (v/v) methanol and 7% (v/v) acetic acid for 1 h and de-stained with 40% (v/v) methanol and 7% (v/v) acetic acid until the backgrounds became clear. The de-stained gels were then transferred into the storage bags containing 25% methanol and were kept at 4°C. The gels were scanned and visualized using a densitometer. The scanned gels were saved as TIFF (Tagged Image File Format) images and exported to PDQuest 2-DE analysis software for further analysis.

To determine differences in protein expression, three gel images from each treatment containing the spots commonly present in all the three gels were analyzed using the PDQuest software. In brief, the software detects all the protein spots with background subtraction and then matches the spots of the control and treated samples. The detection parameters were determined by using reference spots covering small, faint and the largest spots on the master gel. During this process, additional filtering was also carried out to remove the horizontal and vertical streaks. At least three replicates of each gel slice belonging to each treatment were aligned to create a composite image containing only those proteins that are common to all three gels. The images from different treatments were aligned to create a super composite image using the matching tools. The 2-DE images of the salt-treated and control samples were matched with each other to determine the commonly induced proteins.

Identification of the Functional Proteins

The selected induced, upregulated and downregulated leaf and root proteins of *A. paniculata* were trypsin digested and peptides were extracted according to the standard techniques [56]. The peptides were analyzed by automated matrix-associated laser desorption/ionization time-of-flight mass spectrometry (MALDI TOF MS) using a 5800 Proteomics Analyzer. Spectra were analyzed to identify the protein of interest using Mascot sequence matching software (Matrix Science) with Ludwig NR Database.

Statistical analysis

Independent samples t-test was used for comparison of the representing proteins (repressed, induced, upregulated and downregulated proteins) in the control and salt-treated sample.

Acknowledgments

The authors appreciate Universiti Putra Malaysia for supporting this research. Our appreciation is extended to Professor Dr. Soon Guan Tan (Chief Editor of Pertanika Journal of Tropical Agricultural Science) for spending his precious time to perform a final proofreading on this article.

Author Contributions

Conceived and designed the experiments: DT. Performed the experiments: DT AV. Analyzed the data: DT. Contributed reagents/materials/analysis tools: MYR. Wrote the paper: DT AV. Proofreading: AV MM.

References

1. Valdiani A, Mihdzar AK, Tan SG, Talei D, Puad MA, et al. (2012) Nain-e Havandi (*Andrographis paniculata*) present yesterday, absent today: A plenary review on underutilized herb of Iran's pharmaceutical plants. Mol Biol Rep 39: 5409–5424.
2. Ashraf MY, Sarwar G (2002) Salt tolerance potential in some members of Brassicaceae physiological studies on water relations and mineral contents. Prospect Saline Agri: 237–245.
3. Talei D, Valdiani A, Maziah M, Sagineedu SR, Saad MS (2013) Analysis of the Anticancer Phytochemicals in *Andrographis paniculata* Nees. under Salinity Stress. BioMed Res Int 11 pages, doi: 10.1155/2013/319047.
4. Ashraf M (2004) Some important physiological selection criteria for salt tolerance in plants. Flora-Morphology, Distribution, Funct Ecol Plant 199 (5). 361–376.
5. Agarwal S, Pandey V (2004) Antioxidant enzyme responses to NaCl stress in *Cassia angustifolia*. Biol Plantarum 48 (4): 555–560.
6. Sharma P, Jha AB, Dubey RS, Pessarakli M (2012) Reactive oxygen species, oxidative damage, and antioxidative defense mechanism in plants under stressful conditions. J Bot: 26 pages, doi.org/10.1155/2012/217037.
7. Sanchez DH, Lippold F, Redestig H, Hannah MA, Erban A, et al. (2008) Integrative functional genomics of salt acclimatization in the model legume *Lotus japonicus*. Plant J 5353(6): 973–987.
8. Graves PR, Haystead TAJ (2002) Molecular biologist's guide to proteomics. Microbiol Mol Biol Rev 66 (1): 39–63.
9. Kettman JR, Coleclough C, Frey JR, Lefkovits I (2002) Clonal proteomics: One gene–family of proteins. Proteomics 2 (6): 624–631.
10. Malakshah SN, Rezaei MH, Heidari M, Salekdeh GH (2007) Proteomics reveals new salt responsive proteins associated with rice plasma membrane. Biosci Biotechnol Biochem 71 (9): 2144–2154.
11. Parker R, Flowers TJ, Moore AL, Harpham NVJ (2006) An accurate and reproducible method for proteome profiling of the effects of salt stress in the rice leaf lamina. J Exp Bot 57 (5): 1109.
12. Srivastava S, Rahman MH, Shah S, Kav NNV (2006) Constitutive expression of the pea ABA-responsive 17 (ABR17) cDNA confers multiple stress tolerance in *Arabidopsis thaliana*. Plant Biotechnol J 4 (5): 529–549.
13. Saqib M, Zörb C, Schubert S (2006) Salt-resistant and salt-sensitive wheat genotypes show similar biochemical reaction at protein level in the first phase of salt stress. J Plant Nutr Soil Sci 169 (4): 542–548.
14. Salekdeh GH, Siopongco J, Wade L, Ghareyazie B, Bennett J (2002) A proteomic approach to analyzing drought-and salt-responsiveness in rice. Field Crop Res 76 (2–3): 199–219.
15. Hernandez JA, Almansa MS (2002) Short term effects of salt stress on antioxidant systems and leaf water relations of pea leaves. Physiol Plantarum 115 (2): 251–257.
16. Mittova V, Tal M, Volokita M, Guy M (2002) Salt stress induces up-regulation of an efficient chloroplast antioxidant system in the salt-tolerant wild tomato species *Lycopersicon pennellii* but not in the cultivated species. Physiol Plantarum 115 (3): 393–400.
17. Lee DH, Kim YS, Lee CB (2001) The inductive responses of the antioxidant enzymes by salt stress in the rice (*Oryza sativa* L.). J Plant Physiol 158 (6): 737–745.
18. Rossignol M, Peltier JB, Mock HP, Matros A, Maldonado AM, et al. (2006) Plant proteome analysis: a 2004–2006 update. Proteomics 6 (20): 5529–5548.
19. Mohamed AA (2005) Two-dimensional electrophoresis of soluble proteins and profile of some isozymes isolated from maize plant in response to NaCl. Res J Agri Biol Sci 1 (1): 38–44.
20. QingJie G, Lin L, Takano T, ShenKui L (2009) Cloning of an ascorbate peroxidase gene from *Puccinellia tenuiflora* and its expression analysis. Genome Appl Biol 28 (4): 631–639.
21. Chai L, Wang J, Fan Z, Liu Z, Li X, et al. (2011) Ascorbate peroxidase gene from *Brassica napus* enhances salt and drought tolerances in *Arabidopsis thaliana*. Afr J Biotechnol 10 (79): 18085–18091.
22. Zhou J, Wang X, Jiao Y, Qin Y, Liu X, et al. (2007) Global genome expression analysis of rice in response to drought and high-salinity stresses in shoot, flag leaf, and panicle. Plant Mol Biol 63 (5): 591–608.
23. Rabbani MA, Maruyama K, Abe H, Khan MA, Katsura K, et al. (2003) Monitoring expression profiles of rice genes under cold, drought, and high-salinity stresses and abscisic acid application using cDNA microarray and RNA gel-blot analyses. Plant Physiol 133 (4): 1755–1767.
24. Dooki AD, Mayer-Posner FJ, Askari H, Zaiee A, Salekdeh GH (2006) Proteomic responses of rice young panicles to salinity. Proteomics 6 (24): 6498–6507.
25. Shen W, Gómez-Cadenas A, Routly EL, Ho THD, Simmonds JA, et al. (2001) The Salt stress-inducible protein Kinase gene, Esi47, from the salt-tolerant wheat grass *Lophopyrum elongatum* is involved in plant hormone signaling. Plant Physiol 125 (3): 1429–1441.
26. Meskiene I, Baudouin E, Schweighofer A, Liwosz A, Jonak C, et al. (2003) Stress-induced protein phosphatase 2C is a negative regulator of a mitogen-activated protein kinase. J Biol Chem 278 (21): 18945–18952.
27. Mahajan S, Pandey GK, Tuteja N (2008) Calcium-and salt-stress signaling in plants: shedding light on SOS pathway. Arch Biochem Biophys 471 (2): 146–158.
28. Zhu JK (2002) Salt and drought stress signal transduction in plants. Ann Rev Plant Biol 53 (1): 247–273.
29. Alamgir AN, Ali MY (1999) Effect of salinity on leaf pigments, sugar and protein concentrations and chloroplast ATPase activity of rice (*Oryza sativa* L.). Bangl J Bot 28 (2): 145–150.
30. Parida A, Das AB, Das P (2002) NaCl stress causes changes in photosynthetic pigments, proteins, and other metabolic components in the leaves of a true mangrove, *Bruguiera parviflora*, in hydroponic cultures. J Plant Biol 45 (1): 28–36.
31. Prochazkova D, Wilhelmova N (2007) Leaf senescence and activities of the antioxidant enzymes. Biol Plantarum 51 (3): 401–406.
32. Asada K (1999) The water-water cycle in chloroplasts: scavenging of active oxygens and dissipation of excess photons. Ann Rev Plant Biol 50 (1): 601–639.
33. Hernández-Nistal J, Dopico B, Labrador E (2002) Cold and salt stress regulates the expression and activity of a chickpea cytosolic Cu/Zn superoxide dismutase. Plant Sci 163 (3): 507–514.
34. Meloni DA, Oliva MA, Martinez CA, Cambraia J (2003) Photosynthesis and activity of superoxide dismutase, peroxidase and glutathione reductase in cotton under salt stress. Environ Exp Bot 49 (1): 69–76.
35. Lechno S, Zamski E, Tel-Or E (1997) Salt stress-induced responses in cucumber plants. J Plant Physiol 150 (1): 206–211.
36. Hernandez JA, Campillo A, Jimenez A, Alarcon JJ, Sevilla F (1999) Response of antioxidant systems and leaf water relations to NaCl stress in pea plants. New Phytol 141 (2): 241–251.
37. Benavídes MP, Marconi PL, Gallego SM, Comba ME, Tomaro ML (2000) Relationship between antioxidant defence systems and salt tolerance in *Solanum tuberosum*. Funct Plant Biol 27 (3): 273–278.
38. Hernandez JA, Jimenez A, Mullineaux P, Sevilia F (2000) Tolerance of pea (*Pisum sativum* L.) to long-term salt stress is associated with induction of antioxidant defences. Plant Cell Environ 23 (8): 853–862.
39. Comba ME, Benavides MP, Tomaro ML (1998) Effect of salt stress on antioxidant defence system in soybean root nodules. Funct Plant Biol 25 (6): 665–671
40. Asada K (1992) Ascorbate peroxidase–a hydrogen peroxide-scavenging enzyme in plants. Physiol Plantarum 85 (2): 235–241.
41. Noctor G, Veljovic-Jovanovic S, Foyer CH (2000) Peroxide processing in photosynthesis: antioxidant coupling and redox signalling. Philosophical Transactions of the Royal Society of London Series B: Biol Sci 355 (1402): 1465–1475.
42. Sivakumar P, Sharmila P, Pardha Saradhi P (2000) Proline alleviates salt-stress-induced enhancement in ribulose-1, 5-bisphosphate oxygenase activity. Biochem Biophys Res Commun 279 (2): 512–515.
43. Valdiani A, Kadir MA, Saad MS, Talei D, Tan SG (2012) Intra-specific hybridization: Generator of genetic diversification and heterosis in *Andrographis paniculata* Nees. A bridge from extinction to survival. Gene 505(1): 23–36. doi: 10.1016/j.gene. 2012.05.056.
44. Valdiani A, Talei D, Tan SG, Kadir MA, Maziah M, et al. (2014) A classical genetic solution to enhance the biosynthesis of anticancer phytochemicals in *Andrographis paniculata* Nees. PLoS ONE 9(2): e87034. doi:10.1371/journal.-pone.0087034.
45. Valdiani A, Talei D, Kadir MA, Maziah M, Tan SG, (2014) Morpho-molecular analysis as a prognostic model for repulsive feedback of "*Andrographis paniculata*" to allogamy. Gene 542: 156–167. doi: 10.1016/j.gene.2014.03.039.
46. Valdiani A, Kadir MA, Saad MS, Talei D, Omidvar V, et al. (2012) Intra-specific crossability in Andrographis paniculata Nees. A barrier against breeding of the species. Sci World J. 2012: 1–9. doi: 10.1100/2012/297545.
47. Guo B, Chen Y, Zhang G, Xing J, Hu Z, et al. (2013) Comparative proteomic analysis of embryos between a maize hybrid and its parental lines during early stages of seed germination. PLoS ONE 8(6): e65867. doi:10.1371/journal.-pone.0065867.
48. Talei D, Valdiani A, Yusop MK, Abdullah MP (2013) Estimation of salt tolerance in *Andrographis paniculata* accessions using multiple regression model. Euphytica 189: 147–160.
49. Talei D, Valdiani A, Abdullah MP, Hassan SA (2012) A rapid and effective method for dormancy breakage and germination of King of Bitters (*Andrographis paniculata* Nees.) seeds. Maydica 57 (2012): 98–105.
50. Talei D, Saad MS, Khanif YM, Kadir MA, Valdiani A (2011) Effect of different surface sterilizes on seed germination and contamination of King of Bitters (*Andrographis paniculata*). American-Eurasian J Agric Environ Sci 10(4): 639–643.
51. Talei D, Mihdzar AK, Khanif MY, Valdiani A, Puad MA (2012) Response of King of Bitters (*Andrographis paniculata* Nees.) seedlings to salinity stress beyond the salt tolerance threshold. Aust J Crop Sci 6 (6): 1059–1067.
52. Talei D, Valdiani A, Puad M (2013) An effective protein extraction method for two-dimensional electrophoresis in the anticancer herb (*Andrographis paniculata* Nees.). Biotechnol Appl Biochem 2013. doi: 10.1155/2013/319047.
53. Bradford MM (1976) A rapid and sensitive method for the quantitation of microgram quantities of protein utilizing the principle of protein-dye binding. Anal Biochem 72 (1–2): 248–254.
54. O'Farrell PH (1975) High resolution two-dimensional electrophoresis of proteins. J Biol Chem 250 (10): 4007–4021.
55. Laemmli UK (1970) Cleavage of structural proteins during the assembly of the head of bacteriophage T4. Nature 227 (5259): 680–685.
56. Bringans S, Eriksen S, Kendrick T, Gopalakrishnakone P, Livk A, et al. (2008) Proteomic analyses of the venom of *Heterometrus longimanus* (Asian black scorpion). Proteomics. 8: 1081–1096.

21

The *Arabidopsis* PLAT Domain Protein1 Is Critically Involved in Abiotic Stress Tolerance

Tae Kyung Hyun[1,2], Eric van der Graaff[1,3], Alfonso Albacete[1,4], Seung Hee Eom[1,2], Dominik K. Großkinsky[1,3], Hannah Böhm[1], Ursula Janschek[1], Yeonggil Rim[2], Walid Wahid Ali[5], Soo Young Kim[6], Thomas Roitsch[1,3,7]*

1 Institute of Plant Sciences, University of Graz, Graz, Austria, 2 Plant Molecular Biology and Biotechnology Research Center, Gyeongsang National University, Jinju, Korea, 3 Department of Plant and Environmental Sciences, Copenhagen Plant Science Centre, University of Copenhagen, Taastrup, Denmark, 4 Departamento de Nutrición Vegetal, CEBAS-CSIC, Campus Espinardo, Murcia, Spain, 5 Department of Pharmaceutical Biology, University of Würzburg, Würzburg, Germany, 6 Department of Molecular Biotechnology and Kumho Life Science Laboratory, College of Agriculture and Life Sciences, Chonnam National University, Gwangju, Korea, 7 Global Change Research Centre, CzechGlobe AS CR, v.v.i., Drásov, Czech Republic

Abstract

Despite the completion of the *Arabidopsis* genome sequence, for only a relatively low percentage of the encoded proteins experimental evidence concerning their function is available. Plant proteins that harbour a single PLAT (Polycystin, Lipoxygenase, Alpha-toxin and Triacylglycerol lipase) domain and belong to the PLAT-plant-stress protein family are ubiquitously present in monocot and dicots. However, the function of PLAT-plant-stress proteins is still poorly understood. Therefore, we have assessed the function of the uncharacterised *Arabidopsis* PLAT-plant-stress family members through a combination of functional genetic and physiological approaches. *PLAT1* overexpression conferred increased abiotic stress tolerance, including cold, drought and salt stress, while loss-of-function resulted in opposite effects on abiotic stress tolerance. Strikingly, PLAT1 promoted growth under non-stressed conditions. Abiotic stress treatments induced *PLAT1* expression and caused expansion of its expression domain. The ABF/ABRE transcription factors, which are positive mediators of abscisic acid signalling, activate *PLAT1* promoter activity in transactivation assays and directly bind to the ABRE elements located in this promoter in electrophoretic mobility shift assays. This suggests that *PLAT1* represents a novel downstream target of the abscisic acid signalling pathway. Thus, we showed that PLAT1 critically functions as positive regulator of abiotic stress tolerance, but also is involved in regulating plant growth, and thereby assigned a function to this previously uncharacterised PLAT domain protein. The functional data obtained for PLAT1 support that PLAT-plant-stress proteins in general could be promising targets for improving abiotic stress tolerance without yield penalty.

Editor: Vladimir N. Uversky, University of South Florida College of Medicine, United States of America

Funding: This research was supported by the Basic Science Research Program through the National Research Foundation of Korea (NRF) funded by the Ministry of Education (NRF-2014R1A1A2004432), Republic of Korea (T.K.H.), and by the Fundación Séneca, Comunidad Autónoma de la Región de Murcia (A.A.). The funders had no role in study design, data collection and analysis, decision to publish, or preparation of the manuscript.

Competing Interests: The authors have declared that no competing interests exist.

* Email: roitsch@plen.ku.dk

These authors contributed equally to this work.

Introduction

The PLAT domain (PS50095; **P**olycystin-1, **L**ipoxygenase, **A**lpha-toxin and **T**riacylglycerol lipase) forms a β-sandwich composed of two sheets of four strands each and is an intracellular domain. It occurs in a variety of membrane or lipid associated proteins that are multi-domain proteins, but also in proteins harbouring either a single PLAT domain or repeats [1–5]. Because of its similarity to the C2 domain, the PLAT domain was proposed to function in protein-protein interactions as well as protein-membrane interactions [2,4,6]. Indeed, the PLAT domain of *Caenorhabditis elegans* polycystin LOV-1 and human polycystin-1 interact with ATP-2, an ATP synthase F1 subunit [7], while the human polycystin-1L2 interacts with different types of G-proteins [8]. Importantly, the association with membranes is essential for the proper function of PLAT domain proteins [2,7]. Further, the membrane targeted 11*R*-Lipoxygenase from *Gersemia fruticosa* was shown to bind calcium, required to induce its activity [9]. The PLAT domain regulates the catalytic activity in multi-domain proteins, but also in proteins interacting with the PLAT domain [2,6], and was shown to regulate substrate specificity [10]. Whereas substantial experimental data on PLAT domain proteins is available for the animal field, PLAT domain proteins from plants were only poorly studied, despite the fact that genes encoding PLAT domain proteins were isolated from several plant species [3,4,11–16].

Transgenic approaches to improve abiotic stress tolerance often resulted in yield penalties under optimal growth conditions [17,18], while only few studies reported an associated improved plant growth [4,19]. Interestingly, one of these studies addressed the so far only studied PLAT-plant-stress protein *Ca*TIN1 from *Capsicum annuum*, however this protein was only studied using heterologous expression in tobacco [4]. Because both gain-of-function and antisense *CaTIN1* expression promoted abiotic and biotic stress tolerance, *Ca*TIN1 function remained elusive. Proteins that belong to the PLAT-plant-stress protein family (Conserved Domain cd1754) are ubiquitously present in monocot and dicot plant species and harbour a single PLAT domain. Our analyses of the limited *in silico* expression data available for PLAT-plant-stress proteins indicate transcriptional induction by different abiotic and biotic stimuli. This suggests that PLAT-plant-stress proteins in general could promote tolerance towards stress responses, although no data from functional studies in homologous systems are available for these proteins.

The plant hormone abscisic acid (ABA) regulates different aspects of plant development, such as stomatal aperture [20] and seed germination [21]. ABA production is increased by abiotic stresses and ABA regulated genes strongly overlap with those induced under drought, salinity and less prominently cold stress conditions [22–25]. The ABA stimulated stomatal closure has been shown to serve as primary defence mechanism during the initial phase of biotic stress responses [26,27]. In contrast, ABA mostly negatively regulates the subsequent phases in biotic stress responses by repressing the salicylic acid, ethylene, jasmonic acid and cytokinin signalling pathways [26–29]. ABA-deficient mutants showed enhanced defence responses against *Botrytis cinerea* [30], and virulent bacteria in tomato [31] and *Arabidopsis* [32]. Although these findings suggest that ABA is involved in the crosstalk between abiotic and biotic stress responses, no direct link in the antagonistic interaction between these stresses is available.

Based on comparative genomic analysis, we identified three *Arabidopsis* genes (*PLAT1 AT4G39730*, *PLAT2 AT2G22170* and *PLAT3 AT5G65158*) that belong to the PLAT-plant-stress subgroup and submitted this annotation to the TAIR database. PLAT1 and PLAT2 are orthologs of *Ca*TIN1 and *Ca*TIN1-2, respectively [4,16]. Our *in silico* analysis of published experimental data [33] revealed that cold stress induced the expression levels for the *PLAT1* ortholog in *Thlaspi arvense*, which is a close relative of *Arabidopsis*. Based on these findings we hypothesised that the *Arabidopsis* members from this PLAT-plant-stress subgroup, similar to *Ca*TIN1 [4], also promote tolerance towards various stress responses. Here we report on the molecular characterisation and functional analysis of the PLAT-plant-stress subgroup family member AT4G39730 that we designated as *Arabidopsis* PLAT domain protein 1 (PLAT1). We showed that *PLAT1* critically functions as positive regulator of abiotic stress tolerance, also promotes plant growth and is a direct target of the ABF transcription factors, which are positive mediators of the ABA signalling pathway [34,35]. The possible practical application to increase abiotic stress tolerance without yield penalty in crop species is discussed.

Materials and Methods

Plasmid construction and plant transformation

Total RNA isolated from *Arabidopsis* (TRIR reagent from Thermo Fischer Scientific, Germany) was reverse-transcribed using the ReverAid^{TM} First strand cDNA synthesis kit (Thermo Fischer Scientific, Germany). Using this cDNA as template, the full-length *PLAT1* cDNA was amplified by PCR with the PLAT1-F cDNA and PLAT1-R cDNA primers (Table S1 in File S1). The dexamethasone inducible overexpression construct, *35S>> PLAT1* was created by cloning the *PLAT1* cDNA PCR product in the *OP* shuttle vector *pEG647*. The resulting *OP:PLAT1* cassette was transferred to the binary vector *pEG618* harbouring the *35S:LhGR* activator component, resulting in *35S>>PLAT1*. The *PLAT1* rescue/reporter constructs were generated by PCR amplification from genomic DNA with the PLAT1-F and PLAT1-R rescue primers (Table S1 in File S1) to isolate the genomic fragment harbouring 2039 bp *PLAT1* promoter sequence, including the 5′-UTR region, and the *PLAT1* coding region without stop codon. Subsequently, the venus YFP or GUS reporter proteins were fused in frame to the C-terminus of the PLAT1 protein resulting in *PLAT1:PLAT1-YFP* and *PLAT1:-PLAT1 GUS*, respectively. The different *35S:ABF$_{1-4}$* overexpression constructs were created by cloning the *ABF$_{1-4}$* PCR products (ABF$_{1-4}$-F OX and ABF$_{1-4}$-R OX primers, Table S1 in File S1) into the *pPS1* binary vector. The MBP-ABF$_{1-4}$ fusion proteins for the EMSA experiments were created by cloning the ABF$_{1-4}$ PCR products (ABF$_{1-4}$-F and ABF$_{1-4}$-R primers; Table S1 in File S1) into the *pMA-c2xL* vector harbouring the maltose binding protein as tag for protein purification. Binary vectors were introduced into *Agrobacterium tumefaciens* LBA4404 by electroporation and used to transform *Arabidopsis* plants using the floral dip method [36].

Phylogenetic analysis

To identify members of the PLAT-plant-stress family from other plant species, multiple database searches were performed using the Basic Local Alignment Search Tool (BLAST) algorithms BLASTp and tBLASTn available on the public databases, PLAZA 2.0 (bioinformatics.psb.ugent.be/plaza) and Phytozome v8.0 (www.phytozome.net) with cutoff value of $E<10^{-5}$. We used nucleotide and amino acid sequences of PLAT1 from the TAIR database (www.arabidopsis.org) to BLAST all databases. Phylogenetic analysis was performed by using CLUSTALW alignment in PHYLIP format clustal algorithm, and displayed in a phylogram tree format with locus name of each protein. Bootstrap values were presented as a percent of 100 resampled trees at each tree node using default settings of the TreeTop-Phylogenetic Tree (www.genebee.msu.su/services/phtreereduced.html).

Plant materials and growth conditions

Nicotiana benthamiana plants were grown under greenhouse conditions as described previously [37]. *Arabidopsis* plants (Col-0 ecotype) were grown in soil at 8 h light/16 h darkness at 22°C (light intensity: 180 μmol m^{-2} s^{-1}) or on half strength MS medium under continuous light at 22°C (light intensity: 180 μmol m^{-2} s^{-1}) in growth cabinets. T3 homozygous T-DNA insertion lines were obtained for the *PLAT1* gene, *plat1-1* SALK-112728c and *plat1-2* SALK-1283454c, and the *PLAT2* gene, *plat2* SAIL-1171C06. The T-DNA insertions were verified with the primers PLAT1-1-F, PLAT1-1-R and SALK LB2 (*plat1-1*), PLAT1-2-F, PLAT1-1-R and SALK LB2 (*plat1-2*), and PLAT2-F, PLAT2-R and pROK2 LB1 (*plat2*) (Table S1 in File S1). For all plant experiments T3 or T4 homozygous plant lines were employed, based on the segregation of the respective antibiotic selection marker, except for the experiments shown in Figure S4, for which segregating T2 *Arabidopsis* lines were employed.

Abiotic stress conditions

The abiotic stress experiments in soil were performed as 3 biological replicates (cold stress as 2 biological replicates) with at least 10 plants each. For the *Arabidopsis* germination experiment, seeds were directly germinated on half strength MS medium including the respective chemicals as indicated. At least 100 seeds per treatment/genotype were used in 3 independent experiments. For the salt stress tolerance in plates, seeds were germinated and grown for 6 d on half strength MS medium, transferred to half strength MS medium including NaCl (either 0, 150 or 200 mM) and optionally 5 µM dexamethasone (*35S>>PLAT1* lines), and grown for another 4 d. These experiments were performed as 2 biological replicates, each with 2 technical replicates and with >12 Col-0 and >26 mutant/transgenics seedlings per plate. For the tunicamycin (TM) experiments, seeds were directly germinated on half strength MS medium including different TM concentrations. At least 70 seeds per treatment/genotype were used in 3 independent experiments, with 2 technical replicates each.

Biotic stress conditions

To determine pathogen susceptibility, leaves from 8-w-old *Arabidopsis* plants were infected with *Pseudomonas syringae* pv. *tomato* DC3000 with or without the *avrRpm1* gene by infiltration using a needleless syringe as described previously [38]. Visual evaluation of disease symptoms were conducted at 3 to 5 d. For expression analysis, 8-w-old plants were infected with *P. syringae* pv. *tomato* DC3000 or *Sclerotinia sclerotiorum*. Subsequently, the *S. sclerotiorum* infected plants were kept in a clear plastic box under saturating humidity. The biotic stress experiments were performed as 3 biological replicates with at least 10 plants each.

Expression analysis

Total RNA isolation and Northern-blot analysis was carried out as described previously [37]. Filters were exposed to a screen for 4 d, which was scanned with a Phosphor imager (Fuji BAS2000, Ray-test, Germany). The probes for the *PLAT1*, *PLAT2* and *PLAT3* genes were generated by PCR from cDNA with the primers, PLAT1-F probe and PLAT1-R probe, PLAT2-F probe and PLAT2-R probe, and PLAT3-F probe and PLAT3-R probe (Table S1 in File S1), respectively. The RT-PCR analysis was performed essentially as described before [37]. At least 10 seedlings per genotype were grown for 14 d on control plates before transfer to the respective stress and control medium. The optimal cycle number was determined for each primer pair (Table S2 in File S1). Expression values were corrected for the *ACTIN* and *UBIQUITIN* signal intensities and expressed as relative values compared to Col-0. For expression analysis, 3 independent experiments were performed.

ABA determination

The extraction and analysis of ABA was carried out as described previously [39].

PLAT1-YFP localisation

For transient *PLAT1-YFP* expression a single colony of *A. tumefaciens* LBA4404 containing either *PLAT1:PLAT1-YFP* or the *ER-rk CD3-959* ER-marker [40] construct was inoculated into 5 ml induction medium with antibiotics and grown overnight at 28°C. The bacteria were collected by centrifugation and resuspended in 10 mM MES and 10 mM $MgCl_2$ containing 200 µM acetosyringone to an OD_{600} of 1.0. Aliquots (1 ml) of *A. tumefaciens* cells containing *PLAT1:PLAT1-YFP* and ER-marker construct were mixed together, and then a syringe was used to infiltrate the mixture into the lower surface of *N. benthamiana* leaves. YFP and mCherry fluorescence was visualized 48 h post infiltration, using Olympus confocal laser scanning microscope (model FV1000, Tokyo, Japan). For stable *PLAT-YFP* expression following ABA and salt treatment, the *PLAT1:PLAT1-YFP* line *YFP13-1* was employed, 10 plants per treatment.

Transactivation assay

Suspensions of *A. tumefaciens* carrying the respective *35S:ABF₁₋₄* overexpression constructs and the *PLAT1:PLAT1-GUS* rescue/reporter construct were mixed in a ratio of 1:1. The resulting mixed suspensions were used to infiltrate leaves of 6-w-old greenhouse grown *N. benthamiana* in soil. As control, leaves were infiltrated with *A. tumefaciens* carrying the *PLAT1:PLAT1-GUS* rescue construct only, or 10 mM $MgCl_2$. The GUS fluorometric assays were carried out as described previously [41]. Samples were isolated from the infiltrated regions 2 d after infiltration for 5 independent plants and ground in 500 µl of extraction buffer containing 50 mM sodium phosphate, 10 mM EDTA, 10 mM β-mercaptoethanol and 0.1% N-lauroylsarcosine (pH = 7.4). After centrifugation at 4°C for 10 min at 13,000 rpm, the supernatant was used for the determination of GUS enzyme activity. 50 µl of supernatants was transferred into one slot of a black 96-well plate and 50 µl of a 2 mM MUG (methylumbelliferyl-β-D-glucuronide, Sigma) solution was added. For each sample 3 technical replicates were measured. The samples were incubated at 37°C for 30, 60 and 90 min, before the reaction was stopped with 1 M sodium carbonate. A standard curve was prepared with MU (4-methylumbelliferone, Sigma) in a concentration range from 0 to 16 µM. Excitation was measured at 365 nm, emission at 455 nm. Total protein amount was determined by the Bradford assay. GUS enzyme activity was calculated in pmol Mu min^{-1} mg^{-1} protein.

Expression and purification of ABFs

The ABFs were prepared employing a MBP-fusion purification procedure. Five ml of an overnight bacterial culture was incubated with 500 ml rich broth medium containing glucose and ampicillin. The cells were grown to 2×10^8 cells ml^{-1} ($A_{600} = 0.5$). IPTG was added to a final concentration of 0.3 mM and a further incubation at 37°C for 2 h followed. The cells were harvested by centrifugation at 4,000 g for 20 min. The supernatant was discarded and the pellet re-suspended in 25 ml of column buffer (20 mM Tris-HCl, 200 mM NaCl, 1 mM EDTA, 1 mM DTT and 0.1 mM PMSF (pH = 7.4). The pellet was kept at −20°C overnight and thawed in cold water the next morning. The sample was placed in an ice-water bath and sonicated in short pulses of 15 s for at least 2 min. The suspension was centrifuged at 9,000 g for 30 min and the supernatant diluted 1/3 (v/v) with column buffer before purification. Amylase resin was poured in a 5 ml column and washed with 8 column volumes of column buffer. The diluted sample was loaded and slowly ran over the column. The column was then washed with 12 volumes of column buffer and the proteins subsequently eluted with column buffer containing 10 mM maltose. 10 to 15 fractions containing 2 ml each were collected. Proteins were checked via SDS-page.

Electrophoretic mobility shift assay

For EMSA, the 200 bp *PLAT1* promoter, generated by PCR with the primers pPLAT1-F and pPLAT1-R (Table S1 in File S1) was used as positive probe. The mutated version lacking the 2 ABRE elements was generated in 2 steps by PCR using the primers WIP-F1 and WIP-R1, and WIP-F2 and WIP-R2. The 2 PCR fragments were joined to create the negative 200 bp probe WIP1. The 200 bp *PLAT1* promoter fragment was labelled with [γ-^{32}P]ATP using T4 polynucleotide kinase (5′ end labelling). The reaction was incubated for 1 h at 37°C, purified and eluted with 10 mM Tris, pH = 8.0. The labelled probe was incubated 30 min at room temperature with 5 µg of the respective ABFx protein extracts alone, with 100 fold molar excess of "cold" specific competitor (200 bp *PLAT1* promoter), and with 100 fold molar excess of "cold" negative probe (WIP1), including poly-dIdC as nonspecific competitor. The EMSA samples were run on a 5% native poly-acrylamide gel (10×10 cm). Before loading the samples, the gel was pre-run 40 min at 80 V and 4°C and the samples were run at 120 V and 4°C. After electrophoresis, radioactivity was detected in the dried gel as described above. Functionality of the EMSA assay and ABFx preparation using the MBP fusion protein was proven using the ABF1 protein extract and the published [34] positive ABRE-F and ABRE-R, and negative mABRE-F and mABRE-R control primers (Table S1 in File S1).

Statistical analysis

Standard deviations and average values were calculated in excel. Statistical significance for differences between treatments was analysed using the unpaired two sided Student's *t*-test in excel. ***, ** or * indicate statistical significance at $p<0.001$, $p<0.01$ or $p<0.05$, respectively.

Accession numbers

The AGI locus identifiers for the *Arabidopsis* PLAT-plant-stress family members are: PLAT1, AT4G39730; PLAT2, AT2G22170 and PLAT3, AT5G65158.

Results

PLAT1 expression is induced by abiotic stress conditions

The PLAT-plant-stress subgroup of PLAT domain proteins comprises three *Arabidopsis* family members. Phylogenetic analysis of these *Arabidopsis* PLAT-plant-stress proteins using the neighbour-joining method showed that PLAT3 falls outside the other members (Figure S1A). *In silico* expression analysis by the eFP Browser [42] showed *PLAT1* and *PLAT2* expression during many developmental stages, which is affected under different stress conditions, whereas *PLAT3* is not represented on the Affymetrix ATH1 arrays. To corroborate these data, we investigated expression of the *PLAT* family members by northern blot and RT-PCR analysis. This showed that *PLAT1* was indeed expressed throughout development (Figure S1B). In contrast, *PLAT2* expression was only detected in young seedlings (Figure S1E), whereas *PLAT3* expression could not be detected in any of the analysed organs and developmental stages, neither by northern blot nor RT-PCR analysis at 35 cycles. *PLAT1* expression was induced both by salt, following one day of watering with 200 mM NaCl (Figure S1C), and cold treatment, following incubation of 3-w-old plants at 8°C for 2 d (Figure S1D). *PLAT1* expression was also induced following the transfer of young seedlings to medium with 200 mM NaCl (Figure 1). However, both the extent of *PLAT1* induction and the temporal dynamics were different from that following salt stress using older plants grown in soil (Figure S1C), probably because *PLAT1* expression is highest in young seedlings (Figure S1B). *PLAT2* expression was repressed following the transfer of young seedlings to medium with 200 mM NaCl (Figure 1). These results confirmed the *in silico* expression data and suggested that among the *Arabidopsis* PLAT family members mainly PLAT1 is involved in abiotic stress responses.

PLAT1 promotes tolerance towards abiotic stress conditions

To analyse PLAT function in stress responses and plant development, we obtained T-DNA insertion mutants for *PLAT1* and *PLAT2*. Two independent homozygous *Arabidopsis* loss-of-function mutants were obtained for *PLAT1* (*plat1-1* and *plat1-2*) and one promoter insertion mutant was obtained for *PLAT2*

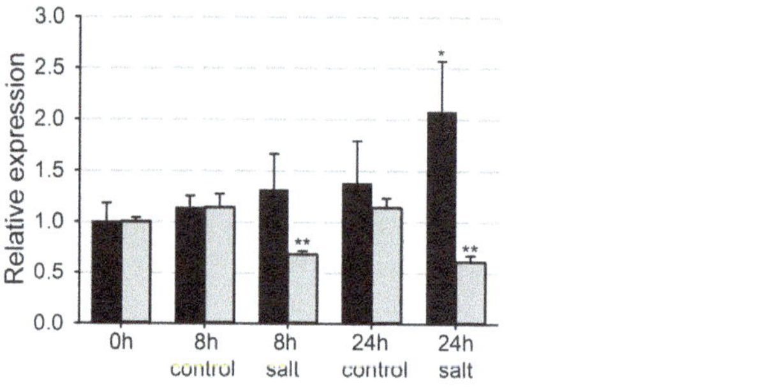

Figure 1. *PLAT1* expression is induced by salt stress conditions. Relative *PLAT1* and *PLAT2* expression in 14-d-old Col-0 seedlings following transfer to salt stress medium compared to control conditions. *PLAT1* black bars, *PLAT2* grey bars. Values are means of 3 replicates ± standard deviation. n≥10 per replicate. ** or * indicate statistical significance calculated using the unpaired Student's *t*-test at $p<0.01$ or $p<0.05$, respectively.

(*plat2*) from the SALK and SAIL mutant collections, respectively [43,44]. Since both *plat1* mutants exhibited similar phenotypic defects, only the characterisation of the *plat1-1* mutant is described in detail. The *plat1-1* and *plat2* mutants exhibited no obvious growth defects under control conditions and most likely represent null alleles because *PLAT1* and *PLAT2* expression was not detected in the respective insertion mutants (Figure S1E).

The *plat1-1* mutant was more sensitive to salt and drought stress, as well as cold stress conditions (Figure 2), evident by a reduction in root length upon growth at 8°C from 9.41±1.25 mm for Col-0 to 7.02±1.27 mm or 6.20±0.94 mm for *plat1-1* and *plat1-2*, respectively ($p<0.001$, $n=14$). In contrast, the *plat2* mutant did not show obvious changes in salt stress tolerance (Figure 3). Together with the differential effect of salt stress on expression of the *PLAT* family members (Figure 1), these data support that only PLAT1 is involved in abiotic stress tolerance.

To analyse whether PLAT1 also plays a role in biotic stress responses, we studied *PLAT1* expression following inoculation of plants with the hemibiotrophic pathogen *P. syringae* pv. *tomato* DC3000 or DC3000 *avrRpm1* (RPM1, Figure S2A), and the necrotrophic fungal pathogen *S. sclerotiorum* (Figure S2B). This showed that *PLAT1* expression was not specifically affected by these pathogens since increased expression was also observed for the respective control treatments (10 mM $MgCl_2$ and PDA medium). Next, disease symptom development following inoculation with *P. syringae* pv. *tomato* DC3000 was investigated. Disease

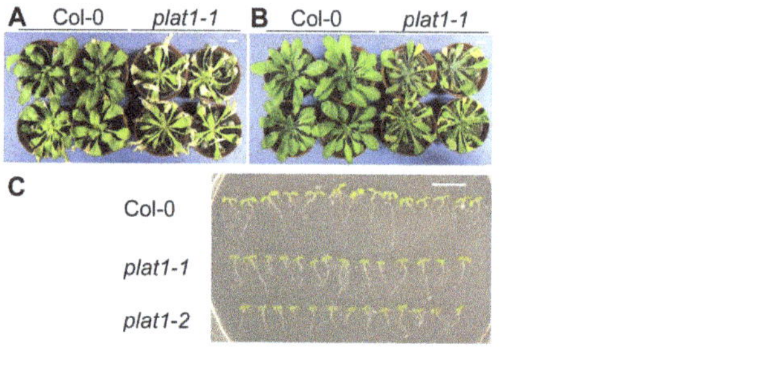

Figure 2. PLAT1 loss-of-function reduces abiotic stress tolerance. (**A**) Salt stress tolerance in wild-type (Col-0) and *plat1-1* seedlings irrigated with 200 mM NaCl for 14 d. n≥10 (**B**) Drought stress tolerance in wild-type and *plat1-1* seedlings, following 14 d without watering. n≥10 (**C**) Cold stress tolerance in 7-d-old wild-type, *plat1-1* and *plat1-2* seedlings following 14 d of incubation at 8°C. n = 14. Scale bar = 1 cm.

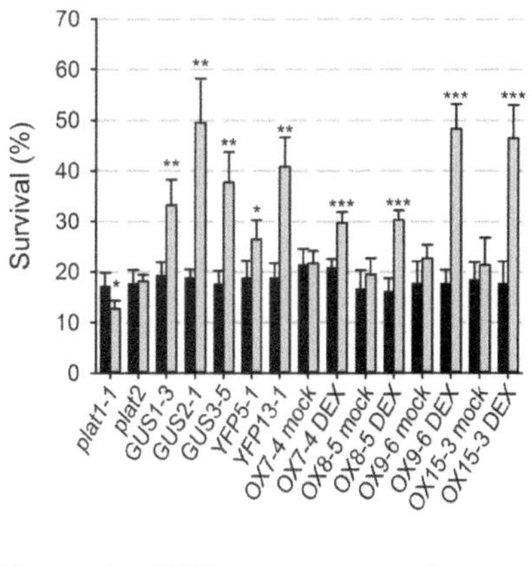

Figure 3. PLAT1 promotes tolerance towards salt stress conditions. Survival, expressed as the percentage of the seedlings transferred at the 6-d-old stage that developed (pale) green leaves during 4 d of salt stress conditions for the different mutant and transgenic lines (grey bars, n≥26 per replicate) compared to Col-0 control plants (black bars, n≥12 per replicate) that were grown on the same plates. Values are means of 3 replicates ± standard deviation. ***, ** or * indicate statistical significance calculated using the unpaired Student's *t*-test at $p<0.001$, $p<0.01$ or $p<0.05$, respectively.

symptom development was not obviously affected in *plat1-1* (Figure S2C), which is in agreement with the fact that biotic stress conditions did not significantly affect *PLAT1* expression.

Because ABA is strongly involved in abiotic stress responses, we analysed whether PLAT1 function is correlated with ABA signalling. One characteristic effect of ABA is the inhibition of seed germination [45]. Further, seed germination frequency is reduced by salt and osmotic stress conditions, resulting from increased ABA signalling. Under control conditions the germination frequency of the *plat1-1* seeds was similar to that of wild-type (Figure 4A). However, under osmotic stress using 300 mM mannitol (Figure 4B) and salt stress conditions employing 200 mM NaCl (Figure 4C), the *plat1-1* seeds exhibited a higher germination frequency compared to wild-type, suggesting that ABA signalling is reduced in *plat1-1*. Indeed, *plat1-1* seed germination is less severely reduced by ABA (1.5 µM) compared to wild-type (Figure 4D).

Since a change in seed dormancy could have influenced the seed germination assays and thus interpretation for ABA sensitivity, we analysed germination for the different PLAT lines on control medium without prior stratification. This showed that seed dormancy was only affected by PLAT2 loss-of-function (Figure S3). Further, ABA levels were not significantly affected in *plat1-1* plants compared to wild-type control (Table S3 in File S1). Together, these results indicate that PLAT1 loss-of-function reduced ABA sensitivity, which could have caused increased sensitivity towards abiotic stress conditions.

We generated genomic *PLAT1* rescue constructs, where the PLAT1 protein was fused to either the YFP or GUS reporter proteins. Following genetic complementation of the *plat1-1* mutant, seed responses towards ABA could be completely restored back to that of wild-type employing the genomic rescue/reporter constructs: 5 out of 6 lines for *PLAT1:PLAT1-GUS* (Figure S4A) and 15 out of 20 lines *PLAT1:PLAT1-YFP* (Figure S4B). This confirmed that the defects in abiotic stress tolerance were indeed caused by the T-DNA insertion in the *plat1-1* mutant and that the PLAT1 fusion proteins are functional in these rescued lines. Transgenic expression levels can be influenced to a large extend by plant sequences flanking the respective T-DNA integration sites, causing variation between individual transgenic lines harbouring the same T-DNA construct. In addition to transformed *plat1-1* lines with restored (wild-type) ABA sensitivity, also transgenics were obtained exhibiting increased ABA sensitivity and thus tissue specific *PLAT1* overexpression phenotype (Figure S4). To study whether *PLAT1* overexpression would confer an opposite phenotype compared to the *plat1* mutant, we selected these lines with tissue specific *PLAT1* overexpression for further analysis and also generated ectopic *PLAT1* overexpression lines. Because we could not obtain transgenics harbouring constitutive *PLAT1* overexpression, we generated lines with inducible *PLAT1* overexpression (*35S>>PLAT1*) employing the dexamethasone inducible *OP-LhGR* two component system [46]. From 17 transgenic *PLAT1* OX lines, the majority exhibited delayed germination on 1.5 µM ABA (Figure S4C).

To compare tissue specific with ectopic overexpression, which could cause considerable differences in overexpression phenotypes [47], we selected the *GUS1-3, GUS2-1, GUS3-5, YFP5-1* and *YFP13-1* genomic rescue lines with increased ABA sensitivity in the seed germination assays for further analysis. In these selected tissue specific (GUS and YFP) *PLAT1* overexpression lines, *PLAT1* transcript levels were increased to a similar extend as the ectopic (OX) overexpression lines, compared to wild-type control (Table S4 in File S1). The different types of *PLAT1* overexpression lines promoted salt stress tolerance after transfer to medium containing both 200 mM NaCl and 5 µM dexamethasone (DEX) (Figure 3). Thus, *PLAT1* overexpression resulted in higher *PLAT1* expression levels and reciprocal phenotypic effects compared to the *plat1-1* mutant, confirming that PLAT1 functions in promoting tolerance towards abiotic stress conditions.

PLAT1 promotes growth

The DEX induced *PLAT1* OX lines from the salt stress experiment shown in Figure 3, appeared to show faster development on control medium including DEX (Figure 5A-E). Therefore, shoot and root growth was analysed in the *plat1-1* mutant and *PLAT1* overexpression lines. This showed that the different *PLAT1* overexpression lines produced more shoot biomass compared to wild-type control (Figure 5F), while the length of the root apical meristem and total root length were not affected (Table S5 in File S1). Interestingly, the number of emerged lateral roots was increased, but only for the ectopic *PLAT1* (OX) overexpression lines (Table S5 in File S1). This effect on lateral root formation was the only difference observed between the tissue specific and ectopic overexpression lines, and could reflect the difference between increased expression at its natural expression site (GUS and YFP) and broader expression in new (ectopic) cell types (OX). In contrast, PLAT1 loss-of-function did not affect either shoot or root growth, except for a reduced number of emerged lateral roots (Figure 5F, Table S5 in File S1).

PLAT1 is expressed in structures related to the regulation of water household

The *PLAT1:PLAT1-GUS* rescue/reporter construct conferred GUS activity (4 out of 6 lines analysed) in vascular tissue, leaf edges, hydathodes and stomata, which are structures related to the regulation of water household. The expression in stomata corresponds to the identification of PLAT1 in guard cells [48]. GUS activity could also be observed in floral organs, root tips, pericycle cells and lateral root primordia (Figure 6A-6H). The basal *PLAT1* expression pattern in the shoot correlated with the abiotic stress tolerance, while that of the root correlated with the changes in lateral root number for the *plat1-1* mutant and *PLAT1* overexpression lines. Both treatment with salt and ABA resulted in

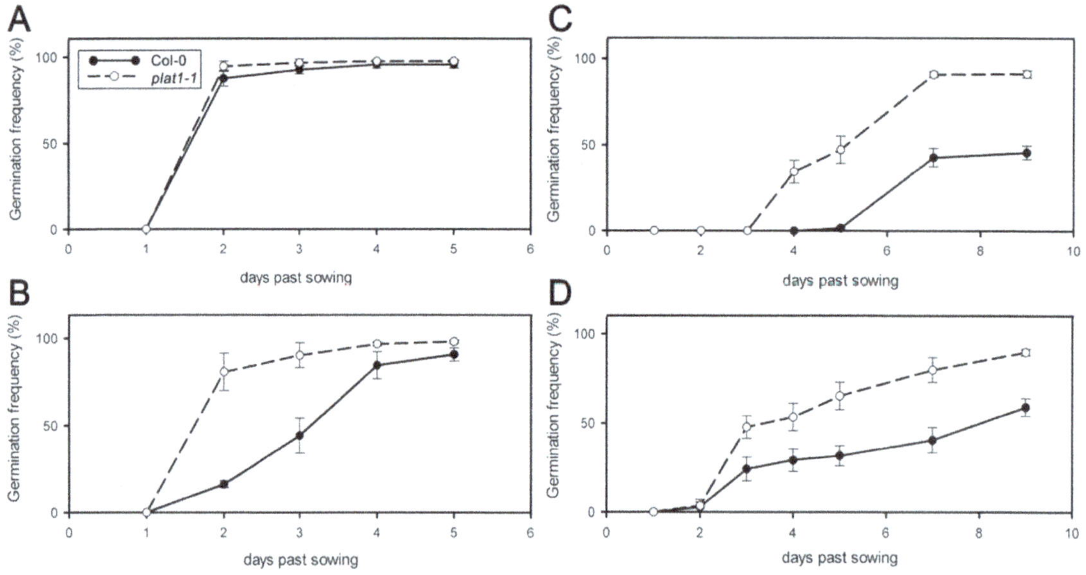

Figure 4. PLAT1 loss-of-function reduces ABA sensitivity during seed germination. (**A**) Seed germination of *plat1-1* and wild-type (Col-0). (**B**) Seed germination of *plat1-1* and wild-type on 300 mM mannitol (osmotic stress). (**C**) Seed germination of *plat1-1* and wild-type on 200 mM NaCl (salt stress). (**D**) Seed germination of *plat1-1* and wild-type on 1.5 μM ABA. Values are means of 3 replicates ± standard deviation. n≥100 per replicate.

an expanded *PLAT1* expression domain in root tips (Figure 6I-6N) and leaves (Figure 6O-6U), evident by the GUS activity in leaf mesophyll cells (Figure 6T and 6U), as well as in root pericycle cells (Figure S5).

To understand protein function, knowledge on its sub-cellular localisation is essential, because this influences access to and availability of interaction partners [49]. According to the Aramemnon membrane protein database [50], PLAT1 was predicted to contain a signal peptide involved in the secretory pathway and one transmembrane spanning domain. To analyse the sub-cellular localisation we transiently transformed the *PLAT1:PLAT1-YFP* construct together with different organelle markers [40] to *N. benthamiana*, which showed that the PLAT1-YFP fusion protein co-localised with the mCherry ER-marker ER-rk CD3-959 (Figure 7A-7C). The analysis of YFP reporter activity in the stable rescued *plat1-1* transformants harbouring the *PLAT1:PLAT1-YFP* construct reflected the expression pattern evident from the GUS rescue reporter lines, confirmed *PLAT1* induction by salt and ABA (Figure 7D-7I), and PLAT1 localisation to the ER (Figure 7J-7M). Further, PLAT1 is localised to rod shaped ER structures that resemble ER bodies (Figure 7M).

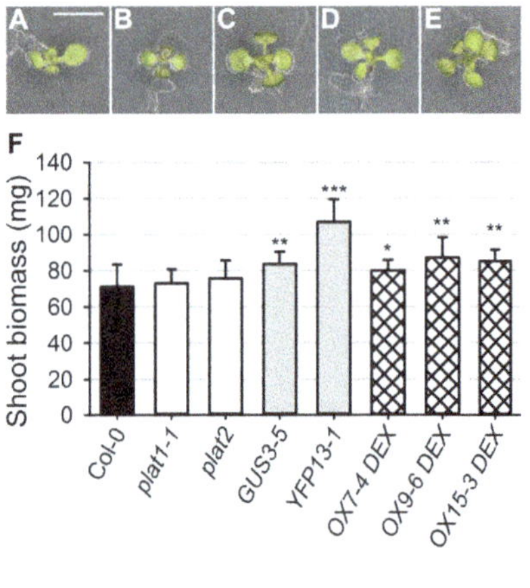

Figure 5. PLAT1 promotes plant growth. Phenotypes of plants from control medium including 5 μM DEX from the salt stress experiment shown in Figure 3. (**A**) Wild-type. (**B**) *plat1-1*. (**C-E**) Three independent *PLAT1* ectopic overexpression lines (OX). Scale bar = 1 cm. (**F**) Shoot biomass production (weight per 5 shoots) for the different PLAT1 lines. Values are means of 8 replicates ± standard deviation. ***, ** or * indicate statistical significance calculated using the unpaired Student's *t*-test at $p<0.001$, $p<0.01$ or $p<0.05$, respectively.

The *PLAT1* promoter is a direct target of the ABF transcription factors

Sequence analysis of the *PLAT1* promoter identified 2 G-boxes/ABRE elements (CACGTG motif) located at positions −165 to −156 and −134 to −126 relative to the transcription start site. ABRE elements are direct binding sites of the ABF/AREB (bZIP) transcription factors, which are positive mediators of the ABA signalling pathway [34,35]. We employed a transactivation assay to investigate whether the *PLAT1* promoter is indeed regulated by these ABF transcription factors. Leaves from wild-type tobacco *N. benthamiana* plants were simultaneously infiltrated with *A. tumefaciens* harbouring either one of the *35S:ABF$_{1-4}$* overexpression constructs (ABF$_{1-4}$) and the *PLAT1:PLAT1-GUS* genomic rescue/reporter construct (pGUS). Each of the double infiltrations resulted in significantly higher GUS activity levels, compared to infiltration with the *PLAT1:PLAT1-GUS* reporter construct only (Figure 8). This indicated that the ABF transcription factors activate the *PLAT1* promoter.

Electrophoretic mobility shift assays (EMSA) confirmed these transactivation results and showed that the ABF1, ABF3 and ABF4 transcription factors directly bind to the *PLAT1* 200 bp promoter fragment (−200 to +1), which contains the 2 ABRE elements. This ABF binding was specifically competed with a 100-fold molar excess of unlabelled 200 bp *PLAT1* promoter fragment (PLAT1; Figure S6), but not with the mutated 200 bp promoter fragment WIP1, where point mutations were introduced in the

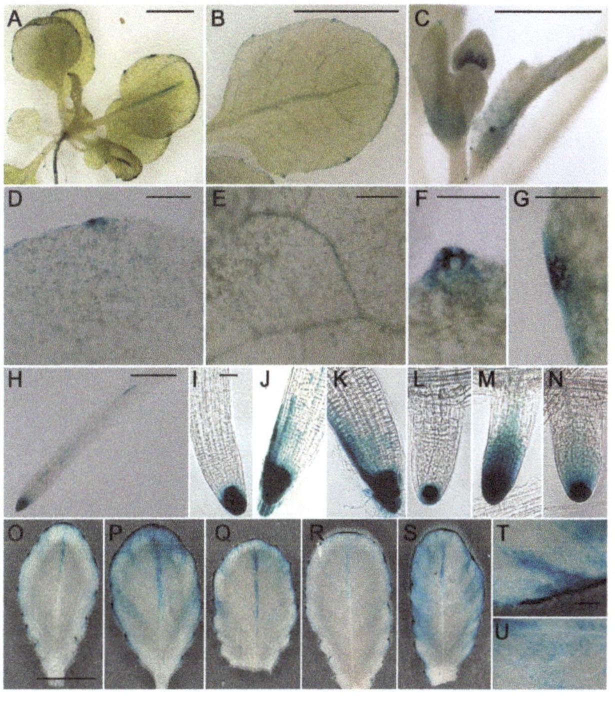

Figure 6. *PLAT1* is expressed in the vasculature, hydathodes and stomata of aerial organs. *PLAT1* expression is reflected by GUS activity in *plat1-1* mutant plants rescued by the *PLAT1:PLAT1-GUS* rescue/reporter construct (line *GUS3-5*). The *PLAT1* promoter confers expression in the leaf vasculature (**A**, **B**, **E**), hydathodes (**B**, **F**, **G**), floral organs (**C**), stomata (**D**) and the primary root tip and root pericycle cells (**H**). *PLAT1* expression in primary root tips, 24 h following transfer to control (**I**), 200 mM NaCl (**J**) or 1.5 μM ABA plates (**K**). *PLAT1* expression in lateral root tips, 24 h following transfer to control (**L**), 200 mM NaCl (**M**) or 1.5 μM ABA plates (**N**). *PLAT1* expression in fully expanded rosette leaf from 4-w-old seedling, control (**O**), following 24 h (**P**), or 48 h (**Q**) of treatment with 1.5 μM ABA and following 24 h (**R**), or 48 h (**S**) watering with 200 mM NaCl. Detail of rosette leaf showing expansion of expression domain in leaf mesophyll following 24 h ABA treatment (**T**), or 48 h watering with NaCl (**U**). Scale bar = 1 cm (A-C) and (O-S), 1 mm (D-H), (T), (U), 0.1 mm (I-N). n≥10.

two ABRE elements (WIP1; Figure S6). Thus, *PLAT1* expression is regulated by the ABA signalling pathway as direct target of the ABF transcription factors.

PLAT1 promotes tolerance towards ER stress elicited by tunicamycin

The PLAT1 subcellular localisation in the ER (Figure 7) suggested that PLAT1 function is related to ER stress. Under stress conditions, misfolded proteins accumulate in the ER that eventually cause ER stress [51], resulting in reduced growth and induction of the unfolded protein response (UPR) to compensate for this increased accumulation of misfolded proteins. TM is used as ER stress agent and interferes with N-linked glycosylation of secreted glycoproteins, which prevents protein folding in the ER [51]. The tolerance of the different *PLAT1* overexpression lines towards TM was improved compared to Col-0 wild-type (Figure 9). In contrast, the *plat1-1* mutant was more sensitive, while the response towards TM was not affected in the *plat2* mutant (Figure 9), which supports a function for PLAT1 in ER stress responses and/or UPR. The basal expression levels for the ER stress markers *BIP1,2*, *CNX1*, *CRT1* and *PDIL* [52] was higher for the different *PLAT1* overexpression lines (Figure S7),

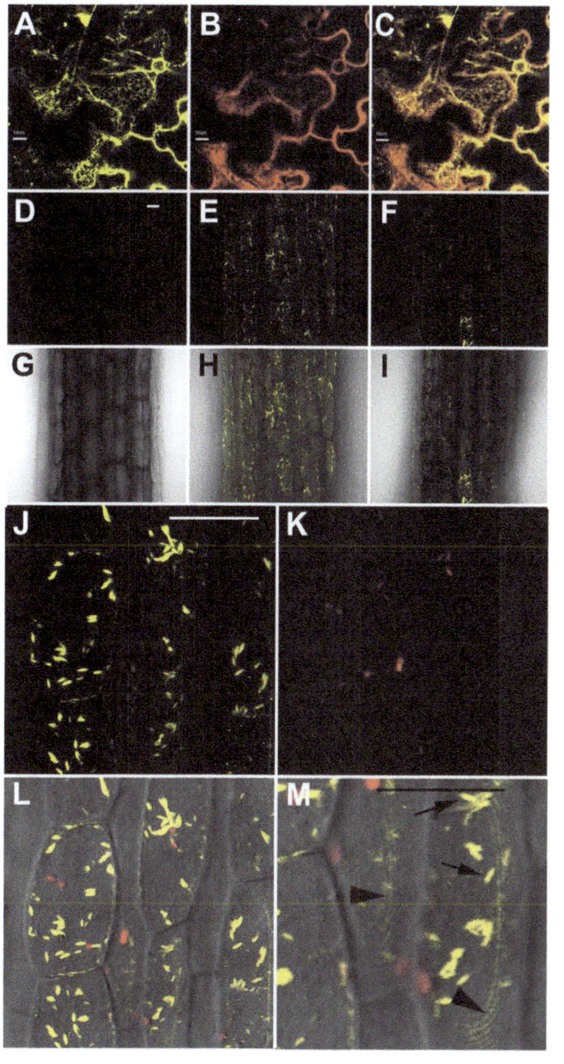

Figure 7. PLAT1 is localised to the ER in *Arabidopsis* and *Nicotiana benthamiana*. Transient transformation of the *PLAT1:PLAT1-YFP* construct to *N. benthamiana*. YFP channel showing *PLAT1-YFP* expression (**A**), RFP channel showing *ER-rk CD3-959* mCherry marker expression (**B**), co-localisation of PLAT1-YFP with the ER-rk CD3-959 marker (C). *PLAT1* expression in stable *plat1-1* transgenics (line *YFP13-1*) rescued by the *PLAT1:PLAT1-YFP* rescue/reporter construct (D-I) 48 h following transfer of 3-d-old seedlings to control, NaCl or ABA plates, with YFP channel (D-F) and merged YFP and bright field channels (G-I). *PLAT1* expression following transfer to control medium (**D**, **G**), following transfer to 200 mM NaCl (**E**, **H**), and expression following transfer to 1.5 μM ABA (**F**, **I**). Hypocotyl section from line *YFP13-1* with YFP channel showing *PLAT1:PLAT1-YFP* reporter activity (**J**), red autofluorescence of chloroplasts (**K**), merged image of bright field, YFP and red autofluorescence (**L**). (**M**) Detail of (L), showing PLAT1 localisation to putative ER bodies (arrows) and the ER (arrow heads). Scale bar = 10 μm (A-C), 0.1 mm (D-L), 0.05 mm (M). n≥10.

which suggests that the capacity of these lines to deal with unprocessed and/or misfolded proteins in the ER is increased, which could contribute to an increased abiotic stress tolerance.

Discussion

The complete *Arabidopsis* genome sequence is available, nevertheless only for a relatively low percentage of *Arabidopsis* proteins experimental evidence concerning their function is

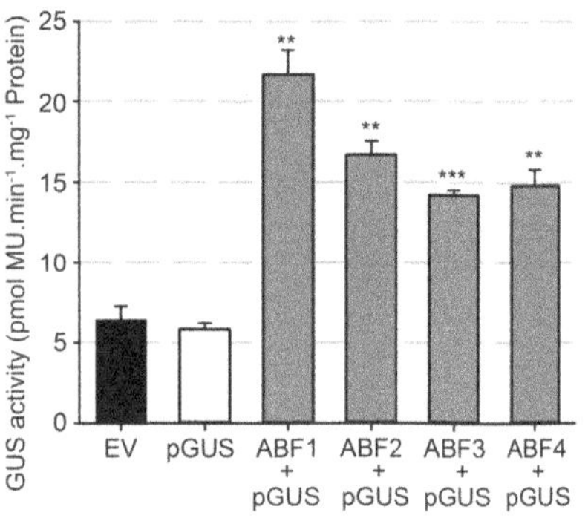

Figure 8. The *PLAT1* promoter is activated by the ABF transcription factors. GUS activity in 8-w-old *N. benthamiana* leaves infiltrated with *A. tumefaciens* harbouring the *PLAT1:PLAT1-GUS* reporter (pGUS) or an empty vector (EV) as negative control, compared to the simultaneous infiltration of the *35S:ABF*$_{1-4}$ and *PLAT1:PLAT1-GUS* constructs. Values are means of 3 replicates ± standard deviation. *** or ** indicate statistical significance calculated using the unpaired Student's *t*-test at $p<0.001$ or $p<0.01$, respectively.

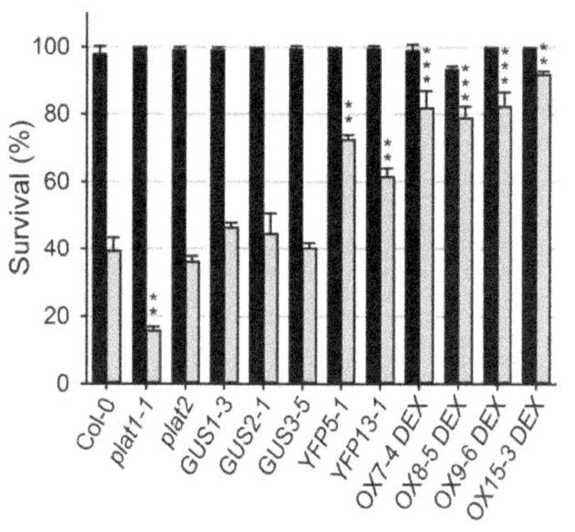

Figure 9. PLAT1 promotes tolerance towards tunicamycin elicited ER stress. Survival is expressed as the percentage of the plated seeds that developed (pale) green seedlings. Control conditions black bars, ER stress (0.05 μg l^{-1} TM) grey bars. Values are means of 3 replicates ± standard deviation. $n \geq 70$ per replicate. *** or ** indicate statistical significance calculated using the unpaired Student's *t*-test at $p<0.001$ or $p<0.01$, respectively.

available. While among the PLAT-plant-stress family only *Ca*TIN1 function has been studied by heterologous expression in tobacco [4], for other family members only limited *in silico* expression data is available. This showed that these members were induced by different abiotic and biotic stimuli, suggesting that PLAT-plant-stress proteins in general could promote tolerance towards stress responses and thus can be of great importance for developing stress tolerant crops. Nevertheless, little information is available on the function of these proteins. We assigned a function to the previously uncharacterised *Arabidopsis* PLAT domain proteins, which are members of this PLAT-plant-stress subgroup. The combination of genetic and physiological approaches supported functional diversification within this *Arabidopsis* PLAT protein family. PLAT1 critically functions as positive regulator of abiotic stress tolerance and also confers increased plant growth. Because *PLAT1* is a direct activated target of the ABF transcription factors and PLAT1 levels affect ABA sensitivity based on seed germination assays, *PLAT1* represents a novel downstream target of the ABA signalling pathway. PLAT2 appears to function specifically in seed dormancy, while *PLAT3* might represent a non-expressed pseudogene.

Analysis of the *plat1-1* loss-of-function mutant, *PLAT1* overexpression lines and genetic complementation of the *plat1-1* mutant, showed that PLAT1 critically functions as positive regulator of abiotic stress tolerance. Abiotic stress treatments induced *PLAT1* expression, PLAT1 loss-of-function resulted in reduced abiotic stress tolerance, whereas *PLAT1* overexpression conferred an inverse response, evident from increased cold, drought and salt stress tolerance. No change in pathogen susceptibility was detected for the *plat1-1* loss-of-function mutant, which correlates with the fact that pathogen infection did not significantly affect *PLAT1* expression. It is however possible that an altered pathogen resistance in *plat1-1* could have been too weak to be detected in the employed pathosystem. Otherwise, functional redundancy among the PLAT family members (partially) could have compensated for PLAT1 loss-of-function in biotic stress tolerance.

High ABA levels confer resistance towards abiotic stress conditions like drought and salinity, which is linked to its role in regulating stomatal aperture, while ABA shows antagonistic interaction with pathogen defence signalling pathways [26,28–31]. The ABF/ABRE transcription factors, which are positive mediators of the ABA signalling pathway [34,35], were shown to be induced by cold (ABF1), salinity (ABF2 and ABF3) and drought stress (ABF4) [34], while *ABF2* or *ABF3* overexpression enhanced abiotic stress tolerance [53]. *PLAT1* overexpression increased and loss-of-function reduced sensitivity towards ABA during seed germination. Because ABA levels were not significantly affected in the different PLAT1 lines, *PLAT1* overexpression resembled the effect of increased ABA signalling, including the differential regulation of abiotic and biotic stress tolerance.

Although it is widely speculated that the effect of ABA signalling on stress tolerance in higher plants is regulated by the complex (antagonistic) interactions with other phytohormones, our understanding of the ABA signalling pathway leading to the adaptation of naturally occurring multi-stress responses remains unclear [54]. Genes harbouring two ABRE elements in their promoter were predicted to be direct (activated) target genes of the ABF/ABRE transcription factors [34] and the expression for a large number of such genes was affected by a triple ABF loss-of-function mutant [35]. However, only the *DREB2C* and *RD29B* genes were functionally shown to be direct ABF target genes [55,56]. The *PLAT1* promoter harbours two ABRE elements, located at positions −165 to −156 and −134 to −126 relative to the transcription start site. In agreement with these findings, transactivation and EMSA experiments showed that the ABF transcription factors directly bind to and activate the *PLAT1* promoter. Analysis of *plat1-1* lines complemented with the *PLAT1:PLAT1-GUS* rescue/reporter construct showed that *PLAT1* is expressed in the leaf vasculature, hydathodes and stomata, which are structures linked to the regulation of water household, but also at specific regions in the root. Together with the induced *PLAT1* expression levels as well as expanded expression domain following ABA treatment and abiotic stress conditions, *PLAT1* expression correlates with PLAT1 function in abiotic stress tolerance and its regulation by ABA signalling. Together, these data showed that *PLAT1* represents a novel component of the ABA signalling as direct target of the ABF transcription factors, which might explain its function in stress tolerance.

ABA in general negatively affects plant growth, mainly through crosstalk with the brassinosteroid pathway [57], and promotes quiescence of stem cells resulting in reduced root growth and lateral root formation [58–60]. In addition to the effects on stress tolerance, which could be correlated with *PLAT1* being a downstream target of the ABA signalling pathway, PLAT1 promotes plant growth, evident by a faster development and consequently increased shoot biomass. Therefore, PLAT1 function exhibits both expected and unexpected ABA related responses. Heterologous *CaTIN1* overexpression in tobacco also resulted in increased abiotic stress tolerance and plant growth, but additionally conferred increased biotic resistance, probably through influencing the redox state. Further, *CaTIN1* expression was induced by ethylene treatment and infection by tobacco mosaic virus, but not following ABA treatment [4]. Thus, despite the fact that PLAT1 and *Ca*TIN1 are orthologs, partially convergent evolution occurred in the different plant species on protein function and transcriptional regulation.

Transient expression experiments and analysis of *plat1-1* lines complemented with the *PLAT1:PLAT1-YFP* rescue/reporter construct showed that PLAT1 is localised to the ER, but also in rod shaped structures resembling ER bodies. This is supported by the induction of these structures with PLAT1-YFP signals following ABA treatment or salt stress conditions. ER bodies are specific to *Brassicaceae* and induced following stress conditions and wounding [61–63], but no direct correlation between ER bodies and abiotic stress responses has been shown. PLAT1 promotes tolerance towards the ER stress elicitor TM and the basal expression levels of ER stress markers, representing chaperonins functioning in ER stress and/or UPR.

Together our results indicate that PLAT1 functions in abiotic stress tolerance, either directly through promoting abiotic stress responses, and/or indirectly through improving basal tolerance/fitness. A direct promotion of abiotic stress responses could result from promoting ABA signaling, which is related to its function as novel direct target gene of the ABA signaling pathway. An indirect effect could result from stimulating ER stress responses for a higher basal tolerance/fitness. ER stress responses were shown to be indispensable for abiotic stress responses [64,65] and the ER appears to play a prominent role in ABA-mediated stress signalling since ABA release from the ER is important for plants coping with stress [66]. The PLAT1 protein essentially harbours one transmembrane spanning domain and one large PLAT domain that covers the rest of the protein sequence and which has been shown to function in protein interaction. Therefore, PLAT1 most likely does not possess enzymatic activity, but rather functions as 'docking site' for interacting proteins with enzymatic or signalling activity functioning in ABA regulated pathways, enabling PLAT1 to regulate their activity.

To obtain plants through biotechnology or breeding approaches with increased tolerance towards adverse conditions, but without yield penalties under optimal growth conditions, it is important to identify all genes involved in stress responses and understand their function. Therefore, the identification and assignment of a function to the previously uncharacterised PLAT-plant-stress family member PLAT1 contributes to this important goal. The improved plant growth associated with the increased tolerance towards cold, drought and salt stress mediated by *PLAT1* overexpression could be an important asset in crop improvement. To enable the application of PLAT1 or other members from the PLAT-plant-stress family in crop improvement, future studies will be needed to address the multifaceted role of these proteins in stress tolerance and plant development.

Supporting Information

Figure S1 *PLAT1* expression patterns under different conditions. (**A**) Phylogenetic tree of the PLAT-plant-stress subgroup. Phylogenetic analysis was carried out using the neighbour-joining method with 100 bootstraps and displayed using TreeTop. *Glycine max* (Glyma), *Zea mays* (GRMZM), *Oryza sativa* (Os), *Populus trichocarpa* (POPTR), *Sorghum bicolor* (Sb). (**B**) *PLAT1* expression in different organs from 3-w-old, 6-w-old and 12-w-old wild-type (Col-0) plants: F, Flower; H, hypocotyl; L, leaf; R, root; S, inflorescence stem and W, whole plant. (**C**) *PLAT1* expression following salt treatment (right) compared to control watering (left). (**D**) *PLAT1* expression following cold treatment. (**E**) *PLAT1* and *PLAT2* expression by RT-PCR in the respective T-DNA insertion mutants *plat1-1* and *plat2*. Bottom panels, rRNA for loading control (A-D).

Figure S2 PLAT1 loss-of-function does not affect biotic stress tolerance. (**A**) *PLAT1* expression following leaf infiltration of 10^7 cfu ml^{-1} of *P. syringae* pv. *tomato* DC3000 or DC3000 *avrRpm1* in 10 mM $MgCl_2$, compared to control treatment ($MgCl_2$). (**B**) *PLAT1* expression following infection with *S. sclerotiorum* compared to control treatment (PDA). (**C**) Leaves from wild-type (Col-0) plants (Top panel) and *plat1-1* plants (Bottom panel), 3 d after infection with 10^7 cfu ml^{-1} *P. syringae* pv. *tomato* DC3000. Scale bar = 1 cm, n≥10.

Figure S3 PLAT2 functions in seed dormancy. Seed germination of *plat1-1*, *PLAT1:PLAT1-GUS* line *GUS3-5*, *35S>>PLAT1* line *OX9-6*, *plat2* and wild-type (Col-0) on control medium including 5 µM DEX without prior stratification. Values are means of 3 replicates ± standard deviation. n≥100 per replicate.

Figure S4 Increased ABA sensitivity by tissue specific or ectopic *PLAT1* overexpression. (**A**, **B**) Seed germination of *plat1-1*, wild-type (Col-0) and *plat1-1* lines complemented with the *PLAT1:PLAT1-GUS* rescue construct (GUS) (A), or *plat1-1* lines complemented with the *PLAT1:PLAT1-YFP* rescue construct (YFP) (B) on medium supplemented with 1.5 µM ABA. (**C**) Seed germination of wild-type and transgenic lines harbouring the *35S>>PLAT1* ectopic overexpression construct (OX) on medium supplemented with 1.5 µM ABA and 5 µM DEX. Values are means of 3 replicates ± standard deviation. n≥100 per replicate.

Figure S5 *PLAT1* expression is induced in adult roots by ABA treatment and salt stress. *PLAT1:PLAT1-GUS* seedlings were monitored for *PLAT1* expression 8 h (A-C) and 24 h (D-F) following transfer to control, NaCl or ABA plates. (**A**, **D**) Detail of 2-w-old adult root with *PLAT1* expression in emerging lateral root primordia following transfer to control medium. (**B**, **E**) Detail of adult root with expanded expression domain following transfer to 200 mM NaCl. (**C**, **F**) Detail of adult root with expanded expression domain following transfer to 1.5 µM ABA. Scale bar = 0.1 mm, n≥10.

Figure S6 PLAT1 functions as direct ABF target in ABA signalling. EMSA assay showing that the ABF transcription factors bind to the 200 bp *PLAT1* promoter region containing two ABRE elements PLAT1 (*). This binding was specifically competed with a 100 molar excess of unlabelled *PLAT1* promoter fragment (PLAT1), but not the negative probe lacking the two

ABRE elements (WIP1). Arrowhead indicates shifted band. Bracket indicates free probe.

Figure S7 ***PLAT1*** **overexpression lines exhibit higher basal expression levels for ER stress markers.** Relative expression levels for ER stress markers in the different *PLAT1* overexpression lines compared to wild-type (Col-0) and the *plat1-1* mutant. (**A**) *BIP1,2* (*HSP70*), (**B**) *CNX1* (*CALNEXIN1*), (**C**) *CRT1* (*CALRETICULIN1*) and (**D**) *PDIL* (*PROTEIN DISULFIDE ISOMERASE*-like). Values are means of 3 replicates ± standard deviation. n≥10 per replicate. ***, ** or * indicate statistical significance calculated using the unpaired Student's *t*-test at $p<0.001$, $p<0.01$ or $p<0.05$, respectively.

File S1 Combined file containing supporting tables. Table S1: List of primers used for cloning, genotyping of T-DNA insertion mutants and EMSA controls. **Table S2**: List of primers used for RT-PCR. **Table S3**: ABA levels are not affected by PLAT1. **Table S4**: *PLAT1* transcript levels in the different overexpression lines. **Table S5**: PLAT1 promotes lateral root formation.

Acknowledgments

We thank A. Luidold for help with the seed germination assays, K. Remele and W. Pfeifhofer for help with ABA determination.

Author Contributions

Conceived and designed the experiments: TKH AA DKG EvdG TR. Performed the experiments: TKH AA SHE DKG HB UJ YR WA EvdG. Analyzed the data: TKH AA DKG EvdG. Contributed reagents/materials/analysis tools: SK EvdG TR. Wrote the paper: TKH AA EvdG TR.

References

1. Minor W, Steczko J, Stec B, Otwinowski Z, Bolin JT, et al. (2006) Crystal structure of soybean lipoxygenase L-1 at 1.4 å resolution. Biochemistry 35: 10687–10701.
2. Bateman A, Sandford R (1999) The PLAT domain: a new piece in the PKD1 puzzle. Curr Biol 9: R588–590.
3. Hong Y, Wang TW, Hudak KA, Schade F, Froese CD, et al. (2000) An ethylene-induced cDNA encoding a lipase expressed at the onset of senescence. Proc Natl Acad Sci U S A 97: 8717–8722.
4. Shin R, An JM, Park CJ, Kim YJ, Joo S, et al. (2004) *Capsicum annuum* tobacco mosaic virus-induced clone 1 expression perturbation alters the plant's response to ethylene and interferes with the redox homeostasis. Plant Physiol 135: 561–573.
5. Tomchick DR, Phan P, Cymborowski M, Minor W, Holman TR (2001) Structural and functional characterization of second-coordination sphere mutants of soybean lipoxygenase-1. Biochemistry 40: 7509–7517.
6. Allard JB, Brock TG (2005) Structural organization of the regulatory domain of human 5-lipoxygenase. Curr Protein Pept Sci 6: 125–131.
7. Hu J, Barr MM (2005) ATP-2 interacts with the PLAT domain of LOV-1 and is involved in *Caenorhabditis elegans* polycystin signaling. Mol Biol Cell 16: 458–469.
8. Yuasa T, Takakura A, Denker BM, Venugopal B, Zhou J (2004) Polycystin-1L2 is a novel G-protein-binding protein. Genomics 84: 126–138.
9. Eek P, Jarving R, Jarving I, Gilbert NC, Newcomer ME, et al. (2012) Structure of a calcium-dependent 11R-lipoxygenase suggests a mechanism for Ca^{2+} regulation. J Biol Chem 287: 22377–22386.
10. Joshi N, Hoobler EK, Perry S, Diaz G, Fox B, et al. (2013) Kinetic and structural investigations into the allosteric and pH effect on the substrate specificity of human epithelial 15-lipoxygenase-2. Biochemistry 52: 8026–8035.
11. Li ST, Zhang M, Fu CH, Xie S, Zhang Y, et al. (2012) Molecular cloning and characterization of two 9-lipoxygenase genes from *Taxus chinensis*. Plant Molecular Biology Reporter 30: 1283–1290.
12. Bhardwaj PK, Kaur J, Sobti RC, Ahuja PS, Kumar S (2011) Lipoxygenase in *Caragana jubata* responds to low temperature, abscisic acid, methyl jasmonate and salicylic acid. Gene 483: 49–53.
13. Inamdar MI, Bhat RS, Kuruvinashetti MS (2010) Cloning a drought-inducible cDNA (*PjPLAT1*) from *Prosopis juliflora*. Karnataka J Agric Sci 23: 350–351.
14. Leal GA Jr, Albuquerque PS, Figueira A (2007) Genes differentially expressed in *Theobroma cacao* associated with resistance to witches' broom disease caused by *Crinipellis perniciosa*. Mol Plant Pathol 8: 279–292.
15. Lee SC, Kim SH, An SH, Yi SY, Hwang BK (2006) Identification and functional expression of the pepper pathogen-induced gene, *CaPIP2*, involved in disease resistance and drought and salt stress tolerance. Plant Mol Biol 62: 151–164.
16. Shin R, Kim MJ, Paek KH (2003) The *CaTin1* (*Capsicum annuum* TMV-induced clone 1) and *CaTin1-2* genes are linked head-to-head and share a bidirectional promoter. Plant Cell Physiol 44: 549–554.
17. Hussain SS, Kayani MA, Amjad M (2011) Transcription factors as tools to engineer enhanced drought stress tolerance in plants. Biotechnol Prog 27: 297–306.
18. Golldack D, Luking I, Yang O (2011) Plant tolerance to drought and salinity: stress regulating transcription factors and their functional significance in the cellular transcriptional network. Plant Cell Rep 30: 1383–1391.
19. Islam S, Griffiths CA, Blomstedt CK, Le TN, Gaff DF, et al. (2013) Increased biomass, seed yield and stress tolerance is conferred in Arabidopsis by a novel enzyme from the resurrection grass *Sporobolus stapfianus* that glycosylates the strigolactone analogue GR24. PLoS One 8: e80035.
20. Joshi-Saha A, Valon C, Leung J (2011) Abscisic acid signal off the STARting block. Mol Plant 4: 562–580.
21. Finkelstein R, Reeves W, Ariizumi T, Steber C (2008) Molecular aspects of seed dormancy. Annu Rev Plant Biol 59: 387–415.
22. Padmalatha KV, Dhandapani G, Kanakachari M, Kumar S, Dass A, et al. (2012) Genome-wide transcriptomic analysis of cotton under drought stress reveal significant down-regulation of genes and pathways involved in fibre elongation and up-regulation of defense responsive genes. Plant Mol Biol 78: 223–246.
23. Hsu FC, Chou MY, Peng HP, Chou SJ, Shih MC (2011) Insights into hypoxic systemic responses based on analyses of transcriptional regulation in Arabidopsis. PLoS One 6: e28888.
24. Seki M, Narusaka M, Ishida J, Nanjo T, Fujita M, et al. (2002) Monitoring the expression profiles of 7000 Arabidopsis genes under drought, cold and high-salinity stresses using a full-length cDNA microarray. Plant J 31: 279–292.
25. Shinozaki K, Yamaguchi-Shinozaki K, Seki M (2003) Regulatory network of gene expression in the drought and cold stress responses. Curr Opin Plant Biol 6: 410–417.
26. Lee SC, Luan S (2012) ABA signal transduction at the crossroad of biotic and abiotic stress responses. Plant Cell Environ 35: 53–60.
27. Ton J, Flors V, Mauch-Mani B (2009) The multifaceted role of ABA in disease resistance. Trends Plant Sci 14: 310–317.
28. Mosher S, Moeder W, Nishimura N, Jikumaru Y, Joo SH, et al. (2010) The lesion-mimic mutant *cpr22* shows alterations in abscisic acid signaling and abscisic acid insensitivity in a salicylic acid-dependent manner. Plant Physiol 152: 1901–1913.
29. Großkinsky DK, van der Graaff E, Roitsch T (2014) Abscisic acid-cytokinin antagonism modulates resistance against *Pseudomonas syringae* in tobacco. Phytopathology 10.1094/PHYTO-03-14-0076-R.
30. Audenaert K, Pattery T, Cornelis P, Hofte M (2002) Induction of systemic resistance to *Botrytis cinerea* in tomato by *Pseudomonas aeruginosa* 7NSK2: role of salicylic acid, pyochelin, and pyocyanin. Mol Plant Microbe Interact 15: 1147–1156.
31. Thaler JS, Bostock RM (2004) Interactions between abscisic-acid-mediated responses and plant resistance to pathogens and insects. Ecology 85: 48–58.
32. Fan J, Hill L, Crooks C, Doerner P, Lamb C (2009) Abscisic acid has a key role in modulating diverse plant-pathogen interactions. Plant Physiol 150: 1750–1761.
33. Sharma N, Cram D, Huebert T, Zhou N, Parkin IA (2007) Exploiting the wild crucifer *Thlaspi arvense* to identify conserved and novel genes expressed during a plant's response to cold stress. Plant Mol Biol 63: 171–184.
34. Choi H, Hong J, Ha J, Kang J, Kim SY (2000) ABFs, a family of ABA-responsive element binding factors. J Biol Chem 275: 1723–1730.
35. Yoshida T, Fujita Y, Sayama H, Kidokoro S, Maruyama K, et al. (2010) AREB1, AREB2, and ABF3 are master transcription factors that cooperatively regulate ABRE-dependent ABA signaling involved in drought stress tolerance and require ABA for full activation. Plant J 61: 672–685.
36. Clough SJ, Bent AF (1988) Floral dip: a simplified method for *Agrobacterium*-mediated transformation of *Arabidopsis thaliana*. Plant Journal 16: 735–743.
37. Großkinsky DK, Naseem M, Abdelmohsen UR, Plickert N, Engelke T, et al. (2011) Cytokinins mediate resistance against *Pseudomonas syringae* in tobacco through increased antimicrobial phytoalexin synthesis independent of salicylic acid signaling. Plant Physiol 157: 815–830.
38. Großkinsky DK, Koffler BE, Roitsch T, Maier R, Zechmann B (2012) Compartment specific antioxidative defense in Arabidopsis against virulent and avirulent *Pseudomonas syringae*. Phytopathology 102: 662–673.

39. Großkinsky DK, Albacete A, Jammer A, Krbez P, van der Graaff E, et al. (2014) A rapid phytohormone and phytoalexin screening method for physiological phenotyping. Mol Plant 7: 1053–1056.
40. Nelson BK, Cai X, Nebenfuhr A (2007) A multicolored set of in vivo organelle markers for co-localization studies in Arabidopsis and other plants. Plant J 51: 1126–1136.
41. Hyun TK, Hoffmann A, Sinha AK, Roitsch T (2009) Tomato mitogen activated protein kinases regulate the expression of extracellular invertase *Lin 6* in response to stress related stimuli. Funct Plant Biol 36: 1088–1097.
42. Winter D, Vinegar B, Nahal H, Ammar R, Wilson GV, et al. (2007) An "Electronic Fluorescent Pictograph" browser for exploring and analyzing large-scale biological data sets. PLoS One 2: e718.
43. Sessions A, Burke E, Presting G, Aux G, McElver J, et al. (2002) A high-throughput Arabidopsis reverse genetics system. Plant Cell 14: 2985–2994.
44. Alonso JM, Stepanova AN, Leisse TJ, Kim CJ, Chen H, et al. (2003) Genome-wide insertional mutagenesis of *Arabidopsis thaliana*. Science 301: 653–657.
45. Barrero JM, Millar AA, Griffiths J, Czechowski T, Scheible WR, et al. (2010) Gene expression profiling identifies two regulatory genes controlling dormancy and ABA sensitivity in Arabidopsis seeds. Plant J 61: 611–622.
46. Craft J, Samalova M, Baroux C, Townley H, Martinez A, et al. (2005) New pOp/LhG4 vectors for stringent glucocorticoid-dependent transgene expression in Arabidopsis. Plant J 41: 899–918.
47. van der Graaff E, den Dulk-Ras A, Hooykaas PJJ, Keller B (2000) Activation tagging of the *LEAFY PETIOLE* gene affects leaf petiole development in *Arabidopsis thaliana*. Development 127: 4971–4980.
48. Zhao Z, Zhang W, Stanley BA, Assmann SM (2008) Functional proteomics of *Arabidopsis thaliana* guard cells uncovers new stomatal signaling pathways. Plant Cell 20: 3210–3226.
49. Scott MS, Calafell SJ, Thomas DY, Hallett MT (2005) Refining protein subcellular localization. PLoS Comput Biol 1: e66.
50. Schwacke R, Schneider A, van der Graaff E, Fischer K, Catoni E, et al. (2003) ARAMEMNON, a novel database for Arabidopsis integral membrane proteins. Plant Physiol 131: 16–26.
51. Howell SH (2013) Endoplasmic reticulum stress responses in plants. Annu Rev Plant Biol 64: 477–499.
52. Liu JX, Srivastava R, Che P, Howell SH (2007) An endoplasmic reticulum stress response in Arabidopsis is mediated by proteolytic processing and nuclear relocation of a membrane-associated transcription factor, bZIP28. Plant Cell 19: 4111–4119.
53. Kim JB, Kang JY, Kim SY (2004) Over-expression of a transcription factor regulating ABA-responsive gene expression confers multiple stress tolerance. Plant Biotechnol J 2: 459–466.
54. Atkinson NJ, Urwin PE (2012) The interaction of plant biotic and abiotic stresses: from genes to the field. J Exp Bot 63: 3523–3543.
55. Uno Y, Furihata T, Abe H, Yoshida R, Shinozaki K, et al. (2000) Arabidopsis basic leucine zipper transcription factors involved in an abscisic acid-dependent signal transduction pathway under drought and high-salinity conditions. Proc Natl Acad Sci U S A 97: 11632–11637.
56. Kim JS, Mizoi J, Yoshida T, Fujita Y, Nakajima J, et al. (2011) An *ABRE* promoter sequence is involved in osmotic stress-responsive expression of the *DREB2A* gene, which encodes a transcription factor regulating drought-inducible genes in Arabidopsis. Plant Cell Physiol 52: 2136–2146.
57. Choudhary SP, Yu JQ, Yamaguchi-Shinozaki K, Shinozaki K, Tran LS (2012) Benefits of brassinosteroid crosstalk. Trends Plant Sci 17: 594–605.
58. Fukaki H, Tasaka M (2009) Hormone interactions during lateral root formation. Plant Mol Biol 69: 437–449.
59. Zhang H, Han W, De Smet I, Talboys P, Loya R, et al. (2010) ABA promotes quiescence of the quiescent centre and suppresses stem cell differentiation in the Arabidopsis primary root meristem. Plant J 64: 764–774.
60. Duan L, Dietrich D, Ng CH, Chan PM, Bhalerao R, et al. (2013) Endodermal ABA signaling promotes lateral root quiescence during salt stress in Arabidopsis seedlings. Plant Cell 25: 324–341.
61. Matsushima R, Hayashi Y, Yamada K, Shimada T, Nishimura M, et al. (2003) The ER body, a novel endoplasmic reticulum-derived structure in Arabidopsis. Plant Cell Physiol 44: 661–666.
62. Hara-Nishimura I, Matsushima R (2003) A wound-inducible organelle derived from endoplasmic reticulum: a plant strategy against environmental stresses? Curr Opin Plant Biol 6: 583–588.
63. Yamada K, Hara-Nishimura I, Nishimura M (2011) Unique defense strategy by the endoplasmic reticulum body in plants. Plant Cell Physiol 52: 2039–2049.
64. Liu L, Cui F, Li Q, Yin B, Zhang H, et al. (2011) The endoplasmic reticulum-associated degradation is necessary for plant salt tolerance. Cell Res 21: 957–969.
65. Liu JX, Howell SH (2010) Endoplasmic reticulum protein quality control and its relationship to environmental stress responses in plants. Plant Cell 22: 2930–2942.
66. Lee KH, Piao HL, Kim HY, Choi SM, Jiang F, et al. (2006) Activation of glucosidase via stress-induced polymerization rapidly increases active pools of abscisic acid. Cell 126: 1109–1120.

Characterization of Arbuscular Mycorrhizal Fungus Communities of *Aquilaria crassna* and *Tectona grandis* Roots and Soils in Thailand Plantations

Amornrat Chaiyasen[1], J. Peter W. Young[2], Neung Teaumroong[3], Paiboolya Gavinlertvatana[4], Saisamorn Lumyong[1]*

1 Department of Biology, Faculty of Science, Chiang Mai University, Chiang Mai, Thailand, **2** Department of Biology, University of York, York, United Kingdom, **3** Schoool of Biotechnology, Institute of Agricultural Technology, Suranaree University of Technology, Nakhon Ratchasima, Thailand, **4** Thai Orchid Labs Co. Ltd., Khannayao, Bangkok, Thailand

Abstract

Aquilaria crassna Pierre ex Lec. and *Tectona grandis* Linn.f. are sources of resin-suffused agarwood and teak timber, respectively. This study investigated arbuscular mycorrhizal (AM) fungus community structure in roots and rhizosphere soils of *A. crassna* and *T. grandis* from plantations in Thailand to understand whether AM fungal communities present in roots and rhizosphere soils vary with host plant species and study sites. Terminal restriction fragment length polymorphism complemented with clone libraries revealed that AM fungal community composition in *A. crassna* and *T. grandis* were similar. A total of 38 distinct terminal restriction fragments (TRFs) were found, 31 of which were shared between *A. crassna* and *T. grandis*. AM fungal communities in *T. grandis* samples from different sites were similar, as were those in *A. crassna*. The estimated average minimum numbers of AM fungal taxa per sample in roots and soils of *T. grandis* were at least 1.89 vs. 2.55, respectively, and those of *A. crassna* were 2.85 vs. 2.33 respectively. The TRFs were attributed to Claroideoglomeraceae, Diversisporaceae, Gigasporaceae and Glomeraceae. The Glomeraceae were found to be common in all study sites. Specific AM taxa in roots and soils of *T. grandis* and *A. crassna* were not affected by host plant species and sample source (root vs. soil) but affected by collecting site. Future inoculum production and utilization efforts can be directed toward the identified symbiotic associates of these valuable tree species to enhance reforestation efforts.

Editor: Zhengguang Zhang, Nanjing Agricultural University, China

Funding: This research was supported by the Thailand Research Fund for Research-Team Promotion Grant (RTA5580007) and the Commission of Higher Education for National Research University (A1) (http://www.trf.or.th/). AC was funded by the Thailand Research Fund; The Royal Golden Jubilee PhD Program (PHD/0150/2550: http://rgj.trf.or.th/indexth.asp). The funders had no role in study design, data collection and analysis, decision to publish, or preparation of the manuscript.

Competing Interests: Paiboolya Gavinlertvatana is employed by a commercial company (Thai Orchid Labs Co. Ltd). The affiliation does not alter the authors' adherence to all PLOS ONE policies on sharing data and materials.

* Email: saisamorn.l@cmu.ac.th

Introduction

Tropical forests are disappearing at the rate of 13.5 million hectares each year owing to logging, burning and clearing for agriculture and shifting cultivation [1]. At present, managed woodlands are required for timber and non-timber products in many countries. *Aquilaria crassna* Pierre ex Lec. (agarwood) and *Tectona grandis* Linn.f. (teak) are perennial plants that are used extensively to provide aromatic resin-infused wood products [2] and good quality teak wood products [3], respectively. The depletion of wild trees from indiscriminate cutting of *Aquilaria* species has resulted in the trees being listed and protected as endangered species. All *Aquilaria* species were listed in Appendix II of the Convention on International Trade in Endangered Species of Wild Fauna and Flora in 2005 [4]; however, a number of countries have outstanding reservations regarding that listing. Plantlets of *A. crassna* and *T. grandis* are produced in Thailand for domestic and foreign markets such as Jamaica, Guatemala, Mozambique, Sri Lanka, Indonesia, Laos, Malaysia, and Australia. Most *T. grandis* plantations in Thailand are planted in the northern provinces such as Chiang Mai, Chaing Rai and Phetchabun, while *A. crassna* plantations are mostly in eastern (Rayong, Trat and Chanthaburi provinces) and central (Nakhon-Nayok) Thailand.

Arbuscular mycorrhizal fungi (AMF) are soil fungi in the phylum Glomeromycota [5] that are mutualistically associated with roots of a wide spectrum of tropical and temperate tree species [6]. AM fungi have major effects on plant growth such as enhance the nutrient uptake by plant roots (especially phosphorus), particularly in low fertility soils [7,8], protected plant against drought stress [9,10], protect plant from soil-borne plant pathogenic infection [11], and improve soil aggregate stability through the action of mycelia and glomalin [12,13,14]. AMF inocula applied to plantlets and plant seedlings increased growth during early tree establishment in the field [2,15,16]. AM fungi have been used to inoculate and enhance growth of *T. grandis* [3,17] and *Aquilaria* spp. [2,18] prior to planting out. Therefore, studying the AM fungal communities of these plants in the field should aid plantation establishment and reforestation efforts. Information about the diversity of AM fungi associated with both plants has been reported mostly from natural forests in India [19,20,21,22,23] and only in *T. grandis* from Thailand [24]. These studies characterized communities based upon spore morphology. However, there are no reports of AM fungal communities of either tree using molecular tools. Identification of AM fungi based on spore morphology inevitably has some limitations, e.g. omission of AM fungi that did not produce spores during the sampling period and inability to identify the AM fungi within the roots.

PCR-based methods have been widely used in AM fungal community studies. Various studies have designed sets of specific primers for AM fungi [25,26,27] to facilitate rapid detection and identification directly from field-grown plant roots. Previously, Terminal restriction fragment length polymorphism (T-RFLP) has been used to study the AM fungi community in roots of arable crops [28], perennial herbs [29], herbaceous flowering plants [30], grass species [31,32], grass species with herbaceous flowering plants [33,34], and temperate deciduous trees [35]. Populations of AM fungi have been well studied in a number of ecosystems around the world, but there is scant information available for tropical forests and plantations of tropical and sub-tropical species.

This study provides the first molecular community analysis of AM fungi associated with field-collected roots and rhizosphere soils of the tropical trees *A. crassna* and *T. grandis*, and is part of a long term goal of optimizing AM fungus inoculation strategies to enhance reforestation efforts with these trees. It also provides an early insight into the biodiversity of AM fungi in Thailand to test the hypothesis that differences in AM fungal communities present in the roots and rhizosphere soils are determined by collecting sites, host plant species, and local environmental factors.

Materials and Methods

Ethics Statement

No specific permits were required to carry out research in the plantations: Chiang Mai (99°15' E, 18°58' N), Chiang Rai (99°29′ E/19°14′ N), Nakhon-Nayok (101°16′ E, 14°9′ N), Phetchabun (100°47' E, 16°2' N) and Thai Orchids Lab Ltd. (101°7′ E, 14°16′ N). The field studies did not involve endangered or protected species in Thailand. *Aquilaria crassna* is defined to be the forbidden forest item in only the forest area as the Forest Act. Therefore, the *A. crassna* planting and deforestation in the land of ownership is legal. All *A. crassna* samples were obtained from privately-owned plantations and are therefore not subject to the restrictions of the Forest Act of Thailand. Permission to sample the *T. grandis* and *A. crassna* were granted by the landowner.

Study sites and sampling

Rhizosphere soils and roots were sampled from plantations of *T. grandis* and *A. crassna* in four provinces of Thailand (Table 1). Two sampling sites were located in Chiang Mai and Chiang Rai provinces in the northern region. These sites are monocultures of *T. grandis* planted at 2 m spacings and left to grow naturally with accumulated leaf litter and negligible understory perennial plants. Only roots attached to the main roots of *T. grandis* were sampled. At the sites in the central region; Nakhon-Nayok and Thai Orchids Lab Ltd., Nakhon-Nayok province, and in the northern region; Phetchabun province, *T. grandis* and *A. crassna* were planted alternately 2 m apart at Thai Orchids Lab Ltd. and Phetchabun. At both sites, weeds were controlled by ploughing and herbicide treatment. Thus, both species were planted without any above-ground vegetation, while in Nakhon-Nayok site, *A. crassna* was left to grow naturally. Paired soil and root samples from each plant species were randomly collected from 3 locations per site at 0–15 cm depth within 50 m^2 and taken to the laboratory. All collections were carried out in July 2010. Root fragments were washed free of soil and air dried on tissue paper. Root fragments and soil samples were stored frozen at −20°C until further analysis.

Soil analyses

Soil pH and electrical conductivity (EC) were determined in a 1:1 soil: water slurry by direct measurement with pH-meter (Waterproof EC Testr, EUTECH instruments). Available phosphorus was measured employing the Bray II method [36]. Total inorganic nitrogen, exchangeable potassium and soil organic carbon were quantified following the methods of soil analysis outlined in Sparks et al. [37].

Molecular analysis

Three replicate rhizosphere soil and root samples from each plant species were used to represent each site of collection. DNA was extracted from rhizophere soils and roots using the PowerSoil DNA isolation kit (MoBio Laboratories, CA) and Nucleospin Plant II (Macherey-Nagel GmbH & Co. KG, Düren), respectively according to the manufacturers' protocols. DNAs were amplified separately by nested PCR and then 20 µl of each of the three replicates from each sampling site were pooled and purified before restriction digestion [38]. The first round of AMF-specific PCR amplification was performed using the unlabelled primers AML1 and AML2 [26] with 30 cycles. In this first PCR, 40 µl reactions were carried out and each mixture contained 10 pmol of each primer, 1 unit of Taq polymerase (Promega) and 25 mM of each dNTP (Invitrogen) in manufacturer's reaction buffer (Promega). PCR was performed on a PTC100 thermocycler (MJ Research) with an initial denaturation at 94°C for 15 min, followed by 30 cycles of denaturation at 94°C for 30 s, annealing at 57°C for 45 s, extension at 72°C for 45 s, followed by a final extension of 72°C for 5 min. PCR products were visualized on a 1% agarose gel containing 0.1 × SybrSafe (Invitrogen). The second round primers, 0.5 unit of Taq polymerase (Promega) and 20 pmol of HEX-labeled NS31 and FAM-labeled AML3 were added directly into 24 µl of each resulting product. Second-round PCR was conducted with 5 additional cycles using the same PCR conditions as the first PCR. The PCR products were purified using the QIAquick PCR purification kit (Qiagen). The purified PCR products were digested separately with the selected restriction enzymes, HinfI, Hsp92II and MboI (Promega) [31,39] for 3 h at 37°C. Digested products were purified as mentioned above. Terminal restriction fragment (TRF) sizes from each sample were determined using the ABI PRISM 3130 Genetic Analyzer System (Applied Biosystems) with GeneScan LIZ-600 (Applied Biosystems) as internal size standards. The GeneMapper software (Applied Biosystems) was used for the analysis of fragment data. To reduce data noise, only fragments containing intensity above a baseline threshold (50 fluorescence units) were recorded. Relative peak heights were calculated and fragments with an average relative abundance <5% were excluded from further analysis.

Screening and DNA sequence analysis

The remainders of the first PCR products were combined and purified using the QIAquick PCR purification kit (Qiagen).

Table 1. Chemical characteristic of soils (mean value ± SEM) in wet season (July 2010) which soils and roots were sampled.

Study plot	Soil pH[a]	Electrical conductivity [a]	Soil organic carbon (%)[a]	Total inorganic N (g kg^{-1} soil)[a]	Available P (mg kg^{-1} soil)[a]	Exchangeable K (mg kg^{-1} soil)[a]	Agricultural management
Phetchabun (PB)	5.77±0.25ab	0.18±0.02a	4.26±0.45ab	222±15a	156±72ab	449±163a	plowing, organic fertilizer, herbicide
Thai Orchid Lab (TO)	6.68± 0.28a	0.14±0.01a	4.15±1.48ab	188±57a	370±158a	296±81a	plowing, organic fertilizer, herbicide
Chiang Mai (CM)	5.70± 0.26bc	0.11±0.05a	6.10 ±0.87a	210±68a	171±133ab	284±10a	No management
Chiang Rai (CR)	5.23±0.11c	0.10±0.04a	2.83±0.25b	140±17a	24±3b	243±28a	No management
Nakhon Nayok (NN)	6.19±0.14ab	0.21±0.07a	3.08±0.66b	200±8a	149±63ab	347±78a	No management

[a]Means of three observations. Means in the same column followed by the same letter are not significantly different ($\alpha = 0.05$).

Purified DNA was cloned into pGEM-T Easy Vector (Promega) and transformed into *Escherichia coli* JM109. One hundred transformants were selected randomly and their insertion checked by PCR using the same primers, AML1 and AML2. The amplified DNAs were digested by the restriction enzymes HinfI and Hsp92II separately. One clone of each RFLP type was screened and sequenced using sequencing primers SP6 and T7 on an ABI PRISM 3130 Genetic Analyzer System (Applied Biosystems). Sequences were trimmed to the NS31-AML3 region and virtually digested with the restriction enzymes HinfI, Hsp92II, and MboI using an online restriction mapping website (RestrictionMapper).

Phylogenetic analysis

Phylogenetic analysis was carried out on the sequences obtained in this study and those corresponding to the closest matches from Genbank, as well as sequences from cultured AMF taxa including representatives of the major groups of Glomeromycota from GenBank. All sequences obtained from this study were aligned by ClustalX using the BioEdit sequence alignment editor [40] along with 28 AMF sequences from GenBank. The aligned SSU rRNA dataset was trimmed to 450 bp by excluding the terminal primer sequences. A neighbour-joining (NJ) phylogeny was constructed using PAUP*4b10 [41] with the Kimura 2-parameter model and 1000 bootstraps. The nucleotide sequences of the clones retrieved in this study have been deposited in GenBank (accession numbers JQ8643324-JQ864355).

Statistical analysis

The total number of TRFs was used as an AM fungal community diversity measurement [31]. The main and interaction effects of collecting sites, host plant species and sample source (root vs. soil) on number of TRFs using three restriction enzymes were tested with two-way factorial ANOVA using SPSS 11.5 for Windows (SPSS Inc., Chicago, IL, USA). Jaccard similarity coefficients were calculated for the T-RFLP patterns of root and soil samples of both plants, which were clustered by the unweighted pair-group average (UPGMA) algorithm with 1000 bootstrap replicates to obtain confidence estimates. These calculations were performed using FreeTree [42] and the results displayed using TreeView [43].

Results

Soil analyses and correlation with TRFs

Chemical characteristics of soil varied among sites (Table 1). Soil pH values ranged from 5.23 to 6.68. No significant different was observed in electrical conductivity, exchangeable potassium, and total inorganic nitrogen. Available phosphorus in soils tended to be highest at the Thai Orchid Lab site (370 mg kg^{-1}soil) and differed significantly from the Chiang Rai site (24 mg kg^{-1}soil). Soil organic carbon was highest at the Chiang Mai site (6.10%) and differed significantly from the Chiang Rai and Nakhon Nayok sites. Pearson correlation analysis between the soil properties measured and TRFs showed that TRFs were positively correlated with available phosphorus, organic matter, and pH (Table S1).

AM fungal community of root and soil samples from *T. grandis* and *A. crassna*

The total number of different TRFs was used as a measure of AM fungal community diversity. Thirty eight TRFs were found in total for the AML3 (FAM-labelled) primer, while the NS31 (HEX-labelled) primer identified 30 TRFs. Since the AML3 primer revealed many more TRFs than the NS31 primer, only the AML3 fragments were used. Overall, in the roots and soils of *T. grandis* and *A. crassna*, we found 13 different AML3 TRFs after digestion with HinfI, 14 after digestion with Hsp92II and 11 after digestion with MboI. The mean number of TRFs in *T. grandis* root and soil samples was 5.67 and 7.67, respectively when the TRF data of the three enzymes were pooled (Figure 1). It is possible to estimate the minimum average number of AM fungi colonizing the *T. grandis* root samples by dividing the average number of TRFs by 3 (three enzymes and one labeled end) [31]. Thus, there were on average at least 1.89 fungal taxa colonizing each *T. grandis* root sample and 2.55 fungal taxa in surrounding soils, respectively. The values for *A. crassna* were at least 2.85 fungal taxa per root sample and 2.33 fungal taxa in surrounding soils. The mean number of TRFs per sample was not significantly affected by source of samples (root and soil) ($F = 0.159$, $P = 0.693$) and host plant ($F = 3.452$, $P = 0.074$) (Table S2), but there was a statistically significant effect of collecting sites ($F = 42.77$, $P<0.01$), and a significant interaction among those three factors (Table S2). The cluster analysis of TRF patterns in roots (R-) and rhizosphere soils (S-) of *A. crassna* and *T. grandis*, based on Jaccard similarities, showed no significant grouping of samples by sites and source of samples (root and soil) (Figure 2a). This suggested that the AM fungal community in roots and rhizosphere soils was almost independent in *A. crassna* (A) and *T. grandis* (T) plots. Some TRF patterns in roots and rhizosphere soils that were collected from the same site were similar, e.g. R-CRT versus S-CRT and R-TOA versus R-TOT. Combining roots and rhizosphere soils of each plant by sampling site (CM: Chiang Mai, CR: Chiang Rai, NN: Nakhon-Nayok, PB: Phetchabun and TO: Thai Orchids Lab) indicated a tendency for *T. grandis* plots to be grouped together (PBT, CMT and TOT) as well as some *A. crassna* plot samples (PBA and TOA) (Figure 2b). This suggests that the AM fungal community associated with each tree species was more similar across plots than were communities for different trees species at the same location. The response for CRT and NNA, however, does not support this.

Occurrence of AM fungi in soils and roots of both plants

Nearly all of the distinct TRFs (31 out of 38) were found in both host plant species (Figure 3). There were some differences in AM fungal communities between *T. grandis* and *A. crassna* because the TRF 329c (TRFs are identified by their relative mobility and a code indicating the restriction enzyme that generated them: a: MboI, b: HinfI and c: Hsp92II) was not found in *T. grandis*, while 5 TRFs (135c, 141b, 158c, 176b, and 435b) were not found in *A. crassna*. Comparison of the population in roots and soils of *T. grandis* (Fig 3a) showed that 6 TRFs (135c, 158c, 176b, 181c, 435b and 438b) were found only in roots, while 141b and 281a were only found in soils. In *A. crassna* (Figure 3b), TRFs 176c, 181c and 438b were only found in root samples.

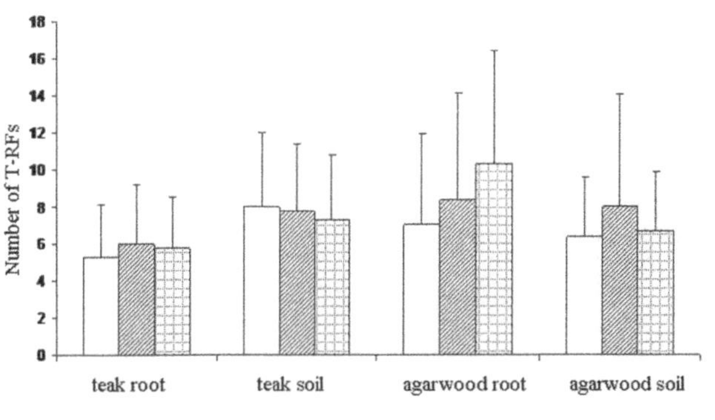

Figure 1. Effects of host plant, *Aquilaria crassna* (agarwood) and *Tectona grandis* (teak), and source of samples (root and soil) on mean number of terminal restriction fragments (TRFs) per sample using three restriction enzymes *MboI* (open bars), *HinfI* (hatched bars) and *Hsp92II* (cross-hatched bars). Values are mean ± SEM (n = 4 for teak and n = 3 for agarwood).

Sequence and phylogenetic analysis

Clones were selected for sequencing on the basis of HinfI and Hsp92II RFLP typing. DNA sequences of 32 selected clones were determined, 7 clones from *A. crassna* and 25 clones from *T. grandis*. Predicted TRFs from the 32 virtually digested clone sequences were compared to observed TRFs from all three restriction enzymes (Table S3). A difference in size of up to 7 nucleotides was accepted as a match, because migration in capillary electrophoresis is sequence-specific, so that mobility (in rmu) is only approximately equivalent to sequence length (in bp). All predicted TRFs were observed, and the great majority of the observed TRFs were represented in the cloned sequences.

Our phylogenetic analysis was based on the new classification of Krüger et al. [44]. The 32 clone sequences were aligned with 23 sequences identified as closely related reference sequences in GenBank and a phylogenetic tree was constructed using the 18S rRNA gene sequences of *Paraglomus occultum* (GenBank accessions AJ276081 and JN687477) as outgroup. This indicated the presence of five AM fungal clades belonging to the families Claroideoglomeraceae, Diversisporaceae, Gigasporaceae, and Glomeraceae (Figure 4), the most frequent sequences corresponding to Glomeraceae. The subclusters contained close matches to taxa previously identified by Singh et al. [22] based on spore morphology of AM fungi in rhizosphere soils of *T. grandis*: TR1-16, TR1-43, TS4-4, AR5-7 and TS6-1 are close to *Rhizophagus intraradices* or *R. irregularis*, while TR1-27 is close to *Redeckera fulvum*. Clone sequences TS4-9 and TS4-32 are similar to *Diversispora aurantia*, while TR3-R10 is probably *Gigaspora margarita*. When sequence data are compared with individual TRFs (Table S3 and Figure 4), it is clear that individual TRFs cannot be used to identify sequence type, because many different species may generate a TRF of the same size. For example, the FAM fragment at 164b could equally well be from *G. indicum*, *Re. fulvum* or *Claroideoglomus etunicatum*.

Discussion

This study examined the AM fungal communities of *A. crassna* and *T. grandis* plantations in Thailand. The estimated numbers of AM fungal taxa in roots and soils of *T. grandis* seedlings were 1.89 and 2.55 respectively, while in roots and soils of *A. crassna* there were 2.85 and 2.33 respectively. The AM fungal diversity was low compared with other plants. Using similar methodologies and definitions, Vandenkoornhuyse et al. [31] reported an average of 6.1 AM fungal taxa colonizing grass roots in a temperate seminatural grassland system, and 5.5 AM fungal taxa were found colonizing each *Solidago virgaurea* L. seedling root sample in low-Arctic meadow habitat [29].

Previous studies quantified the AM fungal diversity in rhizospheres of *T. grandis* and *A. crassna* mainly based on spore morphology and aimed to select efficient AM fungal isolates for growth enhancement. For example, Singh et al. [22] found an average of nine species per 100 g dry soil in a Jhum fallow site at which *T. grandis* was the dominant tree species, and most species belonging to the genus Glomus. Tamuli and Boruah [21] studied the AM fungi association of agarwood (*Aquilaria malaccensis*) plantations in Jorhat District of the Brahmaputra Valley, India. They found that the genus *Glomus* was dominant; among these

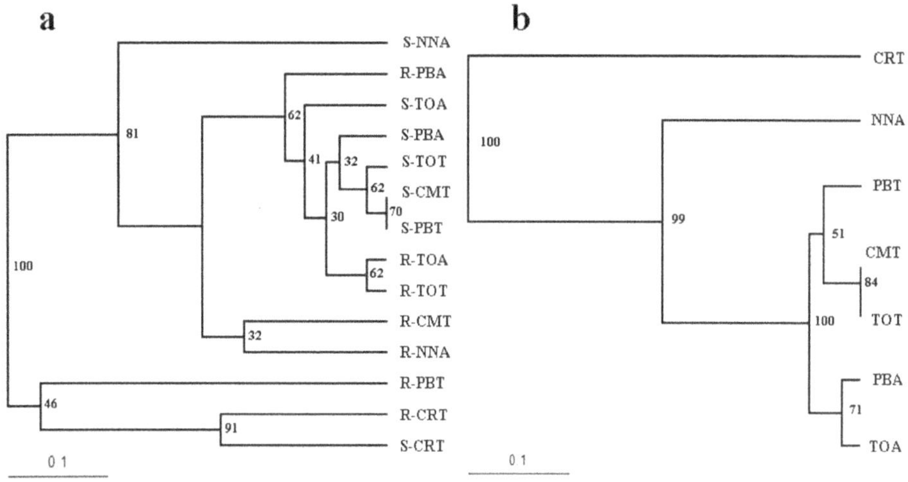

Figure 2. Cluster analysis of terminal restriction fragment length polymorphism patterns from AM fungal communities associated with *Aquilaria crassna* (A) and *Tectona grandis* (T); a) TRFs patterns in roots (R-) and rhizosphere soils (S-) and b) TRFs patterns in five sites (CM: Chiang Mai, CR: Chiang Rai, NN: Nakhon-Nayok, PB: Phetchabun and TO: Thai Orchid Labs). The unweighted pair-group average (UPGMA) algorithm was used to cluster patterns based on Jaccard similarities. Percentage values based on 1000 bootstrap replicates are given at each node.

G. fasciculatum (now known as *Rhizophagus fasciculatus*; [45]) was the most dominant followed by *G. aggregatum*. We are not aware of any information on the diversity of AM fungi on *A. crassna*. According to previous studies, we also found that most sequences belonged to the family Glomeraceae that includes *Glomus* and *Rhizophagus*. This result is consistent with previously

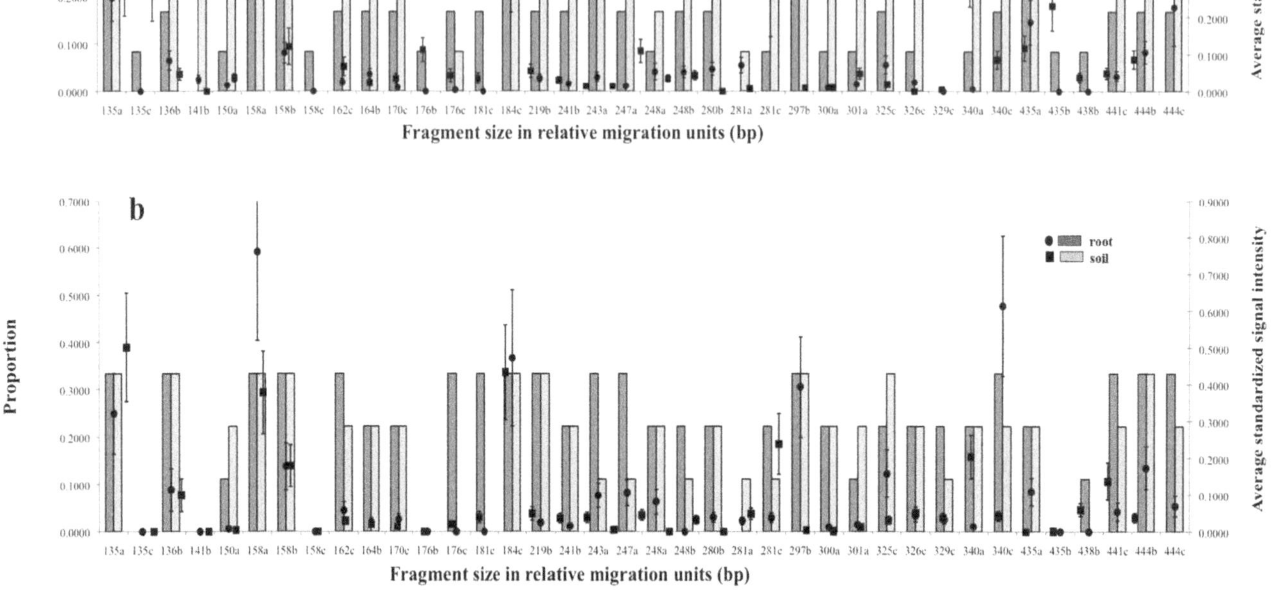

Figure 3. Occurrence of TRFs from roots and soils in (a) *Tectona grandis* and (b) *Aquilaria crassna*. Bars indicate the proportion of samples that yielded each TRF; dots indicate the average intensity of that fragment (± SEM) in those samples. The letters indicate the restriction enzyme involved in each fragment size, a: *MboI*, b: *HinfI* and c: *Hsp92II*.

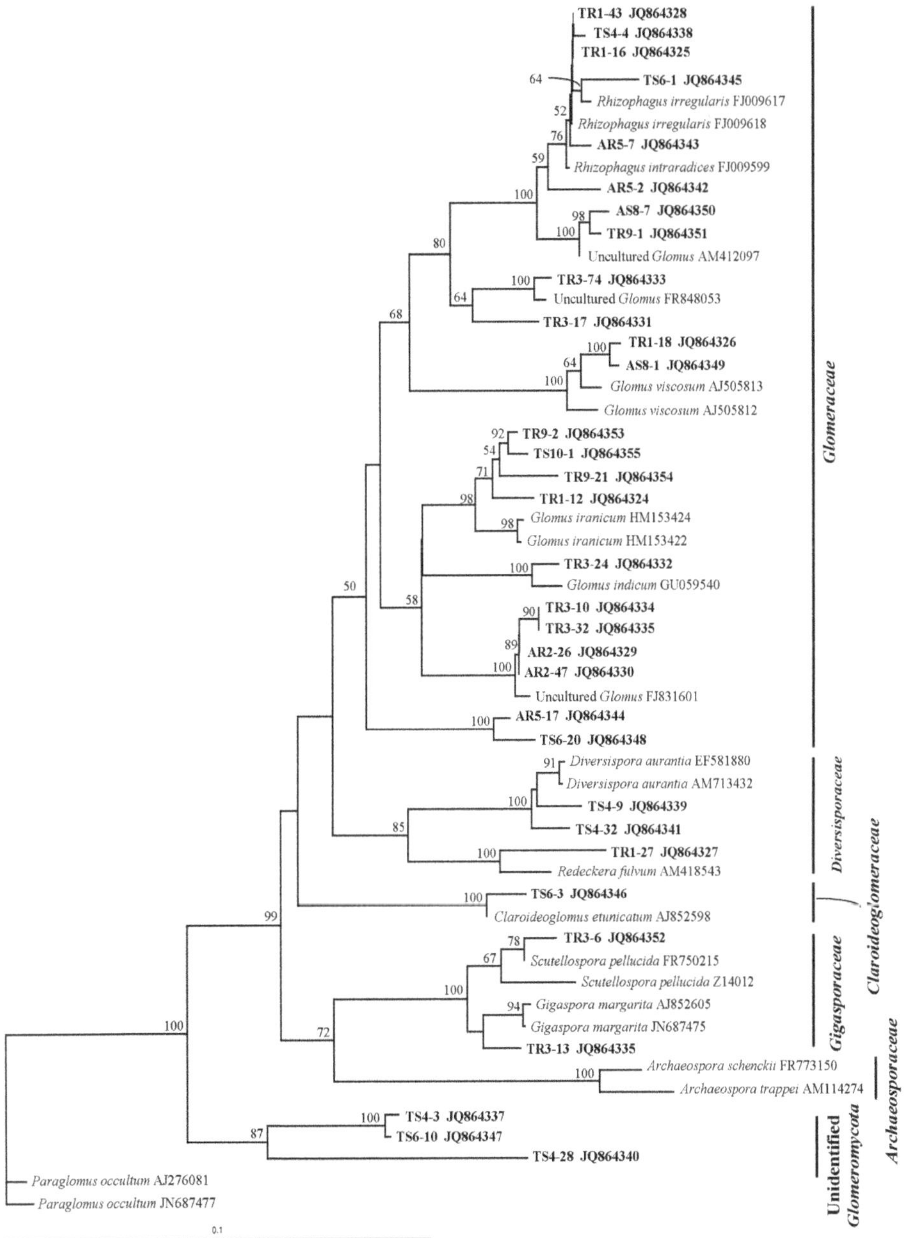

Figure 4. Neighbour-joining (NJ) phylogenetic tree of partial small subunit rRNA gene. Phylogeny was constructed using the region from NS31 to AML3. The percentage support values are based on 1000 bootstraps.

published phylogenies [29,39,46]. The dominance of this family suggests that they able to survive under various agricultural conditions such as soil disturbance from plowing and cultivation and pesticide usage like that used here in the Phetchabun and Nakhon-Nayok sites. Those conditions may be unfavorable for other AM fungi. One possible reason why *Glomus* species have the ability to survive in a disturbed system is related to differences in propagation strategies [29]. Glomeraceae are capable of colonizing via fragments of mycelium, mycorrhizal root pieces, and spores, while Gigasporaceae are only capable of propagation via spores because they do not produce intra-radical vesicles: lipid-rich storage structures which allow for re-growth of hyphae from previously colonized root pieces [46,47,48,49]. This difference can explain the dominance of the Glomeraceae over Gigasporaceae members in an environment with repetitive agricultural disturbance. Oehl et al. [50] revealed a clear seasonal and successional AMF sporulation dynamics and implied that different life strategies of different ecological AMF groups could be defined on the basis of diverging temporal sporulation dynamics.

This study shows that the choice of restriction enzymes (HinfI, Hsp92II, MboI) did not significantly affect AM fungal diversity found per sample. Using a combination of those three restriction enzymes could detect possible species of AM fungi in the samples, even if they resulted in similar-sized fragments. HinfI and Hsp92II were chosen in this study because they showed the highest polymorphism of cleavage sites at the extremities of the amplified DNA fragment [31]. Mummey and Rillig [39] and Wolfe et al. [51] also found that HinfI and MboI can separate different closely-related species of AM fungi identified from phylogenetic analyses. For example, *R. irregularis* and *R. intraradices* are closely related species that group in the same clade (Figure 4). Six clone sequences (TR1-16, TR1-43, TS4-4, AR5-2, AR5-7and TS6-1) that were related to both species were not completely separated using phylogenetic analysis, but virtual digesting with those three enzymes did separate them by using the combination of restriction pattern of each enzyme (Table S3). Clone sequences TR1-16, 1-43, and 6-1 grouped with *R. irregularis* and TS4-4, 5-2, 5-7 grouped with *R. intraradices*.

Some TRFs were only found in roots or only in soils, suggesting that some AM fungi may be rare in soil but produce fungal structures in roots that are rich enough for T-RFLP detection, while some were found only as spores in soils and did not colonize roots. While the majority of TRFs were associated with both *T. grandis* and *A. crassna*, some TRFs were associated with just one plant (i.e. 135c, 141b, 158c, 176b, 329c and 435b). In clustering analysis, samples from each plant species were grouped together even if they were collected from different sites. *A. crassna* samples seemed to group together, but since many AMF taxa were shared by both trees, *A. crassna* shared some AM fungal community patterns with *T. grandis* (Figure 2). Statistical analysis revealed significant effects of collecting sites and the interaction between collecting sites, host plant species and source of samples on TRFs (Table S2). Thus, specific AM taxa in roots and soils of *T. grandis* and *A. crassna* were affected by site but not affected by host plant species and source of samples (root and soil). This is in accordance with the observation of Bever et al. [52] that the host-dependence of the relative growth rates of fungal populations may play an important role in the maintenance of fungal species diversity. Previously, it has been reported that neighboring plants may have a significant impact on the AM fungal colonization and community composition of AM fungi in plant roots [34]. Although *T. grandis* at the Chiang Mai site had other *T. grandis* as closest neighbors with some negligible understory perennial plants, and at the other two sites the closest neighbors were *A. crassna*, the cluster analysis did not reveal any effect of this difference in neighbors. AM fungal community patterns in CMT were grouped with PBT and TOT sites in which weeds were controlled by agricultural management.

In conclusion, we demonstrated here that AM fungal community patterns in rhizosphere soils and roots of *T. grandis* and *A. crassna* were similar even if they were collected from different sites. AM fungal communities of *T. grandis* samples from different sites were similar, as were those in *A. crassna* samples. We also found that most sequences represented Glomeraceae, including *Glomus* spp. and *Rhizophagus* spp. Virtual digestion of sequences using the target sequences of the restriction enzymes HinfI, Hsp92II and MboI yielded expected fragments that mostly matched observed TRFs, linking possible AM fungal species to each TRF. Specific AM taxa in roots and soils of *A. crassna* and *T. grandis* were affected by site but were not affected by host plant species and source of samples (root and soil). Although the T-RFLP technique can provide important information about the AM fungal diversity associated with plant species of interest, trap cultures and cultured spores from the field site are still important in order to assess the ability of the AM fungi to enhance the growth of the plants, and to provide effective candidates for inoculum production targeted for these economically important tree species.

Supporting Information

Table S1 Correlation matrix of soil factors and terminal restriction fragments (TRFs) of study areas in wet season (July 2010) which soils were sampled.

Table S2 Summary of two-way analysis of variance for main and interaction effects of host plants (*Aquilaria crassna* and *Tectona grandis*), sites, and source of samples (root and soil) on AM fungal community diversity measured as the number of different TRFs per sample. Significant P-values are shown in bold.

Table S3 Clone sequences and TRFs derived from roots and rhizosphere soils of *T. grandis* and *A. crassna*. Values in bold indicate TRFs that match the sizes of virtual digest fragments (with differences ranging from 0 to 7 bp).

Author Contributions

Conceived and designed the experiments: SL JY. Performed the experiments: AC JY. Analyzed the data: AC JY. Contributed reagents/materials/analysis tools: JY NT PG SL. Wrote the paper: AC JY SL. Collected samples: PG.

References

1. Kobayashi S (2004) Landscape rehabilitation of degraded tropical forest ecosystems, Case study of the CIFOR/Japan project in Indonesia and Peru. Forest Ecol Manag 201: 13–22.
2. Turjaman M, Tamai Y, Santoso E, Osaki M, Tawaraya K (2006) Arbuscular mycorrhizal fungi increased early growth of two nontimber forest product species *Dyera polyphylla* and *Aquilaria filaria* under greenhouse conditions. Mycorrhiza 16: 459–464.
3. Rajan SK, Reddy BJD, Bagyaraj DJ (2000) Screening of arbuscular mycorrhizal fungi for their symbiotic efficiency with *Tectona grandis*. Forest Ecol Manag 126: 91–95.
4. CITES (2004) Amendments to Appendices I and II of the Convention on International Trade in Endangered Species of Wild Flora and Fauna (CITES). Thirteenth meeting of the conference of the parties, Bangkok, Thailand, 3–14 October, pp. 1–9.
5. Schüßler A, Schwarzott D, Walker C (2001) A new fungal phylum, the *Glomeromycota*: phylogeny and evolution. Mycol Res 105 (12): 1413–1421.
6. Habte M (2000) Mycorrhizal fungi and plant nutrition. In: Silva JA, Uchida R, editors. Plant Nutrient Management in Hawaii's Soils, Approaches for Tropical and Subtropical Agriculture. College of Tropical Agriculture and Human Resources, Manoa: University of Hawaii. pp. 127–131.
7. Brundrett MC, Bougher N, Dell B, Grove T, Malajczuk N (1996) Working with mycorrhizas in forestry and agriculture. Australian Centre for International Agricultural Research Monograph 32, Canberra.
8. Smith S, Read D (2008) Mycorrhizal Symbiosis. New York: Academic Press.
9. Porcel R, Ruiz-Lozano JM (2004) Arbuscular mycorrhizal influence on leaf water potential, solute accumulation, and oxidative stress in soybean plants subjected to drought stress. J Exp Bot 55: 1743–1750.
10. Doubková P, Vlasáková E, Sudová R (2013) Arbuscular mycorrhizal symbiosis alleviates drought stress imposed on *Knautia arvensis* plants in serpentine soil. Plant Soil 370: 149–161.
11. Azcón-Aguilar C, Barea JM (1996) Arbuscular mycorrhizas and biological control of soil-borne plant pathogens – an overview of the mechanisms involved. Mycorrhiza 6: 457–464.
12. Wright SF, Upadhyaya A (1998) A survey of soils for aggregate stability and glomalin, a glycoprotein produced by hyphae of arbuscular mycorrhizal fungi. Plant Soil 198: 97–107.
13. Rillig MC (2004) Arbuscular mycorrhizae, glomalin, and soil aggregation. Can J Soil Sci 84: 355–363.
14. Rillig MC, Ramsey PW, Morris S, Paul EA (2003) Glomalin, an arbuscular-mycorrhizal fungal soil protein, responds to land-use change. Plant Soil 253: 293–299.
15. Habte M, Miyasaka SC, Matsuyama DT (2001) Arbuscular mycorrhizal fungi improve early forest-tree establishment. In: Horst WJ et al., editors. Plant nutrition–Food Security and Sustainability of Agro-ecosystems. Dordrecht: Kluwer Academic Publishers. pp. 644–645.
16. Urgiles N, Loján P, Aguirre N, Blaschke H, Günter S, et al. (2009) Application of mycorrhizal roots improves growth of tropical tree seedlings in the nursery: a

step towards reforestation with native species in the Andes of Ecuador. New Forest 38: 229–239.

17. Swaminathan C, Srinivasan VM (2006) Influence of microbial inoculants on seedling production in teak (*Tectona grandis* L.f.). J Sustain Forest 22 (3): 63–76.
18. Tabin T, Arunachalam A, Shrivastava K, Arunachalam K (2009) Effect of arbuscular mycorrhizal fungi on damping-off disease in *Aquilaria agallocha* Roxb. seedlings. Trop Ecol 50 (2): 243–248.
19. Thapar HS, Khan SN (1988) Seasonal frequency of *Endogone* spores in new-forest soils. In: Khosla PK, Sehgal RN, editors. Trends in Tree Sciences. Solan: Indian Society of Tree Scientists. pp.161–162.
20. Kanakadurga VV, Manoharachary D, Rama RP (1990) Occurrence of endomycorrhizal fungi in teak. In: Bagyaraj DJ, Manjunath A, editors. Mycorrhizal symbiosis and plant growth.Bangalore: University of Agricultural Sciences. pp.17.
21. Tamuli P, Boruah P (2002) Vesicular-abuscular mycorrhizal (VAM) association of agarwood tree in Jorhat District of the Brahmatputra Valley. Indian For 128 (9): 991–994.
22. Singh SS, Tiwari SC, Dkhar MS (2003) Species diversity of vesicular-arbuscular mycorrhizal (VAM) fungi in Jhum fallow and natural forest soils of Arunachal Pradesh, north eastern India. Trop Ecol 44 (2): 207–215.
23. Dhar PP, Mridha MAU (2012) Arbuscular mycorrhizal associations in different forest tree species of Hazarikhil forest of Chittagong, Bangladesh. J For Res 23 (1): 115–122.
24. Ramanwong K (1998) Species diversity of vesicular-arbuscular mycorrhizal fungi of teak (*Tectona grandis* Linn.f.) and their effects on growth of teak seedlings. M.S. Thesis, Kasetsart University, Chatuchak, Bangkok.
25. Helgason T, Daniell TJ, Husband R, Fitter AH, Young JPW (1998) Ploughing up the wood-wide web? Nature 394: 431.
26. Lee J, Lee S, Young JPW (2008) Improved PCR primers for the detection and identification of arbuscular mycorrhizal fungi. FEMS Microbiol Ecol 65: 339–349.
27. Krüger M, Stockinger H, Krüger C, Schüßler A (2009) DNA-based species level detection of Glomeromycota: one PCR primer set for all arbuscular mycorrhizal fungi. New Phytol 183: 212–223.
28. Daniell TJ, Husband R, Fitter AH, Young JPW (2001) Molecular diversity of arbuscular mycorrhizal fungi colonising arable crops. FEMS Microbiol Ecol 36: 203–209.
29. Pietikäinen A, Kytöviita MM, Husband R, Young JPW (2007) Diversity and persistence of arbuscular mycorrhizas in a low-Arctic meadow habitat. New Phytol 176: 691–698.
30. Mummey DL, Rillig MC (2006) The invasive plant species Centaurea maculosa alters arbuscular mycorrhizal fungal communities in the field. Plant Soil 288: 81–90.
31. Vandenkoornhuyse P, Ridgway KP, Watson IJ, Duck M, Fitter AH, et al. (2003) Co-existing grass species have distinctive arbuscular mycorrhizal communities. Mol Ecol 12: 3085–3095.
32. Johnson D, Vandenkoornhuyse PJ, Leake JR, Gilbert L, Booth RE, et al. (2004) Plant communities affect arbuscular mycorrhizal fungal diversity and community composition in grassland microcosms. New Phytol 161: 503–515.
33. van der Heijden MGA, Wiemken A, Sanders IR (2003) Different arbuscular mycorrhizal fungi alter coexistence and resource distribution between co-occurring plant. New Phytol 157: 569–578.
34. Mummey DL, Rillig MC, Holben WE (2005) Neighbouring plant influences on arbuscular mycorrhizal fungal community composition as assessed by T-RFLP analysis. Plant Soil 271: 83–90.
35. Barto EK, Antunes PM, Stinson K, Koch AM, Klironomos JN, et al. (2011) Differences in arbuscular mycorrhizal fungal communities associated with sugar maple seedlings in and outside of invaded garlic mustard forest patches. Biol Invasions 13: 2755–2762.
36. Houba VJG, Van Der Lee JJ, Novozamsky I, Wallinga J (1988) Soil and Plant analysis. Part 5: Soil Analysis Procedure. Agricultural University, Wageningen.
37. Sparks DL, Page AL, Helmke PA, Loeppert RH, Soltanpour PN, et al. (1996) Methods of Soil Analysis. Part 3. Chemical Methods. Madison, Wisconsin: Soil Science Society of America
38. Render C, Weißhuhn K, Kellner H, Buscot F (2006) Rationalizing molecular analysis of field-collected roots for assessing diversity of arbuscular mycorrhizal fungi: to pool, or not to pool, that is the question. Mycorrhiza 16: 525–531.
39. Mummey DL, Rillig MC (2007) Evaluation of LSU rRNA-gene PCR primers for analysis of arbuscular mycorrizal fungal communities via terminal restriction fragment length polymorphism analysis. J Microbiol Meth 70: 200–204.
40. Hall T (1999) BioEdit: a user-friendly biological sequence alignment editor and analysis program for Windows95/98/NT. Nucl Acid S 41: 95–98.
41. Swofford DL (2002) PAUP* Phylogenetic analysis using parsimony (*and other methods). version4b10. Sunderland. MA: Sinauer Associates
42. Hampl V, Pavlícek A, Flegr J (2001) Construction and bootstrap analysis of DNA fingerprinting-based phylogenetic trees with a freeware program Free-Tree: Application to trichomonad parasites. Int J Syst Evol Micr 51: 731–735.
43. Page RDM (1996) TreeView: An application to display phylogenetic trees on personal computers. Comput Appl Biosci 12: 357−358.
44. Krüger M, Krüger C, Walker C, Stockinger H, Schüßler A (2012) Phylogenetic reference data for systematics and phylotaxonomy of arbuscular mycorrhizal fungi from phylum to species-level. New Phytol 193: 970–984.
45. Schüßler A, Walker C (2010) The Glomeromycota: A species list with new families and new genera (Libraries at the Royal Botanic Garden Edinburgh, Edinburgh, UK; The Royal Botanic Garden Kew, Kew, UK; Botanische Staatssammlung Munich, Munich, Germany; and Oregon State University, Corvallis, Oregon, pp.1–56.
46. Helgason T, Fitter AH, Young JPW (1999) Molecular diversity of arbuscular mycorrhizal fungi colonising *Hyacinthoides non-scripta* (bluebell) in a seminatural woodland. Mol Ecol 8: 659–666.
47. Gazey C, Abbott KK, Robson AD (1993) VA mycorrhizal spores from 3 species of *Acaulospora*—germination, longevity and hyphal growth. Mycol. Res. 97: 785–790.
48. INVAM Newsletter 3 (1993) Properties of infective propagules at the suborder level (*Glomineae* versus *Gigasporineae*). West Virginia University Web Services. Available: http//invam.caf.wvu.edu/articles/propagules.htm. Accessed 9 July 2012.
49. Brundrett M, Abbott LK, Jasper DA (1999) Glomalean mycorrhizal fungi from tropical Australia. I. Comparison of the effectiveness and specificity of different isolation procedures. Mycorrhiza 8: 305–314.
50. Oehl F, Sieverding E, Ineichen K, Mäder P, Wiemken A, et al. (2009) Distinct sporulation dynamics of arbuscular mycorrhizal fungal communities from different agroecosystems in long-term microcosms. Agr Ecosyst Environ 134: 257–268.
51. Wolfe BE, Mummey DL, Rillig MC, Klironomos JN (2007) Small-scale spatial heterogeneity of arbuscular mycorrhizal fungal abundance and community composition in a wetland plant community. Mycorrhiza 17: 175–183.
52. Bever JD, Morton JB, Antonovics J, Schultz PA (1996) Host-dependent sporulation and species diversity of arbuscular mycorrhizal fungi in a mown grassland. J Ecol 84: 71–82.

PERMISSIONS

All chapters in this book were first published in PLOS ONE, by The Public Library of Science; hereby published with permission under the Creative Commons Attribution License or equivalent. Every chapter published in this book has been scrutinized by our experts. Their significance has been extensively debated. The topics covered herein carry significant findings which will fuel the growth of the discipline. They may even be implemented as practical applications or may be referred to as a beginning point for another development.

The contributors of this book come from diverse backgrounds, making this book a truly international effort. This book will bring forth new frontiers with its revolutionizing research information and detailed analysis of the nascent developments around the world.

We would like to thank all the contributing authors for lending their expertise to make the book truly unique. They have played a crucial role in the development of this book. Without their invaluable contributions this book wouldn't have been possible. They have made vital efforts to compile up to date information on the varied aspects of this subject to make this book a valuable addition to the collection of many professionals and students.

This book was conceptualized with the vision of imparting up-to-date information and advanced data in this field. To ensure the same, a matchless editorial board was set up. Every individual on the board went through rigorous rounds of assessment to prove their worth. After which they invested a large part of their time researching and compiling the most relevant data for our readers.

The editorial board has been involved in producing this book since its inception. They have spent rigorous hours researching and exploring the diverse topics which have resulted in the successful publishing of this book. They have passed on their knowledge of decades through this book. To expedite this challenging task, the publisher supported the team at every step. A small team of assistant editors was also appointed to further simplify the editing procedure and attain best results for the readers.

Apart from the editorial board, the designing team has also invested a significant amount of their time in understanding the subject and creating the most relevant covers. They scrutinized every image to scout for the most suitable representation of the subject and create an appropriate cover for the book.

The publishing team has been an ardent support to the editorial, designing and production team. Their endless efforts to recruit the best for this project, has resulted in the accomplishment of this book. They are a veteran in the field of academics and their pool of knowledge is as vast as their experience in printing. Their expertise and guidance has proved useful at every step. Their uncompromising quality standards have made this book an exceptional effort. Their encouragement from time to time has been an inspiration for everyone.

The publisher and the editorial board hope that this book will prove to be a valuable piece of knowledge for researchers, students, practitioners and scholars across the globe.

LIST OF CONTRIBUTORS

Yuanyuan Dong, Yepeng Sun, Nan Wang, Fawei Wang, Huan Chen, Na Yao, Lili Guan, Kai Chen, Xiaowei Li and Weican Liu
Ministry of Education Engineering Research Center of Bioreactor and Pharmaceutical Development, Jilin Agricultural University, Changchun, Jilin, China

Junfeng Zhai, and Haiyan Li
Ministry of Education Engineering Research Center of Bioreactor and Pharmaceutical Development, Jilin Agricultural University, Changchun, Jilin, China
College of Life Sciences, Jilin Agricultural University, Changchun, Jilin, China

Xiyan Cui and Meiying Yang
College of Life Sciences, Jilin Agricultural University, Changchun, Jilin, China,

Qi Wang
High School attached to Northeast Normal University, Changchun, Jilin, China

Vokkaliga Thammegowda Harshavardhan, Christiane Seiler and Markus Kuhlmann
Research Group Abiotic Stress Genomics, Interdisciplinary Center for Crop Plant Research (IZN), Halle (Saale), Germany, and Leibniz Institute of Plant Genetics and Crop Plant Research (IPK) Gatersleben, Stadt Seeland, OT Gatersleben, Germany

Le Van Son
Research Group Gene Regulation, Leibniz Institute of Plant Genetics and Crop Plant Research
(IPK) Gatersleben, Stadt Seeland, OT Gatersleben, Germany
National Key Laboratory of Gene Technology, Institute of Biotechnology Vietnam, Academy of Science and Technology, Hanoi, Vietnam

Kathleen Weigelt-Fischer, Astrid Junker and Thomas Altmann
Research Group Heterosis, Leibniz Institute of Plant Genetics and Crop Plant Research (IPK) Gatersleben, Stadt Seeland, OT Gatersleben Germany

Christian Klukas
Research Group Image Analysis, Leibniz Institute of Plant Genetics and Crop Plant Research (IPK) Gatersleben, Stadt Seeland, OT Gatersleben, Germany

Nese Sreenivasulu
Research Group Abiotic Stress Genomics, Interdisciplinary Center for Crop Plant Research (IZN), Halle (Saale), Germany, and Leibniz Institute of Plant Genetics and Crop Plant Research (IPK) Gatersleben, Stadt Seeland, OT Gatersleben, Germany
Grain Quality and Nutrition Center, International Rice Research Institute (IRRI), Metro Manila, Philippines

Helmut Bäumlein
Research Group Gene Regulation, Leibniz Institute of Plant Genetics and Crop Plant Research (IPK) Gatersleben, Stadt Seeland, OT Gatersleben, Germany

Yilin Li and Weiming Shi
State Key Laboratory of Soil and Sustainable Agriculture, Institute of Soil Science, Chinese Academy of Sciences, Nanjing, China

Xingxiang Wang
State Key Laboratory of Soil and Sustainable Agriculture, Institute of Soil Science, Chinese Academy of Sciences, Nanjing, China
Jiangxi Key Laboratory of Ecological Research of Red Soil, Yingtan, Jiangxi, China

Tingting Liu, Chaofeng Shen, Yi Wang, Canke Huang and Jiyan Shi
Institute of Environmental Science and Technology, College of Environmental and Resource Sciences, Zhejiang University Hangzhou, P.R. China

Weiti Cui, Kaikai Zhu, Qijiang Jin, Yanjie Xie, Jin Cui, Yan Xia, Jing Zhang and Wenbiao Shen
College of Life Sciences, Laboratory Center of Life Sciences, Nanjing Agricultural University, Jiangsu Province, Nanjing, China

Huiping Chen
Key Laboratory of Protection and Development Utilization of Tropical Crop Germplasm Resources, Hainan University, Haikou, China

Rupak K. Sarma, Animesh Gogoi, Rajal Debnath, Tarun C. Bora and Ratul Saikia
Biotechnology Division, CSIR-North East Institute of Science and Technology, Jorhat, Assam, India

Budheswar Dehury
Department of Agricultural Biotechnology, Assam Agricultural University, Jorhat, Assam, India

Jeremy D. Long and Laura D. Porturas
Biology Department and Coastal & Marine Institute Laboratory, San Diego State University, San Diego, California, United States of America

Shalabh Dixit, Ma. Teresa Sta Cruz, Paul T. Maturan, Jhon Christian E. Ontoy and Arvind Kumar
International Rice Research Institute (IRRI), Los Baños, Laguna, Philippines

B. Emma Huang
Computational Informatics, CSIRO, Dutton Park, Queensland, Australia

Muhammad Arslan
Soil and Environmental Biotechnology Division, National Institute for Biotechnology and Genetic Engineering, Faisalabad, Pakistan
Earth Sciences Department, King Fahd University of Petroleum and Minerals, Dhahran, Saudi Arabia

Muhammad Afzal, Samina Iqbal and Qaiser M. Khan
Soil and Environmental Biotechnology Division, National Institute for Biotechnology and Genetic Engineering, Faisalabad, Pakistan

Imran Amin
Agricultural Biotechnology Division, National Institute for Biotechnology and Genetic Engineering, Faisalabad, Pakistan

Jing Wu, Lanfen Wang, Long Li and Shumin Wang
Key Laboratory of Crop Germplasm Resources and Utilization, Ministry of Agriculture; The National Key Facility for Crop Gene Resources and Genetic Improvement, Institute of Crop Science, the Chinese Academy of Agricultural Sciences, Beijing, China

Ido Golan, Zvia Konrad, Doron Shkolnik-Inbar and Dudy Bar-Zvi
Department of Life Sciences and Doris and Bertie Black Center for Bioenergetics in Life Sciences, Ben-Gurion University of the Negev, Beer-Sheva, Israel

Pia Guadalupe Dominguez and Fernando Carrari
Instituto de Biotecnología, Instituto Nacional de Tecnología Agropecuaria, Buenos Aires, Argentina

Jian Sun, Qiang Zhang, Jia Zhou and Qinping Wei
Institute of Forestry and Pomology, Beijing Academy of Agriculture and Forestry Sciences, Beijing, China

Tingting Li, Yuanyuan Hu, Hui Tang, Chaohua Shen and Jiasheng Wu
Nurturing Station for the State Key Laboratory of Subtropical Silviculture, Zhejiang A & F University, Li ńan, Hangzhou, P. R. China
School of Forestry & Biotechnology, Zhejiang A & F University, Lin'an, Hangzhou, P. R. China

Xuhua Du
China National Bamboo Research Center, Hangzhou, P. R. China

Deshika Kohli, Ramamurthy Srinivasan, Pradeep Kumar Jain and Amit Atmaram Deokar
NRC on Plant Biotechnology, IARI Campus (PUSA), New Delhi, India

Gopal Joshi, Ankur R. Bhardwaj and Manu Agarwal
Department of Botany, North campus, University of Delhi, Delhi, India

Surekha Katiyar-Agarwal
Department of Plant Molecular Biology, South Campus, University of Delhi, Delhi, India

Renata Velozo Timbo
Department of Cell Biology, University of Brasília, Campus Universitário Darcy Ribeiro, Brasília, Distrito Federal, Brazil
Department of Biological Control, Embrapa Genetic Resources and Biotechnology, Brasília, Distrito Federal, Brazil

Marcelo Hermes-Lima
Department of Cell Biology, University of Brasília, Campus Universitário Darcy Ribeiro, Brasília, Distrito Federal, Brazil

Luciano Paulino Silva, Angela Mehta, Maria Carolina Blassioli Moraes and De´bora Pires Paula
Department of Biological Control, Embrapa Genetic Resources and Biotechnology, Brasília, Distrito Federal, Brazil

Marco Lauteri
Institute of Agro-Environmental and Forest Biology, National Research Council, Porano, Italy

Mauro Centritto and Matthew Haworth
Trees and Timber Institute (IVALSA - Istituto per la valorizzazione del legno e delle specie arboree), National Research Council, Florence, Italy

Rachid Serraj
Consultative Group on International Agricultural Research, Independent Science and Partnership Council Secretariat, Food and Agriculture Organisation of the United Nations, Rome, Italy
International Rice Research Institute, Los Banos, Philippines

Maria Cristina Monteverdi
Forestry Research Centre, National Research Council, Arezzo, ItalyAbstract

Babatunde Bello, Xueyan Zhang, Chuanliang Liu, Zhaoen Yang, Zuoren Yang, Qianhua Wang, Ge Zhao and Fuguang Li
State Key Laboratory of Cotton Biology, Cotton Research Institute, Chinese Academy of Agricultural Sciences, Beijing, China

Yiqing Liu, Yusong Jiang, Jianbin Lan and Yong Zou
Department of Ornamental Horticulture, China Agricultural University, Beijing 100193, China College of Life Science & Forestry, Chongqing University of Art & Science, Yongchuan 402160, China

Junping Gao
Department of Ornamental Horticulture, China Agricultural University, Beijing 100193, China

Xiangzhen Li
Energy Biosciences Institute, University of Illinois at Urbana-Champaign, Urbana, IL, United States of America
Key Laboratory of Environmental and Applied Microbiology, CAS, Environmental Microbiology Key Laboratory of Sichuan Province, Chengdu Institute of Biology, Chinese Academy of Sciences, Chengdu, Sichuan, China

Junpeng Rui
Key Laboratory of Mountain Ecological Restoration and Bioresource Utilization, CAS, Ecological Restoration Biodiversity Conservation Key Laboratory of Sichuan Province, Chengdu Institute of Biology, Chinese Academy of Sciences, Sichuan, PR China

Jingbo Xiong
School of Marine Sciences, Ningbo, Zhejiang, China

Jiabao Li
Key Laboratory of Environmental and Applied Microbiology, CAS, Environmental Microbiology Key Laboratory of Sichuan Province, Chengdu Institute of Biology, Chinese Academy of Sciences, Chengdu, Sichuan, China

Zhili He and Jizhong Zhou
Institute for Environmental Genomics and Department of Microbiology and Plant Biology, University of Oklahoma, Norman, OK, United States of America

Anthony C. Yannarell
Department of Natural Resources and Environmental Sciences, University of Illinois at Urbana-Champaign, Urbana, IL, United States of America

Roderick I. Mackie
Energy Biosciences Institute, University of Illinois at Urbana-Champaign, Urbana, IL, United States of America Resources and Environmental Sciences, University of Illinois at Urbana-Champaign, Urbana, IL, United States of America
Department of Animal Sciences, University of Illinois at Urbana-Champaign, Urbana, IL, United States of America

Daryush Talei
Medicinal Plants Research Center, Shahed University, Tehran, Iran
Department of Cell and Molecular Biology, Faculty of Biotechnology and Biomolecular Sciences, Universiti Putra Malaysia, UPM Serdang, Selangor DE, Malaysia

Alireza Valdiani
Department of Biochemistry, Faculty of Biotechnology and Biomolecular Sciences, Universiti Putra Malaysia, UPM Serdang, Selangor DE, Malaysia

Mohd Yusop Rafii
Institute of Tropical Agriculture, Universiti Putra Malaysia, UPM Serdang, Selangor DE, Malaysia

Mahmood Maziah
Department of Biochemistry, Faculty of Biotechnology and Biomolecular Sciences, Universiti Putra Malaysia, UPM Serdang, Selangor DE, Malaysia
Institute of Tropical Agriculture, Universiti Putra Malaysia, UPM Serdang, Selangor DE, Malaysia
Institute of Bioscience, Universiti Putra Malaysia, UPM Serdang, Selangor DE, Malaysia

Tae Kyung Hyun and Seung Hee Eom
Institute of Plant Sciences, University of Graz, Graz, Austria
Plant Molecular Biology and Biotechnology Research Center, Gyeongsang National University, Jinju, Korea

Eric van der Graaff and Dominik K. Großkinsky
Institute of Plant Sciences, University of Graz, Graz, Austria
Plant Molecular Biology and Biotechnology Research Center, Gyeongsang National University, Jinju, Korea

Alfonso Albacete
Institute of Plant Sciences, University of Graz, Graz, Austria
Departamento de Nutrición Vegetal, CEBAS-CSIC, Campus Espinardo, Murcia, Spain

Hannah Böhm and Ursula Janschek
Institute of Plant Sciences, University of Graz, Graz, Austria

Yeonggil Rim
Plant Molecular Biology and Biotechnology Research Center, Gyeongsang National University, Jinju, Korea

Walid Wahid Ali
Department of Pharmaceutical Biology, University of Würzburg, Würzburg, Germany

Soo Young Kim
Department of Molecular Biotechnology and Kumho Life Science Laboratory, College of Agriculture and Life Sciences, Chonnam National University, Gwangju, Korea

Thomas Roitsch
Institute of Plant Sciences, University of Graz, Graz, Austria
Department of Plant and Environmental Sciences, Copenhagen Plant Science Centre, University of Copenhagen, Taastrup, Denmark
Global Change Research Centre, CzechGlobe AS CR, v.v.i., Drásov, Czech Republic

Amornrat Chaiyasen and Saisamorn Lumyong
Department of Biology, Faculty of Science, Chiang Mai University, Chiang Mai, Thailand

J. Peter W. Young
Department of Biology, University of York, York, United Kingdom

Neung Teaumroong
Schoool of Biotechnology, Institute of Agricultural Technology, Suranaree University of Technology

Paiboolya Gavinlertvatana
Thai Orchid Labs Co. Ltd., Khannayao, Bangkok, Thailand

Index